Biomimetic Protein Based Elastomers
Emerging Materials for the Future

Biomaterials Science Series

Editor-in-chief:
Julian Jones, *Imperial College London, UK*

Series editors:
Cole DeForest, *University of Washington, USA*
Changyou Gao, *Zhejiang University, China*

Titles in the Series:

1: Stimuli-responsive Drug Delivery Systems
2: Biodegradable Thermogels
3: Biofabrication and 3D Tissue Modeling
4: Biomaterial Control of Therapeutic Stem Cells
5: Antimicrobial Materials for Biomedical Applications
6: Decellularized Extracellular Matrix: Characterization, Fabrication and Applications
7: The Chemistry of Medical and Dental Materials, 2nd edition
8: Injectable Hydrogels for 3D Bioprinting
9: Extracellular Vesicles: Applications to Regenerative Medicine, Therapeutics and Diagnostics
10: Biomimetic Protein Based Elastomers: Emerging Materials for the Future

How to obtain future titles on publication:
A standing order plan is available for this series. A standing order will bring delivery of each new volume immediately on publication.

For further information please contact:
Book Sales Department, Royal Society of Chemistry, Thomas Graham House, Science Park, Milton Road, Cambridge, CB4 0WF, UK
Telephone: +44 (0)1223 420066, Fax: +44 (0)1223 420247
Email: booksales@rsc.org
Visit our website at www.rsc.org/books

Biomimetic Protein Based Elastomers
Emerging Materials for the Future

Edited by

Namita Roy Choudhury
RMIT University, Australia
Email: namita.choudhury@rmit.edu.au

Julie C. Liu
Purdue University, USA
Email: julieliu@purdue.edu

and

Naba K. Dutta
RMIT University, Australia
Email: naba.dutta@rmit.edu.au

ROYAL SOCIETY
OF **CHEMISTRY**

Biomaterials Science Series No. 10

Print ISBN: 978-1-78801-078-8
PDF ISBN: 978-1-78801-272-0
EPUB ISBN: 978-1-83916-102-5
Print ISSN: 2397-1401
Electronic ISSN: 2397-141X

A catalogue record for this book is available from the British Library

The Royal Society of Chemistry is a charity, registered in England and Wales, Number 207890, and a company incorporated in England by Royal Charter (Registered No. RC000524), registered office: Burlington House, Piccadilly, London W1J 0BA, UK, Telephone: +44 (0) 20 7437 8656.

For further information see our website at www.rsc.org

Printed in the United Kingdom by CPI Group (UK) Ltd, Croydon, CR0 4YY, UK

Dedication

Dedicated to the brilliant minds of my son, husband, and parents, who had made the most important influence on my life and encouraged me to fly with my dreams.

N. R. C

For my family: Keith and Quentin, thank you for your love, support, and patience during all my late-night conference calls.

J. C. L

In memory of my parents, who taught me the true meaning of Success.

N. K. D

Biomaterials Science Series No. 10
Biomimetic Protein Based Elastomers: Emerging Materials for the Future
Edited by Namita Roy Choudhury, Julie C. Liu and Naba K. Dutta

Published by the Royal Society of Chemistry, www.rsc.org

Preface

'..*One must still have chaos in oneself to be able to give birth to a dancing star.*'
Friedrich Nietzsche, Thus Spoke Zarathustra

The natural world serves as the source of inspiration for our research, and, in editing this book, we were motivated to collate and share the recent findings and knowledge in the field of biomimetic protein/peptide-based elastomers. The traditional view that a stable 3D conformation of a protein is essential to its function has now been challenged by the recent discovery of multistimuli-responsive and multifunctional elastomers based on intrinsically disordered proteins. This paradigm shift has resulted in an intense research focus and the rapid evolution of the field of protein engineering. Over the last two decades, transformational new knowledge has been generated on the structure–property–function relationship of protein-based elastomers. This book attempts to capture the diverse and emerging directions of research on and achievements pertaining to these unusually smart materials. In particular, the book delves into their occurrence and specialized role in nature, their artificial synthesis and characterization, the molecular basis of their function and design principles, and their applications in diverse areas, including biomaterials engineering, template-based synthesis of nanomaterials, scaffolds for tissue engineering, medicine, programmed drug delivery, biomolecule detection and sensing, catalysis, renewable energy, immunoassays, cell bioimaging, and environmental detection and monitoring.

Each of the authors of this volume's chapters is an active researcher who has not only contributed to advancements in the field but is also passionate about the complexity of the science involved. As editors, our strategy for organizing the chapters was to enable the reader to easily navigate the content progressively. Each chapter summarizes the background knowledge, the current state of research, challenges, and future perspectives.

Biomaterials Science Series No. 10
Biomimetic Protein Based Elastomers: Emerging Materials for the Future
Edited by Namita Roy Choudhury, Julie C. Liu and Naba K. Dutta

Published by the Royal Society of Chemistry, www.rsc.org

This book should appeal to students, researchers, scientists, and academics from different fields. The book is a reservoir of resources and ideas that are highly relevant in biology and applicable to biomedicine and a different branch of nano- and nanobiotechnology. We hope it will excite and inspire fellow scientists, students, and many newcomers from both industry and academia to explore this field.

The writing of the book turned out to consume significantly more time than we expected, but, in the end, we found it extremely rewarding. First, finding available authors from a small pool of potential contributors was challenging. Then, the unprecedented situation of COVID-19, which sent shockwaves around the world, occurred in the midst of obtaining chapters and reviews.

We received help on many fronts during this book project, and we appreciate all the time that people took out of their busy schedules. We thank all the outstanding authors for their insightful contributions to this volume and their patience throughout the process. We are grateful to all of our reviewers for their timeliness and their constructive feedback on the chapters. Our appreciation goes to the Royal Society of Chemistry for the opportunity to write this book on a topic that animates our research activities.

Finally, we thank our respective families for their support, as many after-work hours were sacrificed for this project.

Namita Roy Choudhury
Julie C. Liu
Naba K. Dutta

Contents

Biomaterials Science Series No. 10
Biomimetic Protein Based Elastomers: Emerging Materials for the Future
Edited by Namita Roy Choudhury, Julie C. Liu and Naba K. Dutta

Published by the Royal Society of Chemistry, www.rsc.org

CHAPTER 1

Biomimetic Protein-based Elastomers: Emerging Materials for the Future

NAMITA ROY CHOUDHURY,*[a] JULIE C. LIU*[b] AND NABA K. DUTTA*[a]

[a] Chemical and Environmental Engineering, RMIT University, Melbourne, Australia; [b] Davidson School of Chemical Engineering and Weldon School of Biomedical Engineering, Purdue University, USA
*Emails: namita.choudhury@rmit.edu.au; julieliu@purdue.edu; naba.dutta@rmit.edu.au

1.1 Introduction

> *"Those who are inspired by a model other than Nature, a mistress above all masters, are laboring in vain."* – Leonardo da Vinci

Nature has created intriguing materials of amazing functionality with optimal hierarchical structure and architecture through evolution over millions of years. As scientists reveal the intricate workings of natural organisms, they have been inspired to design and develop biomimetic materials that utilize the knowledge gleaned from natural organisms to achieve their outstanding properties. Some exemplars from nature that researchers have imitated include the incredible toughness in nacre that arises from its hierarchical structure, the molecular selectivity in aquaporin water channels,

Biomaterials Science Series No. 10
Biomimetic Protein Based Elastomers: Emerging Materials for the Future
Edited by Namita Roy Choudhury, Julie C. Liu and Naba K. Dutta

Published by the Royal Society of Chemistry, www.rsc.org

toughness of silk fibers, and the high resilience and extreme fatigue lifetime in resilin. The advent of genetic engineering enabled the design of protein-based materials with exquisite control over amino acid sequence and molecular weight. Different amino acid sequences, or domains, can be incorporated to confer specific functionality (*e.g.*, environmental responsiveness) to the material. Compared to synthetic polymers, one clear advantage of biomimetic protein-based materials is the ability to modulate their hierarchical structure, which is essential for self-assembly and self-organization and enables them to be developed for complex functional applications.

An elastomer is a material that has the ability to return to its original size and shape when the external forces that caused the deformation have been removed. Elastomeric proteins are abundant in nature, which has perfected their molecular composition and sequence, conformational ensemble and architecture, crosslinking, and intermolecular interaction over millions of years of evolutionary adaptations to perform their specific biological roles and functions. They are pivotal for the function of many biological machines, including normal cardiac development and function, elasticity in the human arterial wall, and the jumping and flying mechanisms of arthropods.

Many elastic proteins are found in the extracellular matrix. For example, elastin provides elasticity to tissues such as skin, blood vessels, and lungs. In the recent past, significant research has been dedicated to correlating the amino acid sequences responsible for the secondary/tertiary structure/structural ensemble, the protein morphology, and the enviable mechanical properties that arise. In some elastic proteins, a high polypeptide chain entropy drives protein aggregates, which are intrinsically disordered or fuzzy, to form. Significant challenges exist to understand how the amino acid sequence yields disorder yet manages to avoid amyloid formation.

Overall, this book provides an understanding of the relationship between amino acid sequence, structure, and function of elastomeric proteins and demonstrates their use in a wide range of applications. In particular, this book captures the exciting findings on how the sequence modulates disorder, function and responsiveness. Many elastomeric proteins are remarkable as they exhibit not only outstanding elasticity but also a broad range of stimuli-responsive behaviour. Advanced characterization tools provide insights to understanding the complex structure–function relationship of elastic proteins. Understanding the sequence-structure-function relationship of proteins will not only expand scientific knowledge but also undoubtedly advance the rational design and application of biosynthetic elastomers and their analogous polypeptides for medical, commercial, and military applications. Thus, protein engineering methods offer advanced material platforms for specific applications in drug delivery and tissue repair/regeneration or in the synthesis of nanobioconjugates for catalytic, sensor, or imaging applications. Throughout the eleven following chapters, this book reveals the opportunities and challenges of developing advanced functional materials using protein engineering.

Chapter 2, "Native resilin: properties, occurrence and biological functions of a remarkable bio-elastomer" by Appel, Michels, and Gorb presents a striking illustration of the occurrence of native resilin in a variety of strategic locations of a diverse range of insects and crustaceans, including the arthrodial membranes of insects, the jumping system of planthoppers, the flight systems of dragonflies, the discoidal adhesive tarsal setae, the attachment structures of the stinkbug, the orthopteran adhesive pads, the mandibular gnathobases of copepods, the ventral abdominal cuticle of females of the common bed bug, insect mechanoreceptors, and the compound eyes of green lacewing, ants, and beetles. Confocal laser scanning micrographs, scanning electron micrographs, and transmission electron micrographs provide a precise visual description of the existence and morphological distribution of native resilin in different locations in a variety of species. An overlay of bright-field micrographs and wide-field fluorescence micrographs show the presence and distribution of resilin autofluorescence characteristics. Moreover, the precise nature of the graded distribution of mechanical properties of native resilin existing in a specific location has been identified in exquisite detail using atomic force microscopy nanoindentations and numerical models. This chapter demonstrates the importance, spatial correlation, specific role, and functional significance of resilin in locomotion, propulsion, flight, adhesion, and even the reproduction mechanism in insects and crustaceans. This chapter provides a glimpse into the natural origins of native resilins, their general features, their material structure and properties, and their exploitation in the highly effective design found in living organisms. This work thus lays the groundwork for detailed renderings of biomimetic elastic proteins, their structure–property relationships, their structural engineering, their conjugates and bioconjugates, their processing and product development, and the opportunities for their practical applications.

In Chapter 3, Bochicchio, Bracalello, and Pepe present the "Design and production of chimeric resilin-like polypeptides (RLPs)", which describes the development of engineered chimeric polypeptides containing resilin consensus sequences. Chimeric polypeptides are recombinant fusion proteins or biomimetic polypeptides, which are created through the fusion of two or more amino acid consensus sequences inspired by different native proteins. Each amino acid domain adds its distinct function to the resulting polypeptide and/or tunes the final properties. They are appealing as biomaterials because of the multitude of possible designs and the resulting novel physicochemical and biological functionalities in the polypeptide. At the beginning of the chapter, emphasis is devoted to the design strategy and cloning techniques for the production of the recombinant polypeptides, with a specific focus on chimeric RLPs. Subsequently, the potential applications proposed for chimeric RLPs are described. The last section in this chapter is dedicated to the future prospects for chimeric polypeptides in the framework of biomimetic materials and biomaterials science and takes into account the principal advantages and main critical drawbacks to be solved in the future.

In Chapter 4, "Intrinsically disordered and resilin-based protein polymers for high-performance biomaterials applications," the Kiick group describes the properties of an interesting class of proteins: intrinsically disordered proteins (IDPs). IDPs include many structural proteins, such as collagen, silk fibroin, elastin, and resilin. These proteins do not have a well-defined 3D structure and typically have a highly repetitive amino acid sequence that is enriched in certain amino acid residues. The chapter focuses on elastin and, in particular, resilin. Recombinant versions of these proteins recapitulate several key properties, such as their elastomeric nature and thermo-responsive phase behavior, and can be used to fabricate various nano- and microstructures. The chapter concludes by describing a number of tissue engineering applications, including vocal fold engineering, cardiovascular engineering, and cartilage engineering.

As a natural progression, Chapter 5, which is titled "Resilin-mimetic polypeptides and elastomeric modular protein-polymers: amino acid sequence, conformational ensemble and stimuli-responsiveness," by Choudhury *et al.*, continues to discuss the theme of resilin-like polypeptides, which are genetically engineered biopolymers derived from the consensus sequence of native resilin. RLPs exhibit molecular structure analogous to IDPs along with multistimuli (including pH, temperature, ions, and other macromolecules) responsiveness and tunable physicochemical properties. RLPs have thus dramatically changed the structure–function paradigm of proteins over the last two decades. Advances in protein engineering have also led to increasingly available RLP-based modular protein polymers and expanded their potential in diverse biomedical applications, such as tissue engineering, biosensors, and drug delivery. In this chapter, the authors discuss the amino acid sequences, compositions, molecular structures, stimuli-responsive behavior, inter-/intramolecular interactions, and self-assembly of RLPs and RLP-based modular protein polymers. Overall, this chapter provides fundamental knowledge and an understanding of rational design, development, characterization, and applications of resilin-based smart biomaterials.

In Chapter 6, which is entitled "Advanced biomedical applications of self-assembled elastin-like recombinamer structures", Rodríguez-Cabello, García-Arévalo, and Quintanilla-Sierra introduce an outstanding biomimetic elastomer, elastin and elastin-like recombinamers (ELRs). Native elastin is a hydrophobic, fibrous protein in the extracellular matrix. It is embedded in the highly hydrated connective tissue that is critical to many vertebrate tissues. It provides the extensibility and elastic recoil necessary for the function of tissues that are subjected to constant and reversible physiological deformations. Elastins and ELRs are IDPs and display thermo-responsiveness. The highly cooperative and intermolecular nature of the coacervation process is modulated by the ELR sequence, which also modulates growth into different assemblies, stability, interactions, reversibility of the process, and mechanical properties. In this chapter, the concept of self-assembly, morphologies of self-assembled ELR structures, and the parameters controlling the morphology and characteristics of the ELR structure are analyzed in detail. The authors focus on self-assembled structures formed from materials based on ELRs that are intended for biomedical applications, such as tissue engineering, controlled drug delivery, biosensors, or medical devices.

The Montclare group describes self-assembling nanofibrous materials in Chapter 7, "Smart Nanofibrous Materials for Tissue Engineering." Nanofibers not only mimic the fibrous network of extracellular matrix proteins but also result in high-porosity materials, which promote cell viability and function and are particularly useful for tissue engineering applications. The chapter describes the use of nanofibrous proteins such as α-helical coiled coils, elastin-like polypeptides, silk-like proteins, and collagen mimetics in applications including tissue engineering and controlled release of therapeutic agents such as genes or proteins. Strategies to impart smart behavior such as responsiveness to pH, temperature, and mechanical force are described as well. The chapter ends with a description of synthetic polymers that can be designed with a variety of properties, including polymers that are pH- or temperature-responsive, electrically conductive, piezoelectric (*i.e.*, generate electricity in response to mechanical forces), and responsive to magnetic fields.

Chapter 8, "Biomimetic Adhesives for Clinical Applications" by Liu and coworkers, describes the development of biomimetic adhesives for clinical applications. The chapter begins by describing the large, unmet need for surgical glues and delineates the design requirements and FDA regulations that have to be met for a new biomedical adhesive. Nature provides inspiration for the design of new adhesives, and the next section in the chapter describes adhesion mechanisms and biological adhesives that demonstrate these mechanisms. In particular, mussel adhesive chemistry, which has been extensively studied, is described in detail. Examples of synthetic polymers that incorporate lessons learned from mussel adhesive catechol chemistry and sandcastle worm–inspired coacervation are presented. Natural proteins, such as fibrin, and proteins that utilize biomimetic cross-linking are discussed. Finally, the chapter ends with an emphasis on the use of elastomeric proteins, which are particularly suitable for the development of a surgical glue for soft tissues.

Chapter 9 is authored by Ghandehari and his colleagues, who have extensive experience with silk-elastin-like protein-based polymers (SELPs), and is titled "Silk-elastinlike protein-based polymers for controlled delivery applications." SELPs are intriguing because they combine the structural strength of silk fibroin with the temperature-triggered phase behavior of elastin proteins. An overview of SELP structure and the recombinant production of SELPs is provided. The chapter details studies that probe the relationship between amino acid sequence, supramolecular structure, and the resulting function in applications such as liquid embolics and the controlled delivery of macromolecules (*e.g.*, adenoviral gene delivery, synthetic glycosaminoglycans).

Chapter 10, which is entitled "Processing and properties of regenerated *Bombyx mori* silk and recombinant spider silk fibers", is contributed by Lang and Sonnleitner. This chapter discusses native silk materials, their processing, and production. Silk is a highly complex material, and its native processing has been perfected over more than 400 million years of evolution. Silks exhibit a unique combination of high strength, elasticity, and toughness, and these capabilities are unmatched by high-performance synthetic fibers. The chapter highlights various processing methodologies that have been developed and utilized to create a variety of structural forms from regenerated *Bombyx mori* silk and biomimetic spider silks. An account of *B. mori* fibroin regeneration and preconditioning, recombinant spider silk protein creation, silk solution preparation and conditioning for technical processing, and different approaches to manufacture biomimetic silk fibers and nonwovens is provided. The authors also provide an overview of the diversity of different spinning techniques that have recently been used to produce artificial silk fibers, including wet spinning, dry spinning, microfluidic spinning, and electrospinning.

"Protein, biomimetic protein, and designer peptide-directed synthesis of metal nanoparticles, metal nanoclusters and nanobioconjugates, and their potential applications" by Dutta and coworker forms the content of the final chapter. In the natural world, biomineralization is a unique approach employed by living organisms to create hierarchically organized minerals with astonishing characteristics and utility. Recently, protein-/peptide-mediated biomimetic mineralization has been discovered to be a highly effective method for producing stable, biocompatible, highly water dispensable metal nanoparticles (m-NPs), metal nanoclusters (m-NCs), and nanobioconjugates of controlled composition, size, and shape with surface functional groups for bioconjunction. With a focus on noble metals, the chapter begins with a fundamental discussion of how the surface atomic arrangement and electronic structure of metal particles evolve as their size decreases from micro to nano to subnano, and how this affects the physical, photophysical, photoluminescent, catalytic, and other characteristic properties of metal nanostructures. Subsequently, a commentary on the different native protein, recombinant protein, strategically designed polypeptide, and peptoid-directed synthesis of m-NPs and m-NCs is presented. In the process of the formation and structuring of nanocrystals, the roles of individual amino acid residues and the amino acid compositions at the sequence level of peptides are discussed. Additionally, discussions are focused on structural insights at atomic-scale resolution using molecular simulation and new experimental designs to control the shape, composition, and nanoscale structure of m-NPs and m-NCs. Moreover, a specific focus is given to how the noble metal m-NCs, which are made up of only a few to a few tens of metal atoms, bridge the gap between isolated atoms and result in NPs with distinctive properties, including size-dependent fluorescence characteristics. The chapter concludes with a summary of the applications of m-NPs and m-NCs in a wide variety of areas of application, including advanced materials, energy, catalysis, environmental detection and monitoring, biomedicine, cell bioimaging, biomolecule detection and sensing, pharmaceuticals, and their potential future developments as therapeutic nanocomplexes.

CHAPTER 2

Native Resilin: Properties, Occurrence and Biological Functions of a Remarkable Bio-elastomer

ESTHER APPEL,[†] JAN MICHELS*,[†] AND STANISLAV N. GORB

Department of Functional Morphology and Biomechanics, Institute of Zoology, Christian-Albrechts-Universität zu Kiel, Am Botanischen Garten 1–9, D-24118 Kiel, Germany
*Email: jmichels@zoologie.uni-kiel.de

2.1 Native Resilin – The Pliant Protein

Elastomeric proteins occur in a wide range of organisms. They are known, for example, from vertebrates (elastin, titin, and fibrillin), plants (gluten), and bivalve mollusks (abductin, byssus).[1] In arthropods, spider silk proteins and resilin are the most known and common elastomeric proteins. Weis-Fogh[2] was the first to describe resilin on the basis of three resilin-bearing elements of the dragonfly and locust flight apparatus, *i.e.*, the elastic tendon of the pleuro-subalar muscles of the southern hawker (*Aeshna cyanea*) and the wing hinge ligament and prealar arm of the desert locust (*Schistocerca gregaria*) (Figure 2.1A,B). In subsequent studies following shortly thereafter, resilin's molecular structure and mechanical properties were further elucidated.[3–5]

[†]These authors contributed equally.

Biomaterials Science Series No. 10
Biomimetic Protein Based Elastomers: Emerging Materials for the Future
Edited by Namita Roy Choudhury, Julie C. Liu and Naba K. Dutta

Published by the Royal Society of Chemistry, www.rsc.org

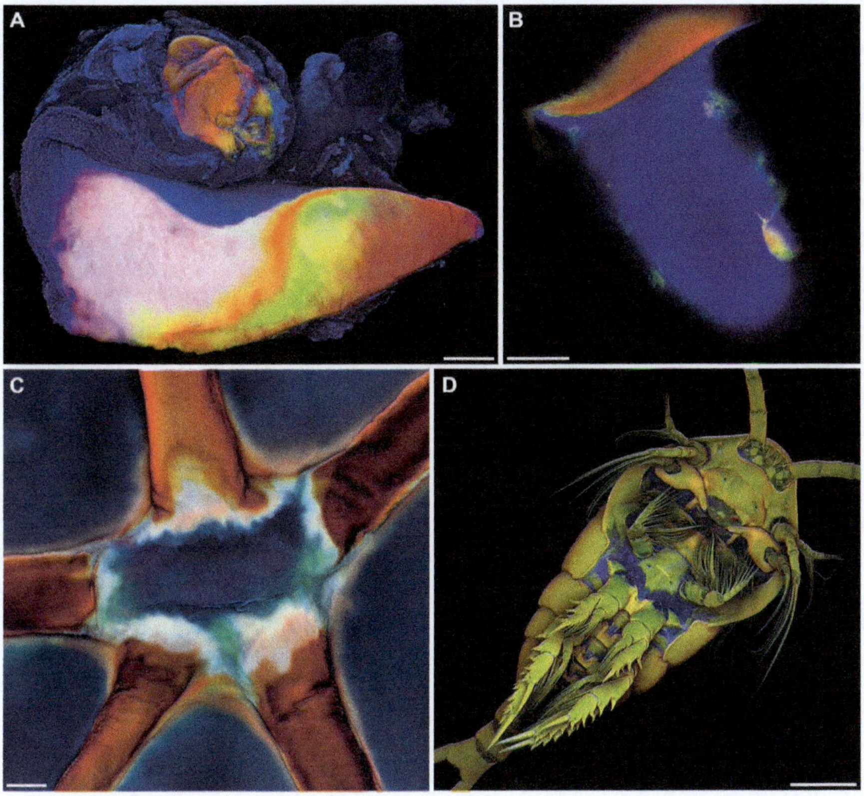

Figure 2.1 Occurrence of resilin in insects and crustaceans. Confocal laser scanning micrographs showing a wing hinge (A) and a prealar arm (B) of the migratory locust (*Locusta migratoria*), a wing vein joint of the common darter (*Sympetrum striolatum*) (C), and the ventral side of a female copepod of the species *Temora longicornis* (D). (A–C) Overlays of four different autofluorescences exhibited by the exoskeleton. Blue colour indicates large proportions of resilin, green exoskeleton structures consist mainly of non-sclerotised or weakly sclerotised chitinous material, and red exoskeleton structures are composed of relatively strongly sclerotised chitinous material. (D) Blue = autofluorescence of resilin, red = Congo red fluorescence of stained chitinous exoskeleton parts, green = mixture of autofluorescence and Congo red fluorescence of stained chitinous exoskeleton parts. (A, C, D) Maximum intensity projections. (B) Optical section. Scale bars = 100 μm (A, B), 20 μm (C), 200 μm (D). (A–D) Adapted from ref. 26 with permission from John Wiley & Sons, Copyright © 2011 The Authors, Journal of Microscopy © 2011 Royal Microscopical Society.

Resilin is a protein of high structural disorder with long polypeptide chains, which show repetitive and uncomplex amino acid sequences with a high amount of glycine (30–40%).[6,7] This composition leads to a low propensity for extended secondary structures (such as α-helices) and instead favors more flexible conformations, thereby decreasing resilin's stiffness and increasing its hydration level, which is beneficial for elastic recoil.[2,5,8–11] Resilin's polypeptide chains are covalently cross-linked to a network by di- and trityrosine, and this ensures that after stretching the polypeptide chains return to their original conformation without irreversible conformational disruption.[12–16] The term resilin is confined to the fully cross-linked protein, whereas the not fully cross-linked protein is called pro-resilin.[12]

As mentioned above, the hydration level of resilin's polypeptide backbone is of high importance for the elastic recoil and, therefore, for its mechanical properties.[5] Haemolymph is the main source for hydration in the living animal, and when fully hydrated, pure resilin exhibits a resilience of 92–97%, which is unmatched by any other elastomeric protein or synthetic rubber.[3,4,8,17] In this state, resilin can be stretched and compressed by a factor of three and recoils without any lasting deformations.[3,4,18] With an endurance limit of over 300 million cycles, resilin is especially suitable for repetitive movements.[19] Furthermore, it has a low stiffness with a Young's modulus of around 0.6–0.7 MPa in the pleuro-subalar tendon of the dragonfly *A. cyanea* and of 0.9 MPa in the locust wing hinge ligament.[17] Finally, as a hydrogel, resilin exhibits a slow viscoelastic response.[20]

These mechanical properties apply for nearly pure, fully hydrated resilin. However, resilin often occurs as a composite, linked to directionally arranged chitin fibres by its chitin-binding domain and to other proteins.[5,13,17] In such composites, the properties of the other components sometimes even dominate the properties of resilin and complicate the identification of resilin based on the criteria of Andersen and Weis-Fogh.[5] Other factors, such as motion frequency, the direction of applied stresses, the ratio and density of di- to trityrosine cross-linking, the proportion of certain amino acids, and the thereupon based detailed structure of the polypeptide chains, as well as the ratio of resilin with and without the chitin-binding domain, also influence resilin's mechanical properties, so that the finally elicited response to a directional stress depends on a variety of environmental and structural components.[12,13,15,20–25]

Despite all these differences, Andersen and Weis-Fogh[5] developed a detailed definition for resilin-bearing structures based on a variety of tests they conducted on the elastic tendon, the prealar arm, and the wing hinge ligament. As the defined visualisation and identification methods are not completely specific for resilin, it is always recommendable to use more than one method to reliably identify resilin.

In some cases, when resilin occurs in the form of large and nearly pure chunks, it can even be discerned with the naked eye, as it has a colourless, amorphous, and transparent appearance.[2,5] This is contrary to structures that contain large proportions of chitin, which are typically pigmented and much less transparent, especially when being strongly sclerotised. Due to the large proportion of proline, glycine, and hydrophilic amino acids, resilin shows an isotropic swelling when coming in contact with aqueous media (and various hydrophilic liquids). The swelling is reversible and depends on the pH, with higher pH values leading to a more pronounced effect.[5] When resilin exists in a composite together with chitin lamellae, the swelling is anisotropic as chitin lamellae directionally constrain the swelling.[2] Tensioning of hydrated resilin makes it positively birefringent in the direction of the extension.[5] Resilin with chitin lamellae, by contrast, is already positively birefringent normally to the lamellae due to its limited extension to only one direction.[2]

When dried, resilin becomes hard and brittle and no longer shows its typical, rubber-like properties.[5] The digestion of resilin with proteolytic enzymes (*e.g.*, pepsin, trypsin) is another test for its existence.[5] For visualisation, resilin can be either stained with conventional dyes or examined by fluorescence microscopy. While Masson and Mallory dyes stain resilin red, aqueous solutions of methylene blue and/or toluidine blue were reported to stain it violet to blue.[5,26–28] Resilin (in its native, isolated, boiled, or hydrolysed form) exhibits a rather pronounced violet/blue autofluorescence due to the cross-linking amino acids di- and trityrosine.[5,26,29,30] The excitation and emission maxima of this autofluorescence are highly pH-dependent and show a reversible and rapid change when transferred to media with changing pH values.[5,29,31] In neutral to alkaline media, the resilin autofluorescence features an excitation maximum at 320 nm and an emission maximum at 415 nm, while at low pH values, the excitation maximum shifts to around 285 nm, and the upper edge of the excitation peak is located at around 330–340 nm.[29]

Sometimes, when not all tests for resilin turn out to be positive, it is hard to determine whether exoskeleton elements actually contain resilin, only a resilin-like protein, or a mixture of both. Therefore, Andersen and Weis-Fogh[5] devised another term for structures with only a few characteristics resembling those of resilin: transitional cuticle.

In the last two decades, both newly developed and improved techniques and methods (*e.g.*, in the areas of mechanical testing, confocal laser scanning microscopy, and atomic force microscopy) have led to progress in the identification, visualisation, and characterisation of resilin, thereby facilitating the processes and leading to more detailed results. For example, the development of the anti-Rec1 antibody, which was raised against (the product of exon 1) of the *Drosophila melanogaster* pro-resilin (rec1-resilin), will probably offer a reliable method to identify resilin in the future.[18,19,32,33] This antibody was shown to be cross-reactive and to label resilin in other insects as well, but only a few species have been tested so far.[18,19,32–34]

Progress has also been made in the utilisation of a wider range of autofluorescences to determine material composition and material gradients and to identify resilin.[26,35–38] By creating overlays of micrographs of different autofluorescences, structures that show a dominance of autofluorescence typically emitted by resilin can be identified.[35–37] In a recently published study, Michels and Gorb[26] used confocal laser scanning microscopy and a combination of four different, separately detected autofluorescences (to which they later on assigned specific colours) to visualise the distribution of resilin and non-sclerotised to strongly sclerotised chitinous exoskeleton material.[26] Thereby, they created very detailed visualisations of the material composition and material gradients within the examined preparations, with resilin-dominated exoskeleton parts being blue, weakly or not sclerotised

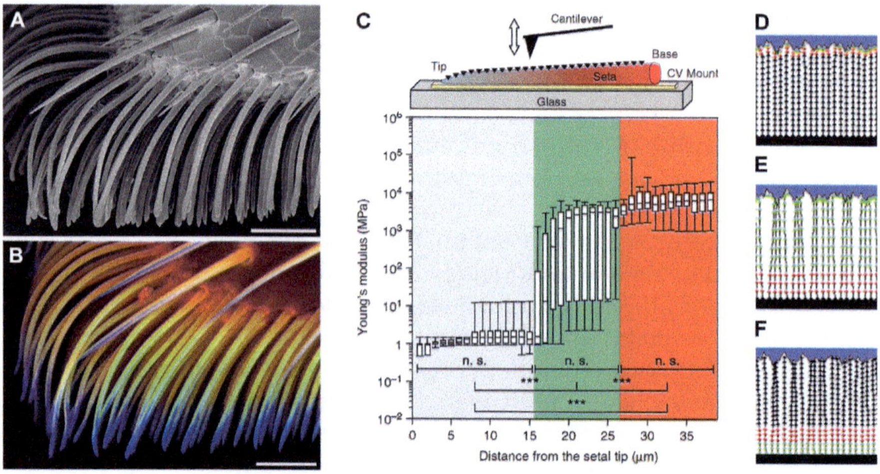

Figure 2.2 Distribution, mechanical properties, and functional significance of resilin in adhesive tarsal setae. (A, B) Ventral part of the second adhesive pad of a female seven-spot ladybird (*Coccinella septempunctata*), lateral view. (A) Scanning electron micrograph. (B) Confocal laser scanning micrograph (maximum intensity projection) showing an overlay of four different autofluorescences exhibited by the exoskeleton. Blue colour indicates large proportions of resilin. (C) Box-and-whisker plots showing the median Young's modulus of fresh adhesive tarsal setae obtained by atomic force microscopy nanoindentations (see the inset above the graph) along each seta. The borders of the boxes define the 25th and 75th percentiles, the median is indicated by a horizontal line, and the error bars define the 10th and 90th percentiles (n. s. = not significant, *** = highly significant). The background colours indicate the different seta sections. (D–F) Numerical model showing typical configurations of a filamentary structure (setal array) attached to a stiff support (black rectangle) in adhesive contact with a random fractal surface (blue region). Three types of fibres were tested: (D) stiff fibres with short elastic ends, (E) long elastic fibres connected to the base by short, stiff roots, and (F) stiff fibres with soft elastic segments near the base. The different stiffnesses of the segments are conditionally shown by circles with different colours. Stiff, medium, and soft segments are marked by black, red, and green circles, respectively. Scale bars = 25 μm (A, B). (A–C) Adapted from ref. 38 with permission from Springer Nature, Copyright 2013. (D–F) Adapted from ref. 146, https://doi.org/10.3762/bjnano.5.95, under the terms of the CC BY 4.0 license https://creativecommons.org/licenses/by/4.0/.

chitinous exoskeleton structures appearing green and relatively strongly sclerotised chitinous material looking red (Figure 2.1A–C).

Material gradients from resilin-dominated to more chitinous and finally mostly sclerotised chitinous cuticle often occur in resilin-containing exoskeleton structures. Such material gradients are not only reflected by a shift in the autofluorescence composition but also by changing mechanical properties. With a combination of autofluorescence analyses and mechanical measurements (atomic force microscopy), Peisker and co-authors[38] showed that the adhesive setae of beetle tarsi are soft at their resilin-bearing tip (Young's modulus: 1.2 ± 0.3 MPa) and rather stiff at their base (6.8 ± 1.2 GPa) (Figure 2.2A–C). When dehydrated, the stiffness of the tip increased to 7.2 GPa, a consequence already mentioned for dried resilin structures above. Another aspect that supports the detected correlation between the results of autofluorescence analyses and mechanical testing and the presence of resilin in the setal tip is the material's behaviour with respect to the removal of load. While the tip showed a purely elastic response, which is typical for resilin, the setal base exhibited an elastic–plastic response due to the presence of a more sclerotised chitinous cuticle.[38]

Such a gradient in the mechanical properties is likely to be found in other exoskeleton structures that show comparable gradients in the autofluorescence composition and, therefore, in the distribution of resilin. Nevertheless, although autofluorescence analyses are beneficial for a variety of reasons and can often be confirmed by mechanical testing, it should always be considered that resilin is not the only compound found in organisms and emitting violet/blue autofluorescence when excited with UV light.[39–42]

2.2 Occurrence and Functions of Resilin in Different Arthropod Exoskeleton Systems

Since its discovery in 1960, resilin has been found in various exoskeleton systems of insects, endowing the respective structures with elasticity and flexibility and serving diverse functions. Amongst others, it has been revealed to occur (1) in leg joints,[31,43] where it can serve energy storage for jumping,[27,44–48] (2) in vein joints and flexion and fold lines of various insect wings, where it helps prevent material damage and supports passive wing deformations in flight and during wing (un-)folding,[35,49–51,53] (3) in tarsal attachment structures, where it facilitates adaptation and improved attachment to different surfaces,[36,38,53,54] and (4) in numerous other structures like food-pumps,[55] poison apparatuses,[56] sound production systems,[28,57] and extensible abdominal cuticles.[58,59]

Apart from insects, resilin has also been found in structures of other arthropods, for example, in mouthparts and various other exoskeleton regions of crustaceans[26,60–63] (Figure 2.1D), in the lamellar alloscutum of mites,[64] in the pedipalp joints of scorpions,[65,66] and in the cuticle of centipedes.[67] Resilin-like proteins that contain the two amino acids di- and trityrosine but show only some of the characteristics described by Andersen and Weis-Fogh[5] have been reported for a handful of non-arthropod organisms, including nematodes,[68] mussels,[69] monogeneans,[70,71] and sea urchins.[72]

In the subsequent sections, the distribution and function of resilin in selected exoskeleton structures are described.

2.2.1 Arthrodial Membranes

Arthrodial membranes are thin, flexible, non-sclerotised cuticular areas that are mostly located between more sclerotised body segments, allowing the stiffer elements to move against each other and the overall structure to increase in length or volume.[73,74] In addition, arthrodial membranes can serve a variety of other purposes, including the prevention of material failure, the tarsal adaptation to the surface profile, protection (with the membrane being sometimes reinforced by tiny outgrowths), and locomotion (*e.g.*, in caterpillars).[37,43,75–77] Often, arthrodial membranes have more than one of these functions.

So far, two different flexibility endowing membrane designs have been reported in insects. The first type stands out for its high extensibility (up to ten times in the locust abdomen),[78–81] which is mainly based on its protein composition, whereas the second type is less extendable, consists of a folded, laminated structure, and has been detected in the abdominal membranes of tsetse flies (*Glossina morsitans*)[81] and assassin bugs of the genus *Rhodnius*.[82,83] Both types of flexible membrane typically contain resilin, but most of them are resilin–chitin composites and, therefore, show directionally changing mechanical properties.

A widespread example of arthrodial membranes can be found in leg joints (Figure 2.3D). These resilin-bearing membranes often work in tension and compression and not only help increase the joint's movability but also serve the prevention of material damage in these frequently used systems. Apart from the rather typical leg joint membranes, the pretarsus of hoverflies comprises an additional resilin-bearing membrane connecting the last tarsomere to the claws and the pulvilli and thereby improving the adaptive behaviour of the latter[37,43] (Figure 2.3A–C).

A further example is the extensible neck membrane of dragonflies and damselflies, which connects the head to the sclerites of the neck region and allows the highly movable head to act as a gravity organ.[84] The membrane exhibits a pronounced three-level folding structure, with the epicuticle forming the first-order folds (Figure 2.3G), the exocuticle and the underlying epidermis generating the second-order folds (Figure 2.3F), and the overall structure forming the third-order macrofolds[84] (Figure 2.3E). In comparison to the sclerotised cuticle of the neighbouring sclerite regions, the exocuticle of the neck membrane is an unusually electron-lucent and thin composite layer of chitin fibres and a flexible resilin-bearing matrix[26,84] (Figure 2.3E).

2.2.2 Legged Locomotion

Insect legs typically consist of six segments, which are flexibly connected by resilin-bearing membranes and mono- or dicondylic articulations. For normal walking locomotion, the segments are flexed and extended by intrinsic and extrinsic muscles, with their contraction pattern being controlled by specialised networks in the central nervous system, which are presumably activated by inputs from central pattern generators, which interact with descending signals

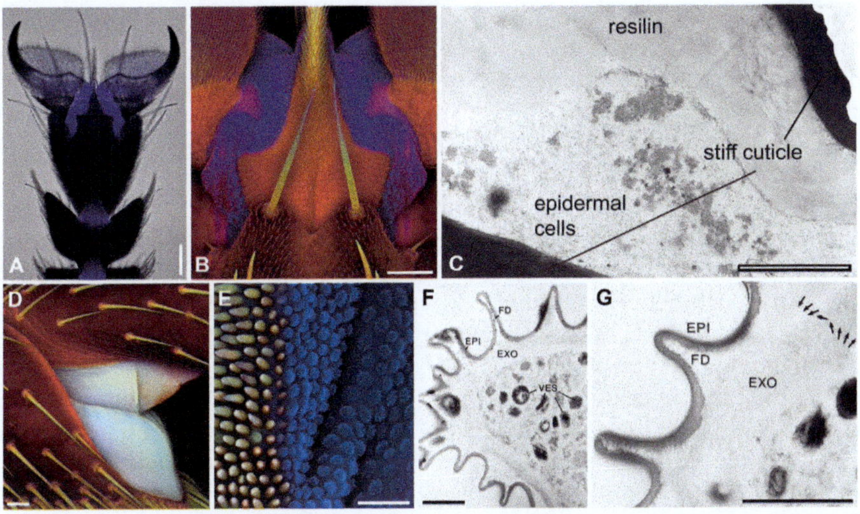

Figure 2.3 Resilin in arthrodial membranes of insects. (A, B) Pretarsus of the drone fly (*Eristalis tenax*), ventral view. (A) Overlay of a bright-field micrograph and a wide-field fluorescence micrograph showing the presence and distribution of resilin autofluorescence (blue). (B) Confocal laser scanning micrograph (maximum intensity projection) revealing large proportions of resilin (shown in blue) in the membranous structures between rigid sclerites of the pretarsus. (C) Pretarsus of the urban bluebottle blowfly (*Calliphora vicina*). Transmission electron micrograph showing an ultrathin section through structures comparable to those shown in blue in B. (D, E) Confocal laser scanning micrographs (maximum intensity projections) showing overlays of different autofluorescences exhibited by the exoskeletons. Blue colour indicates large proportions of resilin. (D) Lateral view of the joint between the tarsomeres 1 and 2 in the third leg of the seven-spot ladybird (*Coccinella septempunctata*). (E) Border between the neck membrane (right side) and a postcervical sclerite (left side) of the broad-bodied chaser (*Libellula depressa*). (F, G) Membranous cuticle in the neck area of the blue-tailed damselfly (*Ischnura elegans*). Transmission electron micrographs showing ultrathin sections. EPI, epicuticle; EXO, exocuticle; FD, folds; VES, vesicles. Arrows: preferential orientation of chitin fibres. Scale bars = 100 μm (A, B), 2 μm (C), 20 μm (D, E), 1 μm (F, G). (A, B, D, E) Adapted from ref. 26 with permission from John Wiley & Sons, Copyright © 2011 The Authors Journal of Microscopy © 2011 Royal Microscopical Society. (F, G) Adapted from ref. 84 with permission from The Zoological Society of London, Copyright 2000.

from certain brain regions (like the gnathal and cerebral ganglia) and sensory information.[85] Except for walking, insect legs can fulfil various other functions and kinds of locomotion, including swimming, grasping, grooming, pollen collection, and jumping. Especially the latter locomotion mode is known to allow very rapid leg movements of less than 1 ms, which outmatch the performance of movements achieved by muscle contraction (normally longer than 10 ms, in extreme cases 2–5 ms).[44,86,87] Jumping movements that partly cover distances of up to 100 times the insect's body length are frequently used to escape from predators and have been detected, amongst others, in fleas,[27,86] snowfleas,[47] Orthoptera,[88,89] beetles,[90–94] and true bugs.[44–46,95] While insects with long legs, like bush crickets, are able to amplify the power produced by direct muscular contraction through long-leg leverage, particularly insects with short legs are facing the problem that the time their propulsive legs are in contact with the ground is relatively short, so that they would normally not be able to reach a high acceleration and the necessary power.[89,96] To overcome this shortcoming, their jumping legs are often modified to catapult-like designs.[27,44–48,86,88,90–97] These include resilin-bearing springs and specialised joints with active latch and trigger muscles that produce tractive forces and release the elastically stored energy[2,27,44–46,86,88,90,91,95,97] (see below).

Spring-actuated jumps are typically conducted in more or less the same way among different insect orders. First, the propulsive legs are moved to a cocked position. With the joints being locked, the power-producing muscles (mostly huge trochanteral depressor muscles or tibial/femoral extensor muscles) are contracted for some milliseconds or seconds.[94,96] In insects like planthoppers, froghoppers, leafhoppers, or jumping plant lice, which sling their legs underneath their body and have relatively short propulsive legs in comparison to their body length, typically metathoracic trochanteral depressor muscles are used[44,95–97] (Figure 2.4B). Insects that move their legs at their body sides (like true fleas, snow fleas, locusts, and flea beetles) and sometimes have rather long legs in comparison to their body length (*e.g.*, locusts, some fleas) use either trochanteral depressor muscles (snow fleas, true fleas) or tibial/femoral extensor muscles (flea beetles, locusts).[27,47,48,89,92,94,96] Instead of moving the legs, the muscle contraction leads to the distortion of either simply modified exoskeleton material (*e.g.*, in froghoppers, grasshoppers, and planthoppers) or specialised springs (*e.g.*, flea beetles) that are both mostly composites of resilin and hard sclerotised cuticle.[27,44–48,89,91–94,96,97] More precisely, flea beetles use a prominent metafemoral spring,[91–94] locusts employ semilunar processes at their distal hind femora,[48] planthoppers and froghoppers use thoracic resilin-supplemented pleural arches[44,97] (Figure 2.4A–C), and snow fleas and fleas use elastic pads at the articulation of the notum (wing hinge) with the dorsal head of the pleuron (pleural ridge).[27,45,47] Bending and twisting of the composite material leads to energy storage, with (1) the soft resilin-dominated part mostly transferring and retransmitting energy to and from the hard part as well as mechanically storing some energy and preventing fracture of the heavily loaded material and (2) the hard, more sclerotised part conducting the major energy storage for the jump.[45,96] This mechanism is, to some extent, comparable to the internal energy storage and release in resilin, which also consists of a soft (exon I) and hard (exon III) segment.[98] With resilin's connection through its Rebers and Riddiford chitin-binding domain (exon II) to a more chitinous and sclerotised cuticle, the described energy-storing effect can be enhanced. The stored energy is suddenly released and results in the elastic recoil of the spring, the extension of the leg within milliseconds, and a catapult-like jump of amplified power.[92,94,96] The sudden release of the stored energy can be triggered by, for example, the contraction of a small trigger

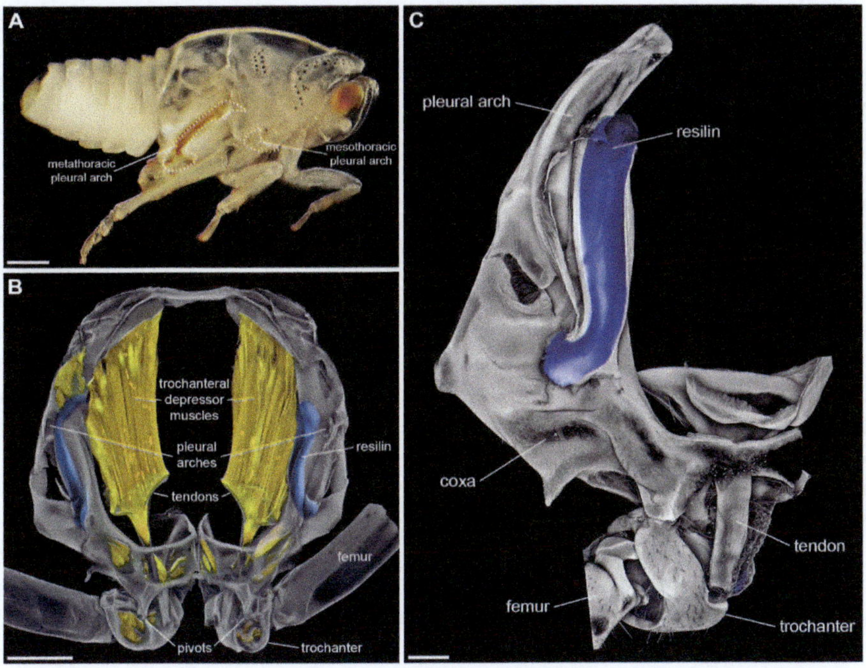

Figure 2.4 Resilin in the jumping system of planthoppers. (A) Bright-field micrograph. Lateral view of a nymph of *Metcalfa pruinosa*, with the right pleural arches of the meso- and metathoracic segments being visible through the external exoskeleton. (B, C) Confocal laser scanning micrographs. Resilin is shown in blue, chitinous cuticle is depicted in grey, and muscles are shown in yellow. (B) Thorax of an adult *Stenocranus acutus* with the resilin-dominated structures in the pleural arches and the large trochanteral depressor muscles, which insert at the trochanters and are responsible for the distortion of the pleural arches. Parts of other, smaller muscles are also visible. (C) Exoskeleton elements of the right side of the metathorax of an adult *Thionia bullata*, clearly showing the pleural arch with its resilin-dominated structures. Scale bars = 500 μm (A), 200 μm (B, C). (A–C) Adapted from ref. 97, https://doi.org/10.7554/eLife.23824, under the terms of the CC BY 4.0 license https://creativecommons.org/licenses/by/4.0/.

muscle being attached to the apodeme of the power muscle (as in the froghopper *Cercopis vulnerata*)[44] or by more complex systems as in flea beetles, where a combination of an elastic plate, a dislodging triangular plate, and muscular tension controls the timing of the jump.[94]

Most of the jumping insects use only the hind legs as propulsive legs during the jump. Snow fleas, by contrast, were shown to use their middle and hind legs, and the respective energy-storing structures were detected in both leg pairs, enabling them to compensate for a lower muscle mass, to jump with lower ground reaction forces from more compliant surfaces, and to reduce airborne spinning due to a more stable take-off.[47]

Apart from the resilin-bearing structures directly associated with jumping movements, locust legs show a specialised resilin-dominated buckling region in the proximal hind leg tibia that acts as shock absorber.[99] When kicking legs miss their target or a leg slips during the jump, the tibia buckles at the buckling region, and the distal tibial part extends, thereby allowing the built-up energy to dissipate.[99]

2.2.3 Flight Systems: Folds, Tendons and Microjoints

Resilin is widespread in the insect flight apparatus. It has already been found in elements of the wing base (*i.e.*, tendons that connect muscles to wing base sclerites), wing hinge ligaments that connect the wings to the thoracic wall, and prealar arms that connect the mesotergum to wing base sclerites.[2,5] In the wing blade, its existence has been reported in wing vein microjoints that connect cross veins to longitudinal veins, in flexion and fold lines, in clearly defined patches within the wing membrane, in the membranous region that establishes a connection to wing veins and even within cuticular layers of wing veins.[35,49,100–102]

As already mentioned above, depending on the purity of resilin and its mixture with chitin fibres of distinct directional arrangement, the materials' mechanical properties vary.[5,13] For example, structures of pure or nearly pure resilin (*e.g.*, the pleuro-subalar dragonfly tendon) have a very low stiffness and behave similarly in extension and compression.[2,5] By contrast, structures consisting of a mixture of chitin fibres and resilin (*e.g.*, the prealar arm) show a highly direction-depending behaviour, with the deformation being either predominantly influenced by the properties of resilin or chitin.[5,17] Nevertheless, all of these structures benefit to a greater or lesser extent from resilin's rubber-like properties, including its low stiffness, large and reversible extensibility, high resilience, long fatigue time, good damping behaviour, and ability to store elastic energy.

The sausage-like swelling of the thoracic dragonfly tendon at the dorsal side of the pleuro-subalar muscle was one of the first resilin-containing structures to be described and characterised.[2] It has a length and width of about 0.7 mm and 0.15 mm, respectively, and consists of virtually pure resilin.[2] On its dorsal side, it is connected to the subalar sclerite, which is located posterior to the pleural wing process and is in turn connected to the axillary sclerites of the wing base[2] (Figure 2.5A,B). On its ventral side, it is connected to the pleuro-subalar muscle, which is attached to the lower part of the pleuron.[2] During flight (*i.e.*, supination), the resilin-bearing tendon, together with the coaxalar muscle, is assumed to control wing twisting.[5,103,104] While the pleuro-subalar muscle contracts tonically and slowly, the tendon is kept at a specific length and tension.[103] This mechanism is particularly important for hovering but might also play a role in other refined flight manoeuvers and in controlling excess wing movements during turbulent flow.[104,105] The tendon shows a large reversible extensibility (over 250% in the forewing of the widow skimmer [*Libellula luctuosa*]) and a long fatigue time, properties that are of key importance for maintaining its function.[103]

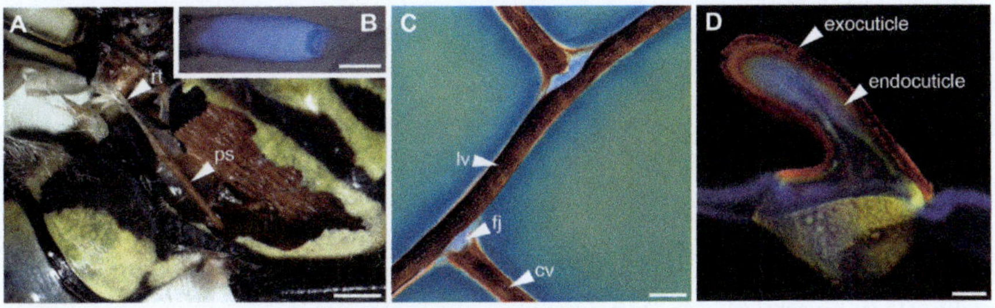

Figure 2.5 Resilin in the flight systems of dragonflies. (A) Stereomicrograph depicting a resilin-bearing tendon (rt) of a pleuro-subalar muscle (ps) of the southern hawker (*Aeshna cyanea*). (B) Overlay of a bright-field micrograph and a wide-field fluorescence micrograph showing autofluorescence (blue) exhibited by resilin in a tendon of the genus *Zyxomma*. (C) Flexible wing vein joints (fj) joining cross veins (cv) to a longitudinal vein (lv) in a wing of the vagrant darter (*Sympetrum vulgatum*). (D) Cross-section through a wing vein of *S. vulgatum*, revealing sclerotised exocuticle and resilin-bearing endocuticle. (C, D) Confocal laser scanning micrographs (maximum intensity projections) showing overlays of different autofluorescences exhibited by the exoskeleton. Structures with large proportions of resilin are shown in blue. Scale bars = 1 mm (A), 100 μm (B), 40 μm (C), 10 μm (D). (B) Adapted from ref. 18 with permission from Springer Nature, Copyright 2005. (D) Adapted from ref. 102 with permission from John Wiley & Sons, Copyright © 2015 Anatomical Society.

A recent study revealed that resilin is not only present in the tendon belonging to the pleuro-subalar muscle but also in all cap tendons of the pterothoracic muscles, which are mostly associated with flight.[106]

Another resilin-bearing structure described by Weis-Fogh in 1960 is the locust wing hinge ligament.[2,5] It connects the (meso-)pleural wing process to a small process at the bottom of the second axillary sclerite.[2,5] The wing hinge ligament is described as a thick pad or cushion and consists of zones with a varying amount of resilin and chitin fibres[2,5,26] (Figure 2.1A). While its ventral part is mainly chitinous and tough, the dorsal part is soft and consists of a part of pure resilin and another part of resilin interspersed with chitin lamellae.[5] The total pad consists of about 86% resilin and 14% chitin (and other proteins).[107] Deposition of resilin and chitin lamellae starts a few days before the final moulting and ends a few days or weeks thereafter.[5,107] The locust wing hinge ligament is assumed to serve wing acceleration during the downstroke by storing kinetic energy at maximum wing deflection during the upstroke.[2,5] During this process, the ligament is able to take up tensile as well as compressive forces.[5,13,108] Apart from Orthoptera, rubber-like wing hinge ligaments have also been found in Odonata and Dictyoptera.[5]

The question of at which frequency range resilin or resilin–chitin composites are still resilient enough to work properly has been examined with different methods and structures (dragonfly tendon, locust wing hinge ligament and prealar arm, cockroach tibia–tarsal joint) and is still discussed controversially.[5,17,20,104,109,110] Although most of the studies suggest that resilin's resilience decreases with increasing frequency, there are pronounced differences regarding the rate of the decrease.[5,17,20,104,109,110] To which extent these differences can be attributed to the different measurement techniques or to the different material composition still has to be elucidated.

The assumption that resilin's resilience might be too small at high wing-beat frequencies is supported by the fact that resilin has been mainly found in the flight system of insects with synchronous flight muscles that fly at low wing-beat frequencies of less than 50 Hz and with inertial forces predominating aerodynamic forces.[5,13,17,109] In representatives of Lepidoptera, some Coleoptera, and some Hymenoptera, wing hinge ligaments are rather tough and inextensible.[5,13] In these systems, elastic energy (for, *e.g.*, wing acceleration) is probably stored in the flight muscles and the rigid cuticle of the thorax.[5,13] However, there is evidence of some small wing hinge ligaments occurring between different sclerites in the genera *Calliphora*, *Bombus*, *Apis*, and *Oryctes*.[5,111]

The locust prealar arm is a hyaline hump and is located at the anterior edge of the mesotergum.[5] Its tip is connected to the first basalar sclerite by a tough and flexible ligament.[5] The prealar arm consists of around 76% resilin and 24% chitin and shows a structure of alternating thick (1–3 μm) resilin and thin (0.2 μm) concentric chitin fibril layers.[5,107] At the bottom, it shows a relatively abrupt transition to a dark sclerotised basis[5] (Figure 2.1B). Owing to the specific arrangement of the chitin fibrils, the deformational response of the prealar arm during stretching is mainly determined by the mechanical behaviour of the chitin fibrils, whereas during bending and compression, it predominantly relies on the properties of resilin.[5,17] During flight, the contraction of the basalar muscle causes a pulling movement on the first basalar sclerite, which is located anterior to the pleural wing process. Thereby, the anterobasal margin (with [sub-]humeral sclerites) of the wing is pulled downwards, causing wing pronation.[112] During this process, the deformation of the prealar arm likely contributes to elastic energy storage, with the energy being released at wing supination.

As already mentioned above, resilin is not only present at the wing base but also at various locations of the wing blade, including vein microjoints, cuticular vein layers, flexion and fold lines, defined patches in the wing membrane, and membranous regions directly abutting on veins.[35,49,100–102,113] In dragonfly and damselfly wings, abundant longitudinal and cross veins create a fine-meshed, corrugated vein network with the cross veins being connected to the longitudinal veins by either stiff, inflexibly fused joints or resilin-bearing joints[50–52,101,113] (Figure 2.5C). Finite element analyses and mechanical testing suggest that resilin-bearing joints are more flexible and enable a larger angular displacement than fused joints and show a decreased stress concentration at the joint itself.[50,52,101,113,114] This probably contributes to the reduction of material failure and provides increased chordwise flexibility, thereby allowing directional passive wing deformations that improve the wing's aerodynamic efficiency during flight.[50,52,101,113,114] Several studies that have examined the distribution of fused and resilin-bearing joints on the ventral and dorsal wing sides in different families of Odonata showed that the distribution pattern is quite diverse but follows phylogenetic trends related to flight behaviour and wing morphology.[50,52,101,113] Furthermore, resilin is not only present in dragonfly vein joints but also in cuticular layers of wing veins, thereby creating a stiffness gradient with a sclerotised, stiff outer layer (mainly exocuticle) and a soft, resilin-bearing inner layer (endocuticle)[102] (Figure 2.5D). Such a stiffness gradient was suggested to improve the vein's damping properties and its load-bearing capacity against failure due to Brazier ovalisation and to reduce the overall vein stiffness.[102,115]

Apart from vein joints and internal vein layers, resilin is also present in the membrane abutting on wing veins[102] (Figure 2.5C,D). Such a flexibly suspended wing membrane can contribute to preventing its tear-off from wing veins by allowing a larger strain.[102] In contrast to these rather diffuse resilin-bearing membrane areas, resilin patches in the wing membrane of the urban bluebottle blowfly (*Calliphora vicina*) are well defined and show an abrupt transition to the more chitinous and sclerotised surrounding wing membrane.[100] These patches are mainly located in close proximity to wing veins and are present in the proximal wing half.[100] The location of the resilin patches coincides with the existence of anterior-posterior-running flexion lines with the occurrence of high spanwise bending stresses during the wing acceleration phase.[100] The presence of resilin patches in these regions was suggested to reduce peak stresses near the wing hinge and, thereby, to prevent failure.[100]

Directional deformation in insect wings is further influenced by the presence of flexion lines and fold lines, which are often supported by resilin.[35,49,116,117] While flexion lines mainly increase the aerodynamic efficiency during flight, fold lines most of all serve wing folding. When flexion and fold lines cross veins, the latter show hyaline, resilin-bearing breaks or gaps, which are called alar fenestrae.[118] In an experimental study, Mountcastle and Combes[118] showed that an artificial stiffening of only one of the flexible, resilin-bearing alar fenestrae in bumblebee wings already affects the overall wing flexibility and maximum vertical aerodynamic force production. These results were supported by a study of Ma and coauthors,[116] who identified five flexion axes for chordwise flexibility (camber formation) on the basis of flexion lines, alar fenestrae, wing deformations, and a flexible hamuli-mediated forewing–hindwing connection in honeybee wings. The morphology, material composition, and functionality of the hook-shaped hamuli in honeybee wings were further examined in detail in additional studies, revealing that the sclerotised hamuli[119] are flexibly suspended in sockets at their bases by resilin-bearing membranes.[120,121] This construction likely enables the hindwing hamuli to maintain a flexible but tight connection to the forewings, which is important for the wings' aerodynamic efficiency and reduces the risk of material failure, thereby increasing the structures' durability.[120,121]

Additional resilin-bearing breaks in longitudinal veins were found at the leading edge of wasp and blowfly wings (*i.e.*, costal and humeral breaks). They not only serve aerodynamic purposes but also mitigate wing wear by increasing wing twisting when the wings hit an obstacle.[122] It is concluded from comparative tests with bumblebee wings, which have a less prominent costal break and longitudinal veins ending before the wing tip, that a reduced venation at the wing tip can already account for collision damage, whereas wings with an increased spanwise rigidity due to full-length veins need alternative mechanisms in the form of costal and humeral breaks to avoid wing fracture.[122]

Wing fold lines enable insects to fold their wings into smaller, often meticulously folded wing packages, thereby allowing the insect to hide in smaller crevices and protect the wing from damage and dessication.[49,117,123,124] Wing folding can be rather simple as in Hymenoptera (longitudinal fold), more elaborated as in Coleoptera (transverse and longitudinal folds), or really complex as in Dermaptera (transverse, longitudinal, and fanwise folds).[49,117,123,124]

In beetle wings, resilin is located along the flexion and fold lines and at the marginal joint, as well as in membranous areas between veins that separate during wing folding.[35] This led to the assumption that resilin not only helps prevent material failure and influence wing deformations but can also support wing unfolding by elastic energy storage.[35] Nevertheless, the latter is probably only true for beetles like *Coccinella septempunctata* that unfold their wings before flapping.[125,126] In beetles like *Pachnoda marginata* that unfold their wings by scissors-like movements of the veins RA and MP1 + 2 (*via* the contraction of the Musculus pleura alaris and basilar muscles) and the involvement of hydraulic haemolymph pressure, elastic energy storage by resilin probably plays only a minor role.[124,127]

Dermapteran wings show a very complex wing folding pattern and achieve a wing area reduction to around 10% up to 5.5% of their original wing size.[49,117] Various structures in dermapteran wings like fold lines, veins, broadened vein patches, membrane areas, and alar fenestrae in cross veins are supported by resilin.[49] Resilin not only helps prevent material failure during the repeated folding and unfolding processes but is also the main driving force for wing folding by storing elastic energy during the active and musculature-driven wing unfolding.[49,117,123] The latter is either driven by wing flapping (*Timomenus lugens*, *Auchenomus* sp.) or by moving the cerci over the wings in a wiping manner (*Forficula auricularia*, *Labia minor*).[49,123] Wing folding is driven by a complex system of (1) alternating convex and concave radiating fold lines, which elicit a fan-wise folding, (2) a ring fold line, which runs through broadened vein patches of anisotropic resilin distribution and folds the fan-wise folded wing package to half of its size, and (3) several other fold lines.[49] Despite their important role in wing folding, fold lines in dermapteran wings are assumed to also support wing flexion during flight, thereby increasing the wings' aerodynamic efficiency.[117,128]

2.2.4 Attachment Systems

Attachment is of key importance for the survival of insects and spiders, allowing them to stick and safely walk or climb on a variety of different surfaces. Except for locomotion, attachment serves additional purposes such as mating, prey capture, escape from predators, or self-grooming. Insects have developed two types of attachment structures to maximise contact formation to a variety of surfaces of different roughness and chemical nature, *i.e.*, hairy and smooth adhesive pads.[129] Hairy adhesive pads are covered with socketed setae or socket-lacking acanthae and maximise their contact by increasing the amount of single contacts at the micro- and nanostructural levels.[76,129–134] Amongst others, they occur in dipterans, megalopterans, and coleopterans. Smooth adhesive pads feature deformable, smooth, sometimes patterned, and fibre-reinforced or foam-like layers, which can be fluid filled and maximise surface contact by adapting to the shape of the substrates' surface.[135] They have been found, for example, in ephemeropterans, hymenopterans, orthopterans, and plecopterans. In general, leg adhesive structures can be found, for example, on derived elements of the pretarsus, on tarsomeres or the tibia, and on claws or transformed thorns.[129,130,135] The attachment efficiency of hairy and smooth adhesive pads relies on the contact they are able to establish with the underlying surface and is influenced by various factors such as (1) pad (and claw) morphology, size, composition, and adaptability, (2) substrate chemistry, wettability, stiffness, and form, (3) surface roughness, and (4) tarsal liquids.[135–140]

Especially on rough surfaces, multiple, rather flexibly adaptable contacts have major benefits.[134] In comparison to one contact of relatively large width, several smaller contacts with widened, spatula-shaped terminal elements show a much higher peeling force (depending on the total peeling line length) and, therefore, attachment ability.[141] Furthermore, when optimal shear force is applied to the thin, spatula-shaped tips, the contact area increases without the presence of slip, especially in the case of previously misaligned spatula.[142] Apart from the widened and thin hair tips, other properties such as hair radius, aspect ratio, Young's modulus, hair density, and flexibility/adaptability play an important role in the contact formation and overall attachment ability of hairy pads.[143] Although tip flexibility and adaptability are very important for optimal contact formation at low normal loads, completely flexible structures with a high aspect ratio possess inferior mechanical stability, as has been shown for plant epicuticular waxes.[144] Buckling and collapse of generally soft setae can lead to clusterisation (or condensation), thereby diminishing the attachment ability of multiple-contact hairy pads.[141,143,145] To overcome this optimisation dilemma of flexibility *versus* stability, thickness and material gradients have evolved in adhesive setae, with the setal thickness and stiffness diminishing towards the tip.[38,130,146] Confocal laser scanning microscopy analyses of the adhesive tarsal setae of the seven-spot ladybird (*C. septempunctata*) (Figure 2.2A,B) and the southern green stink bug (*Nezara viridula*) (Figure 2.7A) revealed that these setae feature a pronounced material gradient with their tips being soft and consisting of resilin-dominated material and their bases exhibiting a relatively high degree of sclerotisation.[38,53] Atomic force microscopical nanoindentation measurements of single adhesive setae of *C. septempunctata* further confirmed the confocal laser scanning microscopy results and showed that Young's modulus ranges from 1.2 MPa at the tip to 6.8 GPa at the base[38] (Figure 2.2C). Numerical simulations demonstrated that such a material gradient with a short, soft tip and a stiff base not only improves the attachment performance of the setae on rough surfaces but also reduces the risk of clusterisation by providing sufficiently stable setal bases to avoid lateral collapse[146] (Figure 2.2D). By contrast, fibres with longer flexible tips and stiff bases show increased clusterisation, though exhibiting good adhesion (Figure 2.2E), and fibres with stiff tips and flexible bases fail due to both inferior adhesion and high clusterisation (Figure 2.2F).

Further studies showed that, apart from the more common thin spatula-shaped, flattened tapered or pointed setal tips, several insects, for example, male beetles, possess adhesive setae with discoidal (or mushroom-shaped) tips[137,147–149] (Figure 2.6A,D). In contrast to the previously mentioned adhesive setae, which feature an overall soft tip, setae with discoidal tips are composed of rather stiff chitinous cuticle at their tip and at most of their central and basal sections[150] (Figure 2.6B,C,E,F). The presence of resilin-dominated cuticles is limited to a joint-like element located between the setal tip and the more proximal section of the seta[150] (Figure 2.6B,C,E,F). It is

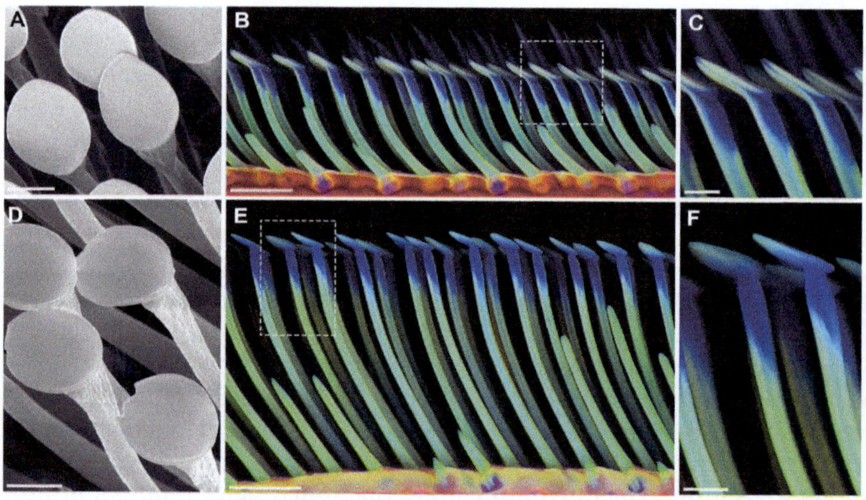

Figure 2.6 Distribution of resilin in discoidal adhesive tarsal setae of male seven-spot ladybirds (*Coccinella septempunctata*) and male Colorado potato beetles (*Leptinotarsa decemlineata*). (A, D) Scanning electron micrographs depicting the tips of single discoidal adhesive tarsal setae of *C. septempunctata* (A) and *L. decemlineata* (D). (B, C, E, F) Confocal laser scanning micrographs (maximum intensity projections) showing overlays of different autofluorescences exhibited by the exoskeleton of discoidal adhesive tarsal setae of *C. septempunctata* (B, C) and *L. decemlineata* (E, F). Blue colour indicates large proportions of resilin. (B, E) Lateral views of complete setae. (C, F) Detailed views of the sections indicated by rectangles in (B) and (E). Scale bars = 5 μm (A, C, D, F) and 20 μm (B, E). Figure reproduced from ref. 150 with permission from the Royal Society of Chemistry.

supposed to increase the adaptability to tilted and uneven surfaces and, thereby, the attachment performance of the tip.[150] For the detachment of discoidal setae from rather stiff and smooth surfaces, high pull-off forces are required so that they are commonly assumed to allow prolonged passive attachment, which, for example, is beneficial for male leaf beetles that attach to female elytra during copulation.[137,151]

Many insect adhesive systems do not operate in dry conditions with their attachment performance mainly relying on Van der Waals forces but instead use tarsal liquids of varying amount and viscosity to improve adhesion by means of capillary actions and self-cleaning processes.[136] In this context, relatively slowly evaporating liquids like the one found on beetle tarsi might also serve the hydration of the soft resilin-bearing setal tips, thereby maintaining the elastic and flexible response of resilin.[38]

Material gradients are also present in smooth adhesive pads.[36,53] As shown for the sac-like pulvilli of the southern green stinkbug (*N. viridula*), the cuticle of the pads' dorsal side is typically rather sclerotised and exhibits a three-dimensional structure of bulges and lamellae, whereas the ventral side is rather smooth, soft, and often dominated by resilin[53] (Figure 2.7B–D). This structure allows the pads to be highly adaptable at the ventral side and, at the same time, facilitates homogenous stress distribution and keeps the pad stable and functional.[53]

Although different pad types (smooth pulvillus, euplantula, arolium) show various species-specific adaptations and differences with respect to the developmental stage and age, their overall ultrastructure is very alike.[36,53,54,135,152] Most of them are composed of a foamy or fibrous material, which differs in thickness and degree of hierarchy, and often show a rather complex cuticle layering with differently developed layers along the proximal–distal and central–lateral axes.[36,53,54,135] Commonly, the pad's ventral cuticle of the fibre-based type consists of four layers: (1) an innermost endocuticle layer, which sometimes is complemented by additional endocuticular layers, (2) an inner exocuticle layer of angled primary rods (called layer of primary rods), which are embedded into a soft matrix layer and/or fluid, (3) an outer exocuticle layer of branching primary rods, generating a more dense layer of secondary fibres (called layer of branching rods; present in the hierarchical pad type), and (4) an electron-dense epicuticle layer, the so-called superficial layer, which keeps the fibres in place and has been shown to be porous at some parts of the pad[36,53,54,135] (Figure 2.8A,C).

Autofluorescence and nanoindentation analyses revealed that, in contrast to hairy adhesive pads that feature setae with a stiff base and a soft tip, smooth pads like the one of the great green bushcricket (*Tettigonia viridissima*) and the migratory locust (*Locusta migratoria*) show a reversed material gradient with a soft core and a stiff outer layer[36,135] (Figure 2.8A–D). More precisely, their filamentous exocuticle was shown to be flexible and to contain resilin, whereas their superficial epicuticular layer (and the most distal part of the dense branching rod layer in *L. migratoria*) is rather stiff.[36] Furthermore, the analyses revealed that the pads of *L. migratoria* are stiffer (higher

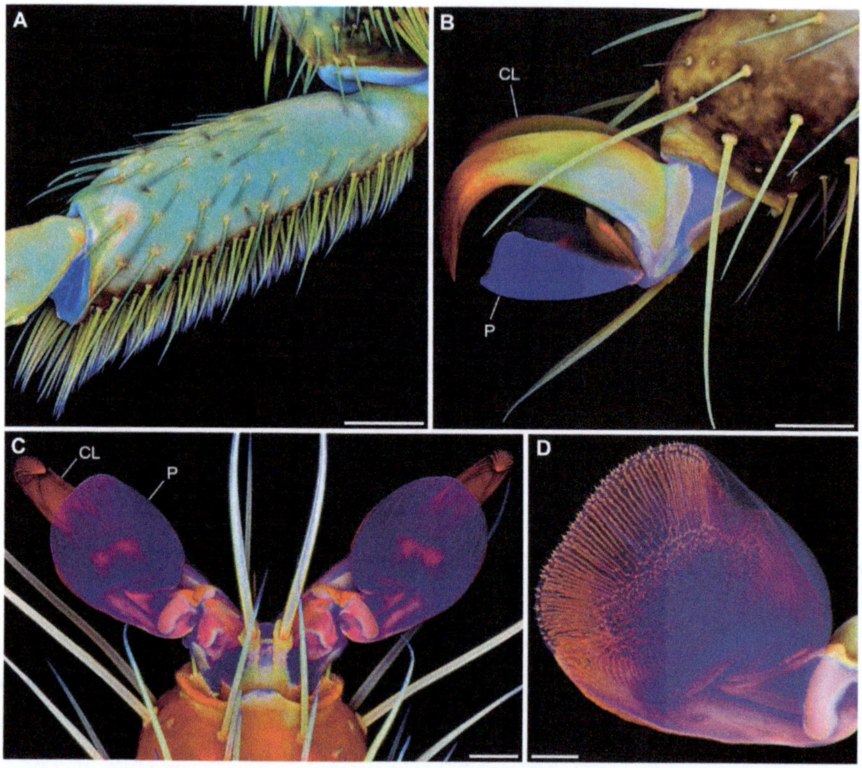

Figure 2.7 Distribution of resilin in attachment structures of the southern green stinkbug (*Nezara viridula*). (A–D) Confocal laser scanning micrographs (maximum intensity projections) showing overlays of different autofluorescences exhibited by the exoskeleton. Blue colour indicates large proportions of resilin, green exoskeleton structures consist mainly of non-sclerotised or weakly sclerotised chitinous material, and red exoskeleton structures are composed of relatively strongly sclerotised chitinous material. (A) Lateral view of the basitarsus with adhesive tarsal setae that exhibit a pronounced longitudinal gradient in the material composition with the tips being mainly composed of resilin. (B, C) Lateral (B) and ventral (C) view of the tip of the distitarsus, showing the pulvilli (P) and the sclerotised claws (CL). (D) Ventral side of a pulvillus featuring large proportions of resilin. Bulges and peripheral ribs that consist of sclerotised chitinous material and are located in the dorsal part of the pulvillus can be seen through the ventral pulvillus part that is transparent because of its large resilin proportion. Scale bars = 150 μm (A), 100 μm (B), 50 μm (C), and 25 μm (D). Figure adapted from ref. 53 with permission from John Wiley & Sons, Copyright © 2018 Wiley Periodicals, Inc.

effective Young's modulus) than those of *T. viridissima* and show a lower work of adhesion[36] (Figure 2.8F,G). This is probably due to the higher rod density and the thicker exocuticle layer of *L. migratoria* pads, making them less adaptable to the substrate[36] (Figure 2.8A–D). Nevertheless, such

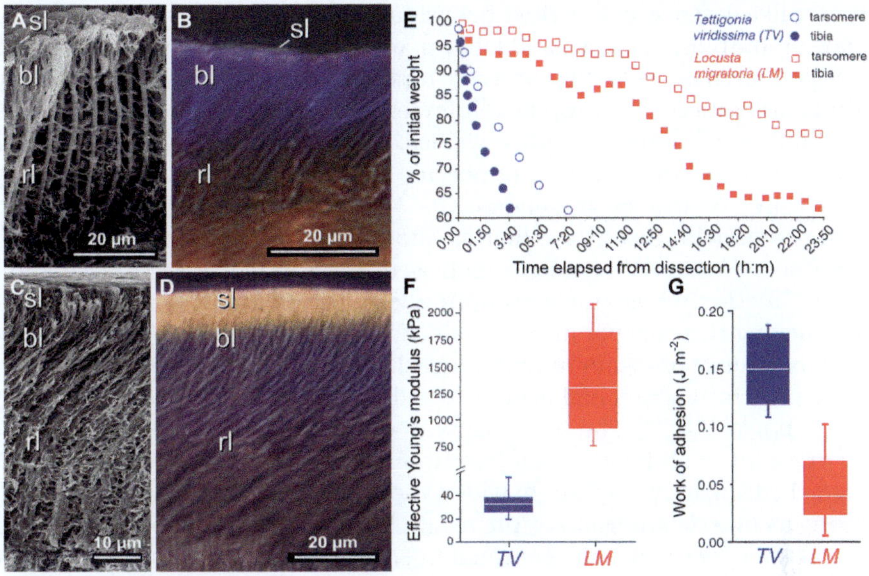

Figure 2.8 Material structure and properties of orthopteran adhesive pads (euplantulae). (A, B) Great green bush-cricket (*Tettigonia viridissima*). (C, D) Migratory locust (*Locusta migratoria*). (A, C) Scanning electron micrographs showing frozen, fractured, substituted, dehydrated, and critical point dried pads. (B, D) Wide-field fluorescence micrographs showing frozen-cut pads. Structures with large proportions of resilin are violet/blue. sl, superficial layer; bl, layer of branching rods; rl, layer of primary rods. (E) Desiccation dynamics of tarsi and pieces of tibiae cut off the body in both species. (F) Effective Young's modulus of attachment pads of both species determined by means of indentation with a spherical tip radius of 250 µm. (G) Work of adhesion of the attachment pads of both species measured by means of indentation with a sphere with a radius of 32 µm. In (F) and (G) the ends of the boxes define the 25th and 75th percentiles, the lines indicate the medians, and the error bars define the 10th and 90th percentiles. (A–G) Adapted from ref. 36 with permission from Springer Nature, Copyright 2006.

a thick layer of epicuticle most likely serves as protection against desiccation, a hypothesis that was supported by weight loss experiments of cut-off pads[36] (Figure 2.8E). Consequently, the structure and the material properties of smooth adhesive pads are probably adapted to the preferred environment, with pad architectures of insects living in dry surroundings representing a compromise between adaptability and stiffness as well as adhesion and resistance to water loss.[36,135]

2.2.5 Mouthparts

Insect mouthparts differ in structure and material composition and are adapted to exploit different food sources. Additionally, they serve other purposes like defence, competitive behaviour, and brood care.[153] They typically include mandibles, maxillae, and labium, which are supplemented by labrum and hypopharynx. Depending on their function, mouthparts can be strongly modified or reduced. They can be reinforced by increased sclerotisation or the incorporation of, amongst others, silica, magnesium, calcium, zinc, and manganese, which is especially important for their tips that directly interact with the prey or ingesta in general.[154,155] At the same time, they are often supplemented with resilin, facilitating shock absorption, joint flexibility, the prevention of material damage, increased adhesion during prey capture, and the functioning as an elastic spring, *e.g.*, for driving the mouthparts of blood-sucking fleas deeper into the skin.[62,63,156–164]

The occurrence of resilin in insect mouthparts was first described in blood-sucking insects, where it had been detected in the salivary and food-pumps of assassin bugs.[55,156–158] The boat-shaped food-pump of *Rhodnius prolixus* features one central and two lateral bands of resilin, which are deformed when the attached muscles contract to widen the volume for blood uptake.[55,157,158] Working as muscle antagonists, the resilin bands spring back to their original shape as soon as the muscle contraction ends, thereby emptying the content into the alimentary tract and allowing the insect to take up large amounts of blood in a short time.[55,157,158] The salivary pump works in more or less the same way. Muscle contraction leads to a sucking of saliva into the pump's cavity, whereas resilin supports the pressurised transport of the saliva into the salivary canal of the mouthparts.[156,164,165] This mechanism is not only used to inject proteolytic proteins underneath the skin of the prey or aggressors but also to spit on predators as an act of defence.[156,157] Apart from food and salivary pumps, resilin has also been found to be involved in the skin-piercing mechanism of blood-sucking insects.[164] For example, fleas use the compression of a resilin pad and a click mechanism to release the stored energy in combination with a lever-like structure at the base of their laciniae to push the latter deeper into the prey's skin.[164]

Dragonfly larvae were shown to use a synchronised dual-catapult mechanism to unfold their prehensile labial mask and accelerate it towards the prey.[166] Similar to, for example, the situation in flea beetles or froghoppers, the power necessary to achieve the recorded acceleration largely surpasses the one achievable by muscles, and morphological experiments revealed that the single catapult mechanisms are comparable to the energy-storing process in jumping insects and include resilin-dominated structures.[166] While the prementum–postmentum joint and the postmentum–head joint are locked in position by a locking muscle and a complex latch mechanism, two power muscles deform resilin-dominated springs.[166] The predatory strike, including the release of both catapults, is triggered by the contraction of a trigger muscle, thereby giving way to the elastically stored energy pushing the labial mask forward and unfolding the prementum–postmentum joint.[166]

Another example of mouthparts benefiting from resilin's elasticity is the butterfly proboscis. The main part of the sucking mouthparts of butterflies comprises (1) modified maxilla with elongated galeae forming the two dorsally and ventrally linked halves of the proboscis, (2) the stipes consisting of a medially sclerotised and laterally membranous tube, and (3) the basal cardo.[159,167] In the resting position, the proboscis is tightly coiled. For food uptake, it is uncoiled and elongated. This process is generated by repeated pumping movements of the stipes compressing the stipital tube, forcing haemolymph into the galeae, and thereby creating enough hydrostatic pressure to completely uncoil the proboscis.[159,160,167,168] During proboscis uncoiling and extension, the dorsal wall or bar of the proboscis, which consists of exo-, meso-, and endocuticle and is supplemented by resilin, is compressed.[160,168] As soon as the stipes valves open and allow the haemolymph to flow back, the proboscis recoils due to the elasticity of the resilin-bearing cuticle, and contractions of the intrinsic galeal muscles further tighten the coil.[159,160,167,169]

Rove beetles of the genus *Stenus* use exceptional adhesive mouthparts for prey capturing.[161–163] Their rod-shaped labium can be extended by several millimetres within milliseconds and is equipped with two sticky pads (paraglossae at the distal part of the prementum).[161–163] The latter show a hierarchical structure with numerous hairs arranged in a hexagonal pattern and possessing terminal spherically shaped ramifications in high density, a hierarchical structure resembling one of dry adhesive systems like gecko feet. The pads are soft and are supplemented by resilin,[163] allowing the overall pad to be expanded by haemolymph before striking the prey and to flexibly adapt to various prey surface topographies of different roughnesses and dimensions.[162] They are covered by a multiphasic adhesive secretion produced in the head capsule and consisting of water-soluble sugars, proteins, and lipids, with only the distal-most tips of the ramifications sticking out. This additional wet adhesive component enables the pads to show high adhesive forces to both hydrophobic as well as hydrophilic surfaces and even more increases the contact area to the prey by pushing the secretion into surface irregularities.[161–163]

Crabs and crayfish use resilin as an antagonist for a single abductor muscle to generate a repeated beating or fanning movement of the flagellum of all three maxillipeds.[170] The joint between the maxillipeds' flagella and exopodits shows a single-plane movement. Repeatedly moving the flagella at 18 to 20 Hz (in crabs) or 10 to 13 Hz (in crayfish) allows the animals to modify exhalant gill currents and guide the water flow over chemoreceptors on their head, thereby contributing to chemoreception and signalling (*e.g.*, by distributing urine odours). During the muscle-controlled abduction of the flagellum, a resilin-bearing spring-like structure at the ventral side of the joint is bent. As soon as the muscle contraction ends, the flagellum is elastically adducted to its original resting position by the energy stored in the resilin–cuticle composite structure.

Copepods occur all over the world, inhabiting virtually every zone of the water column of lakes, rivers, or the sea, no matter their temperature, altitude, or salinity. They account for a large part of the zooplankton[171,172] and feed on smaller zooplankton species or phytoplankton, which largely consists of diatoms, dinoflagellates, cyanobacteria, coccolithophores, and green algae. The diet of marine copepods that feed on phytoplankton commonly consists of a high amount of hard, silica-encased diatoms,[173] which the copepods typically grasp and crush with the gnathobases of their mandibles. The gnathobases of many of these copepods possess stable, relatively short, and compact so-called teeth at their distal end.[174] As shown in two recent studies that examined the diatom-feeding species *Centropages hamatus* and *Rhincalanus gigas*, the structure and material composition of these gnathobase teeth are likely highly adapted to mincing the hard frustules of diatoms without risking material failure.[62,63,154] In *R. gigas*, which is mainly feeding on diatoms, the five biggest of the eight teeth have a large chitinous socket and a siliceous cap-like structure on top.[63,154] The latter is composed of chitin-fibres and aluminium-containing crystalline silica and renders the tip of the teeth stiff and mechanically stable.[63,154] At the bases of their teeth's chitin sockets, the gnathobases feature high proportions of resilin[63] (Figure 2.9F,G). In *C. hamatus*, which is omnivorous but also feeds on considerable amounts of diatoms, only the two biggest of the eight teeth possess an architecture comparable to that of the *R. gigas* gnathobase teeth, with a chitinous socket and a siliceous cap-like structure (Figure 2.9C). The major difference, however, concerns the location of resilin that is found in the form of an additional cap-like structure located in between the silicon cap and the chitinous socket[62,63,154] (Figure 2.9A–E). All other teeth of *C. hamatus* feature a chitinous base with a resilin-dominated cap but lack the siliceous cap (Figure 2.9A–C). Whether located in between the chitinous socket and the siliceous cap or at the bases of the teeth's sockets, resilin is likely compressed during the diatom mincing process and serves the reduction of stress levels and concentrations in the teeth as well as the increase in resistance to failure.[62,63,154] This is especially important for the hard parts of the teeth that probably break more easily under high mechanical load, which has to be applied to be able to crush the mechanically stable frustules.[154] In the smaller teeth of *C. hamatus* that lack the siliceous cap and were assumed to serve the grasping of protists, resilin might additionally help increase the teeth's grip.[62]

As already mentioned above, Odonata larvae possess a modified labium, the prehensile labial mask, which is quickly shot out towards their prey with the help of a catapult.[166] In addition, mandibles and maxillae are used to sense, handle, and crush the food. As in copepods, their tooth-like protrusions, the so-called incisivi, and also the dentisetae of the maxillae have highly sclerotised tips and show material gradients towards their bases and the proximal part of their mouthparts.[175] Resilin occurs at various locations, including dentisetal joints, ridges, or pads, and in larger areas of the mouthparts, likely serving the increase of flexibility and mobility of parts of the mouthparts (either unidirectional or multidirectional) as well as the reduction of material failure and the improvement of shock absorption, especially during the prey-crushing process.[175]

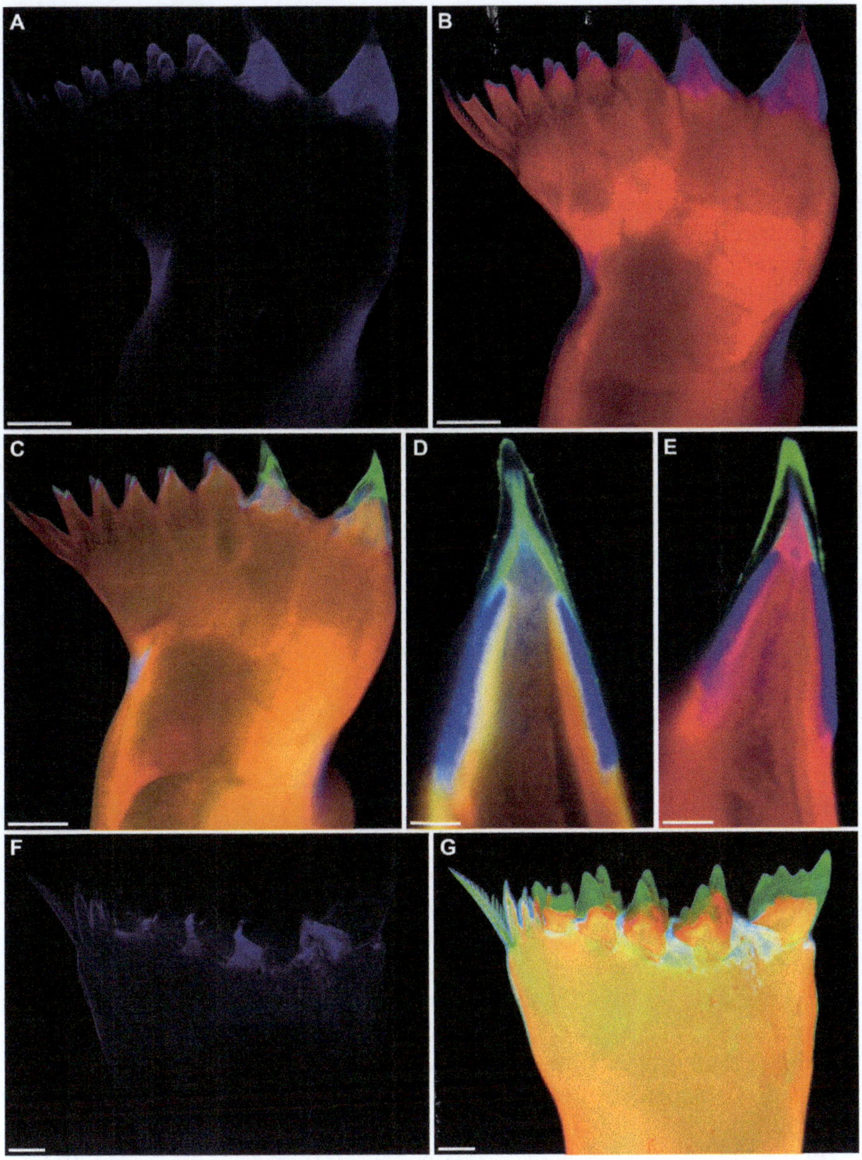

Figure 2.9 Morphology and material composition of mandibular gnathobases of copepods. Confocal laser scanning micrographs showing the gnathobase structures of (A–E) a female of the copepod species *Centropages hamatus* and (F, G) a female of the copepod species *Rhincalanus gigas*. (A–C, F, G) Maximum intensity projections. (D, E) Optical sections of 1 μm thickness through the largest tooth of the gnathobase. (A, F) Distribution of resilin. (B) Chitinous exoskeleton (red) and resilin-dominated structures (blue). (C–E, G) Chitinous exoskeleton (red, orange), resilin-dominated structures (blue, light blue, turquoise), and silica-containing structures (green). Scale bars = 20 μm (A, B, C), 5 μm (D, E), 25 μm (F, G). (A–E) Adapted from ref. 62 with permission from Springer Nature, Copyright 2012. (F, G) Adapted from ref. 63 with permission from Elsevier, Copyright 2015.

2.2.6 Reproductive Organs, Mechanoreceptors and Compound Eyes

Sexual conflicts fuel the continuous coevolution of male and female genitalia as well as behavioural adaptations like sensory exploitation and the female's resistance. As a result, the structure and material composition of male and female genitalia or receptive organs are often arranged in a way that ensures species specificity and a smoother interaction with lower survival costs for the female but at the same time allows sexual selection by, for example, cryptic female choice. A good example for the coevolution of male and female genitalia is the hyper-elongated intromittent organs of male beetles and coiled spermathecal ducts of their female counterparts.[176–178] The female spermathecal duct of *Cassida* leaf beetles is highly spiralled, and its inner wall consists of a thick, homogeneously sclerotised cuticle, whereas its external layer is dominated by resilin.[178] The duct ends in a sperm-storage chamber, the spermatheca, which also controls sperm allocation and usage.[179] A C-shaped part of the spermatheca, which is connected to a gland and the duct, is mostly sclerotised and melanised, and muscle fibres are connected to its C-arms, which enables pumping movements.[179] A region at the centre of the C-arms that is highly stressed during the pumping cycles is dominated by resilin, endowing the site with flexibility and reducing material fatigue.[179] The male intromittent organ is a tube-like flagellum with a length covering several times the beetle's body length.[177,178] For most of its length, it consists of a relatively stiff, sclerotised cuticle.[177,178] The tip region, however, is bent and much softer, with its inner side of the curve being dominated by resilin and the outer curve consisting mainly of non-sclerotised cuticle.[177,178] Being sclerotised at its basal part allows the male flagellum to be pushed inside the female spermathecal duct.[177,178] At the same time, the overall soft, bent tip, with its transverse stiffness gradient with a resilin-dominated inner curve, is able to flexibly adjust to the curvature of the duct and, thereby, to increase the penetration velocity and at the same time reduce the probability of tip fracture.[177,178]

Another famous example of sexual conflict is the traumatic insemination in bed bugs. For mating, the male penetrates the ventral abdominal cuticle of the female with a short, stiff, curved, and pointed intromittent organ and injects sperm and glandular fluid into the female's body cavity, from where these liquids are transported to the ovaries.[180,181] This mating process injures the female and imposes survival costs,[181] and in order to increase mating rates, males typically practice situation exploitation by, for example, increasing their mating attempts at enlarged postfeeding females with less resistance behaviour.[182] Resulting from this sexual conflict, females of the common bed bug (*Cimex lectularius*) have evolved the so-called spermalege. From the outside, this specialised organ is visible as a small notch on the ventral and right side of the posterior edge of the fifth abdominal segment[183] (Figure 2.10A,B). The notch exposes the underlying intersegmental membrane and cuticle of the sixth segment (Figure 2.10A,B). Confocal laser scanning microscopy analyses revealed a high proportion of

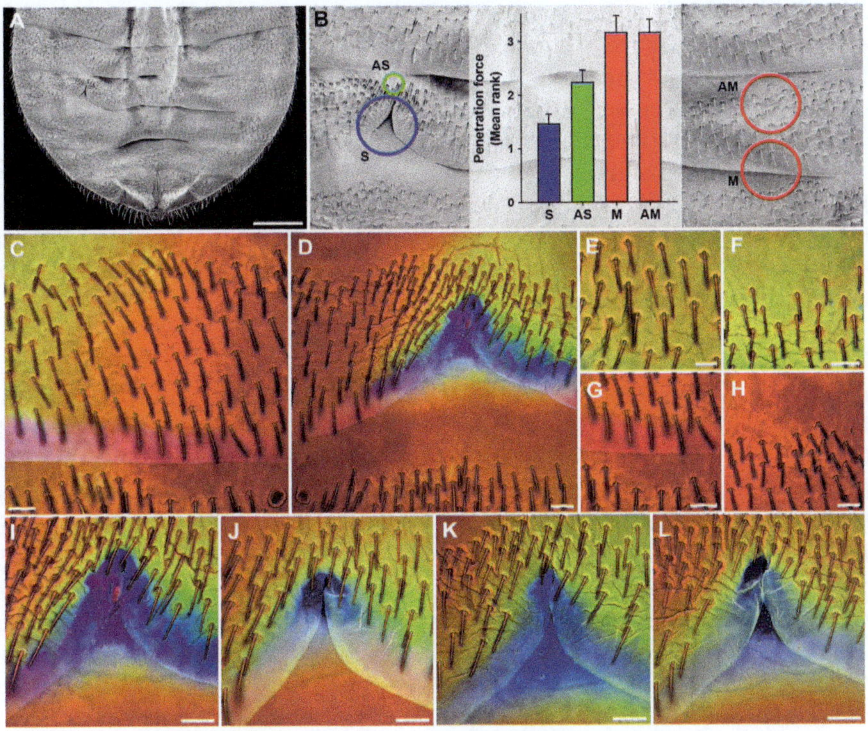

Figure 2.10 Material composition and properties of the ventral abdominal cuticle of females of the common bed bug (*Cimex lectularius*). (A) Abdomen overview (scanning electron micrograph). (B) Section of (A) indicating the locations of the spermalege (S) and three other cuticle areas called AS, M, and AM, and penetration forces (mean ranks and standard errors) determined for these four cuticle sites. (C–L) Confocal laser scanning micrographs (maximum intensity projections) showing overlays of different autofluorescences exhibited by the exoskeletons. Blue colour indicates large proportions of resilin. (C, D) Autofluorescence composition of the cuticle in the left (C) and right (D) abdomen parts. The dominance of violet/blue autofluorescence (shown in blue) is restricted to the spermalege, clearly indicating that only at this site the cuticle contains large proportions of resilin. (E–H) Autofluorescence composition of the cuticle at the sites AS (E, F), M (G), and AM (H). The cuticle at M and AM consists mainly of sclerotised chitinous material, indicated by the dominance of autofluorescence shown in red, while the presence of large proportions of autofluorescence shown in green in the cuticle at AS indicates that the respective material consists mainly of weakly or non-sclerotised chitinous material. (I–L) Autofluorescence composition of the cuticle at the spermaleges of different 1-week-old females, indicating variation of the extent of the resilin-dominated spermalege structures between females. Scale bars = 500 μm (A), 50 μm (C, D, F–L), 25 μm (E). Figure reproduced from ref. 183, https://doi.org/10.1098/rsif.2014.1107, under the terms of the CC BY 4.0 license https://creativecommons.org/licenses/by/4.0/.

resilin in the cuticle at the notch with a relatively abrupt transition to chitinous and in part sclerotised cuticle in the surrounding abdominal areas[183] (Figure 2.10C–L). These results were confirmed by microindentation tests, showing that the necessary force for penetrating the cuticle is much lower at the spermalege than at other tested abdominal sites[183] (Figure 2.10B). In addition, tissue damage and haemolymph loss were lowest at the spermalege.[183] Because of the high proportion of resilin, penetration is likely eased at the spermalege, and resilin's elasticity presumably facilitates a postpenetration sealing of the puncture, thereby reducing haemolymph loss and the risk of bacterial infection and finally reducing the mating costs for the female.[183]

Insects perceive stimuli of their terrestrial or aquatic environment with a variety of different receptors, including numerous types of mechano-, chemo- and thermo-hygroreceptors. Mechanoreceptors occur in various forms, ranging from filiform hairs and bristles to dome-shaped campaniform sensilla and chordotonal organs.[184–187] They allow the insect to detect movements, air currents, sound and vibrations of their surroundings, gravity, and strain in the cuticle that is caused by external or internal forces (*e.g.*, joint deflection). Mechanoreceptive hairs can be found in the form of long, thick bristles, which mainly respond to direct touch, and long, thin filiform hairs, which are used to detect more subtle stimuli like, for example, low-frequency sound and faint air currents.[184,185,187–189] Campaniform sensilla function as proprioceptors and are distributed all over the body but are often found near joints.[184–191] The sensilla typically consist of a hair shaft (in the case of hair sensilla) or a dome-shaped structure (in the case of campaniform sensilla), which originates from a trichogen cell, and a socket with a joint- or cap-membrane, which is formed by a tormogen cell.[76,192] These membranes were found to consist of resilin-dominated cuticle (Figure 2.11) and are assumed to flexibly suspend the hair shaft and the dome-shaped structure to their sockets.[26,34,184,186] Such a flexible suspension of mechanoreceptive hairs not only prevents material damage but also allows the hair shaft to move and bend, thereby stimulating the dendrite of the sensory neuron. The movement is sometimes limited to a certain direction by lateral projections of the hair shaft (articular pegs), fitting into an elliptical counterpart of the septum.[193–195] With the long, thick mechanoreceptive bristles extending beyond the pretarsus and the pulvilli of the drone fly (*Eristalis tenax*) (Figure 2.11A–E), the insects can likely sense the upcoming contact before the pulvilli and the claws actually touch the substrate as well as the deflection of the tarsal segments. By contrast, the filiform hair sensilla (Figure 2.11F) densely covering the cerci of crickets are mainly involved in sensing air current signals from predators and conspecifics.[193]

Insect compound eyes show diverse designs adapted to different light conditions (diurnal or nocturnal activity) and lifestyles. Apart from the type of compound eye (apposition or superposition) with different pigmental distribution, the ommatidia vary with respect to visual pigment, number and size, fusion of rhabdomeres, spatial resolution, and epicuticular and partly antireflective coverage. For their proper functioning, it is essential to transmit and bundle light in order to focus it on the photoreceptor cells.

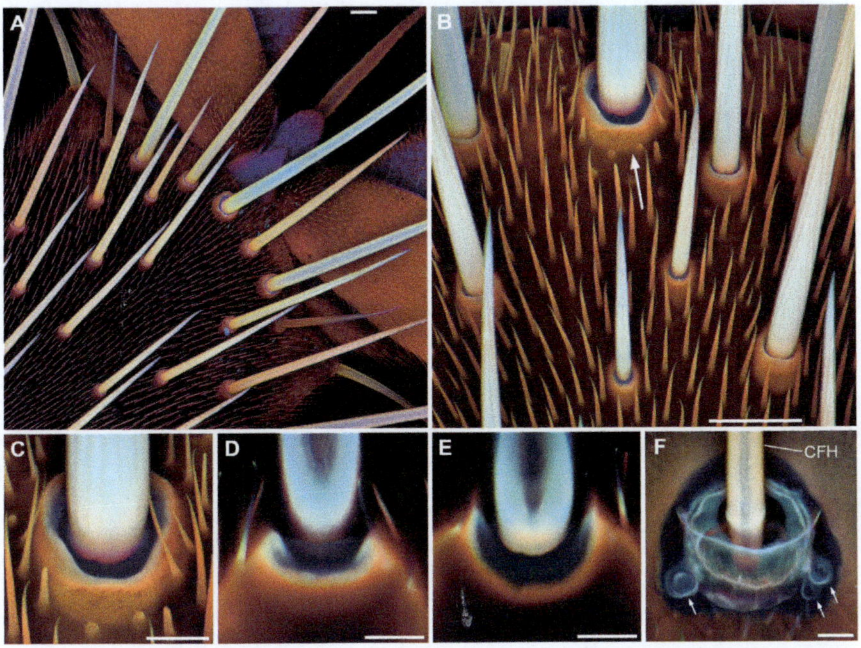

Figure 2.11 Resilin in mechanoreceptors. (A–F) Confocal laser scanning micrographs showing overlays of different autofluorescences exhibited by the exoskeleton. Blue colour indicates large proportions of resilin. (A) Dorsal view of the pretarsus of a third leg of a female drone fly (*Eristalis tenax*). (B) Dorsal view of a section of a second leg's pretarsus of a male *E. tenax*. The arrow highlights a hair plate sensillum. (C) Larger view of the hair plate sensillum highlighted in (B). (D, E) Confocal laser scanning micrographs showing 1 μm thick optical sections through the hair plate sensillum shown in (C). (F) Cercal filiform hair (CFH) and associated campaniform sensilla (highlighted by small arrows) on a cercus of a female house cricket (*Acheta domesticus*). (A–C, F) Maximum intensity projections. Scale bars = 25 μm (A, B), 10 μm (C–F). Figure reproduced from ref. 26 with permission from John Wiley & Sons, Copyright © 2011 The Authors, Journal of Microscopy © 2011 Royal Microscopical Society.

Therefore, apart from sustaining a mechanical resistance and protecting the ommatidia against damage, ensuring an optimal transmittance of light to the crystalline cone is the key role of the corneal lens in diurnal and especially in nocturnal insects. The cuticle of the corneal lens covering the crystalline cone of every single ommatidium is a continuous exocuticle layer of varying thickness (with the maximum thickness found right above the crystalline cones) and consists of numerous cuticular lamellae (*e.g.*, approximately 60 in the scorpionfly *Panorpa dubia*).[196–198] In most insects, this corneal exocuticle layer is devoid of melanin pigment and phenolically tanned protein and is described as hyaline and transparent.[196,197,199] The presence of resilin or resilin-like proteins in the corneal lens has been

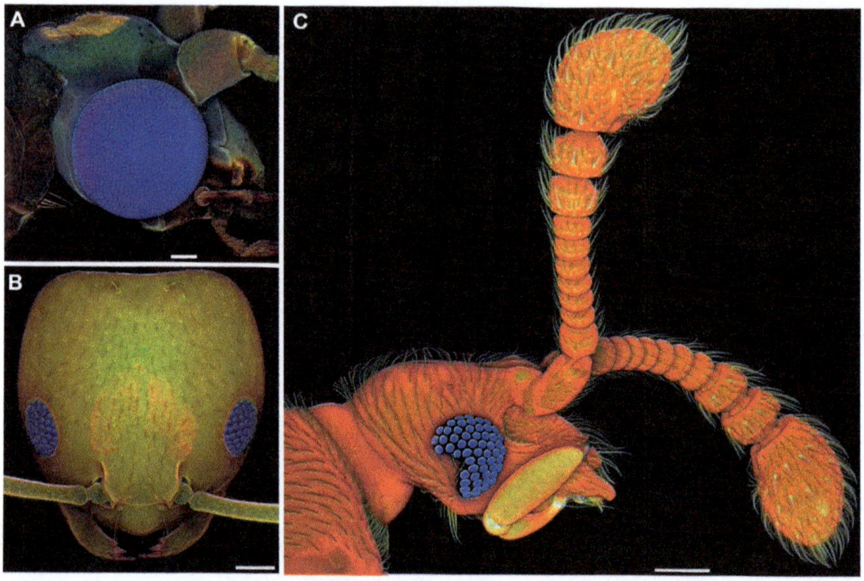

Figure 2.12 Resilin in compound eyes. Confocal laser scanning micrographs (maximum intensity projections) depicting overlays of different autofluorescences exhibited by the exoskeleton. Blue colour indicates large proportions of resilin. (A) Lateral view of the head of a male green lacewing (*Chrysoperla carnea*). (B) Frontal view of the head of a pharaoh ant (*Monomorium pharaonis*) worker. (C) Lateral view of the head of a beetle of the genus *Circocerus*. Scale bars = 100 μm (A, C), 50 μm (B). (A, B) Adapted from ref. 26 with permission from John Wiley & Sons, Copyright © 2011 The Authors, Journal of Microscopy © 2011 Royal Microscopical Society.

described for various arthropods, including centipedes,[200] fireflies,[196] honeybees,[197] dragonflies,[199] lacewings, and ants[26] (Figure 2.12). Such a composite material, consisting of directionally arranged, non-sclerotised chitin lamellae interspersed with amorphous, transparent, colourless, and chemically resistant resilin, guarantees optical transmittance and stability of the corneal lens and, therefore, is considered to be generally very suitable for the construction of optical systems.

2.3 Conclusion

Resilin occurs in a large variety of exoskeleton systems. This intrinsically disordered protein stands out for its high flexibility, resilience, and long-range deformability and exists in either virtually pure form or as a composite, the latter combining diverse properties and advantages of the involved materials. This allows resilin-supplemented systems to efficiently fulfil a broad range of different, highly specialised functions, ranging from the acceleration of catapult-like jumps and the increase in the transparency of compound eyes or the improvement of tarsal attachment ability to the sealing of wounds and the overall reduction of material fatigue and damage. The extraordinary properties of resilin and resilin composites have probably been the basis for the evolution of such highly diverse resilin-supplemented systems in arthropods and are nowadays the reason for the development of resilin-like polypeptide-based biomaterials and resilin-inspired engineering applications.

Acknowledgements

Joseph Parker (Division of Biology and Biological Engineering, California Institute of Technology, Pasadena, CA, United States) made the specimen shown in Figure 2.10C available to us. Igor Siwanowicz (Howard Hughes Medical Institute/Janelia Research Campus, Ashburn, VA, United States) kindly provided the original micrographs of Figure 2.4. Peter Attermeyer, Dario Furlani, and Ingo Bartholomäus (Carl Zeiss Microscopy GmbH, Jena, Germany) permitted the use of a ZEISS LSM 800 and thereby enabled the production of Figure 2.10C. This book chapter is a modification of the publication 'Michels J., Appel E., Gorb S. N. (2016) Functional diversity of resilin in Arthropoda. Beilstein Journal of Nanotechnology 7, 1241–1259'.

References

1. P. R. Shewry, A. S. Tatham and A. J. Bailey, *Elastomeric Proteins: Structures, Biomechanical Properties, and Biological Roles*, Cambridge University Press, Cambridge, 2004.
2. T. Weis-Fogh, *J. Exp. Biol.*, 1960, **37**, 889–907.
3. T. Weis-Fogh, *J. Mol. Biol.*, 1961, **3**, 648–667.
4. T. Weis-Fogh, *J. Mol. Biol.*, 1961, **3**, 520–531.
5. S. O. Andersen and T. Weis-Fogh, *Adv. Insect Physiol.*, 1964, **2**, 1–65.
6. K. Bailey and T. Weis-Fogh, *Biochim. Biophys. Acta*, 1961, **48**, 452–459.
7. S. O. Andersen, Resilin, in *Comprehensive Biochemistry, Vol. 26C: Extracellular and Supporting Structures*, ed. M. Florkin and E. H. Stotz, Elsevier, Amsterdam, 1971, pp. 633–657.
8. S. Rauscher and R. Pomès, Structural Disorder and Protein Elasticity, in *Fuzziness: Structural Disorder in Protein Complexes*. 2012, ed. M. Fuxreiter and P. Tompa, *Adv. Exp. Med. Biol.*, 2012, **725**, pp. 159–183.
9. S. Rauscher, S. Baud, M. Miao, F. W. Keeley and R. Pomès, *Structure*, 2006, **14**, 1667–1676.
10. S. Cheng, M. Cetinkaya and F. Gräter, *Biophys. J.*, 2010, **99**, 3863–3869.
11. A. A. Adzhubei, M. J. Sternberg and A. A. Makarov, *J. Mol. Biol.*, 2013, **425**, 2100–2132.
12. S. O. Andersen, *Insect Biochem. Mol. Biol.*, 2010, **40**, 541–551.
13. S. O. Andersen, Structure and Function of Resilin, in *Elastomeric Proteins: Structures, Biomechanical Properties, and Biological Roles*, ed. P. R. Shewry, A. S. Tatham and A. J. Bailey, Cambridge University Press, Cambridge, 2003, pp. 259–278.
14. T. Weis-Fogh, *Nature*, 1970, **227**, 718–721.

15. S. O. Andersen, *Insect Biochem. Mol. Biol.*, 2004, **34**, 459–466.
16. J. E. Mark and B. Erman, *Rubberlike Elasticity: A Molecular Primer*, Cambridge University Press, Cambridge, 2nd edn, 2007.
17. M. Jensen and T. Weis-Fogh, *Philos. Trans. R. Soc., B*, 1962, **254**, 137–169.
18. C. M. Elvin, A. G. Carr, M. G. Huson, J. M. Maxwell, R. D. Pearson, T. Vuocolo, N. E. Liyou, D. C. C. Wong, D. J. Merritt and N. E. Dixon, *Nature*, 2005, **437**, 999–1002.
19. R. E. Lyons, D. C. Wong, M. Kim, N. Lekieffre, M. G. Huson, T. Vuocolo, D. J. Merritt, K. M. Nairn, D. M. Dudek, M. L. Colgrave and C. M. Elvin, *Insect Biochem. Mol. Biol.*, 2011, **41**, 881–890.
20. A. Kovalev, A. Filippov and S. N. Gorb, *J. Comp. Physiol., A*, 2018, **204**, 409–417.
21. G. Qin, S. Lapidot, K. Numata, X. Hu, S. Meirovitch, M. Dekel, I. Podoler, O. Shoseyov and D. L. Kaplan, *Biomacromolecules*, 2009, **10**, 3227–3234.
22. L. Li, S. Teller, R. J. Clifton, X. Jia and K. L. Kiick, *Biomacromolecules*, 2011, **12**, 2302–2310.
23. L. Li, Z. Tong, X. Jia and K. L. Kiick, *Soft Matter*, 2013, **9**, 665–673.
24. M. Y. Truong, N. K. Dutta, N. R. Choudhury, M. Kim, C. M. Elvin, A. J. Hill, B. Thierry and K. Vasilev, *Biomaterials*, 2010, **31**, 4434–4446.
25. R. Balu, J. Whittaker, N. K. Dutta, C. M. Elvin and N. R. Choudhury, *J. Mater. Chem. B*, 2014, **2**, 5936–5947.
26. J. Michels and S. N. Gorb, *J. Microsc.*, 2012, **245**, 1–16.
27. H. C. Bennet-Clark and E. C. A. Lucey, *J. Exp. Biol.*, 1967, **47**, 59–76.
28. D. Young and H. C. Bennet-Clark, *J. Exp. Biol.*, 1995, **198**, 1001–1019.
29. S. O. Andersen, *Biochim. Biophys. Acta*, 1963, **69**, 249–262.
30. S. O. Andersen, *Acta Physiol. Scand.*, 1966, **66**(Suppl. 263), 1–81.
31. D. Neff, S. F. Frazier, L. Quimby, R.-T. Wang and S. Zill, *Arthropod Struct. Dev.*, 2000, **29**, 75–83.
32. M. Burrows, J. A. Borycz, S. R. Shaw, C. M. Elvin and I. A. Meinertzhagen, *PLoS One*, 2011, **6**, e28456.
33. D. C. C. Wong, R. D. Pearson, C. M. Elvin and D. J. Merritt, *Dev. Dyn.*, 2012, **241**, 333–339.
34. S. Lerch, R. Zuber, N. Gehring, Y. Wang, B. Eckel, K.-D. Klass, F.-O. Lehmann and B. Moussian, *BMC Biol.*, 2020, **18**, 195.
35. F. Haas, S. Gorb and R. Blickhan, *Proc. R. Soc. Lond. B*, 2000, **267**, 1375–1381.
36. P. Perez Goodwyn, A. Peressadko, H. Schwarz, V. Kastner and S. Gorb, *J. Comp. Physiol., A*, 2006, **192**, 1233–1243.
37. S. Niederegger and S. Gorb, *J. Insect Physiol.*, 2003, **49**, 611–620.
38. H. Peisker, J. Michels and S. N. Gorb, *Nat. Commun.*, 2013, **4**, 1661.
39. E. Fujimori, *Biochim. Biophys. Acta*, 1978, **534**, 82–88.
40. S. Garcia-Castineiras, J. Dillon and A. Spector, *Exp. Eye Res.*, 1978, **26**, 461–476.
41. R. Gast and J. Lee, *Proc. Natl. Acad. Sci. U. S. A.*, 1978, **75**, 833–837.

42. A. Giurginca, V. Šustr, K. Tajovský, M. Giurginca and I. Matei, *ZooKeys*, 2015, **515**, 111–125.
43. S. N. Gorb, *J. Morphol.*, 1996, **230**, 219–230.
44. S. N. Gorb, *Arthropod Struct. Dev.*, 2004, **33**, 201–220.
45. M. Burrows, S. R. Shaw and G. P. Sutton, *BMC Biol.*, 2008, **6**, 41.
46. M. Burrows, *J. Exp. Biol.*, 2010, **213**, 469–478.
47. M. Burrows, *J. Exp. Biol.*, 2011, **214**, 2362–2374.
48. M. Burrows and G. P. Sutton, *J. Exp. Biol.*, 2012, **215**, 3501–3512.
49. F. Haas, S. Gorb and R. J. Wootton, *Arthropod Struct. Dev.*, 2000, **29**, 137–146.
50. S. N. Gorb, *Naturwissenschaften*, 1999, **86**, 552–555.
51. E. Appel and S. N. Gorb, *Bioinspiration Biomimetics*, 2011, **6**, 046006.
52. S. Donoughe, J. D. Crall, R. A. Merz and S. A. Combes, *J. Morphol.*, 2011, **272**, 1409–1421.
53. M. Rebora, J. Michels, G. Salerno, L. Heepe, E. Gorb and S. N. Gorb, *J. Morph.*, 2018, **279**, 660–672.
54. M. Rebora, S. Salerno, S. Piersanti, E. V. Gorb and S. N. Gorb, *Zoomorphology*, 2021, **140**, 85–102.
55. H. A. Edwards, *J. Exp. Biol.*, 1983, **105**, 407–409.
56. A. Raghu Varman and H. R. Hermann, *J. Anim. Morphol. Physiol.*, 1982, **29**, 284–285.
57. H. C. Bennet-Clark, *J. Exp. Biol.*, 1997, **200**, 1681–1694.
58. A. Raghu Varman, *J. Ga. Entomol. Soc.*, 1981, **16**, 11–13.
59. A. Raghu Varman, *Experientia*, 1980, **36**, 564.
60. T. Kannupandi, *Acta Histochem.*, 1976, **56**, 73–79.
61. S. O. Andersen, Structure and Function of Resilin, in *Elastomeric Proteins: Structures, Biomechanical Properties, and Biological Roles*, ed. P. R. Shewry, A. S. Tatham and A. J. Bailey, Cambridge University Press, Cambridge, 2003, pp. 259–278.
62. J. Michels, J. Vogt and S. N. Gorb, *Sci. Rep.*, 2012, **2**, 465.
63. J. Michels, J. Vogt, P. Simon and S. N. Gorb, *Zoology*, 2015, **118**, 141–146.
64. S. C. G. Dillinger and A. B. Kesel, *Arthropod Struct. Dev.*, 2002, **31**, 95–101.
65. S. Govindarajan and G. S. Rajulu, *Experientia*, 1974, **30**, 908–909.
66. A. T. Sensenig and J. W. Shultz, *J. Arachnol.*, 2004, **32**, 1–10.
67. G. Sundara Rajulu, *Indian J. Exp. Biol.*, 1971, **9**, 122–123.
68. L. V. Lopez-Llorca and S. C. Fry, *Nematologica*, 1989, **35**, 165–179.
69. D. P. DeVore and R. Gruebel, *J. Biochem. Biophys. Res. Commun.*, 1978, **80**, 993–999.
70. K. Ramalingam, *Parasitology*, 1973, **66**, 1–7.
71. W. L. Wong, J. Michels and S. N. Gorb, *Parasitology*, 2013, **140**, 95–98.
72. C. A. Foerder and B. M. Shapiro, *Proc. Natl. Acad. Sci. U. S. A.*, 1977, **74**, 4214–4218.
73. J. F. V. Vincent Design for Living: The Elastic-sided Locust, in *The Insect Integument*, ed. H. R. Hepburn, Elsevier, Amsterdam, 1976, pp. 401–419.
74. R. H. Hackman and M. Goldberg, *Int. J. Parasitol.*, 1985, **15**, 249–254.
75. H. R. Hepburn and P. Levy, *J. Entomol. Soc. South. Afr.*, 1975, **38**, 131–141.

76. A. G. Richards and P. A. Richards, *Int. J. Insect Morphol. Embryol.*, 1979, **8**, 143–157.
77. H. R. Hepburn Structure of the Integument, in *Comprehensive Insect Physiology, Biochemistry and Pharmacology, Vol. 3: Integument, Respiration and Circulation*, ed. G. A. Kerkut and L. I. Gilbert, Pergamon Press, Oxford, 1985, pp. 1–58.
78. J. F. V. Vincent and S. D. E. Wood, *Nature*, 1972, **235**, 167–168.
79. J. F. V. Vincent, *Proc. R. Soc. Lond. B*, 1975, **188**, 189–201.
80. J. F. V. Vincent, *Tissue Cell*, 1981, **13**, 18–31.
81. R. H. Hackman and M. Goldberg, *J. Insect Physiol.*, 1987, **33**, 39–50.
82. R. H. Hackman, *J. Insect Physiol.*, 1975, **21**, 1613–1623.
83. S. E. Reynolds, *J. Exp. Biol.*, 1975, **62**, 69–80.
84. S. N. Gorb, *J. Zool.*, 2000, **250**, 479–494.
85. S. S. Bidaye, T. Bockemühl and A. Büschges, *J. Neurophysiol.*, 2018, **119**, 459–475.
86. M. Rothshild, Y. Schlein, K. Parker, C. Neville and S. Sternberg, *Philos. Trans. R. Soc., B*, 1975, **271**, 499–515.
87. W. Gronenberg, *J. Comp. Physiol., A*, 1996, **178**, 727–734.
88. H. C. Bennet-Clark, *J. Exp. Biol.*, 1975, **63**, 53–83.
89. M. Burrows and O. Morris, *J. Exp. Biol.*, 2003, **206**, 1035–1049.
90. M. E. G. Evans, *J. Zool.*, 1973, **169**, 181–194.
91. D. G. Furth, W. Traub and I. Harpaz, *J. Exp. Zool.*, 1983, **227**, 43–47.
92. D. G. Furth, The Jumping Apparatus of Flea Beetles (Alticinae) – The Metafemoral Spring, in *Chrysomelidae Biology*, ed. P. Jolivet, E. Petitpierre and T. H. Hsiao, Kluwer Academic Publishers Group, Dordrecht, 1988, pp. 287–297.
93. D. Furth and K. Suzuki, *Syst. Entomol.*, 1992, **17**, 341–349.
94. Y. Ruan, A. S. Konstantinov, G. Shi, Y. Tao, Y. Li, A. J. Johnson, X. Luo, X. Zhang, M. Zhang, J. Wu, W. Li, S. Ge and X. Yang, *Zookeys*, 2020, **915**, 87–105.
95. M. Burrows, *J. Exp. Biol.*, 2012, **215**, 3612–3621.
96. G. P. Sutton and M. Burrows, *Curr. Biol.*, 2018, R135–R147.
97. I. Siwanowicz and M. Burrows, *eLife*, 2017, **6**, e23824.
98. G. Qin, X. Hu, P. Cebe and D. L. Kaplan, *Nat. Commun.*, 2012, **3**, 1003.
99. T. G. Bayley, G. P. Sutton and M. Burrows, *J. Exp. Biol.*, 2012, **215**, 1151–1161.
100. F. O. Lehmann, S. Gorb, N. Nasir and P. Schützner, *J. Exp. Biol.*, 2011, **214**, 2949–2961.
101. E. Appel and S. N. Gorb, *Zoologica*, 2014, **159**, 1–159.
102. E. Appel, L. Heepe, C. P. Lin and S. N. Gorb, *J. Anat.*, 2015, **227**, 561–582.
103. A. C. Neville, *J. Exp. Biol.*, 1960, **37**, 631–656.
104. R. King, *Dynamic Mechanical Properties of Resilin*, Master's thesis, Virginia Polytechnic Institute and State University, Blacksburg, VA, U.S.A., 2010.
105. S. O. Andersen, *Biochim. Biophys. Acta*, 1964, **93**, 211–213.

106. F. Bäumler and S. Büsse, *Biol. Lett.*, 2019, **15**, 20190127.
107. A. C. Neville, *J. Insect Physiol.*, 1963, **9**, 265–278.
108. T. Weis-Fogh, *J. Exp. Biol.*, 1973, **59**, 169–230.
109. J. Gosline, M. Lillie, E. Carrington, P. Guerette, C. Ortlepp and K. Savage, *Philos. Trans. R. Soc., B*, 2002, **357**, 121–132.
110. U. Choudhury, *Dynamic Mechanical Properties of Cockroach (Periplaneta americana) Resilin*, Master's thesis, Virginia Polytechnic Institute and State University, Blacksburg, VA, U.S.A., 2012.
111. W. Nachtigall, A. Wisser and D. Eisinger, *J. Comp. Physiol., B*, 1998, **168**, 323–344.
112. R. J. Wootton, K. E. Evans, R. Herbert and C. W. Smith, *J. Exp. Biol.*, 2000, **203**, 2921–2931.
113. D. J. S. Newman, *The Functional Wing Morphology of Some Odonata*, PhD thesis, University of Exeter, Exeter, 1982.
114. H. Rajabi, N. Ghoroubi, A. Darvizeh, J.-H. Dirks, E. Appel and S. N. Gorb, *Bioinspiration Biomimetics*, 2015, **10**, 056003.
115. H. Rajabi, A. Shafiei, A. Darvizeh, J.-H. Dirks, E. Appel and S. N. Gorb, *R. Soc. Open Sci.*, 2016, **3**, 160006.
116. Y. Ma, J. G. Ning, H. L. Ren, P. F. Zhang and H. Y. Zhao, *J. Exp. Biol.*, 2015, **218**, 2136–2142.
117. J. Deiters, W. Kowalczyk and T. Seidl, *Biol. Open*, 2016, **5**, 638–644.
118. A. M. Mountcastle and S. A. Combes, *Proc. R. Soc. Lond. B*, 2013, **280**, 20130531.
119. Y. Ma, H. Ren, H. Rajabi, H. Zhao, J. Ning and S. N. Gorb, *J. Insect Physiol.*, 2019, **118**, 103936.
120. A. Toofani, S. H. Eraghi, M. Khorsandi, A. Khaheshi, A. Darvizeh, S. N. Gorb and H. Rajabi, *Acta Biomater.*, 2020, **110**, 188–195.
121. J. Michels, E. Appel and S. N. Gorb, *Arthropod Struct. Dev.*, 2021, **60**, 101008.
122. A. M. Mountcastle and S. A. Combes, *J. Exp. Biol.*, 2014, **217**, 1108–1115.
123. F. Haas, J. T. C. Hwen and H. B. Tang, *Arthropod Syst. Phyl.*, 2012, **70**, 95–105.
124. F. Haas and R. G. Beutel, *Zoology*, 2001, **104**, 123–141.
125. K. Saito, S. Nomura, S. Yamamoto, R. Niyama and Y. Okabe, *Proc. Natl. Acad. Sci. U. S. A.*, 2017, **114**, 5624–5628.
126. H. V. Phan and H. C. Park, *Science*, 2020, **370**, 1214–1219.
127. J. Sun, M. Ling, W. Wu, B. Bhushan and J. Tong, *Int. J. Mol. Sci.*, 2014, **15**, 6009–6018.
128. R. J. Wootton, *Annu. Rev. Entomol.*, 1992, **37**, 113–140.
129. S. N. Gorb and R. G. Beutel, *Naturwissenschaften*, 2001, **88**, 530–534.
130. S. N. Gorb, *Attachment Devices of Insect Cuticle*, Kluwer Academic Publishers, Dordrecht, 2001.
131. S. N. Gorb, R. G. Beutel, E. V. Gorb, Y. Jiao, V. Kastner, S. Niederegger, V. L. Popov, M. Scherge, U. Schwarz and W. Vötsch, *Integr. Comp. Biol.*, 2002, **42**, 1127–1139.
132. C. Creton and S. N. Gorb, *MRS Bull.*, 2007, **32**, 466–472.

133. D. Voigt, J. M. Schuppert, S. Dattinger and S. N. Gorb, *J. Insect. Physiol.*, 2008, **54**, 765–776.
134. C.-Y. Hui, N. J. Glassmaker, T. Tang and A. Jagota, *J. R. Soc., Interface*, 2004, **1**, 35–48.
135. S. N. Gorb, Smooth Attachment Devices in Insects, in *Advances in Insect Physiology: Insect Mechanics and Control*, ed. J. Casas and S. J. Simpson, Elsevier (Academic Press), London, 2008, pp. 81–116.
136. H. Peisker, L. Heepe, A. E. Kovalev and S. N. Gorb, *J. R. Soc., Interface*, 2014, **11**, 20140752.
137. L. Heepe, D. S. Petersen, L. Tölle, J. O. Wolff and S. N. Gorb, *Appl. Phys. A: Mater. Sci. Process.*, 2017, **123**, 34.
138. D. Voigt, P. Perez Goodwyn, M. Sudo, K. Fujisaki and M. Varenberg, *Entomol. Sci.*, 2019, **22**, 105–118.
139. E. V. Gorb and S. N. Gorb, *J. Insect Physiol.*, 2020, **121**, 104011.
140. G. Salerno, M. Rebora, S. Piersanti, E. V. Gorb and S. N. Gorb, *Zoology*, 2020, **139**, 125748.
141. M. Varenberg, N. M. Pugno and S. N. Gorb, *Soft Matter*, 2010, **6**, 3269–3272.
142. A. E. Filippov, V. L. Popov and S. N. Gorb, *J. Theor. Biol.*, 2011, **276**, 126–131.
143. R. Spolenak, S. N. Gorb and E. Arzt, *Acta Biomater.*, 2005, **1**, 5–13.
144. F. M. Borodich, E. V. Gorb and S. N. Gorb, *Appl. Phys. A: Mater. Sci. Process.*, 2010, **100**, 63–71.
145. A. Jagota and S. Bennison, *J. Integr. Comp. Biol.*, 2002, **42**, 1140–1145.
146. S. N. Gorb and A. E. Filippov, *Beilstein J. Nanotechnol.*, 2014, **5**, 837–846.
147. N. Stork and M. Evans, *Int. J. Insect Morphol. Embryol.*, 1976, **5**, 219–221.
148. N. E. Stork, *J. Linn. Soc. London, Zool.*, 1980, **68**, 173–306.
149. Y. Pelletier and Z. Smilowitz, *Can. Entomol.*, 1987, **119**, 1139–1142.
150. L. Heepe, S. Höft, J. Michels and S. N. Gorb, *Soft Matter*, 2018, **14**, 7026.
151. S. N. Gorb and M. Varenberg, *J. Adhes. Sci. Technol.*, 2007, **21**, 1175–1183.
152. G. Salerno, M. Rebora, S. Piersanti, Y. Matsumura, E. Gorb and S. N. Gorb, *J. Insect Physiol.*, 2020, **127**, 104117.
153. H. W. Krenn, Form and Function of Insect Mouthparts, in *Insect Mouthparts. Form, Function, Development and Performance*, ed. H. Krenn, Zoological Monographs, Springer Nature Switzerland AG, Cham, 2019, vol. 5, pp. 9–46.
154. J. Michels and S. N. Gorb, *Beilstein J. Nanotechnol.*, 2015, **6**, 674–685.
155. B. W. Cribb, A. Stewart, H. Huang, R. Truss, B. Noller, R. Rasch and P. Zalucki, *Naturwissenschaften*, 2008, **95**, 17–23.
156. J. S. Edwards, *Predation and Digestion in Assassin Bugs (Heteroptera, Reduviidae)*, PhD Thesis, University of Cambridge, Cambridge, 1960.
157. H. C. Bennet-Clark, *J. Exp. Biol.*, 2007, **210**, 3879–3881.
158. H. C. Bennet-Clark, *J. Exp. Biol.*, 1963, **40**, 223–229.
159. H. W. Krenn, *Zoomorphology*, 1990, **110**, 105–114.
160. H. R. Hepburn, *J. Insect Physiol.*, 1971, **17**, 637–656.
161. O. Betz, L. Koerner and S. N. Gorb, *Adhes. Adhes. Sealants*, 2009, **6**, 32–35.

162. L. Koerner, S. N. Gorb and O. Betz, *J. Insect Physiol.*, 2012, **58**, 155–163.
163. L. Koerner, S. N. Gorb and O. Betz, *Zoology*, 2012, **115**, 117–127.
164. H. Krenn and H. Aspöck, *Arthropod Struct. Dev.*, 2012, **41**, 101–118.
165. P. Wenk, S. Lucic and O. Betz, *Zoomorphology*, 2010, **129**, 225–234.
166. S. Büsse, A. Koehnsen, H. Rajabi and S. N. Gorb, *Sci. Robot.*, 2021, **6**, eabc8170.
167. L. E. S. Eastham and Y. E. E. Eassa, *Philos. Trans. R. Soc., B*, 1955, **239**, 1–43.
168. H. Krenn, *Annu. Rev. Entomol.*, 2010, **55**, 307–327.
169. E. Pinterich and H. Krenn, *Entomologica Austriaca*, 2020, **27**, 91–105.
170. M. Burrows, *BMC Biol.*, 2009, **7**, 27.
171. A. R. Longhurst, *Prog. Oceanogr.*, 1985, **15**, 1–35.
172. R. Huys and G. A. Boxshall, *Copepod Evolution*, The Ray Society, London, 1991.
173. A. F. Pasternak and S. B. Schnack-Schiel, *Hydrobiologica*, 2001, **453/454**, 25–36.
174. J. Michels and S. B. Schnack-Schiel, *Mar. Biol.*, 2005, **146**, 483–495.
175. S. Büsse and S. N. Gorb, *R. Soc. Open Sci.*, 2018, **5**, 172117.
176. Y. Matsumura, J. Michels, E. Appel and S. N. Gorb, *Zoology*, 2017, **120**, 1–14.
177. Y. Matsumura, A. E. Kovalev and S. N. Gorb, *Sci. Adv.*, 2017, **3**, eaao5469.
178. A. E. Filippov, Y. Matsumura, A. E. Kovalev and S. N. Gorb, *Sci. Rep.*, 2016, **6**, 27608.
179. Y. Matsumura, S. Gürke, H. T. Tramsen and S. N. Gorb, *BMC Zool.*, 2020, **5**, 9.
180. J. Carayon, Traumatic Insemination and the Paragenital System, in *Monograph of Cimicidae (Hemiptera-Heteroptera)*, ed. R. L. Usinger, Entomological Society of America, Lanham, 1966, pp. 81–166.
181. A. D. Stutt and M. T. Siva-Jothy, *Proc. Natl. Acad. Sci. U. S. A.*, 2001, **98**, 5683–5687.
182. K. Reinhardt, R. Naylor and M. T. Siva-Jothy, *Evolution*, 2009, **63**, 29–39.
183. J. Michels, S. N. Gorb and K. Reinhardt, *J. R. Soc., Interface*, 2015, **12**, 20141107.
184. U. Thurm, *Science*, 1964, **145**, 1063–1065.
185. R. Heusslein and W. Gnatzy, *Cell Tissue Res.*, 1987, **247**, 591–598.
186. T. A. Keil, *Microsc. Res. Tech.*, 1997, **39**, 506–531.
187. M. Rebora, G. Salerno and S. Piersanti, Aquatic Insect Sensilla: Morphology and Function, in *Aquatic Insects. Behaviour and Ecology*, ed. K. Del-Claro and R. Guillermo, Springer Nature Switzerland AG, Cham, 1st edn, 2019, pp. 139–166.
188. A.-C. Joel, H. Adamova and P. Bräunig, *J. Morphol.*, 2018, **279**, 1654–1664.
189. J. C. Tuthill and R. I. Wilson, *Curr. Biol.*, 2016, **26**, R1022–R1038.
190. W. Gnatzy, U. Grünert and M. Bender, *Zoomorphology*, 1987, **106**, 312–319.
191. T. Ichikawa, Y. Toh, K. Ohkubo and H. Nishino, *Zoomorphology*, 2014, **133**, 273–284.

192. S. L. Winterton, Scales and Setae, in *Encyclopedia of Insects*, ed. V. H. Resh and R. T. Cardé, Academic Press, Burlington, 2nd edn, 2009, pp. 901–904.
193. J. P. Miller, S. Krueger, J. J. Heys and T. Gedeon, *PLoS One*, 2011, **6**, e27873.
194. W. Gnatzy and J. Tautz, *Cell Tissue Res.*, 1980, **213**, 441–463.
195. J. S. Edwards and J. Palka, *Proc. R. Soc. Lond. B*, 1974, **185**, 83–103.
196. A. Sannasi, *Experientia*, 1970, **26**, 154.
197. S. Dey and A. Raghuvarman, *Indian Zool.*, 1983, **7**, 31–34.
198. Q.-X. Chen and B.-Z. Hua, *PLoS One*, 2016, **11**, e0156970.
199. S. Viswanathan and G. Varadaraj, *Odonatologica*, 1985, **14**, 155–157.
200. R. Jaganathan and G. Sundara Rajulu, *Indian Zool.*, 1979, **3**, 133–134.

CHAPTER 3

Design and Production of Chimeric Resilin-like Polypeptides (RLPs)

B. BOCHICCHIO, A. BRACALELLO AND A. PEPE*

Laboratory of Bioinspired Materials, Department of Science, University of Basilicata, Via dell'Ateneo Lucano 10, 85100 Potenza, Italy
*Email: antonietta.pepe@unibas.it

3.1 Introduction

Chimera, in genetics, is an organism or tissue that contains at least two different sequences of DNA. The term is derived from the *Chimera* of Greek mythology, a fire-breathing monster that was part lion, part goat, and part dragon. In analogy with this definition, also "engineered" recombinant polypeptides coded by DNA sequences of different organisms or proteins can be considered chimeras.[1] Thanks to the great advancement in molecular biotechnology, a plethora of examples of biologically functional molecular chimeras have been developed in different research fields. A paradigmatic example can be considered the chimeric green fluorescent proteins (GFPs) in heterogeneous systems. The chimeric GFP systems have revolutionized cell biology, providing unprecedented insights into the subcellular localization, mobility, transport routes, and binding interactions of proteins in living cells.[2]

More generally speaking, the almost endless numbers of possible combinations of fusion partners have turned this technique into a versatile and valuable tool within many areas of biochemistry and biotechnology. Chimeric polypeptides can easily be prepared *in vitro* using recombinant DNA

Biomaterials Science Series No. 10
Biomimetic Protein Based Elastomers: Emerging Materials for the Future
Edited by Namita Roy Choudhury, Julie C. Liu and Naba K. Dutta

Published by the Royal Society of Chemistry, www.rsc.org

technology by combining DNA sequences belonging to the genes of the proteins of interest in an appropriate expression vector. The fused DNA molecules can be short synthetic oligonucleotides coding for peptide sequences or full-length structural genes, giving rise to a single polypeptide chain with the properties of both of the original gene products.[3] This type of chimeric protein is typically multifunctional; that is, each sequence adds its own separate function to the polypeptide (*e.g.*, mechanical properties, binding activity, enzymatic activity, protease susceptivity, improved solubility, and self-assembling characteristics).[4]

In other cases, as for the *de novo* design of chimeric proteins, sequences of different, related, or unrelated proteins are combined in a sequential or alternating manner, often providing a novel polypeptide with unpredictable properties. In this chapter, we will present different design and production schemes of chimeric resilin-like polypeptides (RLPs). The common variable is the presence of resilin-related sequences combined with a various number of versatile polypeptide/protein sequences.

3.2 Why Chimeric RLPs?

RLPs are polypeptides inspired by the primary structure of the protein resilin and are proposed as promising biomaterials in many biomedical and technological applications for their remarkable mechanical properties mimicking the native protein. Resilin is a structural protein of arthropods occurring in specialized regions of insect cuticle that plays a key role in insect flight, jumping mechanism of fleas, and vocalization of cicadas.[5] The amazing mechanical properties were described by Weis-Fogh in 1960, who showed that resilin could be stretched to 250–300% of its length without breaking and without creep or relaxation even after a long bearing load.[6,7]

Distinctive mechanical properties of resilin can be responsible for a number of multiple functions that are assigned to resilin in arthropods. For example, high deformability and flexibility are useful in membrane and joint systems, while the storage of elastic energy is important in jumping and catapulting systems. Also, a passive action on stabilizing the wings during flight is assigned as a biological function to resilin. Resilin is also able to work at a wide range of frequencies, as observed in the case of salivary pumps of the assassin bugs, where resilin is a spring antagonist of a muscle working at 3–10 Hz,[8] or in the wings of locusts or dragonflies, where resilin is present in the joint system of the wings beating at 15–50 Hz.[9] The highest frequencies are reached in the case of cicadas, where resilin is used to create a resonant sound pulse at 13 kHz.[10] Finally, the high resilience of fully hydrated resilin approaching 99% at 15 Hz and remaining above 80% at 1000 Hz is an added value for resilin that attracted great interest for this protein in the (bio)material field.

The mechanical properties are closely dependent on its hydration state. Water-swollen resilin exhibits a long-range elasticity and complete recovery after deformation[11] showing a near-perfect resilience of up to 92–97%, while dehydrated resilin is extremely brittle. Furthermore, fully hydrated resilin shows high extensibility, stretchable to more than three times its original length, and extraordinary fatigue lifetime.[7,12,13]

The rubber-like elasticity of resilin is of entropic origin and is based on the continuously changing form of long disordered polypeptide chains. Based on their molecular studies, a mechanism for resilin elasticity was proposed by Tamburro and coworkers in analogy with the elastin mechanism.[14] As a matter of fact, molecular studies have shown high conformational flexibility for both resilin sequences of *Drosophila melanogaster*, the 15-residue sequence (GGRPSDTYGAPGGGN) encoded by exon 1 and the 13-residue sequence (GYSGGRPGGQDLG) encoded by exon 3, which adopt two main conformational structures: folded β-turns and (quasi)extended structures (*e.g.*, poly-L-proline II conformation) in a dynamic interconversion. On the basis of their molecular studies, it has been presumed that resilin behaves like an entropic spring with the elasticity mechanism governed by its conformational entropy. The high entropy of the relaxed state originates from the conformational equilibria between open/folded structures, as also proposed for elastin and other elastomeric proteins.[14] On stretching, a shift of the equilibria toward the open, extended conformations reduces the entropy of the system, thus allowing the return to the relaxed state once the applied force is removed.[15] Of fundamental importance for elastomers is the presence of crosslinks, which in resilin were identified by Andersen to be dityrosine and trityrosine interconnecting the polypeptide chains and thus preventing the chains from sliding apart during deformation.[16]

The amino acid composition of resilin is usually dominated by the presence of a high number of glycine (about 40% in locust resilin) and proline residues (about 8%). Many native resilins from other arthropods have similar amino acid composition, while others show significant differences, such as the rhinoceros beetle containing more than 20% of asparagine residues and only 33% of glycine residues. Even if resilin is widespread in arthropods, it is present in limited tissue districts of small size, like in the wing joint of locusts or dragonflies or fleas. Often it is intimately connected to other molecules, like chitin, forming a composite material. Furthermore, it is very sensible to enzyme degradation, thus hampering the extraction of the protein in a massive way. For this reason, structural and mechanical studies that were dedicated to resilin were limited or absent since alternative ways to obtain resilin/resilin-like polypeptides were developed based on the gene sequences of putative pro-resilins.[17,18] A search in the complete genome of the fruit fly, *D. melanogaster*, by Ardell and Andersen revealed the presence of two genes coding for putative pro-resilin protein sequences. They based their search on the tryptic peptide sequences of cockroach resilin[19] and locust resilin, recognizing a putative translational product of the *Drosophila* gene (CG15920), considered nowadays as the resilin precursor of *D. melanogaster.*[20] The putative protein comprises three domains; the N-terminal domain, coded by exon 1, and containing a 17-residue-long signal peptide, is dominated by 18 slightly varying copies of a 15-residue repeat sequence, while the C-terminal domain presents 11 variants of a 13-residue repeat. The central region, comprising 62 residues, shows high sequence similarity with the Rebers–Riddiford consensus sequence present in cuticular proteins. This region was shown to be involved in the interaction with chitin.[21] Since the discovery of the *D. melanogaster* resilin, many other putative resilin-like proteins have been reported.[22–24]

Based on the resilin gene of *D. melanogaster*, the production of resilin and RLPs was developed by recombinant DNA technologies. Elvin and coworkers were the first who expressed a polypeptide encoded by exon 1 of the *Drosophila* CG15290 gene (called Rec-1 resilin) in *Escherichia coli*, paving the way for the recombinant production of RLPs.[17] Full-length recombinant resilin was produced by Qin *et al.* and contained all three exons of *D. melanogaster* resilin.[21]

The presence of repetitive sequences in resilin was supposed to be responsible for the remarkable mechanical properties of the native protein and has motivated the engineering of polypeptides containing a variable number of resilin-inspired repeated units, called RLPs, by many research groups. The most used repeat sequence is GGRPSDTYGAPGGGN, with the number of repeats varying from 4 to 16.[15,23] Kiick and coworkers produced a great variety of RLPs, with the repetitive sequence GGRPSDT**F/M**GAPGGGN, where Y is replaced by F and/or M in the *Drosophila* consensus sequence.[25,26]

Also, RLPs inspired by repetitive sequences of other arthropods were produced based on the consensus sequence of African malaria mosquito resilin (AQTPSSQYGAP),[23,27] while Chilkoti and coworkers produced RLPs containing a simplified resilin-like sequence (QYPSDGRG) derived from their studies about intrinsically disordered polypeptide (IDP) sequences able to undergo upper critical solution temperature (UCST) transition in a similar way to resilin and RLPs.[28]

The introduction of new functionalities into the RLPs by inserting or appending specific sequences represents a step forward in the design of chimeric RLPS. The main aim of the development of chimeric RLPs is the production of biomaterials with tailored properties able to be responsive to the particular needs of the final (bio)materials characterized by tunable mechanical properties. Additionally, it is possible to improve the interaction of RLPs with cells and, in general, with biological systems. In the latter case, the following aspects are considered: introduction of sequences that favor cell adhesion or migration or that improve interaction with other components of the extracellular matrix (ECM), not neglecting sequences that promote degradation.

3.3 Different Design Schemes of Chimeric RLPs

The introduction of sequences from other proteins or *de novo* designed in the RLPs was performed with the aim to produce materials with new properties and/or responsiveness. Different design schemes were presented where chimeric RLPs were distinguished according to the category of sequences introduced (Figure 3.1).

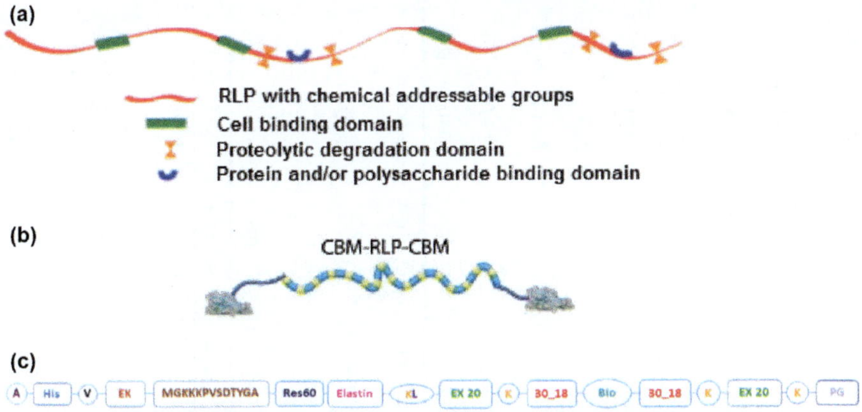

Symbol	Module name	Amino acid sequence
His	*Histidine Tag*	HHHHHH
EK	*Enterokinase cleavage site*	DDDDK
Res60	*Resilin*	$(PGGGNGGRPSDTYGA)_4$
Elastin		$(LGGVG)_3$
Elastin EX20	*Esone 20*	GARPGVGVGGIPTYGVGAGGFPGFGVGVGGIPGVAGVPSVGGVPGVGG VPGVGIS
Elastin (30_18)	*Esone 30_18*	LGGLGVGGLGVPGVGGLG
Bio	*Cell-adhesive*	GRGDSP
Elastin (30_18)	*Esone 30_18*	LGGLGVGGLGVPGVGGLG
Elastin EX20	*Esone 20*	GARPGVGVGGIPTYGVGAGGFPGFGVGVGGIPGVAGVPSVGGVPGVGG VPGVGIS

Figure 3.1 Schematic representations of (a) **RLP12**, a chimeric RLP containing bioactive domains. Adapted from ref. 25 with permission from the Royal Society of Chemistry; (b) **CBM-RLP-CBM**, a chimeric RLP comprising peptide sequences for tuning mechanical properties. Adapted from ref. 51 with permission from Elsevier, Copyright 2018. (c) **RE** polypeptide, a chimeric RLP containing elastin-like domains. Adapted from ref. 67, https://doi.org/10.3390/nano9111613, under the terms of the CC BY 4.0 license https://creativecommons.org/licenses/by/4.0/.

The advantage in the insertion of different polypeptide sequences in the design of RLP-based biopolymers is represented by the improvement of some of the properties necessary for the development of new materials. With this aim, bioactive sequences (Table 3.1) were introduced in order to favor interaction with cells, while other specific sequences were added to optimize mechanical properties of the produced biomaterial (Table 3.2) or by combining RLP sequences with other structural proteins (Table 3.3) like elastin-like, silk-like, or collagen-like sequences.

3.3.1 Chimeric RLPs Containing Bioactive Peptide Sequences

The engineering of bioelastomers based on mechanically active polypeptides has emerged in the biomaterial field due to the possibility to add specific biological active sequences in order to increase the opportunities in the production of scaffolds for tissue engineering or, more generally, biomedical applications. Many research groups took advantage of this opportunity in

Table 3.1 Design and sequences of chimeric RLPs containing bioactive peptide sequences.

Label	Resilin sequence (R)	Protein/polypeptide (P)[a]	Design scheme[b]	Reference
RLP12	(GGRPSDT**F**GAPGGGN)	GGGRGDSPAEDL (**C**) GPQG-IWGQ (**M**) CKAAKRPKAAKDKQTK (**H**)	GGG[GDQK(**R**$_3$-GGK)$_2$-**C**]$_2$-**M**-GGRGG-**H**-G	Charati *et al.*[25]
RLP-RGD	(GGRPSDT**X**GAPGGGN) where **X** = **F** and/or **M**	GGKGGKGGKGG (**K**) T-GRGDSPG-G (**C**)	**K**-(**R**$_3$-**K**)$_4$-**C**	Li *et al.*[36]
RLP-MMP	(GGRPSDT**X**GAPGGGN) where **X** = **F** and/or **M**	GGKGGKGGKGG (**K**) GPQGIWGQP-V- (**M**)	**K**-(**R**$_3$-**K**)$_4$-**K**-**M**	Li *et al.*[36]
RZ30-QK	(AQTPSSKQFGAPAQTPSSQFGAP)	KWADRHGGMR (**L**) GGGKLTWQELYQLKYKGIGG (**V**)	(**R**$_5$-**L**-**V**)$_3$	Renner *et al.*[27]
RZ30-RGD	(AQTPSSKQFGAPAQTPSSQFGAP)	KWADRHGGMR (**L**) TVYAVTGRGDSPASSKPISI (**C**)	(**R**$_5$-**L**-**C**)$_3$	Renner *et al.*[27]
RZ10-RGD	(AQTPSSKQFGAPAQTPSSQFGAP)	KWADRHGGMR (**L**) TVYAVTGRGDSPASSKPISI (**C**)	**R**$_5$-**L**-**C**	Renner *et al.*[37]
RZ-BMP	(AQTPSSKQFGAPAQTPSSQFGAP)	KWADRHGGMR (**L**) KIPKASSVPTELSAISTLYL (**B**)	**R**$_5$-**L**-**B**	Renner *et al.*[39]
RZ-RGD	(AQTPSSKQFGAPAQTPSSQFGAP)	KWADRHGGMR (**L**) GGTVYAVTGRGDSPASSGGG (**C**)	**R**$_5$-**L**-**C**	Renner *et al.*[39]
RES-MAG	Exon 1–coded pro-resilin (CG15920)	GIGKFLHSAGKFGKAFVGEIMKS (**A**)	**S-R-F-A**	Azam *et al.*[47]

[a]**C**: cell-binding domain; **M**: matrix metalloprotease sensitive domain; **H**: heparin-binding domain; **K**: crosslinking domain; **L**: linker sequence; **B**: bone morphogenic protein-2 domain; **S**: secretion tag; **F**: FLAG-tag; **A**: antibiotic peptide sequence.
[b]Letters in bold indicate domains described in the preceding columns; the other letters indicate amino acids by one-letter code.

Table 3.2 Design and sequences of chimeric RLPs comprising peptide sequences for improving mechanical properties.

Label	Resilin sequence (R)	Protein/polypeptide (P)[a]	Design scheme[b]	Reference
RLP	(GGRPSDT**X**GAPGGGN) where **X** = **F** and/or **M**	GGKGGKGGKGG (**K**)	**K**-(**R**$_3$-**K**)$_4$	Li *et al.*[36,48]
CBM-RLP-CBM	Rec1-resilin (Exon 1)[17]	QACSSVWGQCGGQNWSGPTCCASGSTCVYSNDYYSQL (**C**)	**C-R-C**	Fang *et al.*[49,51]
dCBM-RLP-HFBI	Rec1-resilin (Exon 1)[17]	QACSSVWGQCGGQNWSGPTCCASGSTCVYSNDYYSQL (**C**) SNGNGNVCPPGLFSNPQCCATQVLGLIGLDCKVPSQNV YDGTDFRNVCAKTGAQPLCCVAPVAGQALLCQTAVGA (**H**) WSHPQFEK (**S**)	**C-L-C-L-R-L-H-S**	Fang *et al.*[51] Griffo *et al.*[86]
Res-CBD	Rec1-resilin (Exon 1)[17]	CBD sequence of *C. Cellulovorans*[87](**C**)	**R-C**	Rivkin *et al.*[50] Verker *et al.*[53]

[a]**C**: cellulose-binding domain; **H**: hydrophobin domain; **K**: crosslinking domain; **L**: linker sequence; **S**: strep-tag.
[b]Letters in bold indicate domains described in the preceding columns; the other capital letters indicate amino acids by one-letter code.

Table 3.3 Design and sequences of chimeric RLPs comprising structural protein sequences.

Label	Resilin sequence (R)	Protein/polypeptide (P)[a]	Design scheme[b]	References
(GR)$_4$	GGRPSDSYGAPGGGN	GB1 domain (**G**)	**(G-R)$_4$**	Lv *et al.*[56]
GRG$_5$RG$_4$R			**G-R-G$_5$-R-G$_4$-R**	Duan *et al.*[61]
V$_n$-(GR)$_4$		(VPGVG)$_n$ $n=24, 48, 72$ (**V$_n$**)	**V$_n$-(GR)$_4$**	
FRF$_4$RF$_4$R	GGRPSDSYGAPGGGN	TNfn3 (**F1**)	**F1-R-F1$_4$-R-F1$_4$-R**	Lv *et al.*[60]
RS	GGRPSDSYGAPGGGN	GAGAGS (**S**)	**(R$_4$S$_4$)$_5$**	Huang *et al.*[59]
R$_4$C	GGRPSDSYGAPGGGN	C-terminal domain of spider *N. clavipes* major ampullate spidroin 1 (**CT**)	**R$_4$-CT**	Luo *et al.*[62]
R$_8$C			**R$_8$-CT**	
REC	PGGGNGGRPSDSYGA	MAHHHHHHVDDDDK (**H**) (LGGVG)$_3$ (**E^1**) PKG-(PGG)$_3$-K-(PAG)$_3$-PKG (**C**)	**H-R-E^1-K-E^1-C**	Bracalello *et al.*[57] Sbrana *et al.*[75]
RE	SDTYGAPGGGNGGRP	(LGGVG)$_3$ (**E^1**) (GARPGVGVGGIPTYGVGAGGFPGFGVGVGGIPGV AGVPSVGGVPGVGGVPGVGIS) (**E^2**) LGGLGVGGLGPGVGGLG) (**E^3**) GRGDSP (**C^I**)	**R$_4$-E^1-KL-E^2-K-E^3-C^b-E^3-K-E^2-KPG**	Bracalello *et al.*[67]
(RLP$_y$-ELP$_y$)	(QYPSDGRG)	VPGXG (X = A, G, V, S) (**ELP**)	**G-R$_n$-ELP$_m$**-Y $n=20, 40, 60, 80, 100$ $m=40, 80, 160$	Weitzhandler *et al.*[58]
(RLP$_y$-ELP$_{80}$-Fn3)	(QYPSDGRG)	VPGXG (X = A, G, V, S) (**ELP**) Fn3 (**F2**)	**G-R$_n$-ELP$_m$-F2**-Y $n=20, 40, 80$ $m=40, 80, 160$	Dzuricky *et al.*[68]
RES-2XTE	Exon 1–coded pro-resilin (CG15920)	GLGAFPAVTFPGALVPGGVADAAAAYKAAKA (**E**)	**S-R-F-E$_2$**	Azam *et al.*[47]
RES-8XTE			**S-R-F-E$_8$**	

[a]**C**: collagen-like cell-binding domain; **C^I**: cell-binding domain; **E^1**: glycine-rich elastin domain; **E^2**: exon 20–coded domain of human tropoelastin; **E^3**: N-terminal part of exon 30–coded domain of human tropoelastin; **F1**: third fibronectin type III domain of tenascin-C; **F2**: tenth type III domain of human fibronectin; **G**: streptococcal B1 immunoglobulin-binding domain of G protein; **H**: Histidine tag.

[b]Letters in bold indicate domains described in the preceding columns; the other letters indicate amino acids by one-letter code.

the elastin-like[16,30–33] and silk-like polypeptide[34] design. However, also in the design of RLPs, the insertion of several bioactive domains has emerged with the aim to improve interactions with cells and molecules of the ECM, to favor biodegradation, and to insert biological cues.

A highly complex polypeptide design was first developed by Kiick and coworkers. The recombinantly produced chimeric **RLP12** comprised a total of 12 GGRPSDS**F**GAPGGGN repeats, which differs from the consensus repeat sequences by substitution of Y residues with the F residues, in order to plan further incorporation of nonnatural amino acids[35] that would facilitate photocrosslinking without compromising the biological activity of the other domains.[25,26] For crosslinking purposes, six lysine residues were interspersed between the RLP repeats, while for improving the cell attachment of human cells to the scaffold, the RGDSP sequence, derived from fibronectin, was inserted. Also, a matrix metalloprotease (MMP) susceptive sequence, GPQG-IWGQ, derived from human α1(I) collagen for promoting proteolytic degradation and a heparin-binding domain CKAAKRPKAAKDKQTK for the immobilization of both heparin and growth factors were added.[25] The chimeric **RLP12**, containing several bioactive domains, was shown to possess outstanding mechanical properties in both low- and high-frequency regimes together with favorable cell adhesion behavior[26] (Figure 3.1a). In order to independently tune the concentration of the bioactive domains in the RLP constructs, an alternative approach was proposed by Li *et al.* Rather than a multimodular RLP (**RLP12**), they proposed the combination in different proportions of distinct bimodular RLPs consisting of 12 repeats of the *Drosophila* resilin consensus sequence followed by a single biological domain (for example, cell binding in **RLP-RGD**, protease susceptivity in **RLP-MMP**, and heparin-binding in **RLP-HBD**) to finely tune the biological activities of the scaffold.[36]

Bioactive chimeric RLPs based on the consensus sequence derived from *Anopheles gambiae* (mosquito) genes were developed by Liu and coworkers.[27,37–39] The protein design is based on an elastomeric region composed of ten repeats of the *A. gambiae* resilin motif (AQTPSSQFGAP) linked to a bioactive domain. The bioactive domains were, alternatively, i) an RGD cell-binding sequence, GGTV**YAVTGRGDSPASS**GGG, derived from fibronectin[40–42] (RZ-RGD);[37,43] ii) a bioactive domain, GGG**KLTWQE-LYQLKYKGI**GG, that induces angiogenesis by mimicking vascular endothelial growth factor (VEGF)[44] (RZ-QK);[27,45] and iii) a bioactive domain, **KIPKASSVPTELSAISTLYL**,[46] derived from the bone morphogenetic protein-2 (BMP-2) growth factor, for favoring osteoinduction (**RZ-BMP**).[39] Furthermore, some lysine residues were added as potential chemical crosslinking sites. With the aim to improve the production yield and purity of the chimeric RLPs, Azam *et al.* designed and produced a chimeric RLP containing an antimicrobial peptide sequence. The resilin block comprised the exon 1–encoded domain of the native resilin gene of *D. melanogaster* (CG15920), while the bioactive sequence introduced was the peptide sequence of magainin-1, an antimicrobial peptide produced by the frog *Xenopus laevis.*

They used a bacterial expression system based on *Salmonella* cells, taking advantage of the type III secretion system for protein secretion. For this reason, a secretion domain at the N-terminus was added to the polypeptide sequence.[47] Bioactive sequences in chimeric RLPs were inserted in the polypeptide sequence at different locations. They were interspersed between RLP sequences or were appended at the end of the RLP sequences. The rationale for choosing alternative locations along the polypeptide chains was not discussed, but it is probably due to DNA or polypeptide design considerations. A systematic study examining in depth if and how the location of a non-RLP sequence affects the biological, mechanical, and physicochemical properties of the produced polypeptides could furnish additional indications useful in the general RLP design.

3.3.2 Chimeric RLPs Comprising Peptide Sequences for Tuning Mechanical Properties

Chimeric RLPs offer enormous advantages in the engineering of new elastomeric and highly mechanically active materials. In order to tailor the mechanical performance to the potential applications of the produced bioelastomers, several strategies have been developed (Table 3.2). For example, Kiick and coworkers inserted a *de novo* designed crosslinking domain, GG**K**GG**K**GG**K**GG**K**GG, containing three lysine residues evenly spaced in the RLP sequence. The aim was to increase the crosslinking sites by modulation of the crosslink density affecting the mechanical properties.[36,48] The hydrogels were crosslinked by [tris(hydroxymethyl phosphine)] propionic acid (THPP) and produced hydrogels by fast gelation kinetics compatible with cell incapsulation and subcutaneous injection.

Recently, RLPs have been engineered to include cellulose-binding domains with the aim to favor interaction with cellulose-based nanomaterials like nanocrystalline cellulose (NCC) or cellulose nanofibrils (CNF).[49–51] As a matter of fact, in recent years, nanocellulose has been emerging as an eco-compatible, biodegradable, cost-effective material with excellent mechanical properties and has been proposed as a component in nanocomposites in the development of new technologically advanced materials.[52] The principal aim of the proposed design is to develop new nanocomposite biomaterials by combining the soft biopolymer with high elasticity and low stiffness (RLP) with stiffer components like NCC and CNF, obtaining structures with mechanically performant properties. The reinforcement of two chemically and mechanically different components requires an intimate interaction of them. In order to fulfill this requirement in RLPs, the insertion of adhesive cellulose-binding domains (CBDs) able to interact with NCC or CNF was performed, producing chimeric RLP-CBD biopolymers. The CBD sequences introduced by Fang *et al.* belong to the noncatalytic domains of the fungal cellobiohydrolase enzyme[49,51] (Figure 3.1b), while the CBD domain inserted by Shoseyov and coworkers[50,53] belongs to the cellulose-binding protein of *Clostridium cellulovorans*.[54]

With the aim to add improved adhesive functionality to the RLPs, Fang *et al.* added together with CBD domains also a hydrophobin (HFB) domain and studied the influence of the fusion proteins on the liquid–liquid phase transition of RLPs.[51] Remarkably, the introduction of CBD domains did not change the coacervation process. RLP and CBD–RLP–CBD polypeptides were affected by pH and concentration in the same way, producing particles in the micrometer size. The presence of the HFB domain, in turn, confers a higher tendency to coacervate at a lower concentration, with the formation of smaller size particles that cluster together. The dissimilar behavior could be ascribed to differences in the size of the HFB domain or, more probably, to differences in the amino acid composition, enriched of hydrophobic residues that favor coacervation.[55]

3.3.3 Chimeric RLPs Comprising Structural Protein Sequences

In an effort to develop RLP-based biomaterials with a great variety of chemical, mechanical, and biological properties, recombinant chimeric RLPs have been engineered by the arrangement of resilin sequences with other sequences inspired by structural proteins. Chimeric RLPs have been designed to combine different sequences in a diblock, triblock, or multiblock elastomeric protein structure, where resilin domains are interspersed with other domains to mimic the passive elasticity mechanical properties of muscle protein titin,[56] to induce assembly,[57–59] and to ensure the biophysical and biochemical characteristics of native ECM microenvironments[60] (Table 3.3).

Engineering titin-mimicking elastomeric proteins was the main aim of Lv *et al.*[56] Titin is a giant muscle protein composed of a combination of folded domains, immunoglobulin-like sequences, and highly unstructured domains. The combination of these two elements ensures the passive elasticity typical of the muscle. In order to design a muscle titin-mimicking polypeptide, they produced a recombinant chimeric polypeptide by combining the GB1 domain (**G**) from the streptococcal B1 immunoglobulin-binding domain of G protein as mimicking the structure of the immunoglobulin-like fold of titin, with a short resilin sequence, corresponding to the canonical *Drosophila* consensus motif of exon 1 as an unstructured domain. The complex molecular architecture of titin was simulated by two chimeric polypeptides (**G-R**)$_4$ and **GRG$_5$RG$_4$R**, where **R** = GGRPSDSYGAPGGGN with different arrangements of the two domains.[60] Both polypeptides were shown to behave like rubber-like biomaterials with high resilience at low strain and as shock adsorbent materials at high strain. These features confer to the obtained material passive elastic properties close to muscles but are significantly different from those of resilin-based biomaterials. The GB1 domain dominates the mechanical behavior at high tensile strain, while at low strain, the entropic-based elasticity of resilin segments prevails.

Further evolution of the molecular design determined the introduction of the canonical elastin sequence (VPGVG)$_n$, obtaining a series of protein block copolymers V_n-$(GR)_4$, with $n = 24, 48, 72$. The addition of elastin sequences endowed the polypeptide with thermoresponsive properties. Hydrogel produced from the ruthenium-catalyzed photochemical crosslinking of V_n-$(GR)_4$ is thermoresponsive, the mechanical properties of which are shown to be correlated to the working temperature.[61]

Other chimeric RLPs belonging to this class (Table 3.3) have a higher contribution of resilin sequences and are combined essentially with structural proteins having crucial mechanical functions as well as peculiar self-assembled supramolecular structures like silk-, elastin-, and collagen-like sequences.

In this framework, Xia and coworkers have designed, produced, and investigated the self-assembling properties of a chimeric resilin-silk polypeptide.[59] The polypeptide design has been based on the repetition of a monomer comprising the four-fold repeat of the canonical *Drosophila* consensus sequence, $(GGRPSDSYGAPGGGN)_4$, and the four-fold repeat of the silk-like sequence, $(GAGAGS)_4$, from the silkworm *Bombyx mori.* The multimer **RS** derived from five-fold multimerization of the monomeric resilin-silk sequence produced a recombinant chimeric polypeptide of 37.8 kDa. In addition, an RLP containing a 32-times-repeated GGRPSDSYGAPGGGN sequence (**R32**) was produced in order to compare the self-assembling and mechanical properties.[59] The obtained chimeric RLP has been shown to have thermoresponsiveness at low and high temperatures. Further studies on thermoresponsive structural proteins pushed the researchers to combine other proteins to RLPs, like the carboxy-terminal (**CT**) domain of the major ampullate spidroin 1 (MaSp1) of the spider *Nephila clavipes*. The design is based again on either four or eight resilin consensus motifs followed by the *ca.* 110 amino acid–long CT domain (**R4C** and **R8C**).[62]

The design and production of complex block polypeptides containing repetitive sequences belonging to different structural proteins, like resilin, elastin, and collagen (**REC**), was undertaken by Bracalello *et al.* in order to obtain a highly performant biomaterial with tailored mechanical and biological properties for tissue engineering.[57] The choice of the amino acid sequence of the three blocks is based on previous studies on repetitive sequences of elastomeric proteins.[14,15,63,64] The RLP block comprised four repetitions of the *Drosophila* consensus sequence (SDTYGAPGGGNGGRP), while the ELP domain contained the seven-times-repeated hydrophobic sequence (LGGVG) with one lysine residue in the middle of the domain as a chemical crosslinking site. The collagen-like sequence contained three lysine residues interspersed in the GPZ ($Z = G$, A) triplets. Although RLP and ELP blocks are highly elastic polypeptides, they have very different physicochemical properties, with the resilin-like domain being hydrophilic and the elastin-like block being hydrophobic, as well as different self-assembling patterns. Surprisingly, the **REC** polypeptide exhibited a great tendency to self-assemble into an extremely peculiar supramolecular structure, constituted of higher-order fibrillar morphology, definitively dissimilar from twisted-rope helical structures previously observed for glycine-rich elastin-like polypeptides.

Other chimeric sequences were designed and produced by the authors' research group based on the combination of RLPs and ELPs (**RE**), integrated with bioactive sequences. The RLP domain is analogous to the resilin domain contained in **REC** polypeptides, while the ELP domain is made of hydrophobic elastin sequences encoded by exon 20 and exon 30 of human tropoelastin. These elastin sequences were previously shown to undergo coacervation, a thermally triggered phase separation that favors fiber formation.[65,66] Finally, to improve cell interaction, the RGD cell-binding domain from fibronectin was inserted in the **RE** polypeptide[67] (Figure 3.1c).

A different family of chimeric resilin-elastin diblock polypeptides was designed and recombinantly produced by Chilkoti and coworkers. They based the copolypeptide design on simple principles of block polymer physics. The RLP blocks contained systematically varied lengths of a simplified resilin-like sequence (QYPSDGRG) derived from studies on IDP sequences,[28] while the elastin-like blocks were composed of 80, 160, or 40 repeats of the VPGXG pentapeptide motif, where X = A, G, V, S denoted the guest residue (**RLP$_y$-ELP$_y$**). The variation of the length of the resilin-like and/or the elastin-like blocks affects the hydrophilicity of the polypeptide chains influencing the self-assembling process. Results have shown that the amphiphilic diblock polypeptides were able to self-assemble into spherical or worm-like micelles based on the hydrophilic contribution of each block.[58] The produced polypeptides, manufactured as stable micelles, were considered highly attractant as drug delivery systems and imaging agents. In order to improve tumor targeting and the accumulation of the nanoparticles, the **RLP-ELP** polypeptides were fused to a protein ligand.[68] As the targeting domain, the 10th type III domain from human fibronectin (Fn3) that binds to human $\alpha_v\beta_3$ integrin, a receptor that is highly overexpressed in many solid tumor cells, was selected.

Different recombinant resilin–elastin coblock polypeptides were designed and produced by combining the exon 1–coded pro-resilin sequence of *D. melanogaster* with a two- or eightfold sequence, belonging to the exon 6–coded domain of human tropoelastin.[47]

Recently, a new chimeric RLP was designed and produced by Mikael *et al.*[69] The resilin block comprised the exon 1–encoded domain of the native resilin gene of *D. melanogaster* (CG15920), fused with a collagen-binding domain ColBD (**Res-ColBD**) from ColH collagenase from *Clostridium histolyticum*.[69] The main aim of **Res-ColBD** was to favor the interface among resilin polypeptide and the collagen-rich tissues present in the ligaments, tendons, bone, and other tissue districts.

3.4 Production of Chimeric RLPs

In recent decades, the genetic engineering approach based on DNA recombinant technology has promoted the design and production of chimeric polypeptide polymers with defined composition, sequence, and tailored properties. Also, in the case of chimeric RLPs, the recombinant expression strategy was the principal production technology because of the facilitated control of the primary sequence, the ease of combining different DNA sequences to produce a chimeric gene, and the potential for large-scale production. The principal expression systems for the production of chimeric RLPs have been bacteria, mainly *E. coli*, even if other expression systems, like yeast and plants, have also been recently explored.[69,70]

The steps toward the production of chimeric RLPs can be schematized into the following five steps:

1. Synthesis of the DNA sequence encoding the chimeric polypeptide;
2. Insertion of the DNA sequence in an appropriate expression plasmid vector;
3. Cloning into a host cell;
4. Production in prokaryotic or eukaryotic cells; and
5. Purification of the chimeric polypeptide

One of the most demanding steps in the production of chimeric RLPs is the construction of the desirable gene sequence by the accurate choice of the cloning strategy. Different methods have been developed, which can be summarized as follows: i) concatemerization strategy; ii) convergent assembly of modular blocks domains; and iii) recursive directional ligation (RDL).[71] In several cases, more than one strategy is applied. Concatemerization is a common strategy for the production of multimeric genes, starting from a monomer gene.[72] This approach allows obtaining a set of multimer DNA sequences by ligation of a monomer gene unit, creating a library of oligomers of different lengths (Figure 3.2a). Concatemerization is a useful method when a library of genes of different sizes is desired but has limitations in the preparation of genes with one specific size. While this cloning strategy was frequently employed for the synthesis of elastin-like genes, it was not so common for chimeric resilin genes.[73] To the best of the authors' knowledge, the only example was described by Huang *et al.*[59] for the production of the chimeric resilin-silk gene encoding a multimeric polypeptide.

Convergent assembly can be used to combine different modular polypeptide blocks by assembling different DNA sequences encoding for the individual domains in order to produce chimeric RLPs. A convergent assembly from plasmids containing individual blocks is performed to generate diblock, triblock, or copolymer genes with more blocks.

As an example, some chimeric RLPs produced by Kiick and coworkers were obtained by using this strategy. The pQE80-RLP plasmid containing the RLP-related DNA sequence was made to be digested by the appropriate restriction enzymes (RE). The linearized vectors were ligated with DNA oligos that encoded the bioactive domain (RGD, HBD, MMP).[36] The chimeric **RE** and **REC** genes were obtained by convergent assembly from plasmids containing individual blocks generating the di- and triblock copolymer genes[57,67] (Figure 3.2b).

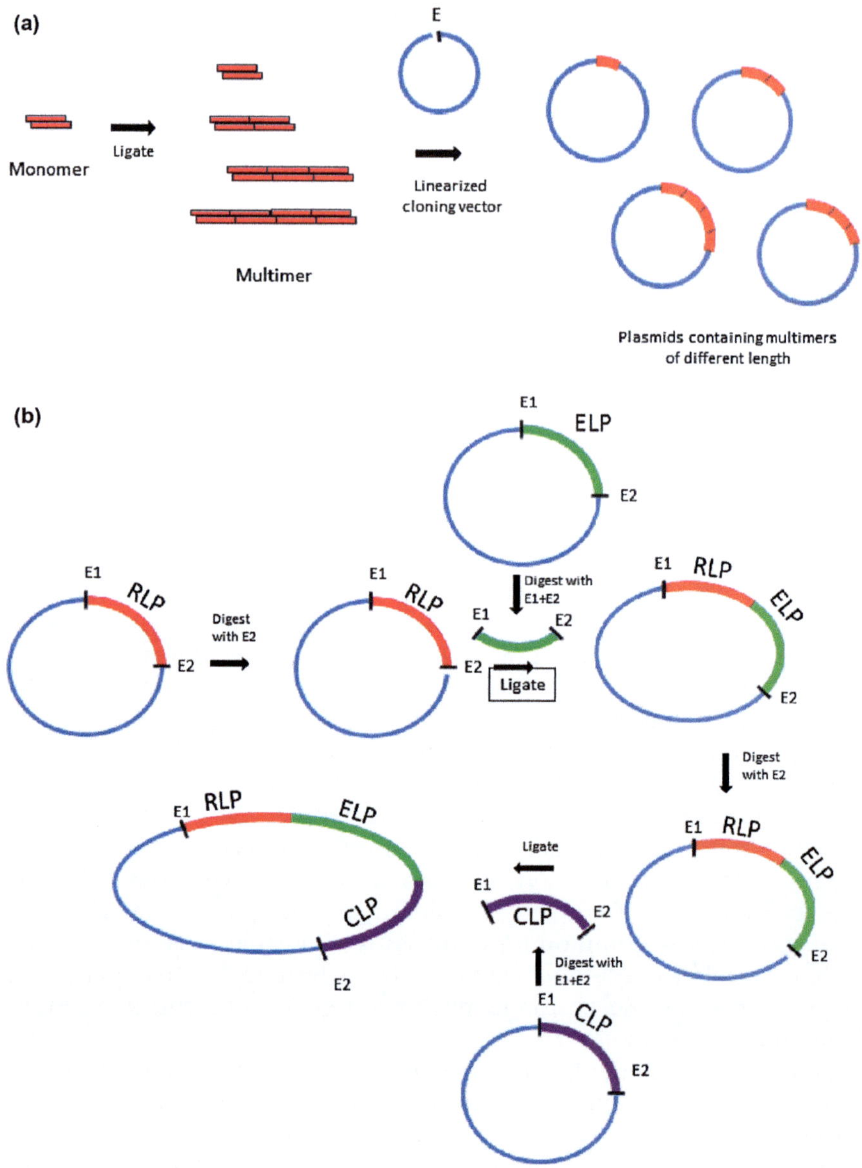

Figure 3.2 Schematic representation of the cloning strategy for producing chimeric RLPs genes. (a) Concatemerization of the DNA monomers under the action of DNA ligase. The nonpalindromic overhanging cohesive ends promote head-to-tail concatemerization. The multimers were inserted in an appropriate plasmid; (b) convergent assembly of resilin-like (RLP), elastin-like (ELP), and collagen-like (CLP) modular block domains by the use of restriction enzyme E1 and E2.

Liu and coworkers have developed a cloning scheme based on the RDL technique adapted for the precise control of the number of resilin repeats using REs, AleI, and AfeI. RDL is a seamless cloning method that leaves no extraneous nucleotides in the ligation junction. The obtained vectors, containing 10 resilin repeats, were further assembled with DNA sequences encoding bioactive (BA) domains to obtain the recombinant plasmids for the expression of chimeric RLPs containing VEGF-like, BMP-2 peptide, or cell-binding domains[27] (Figure 3.3a). A step-wise construction scheme that makes use of the iterative cloning of DNA sequences encoding titin-like or Fn3 domains with resilin-like domains was also employed by Lv *et al.*[56,60]

Chilkoti and coworkers have extensively used the recursive directional ligation by plasmid reconstruction (pre-RDL) strategy to clone highly repetitive chimeric RLPs constituted of resilin- and elastin-like repeats of defined sequence and specified length over a large range of molecular weights.[58,68,74] This strategy can be used to clone highly repetitive polypeptides of any sequence and specified length by ligating two halves of a parent plasmid, each containing a copy of an oligomer, thereby dimerizing the oligomer and reconstituting a functional plasmid (Figure 3.3b). It has also been used to assemble genes that encode different sequences in a predetermined order and to append bioactive sequences.[68]

In Table 3.4, a summary of the main information regarding chimeric RLPs expression systems is provided.

3.5 Mechanical Properties of Chimeric RLPs

The outstanding material properties of high resilience and high fatigue lifetime of resilin have driven great research interest into resilin-like biomaterials that offer potential to be used in, but not limited to, the biomedical device field. Consequently, RLPs and chimeric RLPs were mainly produced with the aim of reproducing the astonishing mechanical properties of the native resilin protein. Accordingly, a particular focus on mechanical studies and their relationship with polypeptide sequences is presented.

The mechanical studies presented are distinguished as nanostructural analysis or single-molecule analysis commonly performed by using atomic force microscopy (AFM) and bulk material analyses performed by dynamic oscillatory rheology and/or tensile testing. In the first approach, nanoindentation measurements are commonly used to evaluate the stiffness of the material by recording force–distance curves while approaching the material in a compressive mode. Nanoindentation data obtained by measuring hydrogels obtained from resilin proteins derived from different insects indicated that flea-derived hydrogels are softer than hydrogels derived from buffalo fly and fruit fly, as evidenced by the observed penetration depths, respectively, of 654, 347, and 324 nm.[22] Further on, a measure of the reversibility and resilience of the polypeptide is obtained by acquiring repeated single-molecule stress-elongation cycles and evaluating the hysteresis among the stress and relaxation curves. The results on recombinant resilin hydrogels showed a resilience of 90–92% for

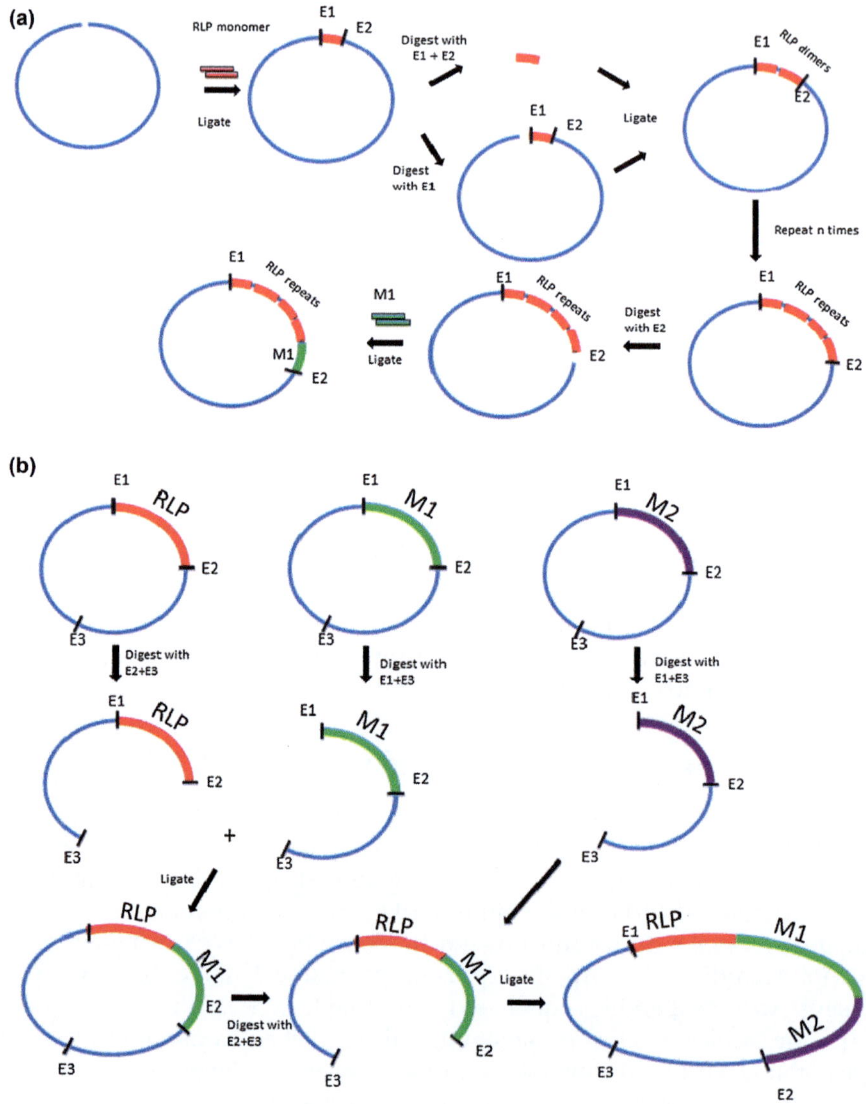

Figure 3.3 Schematic representation of the cloning strategies for producing chimeric RLPs genes by (a) recursive directional ligation (RDL) allowing to obtain RLPs with a defined number of repeats and insertion of the DNA sequence encoding bioactive or other polypeptide domains; (b) RDL by plasmid reconstruction (pre-RDL).

Rec-1, while exon 3–encoded RLP hydrogels showed a reduced value of 63%, which increases to 83% after photochemical crosslinking. Flea- and buffalo fly-derived resilin show resilience values of 88 and 87%, respectively, while RLPs derived from repeated sequences of *D. melanogaster* (Dros16) and *A. gambiae* (An16) show resilience comparable to that of native resilin Rec-1.

Table 3.4 Summary of the experimental conditions of expression and purification of chimeric RLPs.

Chimeric RLPs	Plasmid/ expression vector	Promotor	Tag	Host	Protein expression conditions	Purification methods	Yield	References
RLP12	pET-28a	T7	6xHis	BL21 Star™ (DE3)	Studier autoinduction	Ni-NTA[a]	70–80 mg L^{-1}	Charati *et al.*[25]
RLP-RGD **RLP-MMP**	pQE80	T5	6xHis	M15[pREP4]	IPTG[e]	Ni-NTA[a]	20–30 mg L^{-1}	Li *et al.*[36]
RZ30-QK **RZ30-RGD** **RZ-QK**	pUC19	Lac	T7-tag 7xHis	BL21(DE3)pLysS	Studier autoinduction, 20 °C, 0.5% glucose	Salting out and heating	n.d. n.d. 13.8 mg L^{-1}	Renner *et al.*[27] Kim & Liu,[45]
RZ10-RGD	pUC19	Lac	T7-tag/7xHis	BL21-CodonPlus-(DE3)-RIPL	IPTG, 37 °C	Salting out and heating	22 mg L^{-1}	Renner *et al.*[27,37]
RZ-BMP	pUC19		T7-tag/7xHis	BL21-Codon Plus-(DE3)-RIPL	IPTG induction 37 °C	Salting out and heating	10 mg L^{-1}	Kim *et al.*[39]
RES-MAG	pSicA	pSicA	6xHis/FLAG-tag	*S. enterica* SL1344	T3SS+/HilA overexpression conditions	Ni-NTA[a]	15 mg L^{-1}	Azam *et al.*[47]
RES-2XTE **RES-8XTE**	pSicA	pSicA	6xHis/FLAG-tag	*S. enterica* SL1344	T3SS+/HilA overexpression conditions	Ni-NTA[a]	10 mg L^{-1} 20 mg L^{-1}	Azam *et al.*[47]
CBM-RLP-CBM	pML124	*Cbh*1	8xHis/Strep-tag	*Trichoderma reesei* strain M1230	28 °C	Ni-NTA[a]	n.d.	Fang *et al.*[49]

Table 3.4 (*Continued*)

Chimeric RLPs	Plasmid/ expression vector	Promotor	Tag	Host	Protein expression conditions	Purification methods	Yield	References
dCBM-RLP-HFBI	pML124	cbh1		*T. reesei* strain M1230	28 °C, pH 4.0	ATPS[b] + RP-HPLC[f]	60 mg L^{-1}	Fang *et al.*[51]
Res-CBD	pHis-parallel3_17	T7	6xHis	BL21 (DE3)	IPTG, 36 °C	Ni-NTA[a]	n.d.	Rivkin *et al.*[50]
$(GR)_4$	pQE80L/	T5	6xHis	DH5α	IPTG, 37 °C, 4–5 h	Ni-NTA[a]	n.d.	Lv *et al.*[56]
GRG_5RG_4R	pET29b+			BL21 (DE3)		Co^{2+} AC [c]	30–60 mg L^{-1}	Duan *et al.*[61]
V_n-$(GR)_4$	pET25+							
FRF_4RF_4R	pQE80L	T5	6xHis	DH5α	IPTG, 37 °C, 5 h	Co^{2+} AC [c]	40 mg L^{-1}	Lv *et al.*[60]
RS	pET-19b3	T7	10xHis	BL21 (DE3)	IPTG, 16 °C, 12–14 h	Ni-NTA[a]	50–100 mg L^{-1}	Huang *et al.*[59]
R_4C/R_8C	pET-19b	T7	10xHis		IPTG, 30 °C, 6 h	Ni-NTA[a]	0.6–0.8 g L^{-1}	Luo *et al.*[62]
REC	pET46EK/Lic	T7	6xHis	BL21 (DE3)	IPTG, 37 °C, 3 h	Ni-NTA[a] + RP-HPLC	40–50 mg L^{-1}	Bracalello *et al.*[57]
RE	pET46EK/Lic	T7	6xHis	BL21 Star™ (DE3)	IPTG, 37 °C, 1 h	Ni-NTA[a] + RP-HPLC	35–40 mg L^{-1}	Bracalello *et al.*[58,67]
(RLP_y-ELP_y)	pET24+		T7-tag/6xHis	BL21 (DE3)	37 °C, 24 h	ITC[d]	20 mg L^{-1}	Weitzhandler *et al.*[58]
(RLP_y-ELP_{80}-Fn3)	pET24+		T7-tag/6xHis	BL21 (DE3)	37 °C, 24 h	ITC[d]	20 mg L^{-1}	Dzuricky[68]

[a]Ni-NTA affinity chromatography.
[b]Aqueous two-phase separation.
[c]Co^{2+} affinity chromatography.
[d]Inverse transition Cycling.
[e]Isopropyl β-D-1-thiogalactopyranoside.
[f]Reverse phase – high performance liquid chromatography.

Other types of AFM-based experiments, such as single-molecule force-spectroscopy analysis, can be performed. In this experimental setup, the tip of the AFM probe catches the single molecule of polypeptide immobilized on the substrate and stretches the polypeptide chain by pulling it until detaching it from the substrate. The forces involved in the experiment describe the response of the polypeptide to the manipulation and give useful insights into the mechanical signature of the molecules. However, a very limited number of chimeric RLPs were studied by using this technique and the results show peculiar results according to the type of investigated chimeric RLP. Lv *et al.* observed that stretching of the $(GR)_4$ polypeptide by AFM single-molecule force spectroscopy resulted in a peculiar sawtooth-like force–elongation curve, corresponding to progressive unfolding of each globular GB1 domain present in chimeric RLP. The presence of resilin-like linker sequences determined a featureless spacer length detectable in the force–spectroscopy curve.[56] In the case of REC polypeptide, the single-molecule stretching curve showed that the unrolling behavior of the polypeptide can be well described by the worm-like chain model.[75] Further stress-relaxation cycles highlighted the quasi-ideal elastomeric behavior of REC. The elasticity modulus was evaluated on the self-assembled fibers and resulted in being in the range of 0.1–3 MPa, while the elastic reversibility was determined to be near 100%, showing a quasi-ideal elastomeric behavior.[57,75]

Studies at the single-molecule level can give useful insight into the intrinsic mechanical properties of the polypeptide chain at the nanoscale; however, material properties are more commonly described by evaluating the manufactured (bio)material as bulk. In this way, also other effects arising from the production process or due to local defects of the bulk material are taken into account.

Hydrogels obtained from chimeric RLPs were commonly characterized from a mechanical point of view by oscillatory rheology. Measurements of the storage modulus and loss modulus are obtained as a function of the concentration of the polypeptides, crosslinking conditions, time, and temperature.

In all the classes of investigated hydrogels, the mechanical properties are mainly correlated to the polypeptide and the crosslinker. It has been shown that it is possible to produce hydrogels with a wide range of storage modulus by tuning the concentration of chimeric RLPs, and the stoichiometric crosslinking ratio. A summary of the rheological results is presented in Table 3.3. A wide range of storage moduli was described by Kiick and coworkers for the **RLP12**-based hydrogels. By changing the concentration of the RLPs (from 10 to 25% wt) and the crosslinking ratio, storage modulus (G') values of 500 Pa to 10 kPa were recorded. For **RLP12** hydrogels with potential application as vocal fold tissues, a working setup that mimics the frequency conditions during human phonation was performed. The mechanical properties were recorded by torsional wave amplification experiments in the range of 30 to 150 Hz. They resulted in being compatible with those observed for human vocal fold tissues.[26] In a different polypeptide design, RLPs containing crosslinking sequences that comprise five lysine residues evenly incorporated in the polypeptide chain have shown higher storage modulus values due to a higher number and more uniformly distributed presence of crosslinking points (1 kPa–20 kPa).[36] The oscillatory shear strain-to-break values were in the range of 160% to 240%, with increased strain for hydrogels at lower concentrations. The presence of bioactive sequences did not affect the observed mechanical properties. Further dynamical mechanical data were recorded in order to explore the stress relaxation behavior of the RLP hydrogels, central for materials proposed for vocal fold tissue regeneration. RLP hydrogels have shown a rapid elastic response to the shear stress and a low dissipation of energy during stress relaxation test.[76] Hydrogels containing different resilin repeat sequences derived from *A. gambiae* resilin sequence were produced by Liu and coworkers. **RZ10** and **RZ10-RGD** polypeptide hydrogels showed a wide range of storage modulus values varying from 75 Pa to 20 kPa, as a function of concentration from 8 wt% to 16 wt%.[77] These values matched the stiffness of soft tissues, such as brain tissues, and stiffer tissues, such as muscle, offering high versatility in the application as scaffolds in tissue engineering. Again, the presence of bioactive sequences did not affect the mechanical properties. This finding highlights the independent control over mechanical properties, thus suggesting opportunities to endow the artificial extracellular matrix materials with significant biological activities, which could improve the interaction inside the tissues and organs without affecting the mechanical properties significantly. In the case of chimeric RLPs constituted of sequences belonging to different structural proteins, the presence of two mechanical active sequences can have an interesting impact on the mechanical behavior of the hydrogel. Luo *et al.* showed that a resilin-silk chimeric polypeptide formed hydrogels in which the stiffness is correlated with the incubation time.[59,62] Taking advantage of self-assembling properties of the silk domain, **RS** polypeptide formed a soft gel with a storage modulus of 195 Pa by incubation of 37 °C for 10 hours. Photochemical crosslinking of the **RS** gel produced a hydrogel whose stiffness was related to the preincubation time. After photocrosslinking, the preincubated **RS** hydrogel showed a G' modulus of 7 kPa, while the non-preincubated revealed a storage modulus of 2.9 kPa.

A summary of the rheological results on chimeric RLPs is presented in Table 3.5.

A complete mechanical characterization of the chimeric RLPs was usually accomplished by standard tensile testing. Chimeric RLP-based hydrogel samples were exposed to repeated loading and unloading strain cycles in order to define the longitudinal elasticity, to determine hysteresis and calculate the resilience of hydrated RLP-based hydrogel films, and to record the extension-to-break values. Furthermore, stress-relaxation data are not to be neglected. Tensile testing results on chimeric RLPs described in this section are summarized in Table 3.5.

On the whole, the mechanical characterization of diverse chimeric RLP hydrogels suggests that variations in the stiffness and elasticity, resilience, and extensibility could be related to a number of factors, such as differences in resilin repeat amino acid composition, and consequently in polypeptide

Table 3.5 Summary of the mechanical properties of RLPs and chimeric RLPs.

Chimeric RLPs	Concentration (crosslinking ratio)	Storage modulus (G')	E modulus	Extension to break	Resilience	Other results	References
Rec-1	20 wt%		26 kPa	250 (±43)%	97%	92%[m]; 325 (±23) nm[a]	Lyon *et al.*[29]
Dros16	20 wt%		–	347 (±61)%	94%	91%; 235 (±39) nm	
An16	20 wt%		5.7 kPa			98%; 495 (±199) nm	
RLP12	10–25 wt% (x1-x5)	600 Pa–10 kPa	14–34 kPa	250–340%	90–98%		Charati *et al.*[25]/Li *et al.*[26]
RLP	5–20 wt%	1 kPa–20 kPa	30 kPa	190%	95–100%		Li *et al.* 2013
RLP-RGD[b]	20 wt% (1x)	20 kPa					Li *et al.* 2014
RLP-MMP[c]	10–20 wt%	10–20 kPa					
	10–20 wt%	10–20 kPa					
RZ10-RGD	16 wt% (1x)	12.9 kPa				2.4 MPa[d]	Renner *et al.*[37]
RZ-RGD[e]	8–16 wt% (5x)	7 Pa–22 kPa				1 kPa–1.7 kPa; 29–70%[d]	Su *et al.*[77]
	6–12 wt%	44 Pa–7 kPa					Kim *et al.*[43]
CBM-RLP-CBM	CNF/RLP[f]	0–9 Pa	4.5–7 GPa	10–16%		35–65 MPa yield stress	Fang *et al.* 2017
	CNF/RLP[g]						
Res-CBD	CNF/RLP[g]		3 GPa	3%		70 MPa tensile strength	Rivkin *et al.* 2015
$(GR)_4$	20 wt%		70 kPa	40%		High stress relaxation	Lv *et al.* 2010
GRG_5RG_4R	20 wt%		50 kPa	135%		at strain >35%	Duan *et al.*[61] 2020
V_n-$(GR)_4$/$(GR)_4$	15 wt%		33 kPa (25 °C)				
			40 kPa (50 °C)				
FRF_4RF_4R	20 wt%	25 kPa	24 kPa		80%		Lv *et al.* 2013
RS	20 wt%	200 Pa/2.9 kPa[h]/ 7 kPa[i]					Huang *et al.* 2017
R_4C	20 wt%	11.0 kPa					Luo *et al.*[62] 2018
R_8C	20 wt%	7.1 kPa					
REC						0.1–3 MPa	Bracalello *et al.*[57]
						90(±8)%[j]	Sbrana *et al.*[75]

[a]Penetration depth determined by AFM.
[b]50 wt% RLP with 50 wt% RLP-RGD.
[c]50 wt% RLP with 50 wt% RLP-MMP.
[d]Unconfined compressive modulus.
[e]Transglutaminase mediated crosslinking.
[f]Hybrid hydrogel.
[g]Hybrid film.
[h]After photochemical crosslinking.
[i]Preincubation and photochemical crosslinking.
[j]Resilience determined by AFM.

hydrophobicity, molecular weight and crosslinking efficiency. However, a direct comparison of the results is difficult to realize in consideration of the evident differences in the experimental conditions. Some useful insights were acquired from molecular dynamics (MD) simulations on resilin-based polypeptide chains. Cohen *et al.* investigated, in particular, the relationship between the amino acid composition of the RLPs and the mechanical properties, pointing their attention to the polarity of the amino acid side chains.[78]

Their results revealed a strong correlation between the presence of many highly polar amino acids of resilin sequence and its mechanical properties. High resistance to fracture and exceptionally low stiffness was attributed to a high percentage of hydrophilic residues within the resilin motif. A fundamental role in resilin elasticity was assigned to the plasticizer effect of water, able to break strong hydrogen bonds that behave like noncovalent crosslinks. Their reduction leads to additional chain mobility, a decrease in enthalpic energy, and a subsequent increase in entropy that finally results in the improved mechanical properties of the swollen hydrophilic resilin domains. Accordingly, differences in the hydrophobic/hydrophilic balance in the different resilin sequences could account for the broad range of mechanical properties of RLPs and chimeric RLPs.

3.6 Potential Applications of Chimeric RLPs

As of now, biomedical applications are the main interest of researchers producing chimeric RLPs related to tissue engineering (TE), regenerative medicine, or drug delivery applications. This is also the main motivation for introducing bioactive domains in the RLP polypeptide sequences in order to favor the interaction, proliferation, spreading, or differentiation of the seeded cells. Also, the mechanical properties are fundamental for the production of scaffolds for tissue engineering applications. Several studies have focused on the impact of the stiffness of the matrix on the physiological development of tissue during embryogenesis and growth, highlighting the influence of the mechanical properties of the substrates on cell behavior.[79,80] In this context, the possibility to control the mechanical properties of the scaffold is of paramount importance in order to realize scaffolds directed at the proliferation of different cell types. Numerous studies on chimeric RLPs have reported the possibility to tune the mechanical properties of the scaffolds, thus allowing to set up a powerful platform for studying cell behavior and differentiation in a more "native" environment. Given that stiffness and resilience are mainly regulated by the chemical or physical crosslinks present in the scaffold, it is possible to modulate and adapt the mechanical properties to specific needs inherent to variable biomedical applications by varying the crosslinker/substrate ratio. As an example, biomaterials with a storage modulus in the range of 75 Pa–25 kPa were produced by regulating the polypeptide and crosslinker concentrations, thus demonstrating their potential use as scaffolds for very soft tissues (brain tissue) to stiffer tissues (muscle) in tissue engineering applications.[77] The high compressive modulus recorded for crosslinked chimeric RLP **RZ10-RGD** points to a potential application in cartilage tissue engineering.[38] Muscle tissue engineering applications were also proposed for the titin-mimicking chimeric RLPs exhibiting high resilience at low strain and high toughness at high strain. These features confer to the produced material passive elastic properties close to muscles.[56]

The peculiar mechanical properties reported for some chimeric RLPs, such as low modulus, high toughness, excellent resilience, and high-frequency responsiveness, were proposed to be useful also for other mechanically demanding tissue engineering applications, like vocal fold tissue regeneration.[36] Also, in these chimeric RLPs, the presence of biological active domains favored the interaction of cells with the scaffold and the *in vivo* biocompatibility with surrounding tissues.[48]

Self-aggregation of RE and REC into higher-order fibrillary structures, together with the ability to guide cell adhesion and growth of mesenchymal stem cells (MSCs) along the fibers, has suggested a potential application in the directed growth of cells and guided differentiation.[57,67,75]

Some authors proposed chimeric RLPs as drug delivery systems, in particular when they were designed for stimuli-responsiveness.[62,81] Thermal and pH responsiveness was thoroughly exploited in hydrogels constituted of chimeric RLPs,[59,62] and the pH-dependent delivery of rhodamine B from the hydrogel was reported as well. In this framework, chimeric RLPs were exploited to develop nanoparticle carriers for drug delivery with characteristics useful for passive targeting. The passive targeting approach in drug delivery is possible if the molecules or nanoparticles have an extended blood circulation time and a dimension smaller than 100 nanometers, which allows them to escape from the tumor blood vessel through the leaky vasculature and be captured by tumor cells. In order to favor the last step of tumor cell capturing, Dzuricky *et al.*[68] proposed integrin-targeting chimeric RLP able to form micelles and studied their internalization. Their preliminary studies are very encouraging, even if they require further investigation on the drug delivery aspect.

Other nonbiomedical but technological applications were proposed by Shoseyov and coworkers, who suggested the production of a high-performance adhesive by combining epoxy resin with nanocellulose and as binder to a chimeric RLP (**Res-CBD**) that contained cellulose-binding domains.[53] The role of the **Res-CBD** is to act as a surfactant-like agent able to improve and reinforce the interactions between epoxy resin and NCC.[53] Other studies of the same research group proposed the production of chimeric RLPs in transgenic plants to manipulate the characteristics of the plant cell walls and modify the mechanical properties of the woody tissues.[70] They succeeded in producing transgenic eucalyptus with chimeric RLPS expressed in the cell wall, creating novel materials for wood and paper industries. Further research is needed in this breakthrough field; however, these and many other research works are paving the way to produce new materials by exploiting *de novo* designed genes for innovative materials.

3.7 Future Perspectives

Starting from the observation of the jump of the fleas able to reach hundreds of times their height, researchers were pushed to investigate the material responsible for this property. The discovery of the resilin protein in charge of the exceptional mechanical properties built the basis for investigating nature-produced rubber-like material. Later studies were directed on one side in unraveling the mechanism of elasticity by studying the sequence–structure–function relationship of resilin and on the other side in mimicking the resilin protein by producing recombinant RLPs. The great potential of chimeric RLPs is related to the preserved mechanical properties, evidenced by many research groups together with additional functionalities that improve one or more characteristics of the produced material. The potential of producing different chimeric RLPs is endless, considering the myriads of sequences that could be added to RLPs with the aim of introducing novel properties to the material. Specific sequences that favor interactions with cells or with other components of the ECM, like glycosaminoglycans (GAGs), proteoglycans, receptors, collagen, elastin, or hydroxyapatite, could push new designs for tissue engineering applications.

Until now, the principal applications claimed for chimeric RLPs are mainly in the biomedical field. A search in the patent database highlighted some recent claims for RLPs and chimeric RLPs in several unexpected fields, like in the footwear production application[82] or in golf ball production.[83] In the future, new functionalities can be added to RLPs with the aim to render the produced material of potential interest to a broad range of functions.

Another appealing property of recombinant produced chimeric RLPs is their environmentally friendly production process and the possibility to be degraded by enzymatic processes, which is in line with the objectives of the new "green deal" plans aiming to produce high-performance materials with a sustainable life cycle. However, the main problem in the massive use of new chimeric RLPs is related to the low-yield expression systems developed until now. As a matter of fact, yields in the range of 10–70 $mg\,L^{-1}$ were reported, with great variability due to the polypeptide sequence and the expression systems employed.[25,37,45,57,60,67,77] Some other researchers improved the yield to 200–300 $mg\,L^{-1}$ with the accurate choice of the inducing method, expression system, and expression temperature; however, these results, even if encouraging, are far from making recombinant expression systems competitive.[29,84,85] Accordingly, improved systems for expressing and purifying chimeric RLPs with greater productivity are desired to provide more efficient protein production at larger scales. Some recent reports proposed an alternative expression system, taking advantage of the type III secretion system (T3SS) found in *Salmonella enterica* and many other gram-negative bacteria. Even if the proposed protocol is promising regarding the purity of the secreted polypeptide (>90%), a significant portion of the expressed chimeric polypeptide remains in the cytosol together with truncated or degraded RLPs.[47]

Definitively, the current costs of chimeric RLPs are still not competitive with the analogous petroleum-based polymers, but research on this aspect poses a challenge that could be solved in the near future. Alternative expression systems based on plants (tobacco, eucalyptus) could be the most efficient solution?

Over 3 billion years, nature has created astonishing materials that fit perfectly the function they are planned for. Researchers in many laboratories are trying to imitate nature with very poor tools available; however, great advances have now been achieved and further improvements in the design and production of chimeric RLPs for many applications are in progress. We are confident that in the near future, the production of high-performance, cost-effective chimeric RLPs will not be a *Chimaera*.

References

1. J. S. Bonde and L. Bülow, in *Brenner's Encyclopedia of Genetics*, ed. S. Maloy and K. Hughes, Academic Press, San Diego, 2nd edn, 2013, DOI: 10.1016/B978-0-12-374984-0.00229-1, pp. 519–520.
2. J. Lippincott-Schwartz, E. Snapp and A. Kenworthy, *Nat. Rev. Mol. Cell Biol.*, 2001, **2**, 444–456.
3. W. Strohl, in *Reference Module in Life Sciences*, Elsevier, 2017, DOI: 10.1016/B978-0-12-809633-8.06199-9.
4. R. L. DiMarco and S. C. Heilshorn, *Adv. Mater.*, 2012, **24**, 3923–3940.
5. H. Bennet-Clark, *J. Exp. Biol.*, 2007, **210**, 3879–3881.
6. T. Weis-Fogh, *J. Exp. Biol.*, 1960, **37**, 889–907.
7. T. Weis-Fogh, *J. Mol. Biol.*, 1961, **3**, 648–667.
8. H. Bennet-Clark, *J. Exp. Biol.*, 1963, **200**, 1681–1694.
9. C. M. Jensen and O. Weis-Fogh, *Philos. Trans. R. Soc., B*, 1962, **245**, 137–169.
10. P. J. Fonseca and H. Bennet-Clark, *J. Exp. Biol.*, 1998, **201**, 717–730.
11. K. Bailey and T. Weis-Fogh, *Biochim. Biophys. Acta*, 1961, **48**, 452–459.
12. J. Michels, E. Appel and S. N. Gorb, *Beilstein J. Nanotechnol.*, 2016, **7**, 1241–1259.
13. S. O. Andersen and T. Weis-Fogh, in *Advances in Insect Physiology*, ed. J. W. L. Beament, J. E. Treherne and V. B. Wigglesworth, Academic Press, 1964, vol. 2, pp. 1–65.
14. B. Bochicchio, A. Pepe and A. M. Tamburro, *Chirality*, 2008, **20**, 985–994.
15. A. M. Tamburro, S. Panariello, V. Santopietro, A. Bracalello, B. Bochicchio and A. Pepe, *Chembiochem: A Eur. J. Chem. biol.*, 2010, **11**, 83–93.
16. S. O. Andersen, *Biochim. Biophys. Acta*, 1963, **69**, 249–262.
17. C. M. Elvin, A. G. Carr, M. G. Huson, J. M. Maxwell, R. D. Pearson, T. Vuocolo, N. E. Liyou, D. C. C. Wong, D. J. Merritt and N. E. Dixon, *Nature*, 2005, **437**, 999–1002.
18. R. S. C. Su, Y. Kim and J. C. Liu, *Acta Biomater.*, 2014, **10**, 1601–1611.
19. E. C. Lombardi and D. L. Kaplan, *Mater. Res. Soc. Symp. Proc.*, 1993, **292**, 3–7.
20. D. H. Ardell and S. O. Andersen, *Insect Biochem. Mol. Biol.*, 2001, **31**, 965–970.

21. G. K. Qin, S. Lapidot, K. Numata, X. Hu, S. Meirovitch, M. Dekel, I. Podoler, O. Shoseyov and D. L. Kaplan, *Biomacromolecules*, 2009, **10**, 3227–3234.
22. R. E. Lyons, D. C. C. Wong, M. Kim, N. Lekieffre, M. G. Huson, T. Vuocolo, D. J. Merritt, K. M. Nairn, D. M. Dudek, M. L. Colgrave and C. M. Elvin, *Insect Biochem. Mol.*, 2011, **41**, 881–890.
23. R. E. Lyons, E. Lesieur, M. Kim, D. C. Wong, M. G. Huson, K. M. Nairn, A. G. Brownlee, R. D. Pearson and C. M. Elvin, *Protein Eng., Des. Sel.*, 2007, **20**, 25–32.
24. S. O. Andersen, *Insect Biochem. Mol. Biol.*, 2010, **40**, 541–551.
25. M. B. Charati, J. L. Ifkovits, J. A. Burdick, J. G. Linhardt and K. L. Kiick, *Soft Matter*, 2009, **5**, 3412–3416.
26. L. Li, S. Teller, R. J. Clifton, X. Jia and K. L. Kiick, *Biomacromolecules*, 2011, **12**, 2302–2310.
27. J. N. Renner, Y. Kim, K. M. Cherry and J. C. Liu, *Protein Expression Purif.*, 2012, **82**, 90–96.
28. F. Garcìa Quiroz and A. Chilkoti, *Nat. Mater.*, 2015, **14**, 1164–1171.
29. R. E. Lyons, K. M. Nairn, M. G. Huson, M. Kim, G. Dumsday and C. M. Elvin, *Biomacromolecules*, 2009, **10**, 3009–3014.
30. J. C. Liu and D. A. Tirrell, *Biomacromolecules*, 2008, **9**, 2984–2988.
31. L. Martin, M. Alonso, M. Moller, J. C. Rodriguez-Cabello and P. Mela, *Soft Matter*, 2009, **5**, 1591–1593.
32. A. Girotti, J. Reguera, J. C. Rodriguez-Cabello, F. J. Arias, M. Alonso and A. M. Testera, *J. Mater. Sci.: Mater. Med.*, 2004, **15**, 479–484.
33. W. Hassouneh, K. Fischer, S. R. MacEwan, R. Branscheid, C. L. Fu, R. H. Liu, M. Schmidt and A. Chilkoti, *Biomacromolecules*, 2012, **13**, 1598–1605.
34. E. Bini, C. W. P. Foo, J. Huang, V. Karageorgiou, B. Kitchel and D. L. Kaplan, *Biomacromolecules*, 2006, **7**, 3139–3145.
35. K. Kirshenbaum, I. S. Carrico and D. A. Tirrell, *Chembiochem: Eur. J. Chem. biol.*, 2002, **3**, 235–237.
36. L. Q. Li, Z. X. Tong, X. Q. Jia and K. L. Kiick, *Soft Matter*, 2013, **9**, 665–673.
37. J. N. Renner, K. M. Cherry, R. S. C. Su and J. C. Liu, *Biomacromolecules*, 2012, **13**, 3678–3685.
38. J. N. Renner, Y. Kim and J. C. Liu, *Tissue Eng. Part A*, 2012, **18**, 2581–2589.
39. Y. Kim, J. N. Renner and J. C. Liu, *Biomater Sci-Uk*, 2014, **2**, 1110–1119.
40. M. Pierschbacher, E. G. Hayman and E. Ruoslahti, *Proc. Natl. Acad. Sci. U. S. A.*, 1983, **80**, 1224–1227.
41. M. D. Pierschbacher and E. Ruoslahti, *Nature*, 1984, **309**, 30–33.
42. M. D. Pierschbacher and E. Ruoslahti, *Proc. Natl. Acad. Sci. U. S. A.*, 1984, **81**, 5985–5988.
43. Y. Kim, E. E. Gill and J. C. Liu, *Biomacromolecules*, 2016, **17**, 2530–2539.
44. L. D. D'Andrea, G. Iaccarino, R. Fattorusso, D. Sorriento, C. Carannante, D. Capasso, B. Trimarco and C. Pedone, *Proc. Natl. Acad. Sci. U. S. A.*, 2005, **102**, 14215–14220.
45. Y. Kim and J. C. Liu, *Biomater Sci-Uk*, 2016, **4**, 1761–1772.

46. A. Saito, Y. Suzuki, S.-I. Ogata, C. Ohtsuki and M. Tanihara, *Biochim. Biophys. Acta, Proteins Proteomics*, 2003, **1651**, 60–67.
47. A. Azam, C. Li, K. J. Metcalf and D. Tullman-Ercek, *Biotechnol. Bioeng.*, 2016, **113**, 2313–2320.
48. L. Li, J. M. Stiadle, E. E. Levendoski, H. K. Lau, S. L. Thibeault and K. L. Kiick, *J. Biomed. Mater. Res., Part A*, 2018, **106**, 2229–2242.
49. W. Fang, A. Paananen, M. Vitikainen, S. Koskela, A. Westerholm-Parvinen, J. J. Joensuu, C. P. Landowski, M. Penttila, M. B. Linder and P. Laaksonen, *Biomacromolecules*, 2017, **18**, 1866–1873.
50. A. Rivkin, T. Abitbol, Y. Nevo, R. Verker, S. Lapidot, A. Komarov, S. C. Veldhuis, G. Zilberman, M. Reches, E. D. Cranston and O. Shoseyov, *Ind. Biotechnol.*, 2015, **11**, 44–58.
51. W. Fang, Nonappa, M. Vitikainen, P. Mohammadi, S. Koskela, M. Soikkeli, A. Westerholm-Parvinen, C. P. Landowski, M. Penttila, M. B. Linder and P. Laaksonen, *Colloids Surf., B*, 2018, **171**, 590–596.
52. E. Kontturi, P. Laaksonen, M. B. Linder, Nonappa, A. H. Gröschel, O. J. Rojas and O. Ikkala, *Adv. Mater.*, 2018, **30**, e1703779.
53. R. Verker, A. Rivkin, G. Zilberman and O. Shoseyov, *Cellulose*, 2014, **21**, 4369–4379.
54. M. A. Goldstein, M. Takagi, S. Hashida, O. Shoseyov, R. H. Doi and I. H. Segel, *J. Bacteriol.*, 1993, **175**, 5762–5768.
55. D. W. Urry, D. C. Gowda, T. M. Parker, C. H. Luan, M. C. Reid, C. M. Harris, A. Pattanaik and R. D. Harris, *Biopolymers*, 1992, **32**, 1243–1250.
56. S. Lv, D. M. Dudek, Y. Cao, M. M. Balamurali, J. Gosline and H. B. Li, *Nature*, 2010, **465**, 69–73.
57. A. Bracalello, V. Santopietro, M. Vassalli, G. Marletta, R. Del Gaudio, B. Bochicchio and A. Pepe, *Biomacromolecules*, 2011, **12**, 2957–2965.
58. I. Weitzhandler, M. Dzuricky, I. Hoffmann, F. Garcia Quiroz, M. Gradzielski and A. Chilkoti, *Biomacromolecules*, 2017, **18**, 2419–2426.
59. S. C. Huang, Z. G. Qian, A. H. Dan, X. Hu, M. L. Zhou and X. X. Xia, *ACS Biomater. Sci. Eng.*, 2017, **3**, 1576–1585.
60. S. S. Lv, T. J. Bu, J. Kayser, A. Bausch and H. B. Li, *Acta Biomater.*, 2013, **9**, 6481–6491.
61. T. Duan and H. Li, *Biomacromolecules*, 2020, **21**, 2258–2267.
62. F. Luo, Z. G. Qian and X. X. Xia, *Polymers*, 2018, **10**, 915.
63. A. M. Tamburro, A. Pepe, B. Bochicchio, D. Quaglino and I. P. Ronchetti, *J. Biol. Chem.*, 2005, **280**, 2682–2690.
64. M. Martino, A. Coviello and A. M. Tamburro, *Int. J. Biol. Macromol.*, 2000, **27**, 59–64.
65. A. Pepe, D. Guerra, B. Bochicchio, D. Quaglino, D. Gheduzzi, I. Pasquali Ronchetti and A. M. Tamburro, *Matrix Biol. J. Int. Soc. Matrix Biol.*, 2005, **24**, 96–109.
66. B. Bochicchio, M. R. Armenante, M. A. Crudele and A. Pepe, *Conf. Pap. Sci.*, 2014. Article ID 214235.
67. A. Bracalello, V. Secchi, R. Mastrantonio, A. Pepe, T. Persichini, G. Iucci, B. Bochicchio and C. Battocchio, *Nanomaterials*, 2019, **9**, 1613.

68. M. Dzuricky, S. Xiong, P. Weber and A. Chilkoti, *Nano Lett.*, 2019, **19**, 6124–6132.
69. P. E. Mikael, R. Udangawa, M. Sorci, B. Cress, Z. Shtein, G. Belfort, O. Shoseyov, J. S. Dordick and R. J. Linhardt, *Regener. Eng. Transl. Med.*, 2019, **5**, 362–372.
70. I. Preis, M. Abramson and O. Shoseyov, *Mol. Biotechnol.*, 2018, **60**, 310–318.
71. W. Kim, *Front. Biosci.-Landmark*, 2013, **18**, 289–304.
72. S. Ding, X. Wang and A. E. Barron, *Nat. Mater.*, 2011, **10**, 83–84.
73. R. E. Sallach, V. P. Conticello and E. L. Chaikof, *Biotechnol. Prog.*, 2009, **25**, 1810–1818.
74. J. R. McDaniel, J. A. MacKay, F. G. Quiroz and A. Chilkoti, *Biomacromolecules*, 2010, **11**, 944–952.
75. F. Sbrana, C. Fotia, A. Bracalello, N. Baldini, G. Marletta, G. Ciapetti, B. Bochicchio and M. Vassalli, *Bioinspiration Biomimetics*, 2012, **7**, 046007.
76. L. Li and K. L. Kiick, *Front. Chem.*, 2014, **2**, 21.
77. R. S. C. Su, E. E. Gill, Y. Kim and J. C. Liu, *J. Mech. Behav. Biomed.*, 2019, **91**, 68–75.
78. N. Cohen and C. D. Eisenbach, *Acta Biomater.*, 2019, **96**, 303–309.
79. I. G. Levental, P. C Georges and P. A. Janmey, *Soft Matter*, 2006, **3**, 299–306.
80. R. G. Wells, *Hepatology*, 2008, **47**, 1394–1400.
81. R. S. C. Su, R. J. Galas, C. Y. Lin and J. C. Liu, *Macromol. Biosci.*, 2019, **19**, 1900122.
82. M. J. Smith, M. E-S. Lee, M. J. H.P. J. Jehlen. WO patent 2020016810, 2020.
83. W. E. Morgan. US patent 20080305889, 2008.
84. M. Kim, C. Elvin, A. Brownlee and R. Lyons, *Protein Expression Purif.*, 2007, **52**, 230–236.
85. A. Griffo, H. Hähl, S. Grandthyll, F. Müller, A. Paananen, M. Ilmén, G. R. Szilvay, C. P. Landowski, M. Penttilä, K. Jacobs and P. Laaksonen, *ACS Omega*, 2017, **2**, 6906–6915.
86. A. Griffo, H. Hahl, S. Grandthyll, F. Muller, A. Paananen, M. Ilmen, G. R. Szilvay, C. P. Landowski, M. Penttila, K. Jacobs and P. Laaksonen, *ACS Omega*, 2017, **2**, 6906–6915.
87. M. A. Goldstein, M. Takagi, S. Hashida, O. Shoseyov, R. H. Doi and I. H. Segel, *J. Bacteriol.*, 1993, **175**, 5762–5768.

CHAPTER 4

Intrinsically Disordered and Resilin-based Protein Polymers for High-performance Biomaterials Applications

SAI S. PATKAR, CRISTOBAL GARCIA GARCIA AND KRISTI L. KIICK*

Department of Materials Science and Engineering, University of Delaware, Newark, DE 19716, USA
*Email: kiick@udel.edu

4.1 Intrinsically Disordered Proteins (IDPs)

Traditional understanding about protein function has historically been based on the established structure–function paradigm,[1] and, in globular proteins, there is a loss of function due to denaturation at high temperatures and nonneutral pH. In contrast, intrinsically disordered proteins (IDPs) comprise intrinsically disordered regions (IDRs) that do not adopt a specific 3D structure;[1] these regions in IDRs are repetitive, have low sequence complexity, and are enriched in amino acid residues such as Gly, Gln, Asn and Ser; charged residues Arg, Lys, Asp, and Glu; and aromatic residues Phe and Tyr.[2] The presence of structure-breaking Pro and Gly residues, patterning of charge, and residues with low hydropathy prevents the formation of defined tertiary structures.[1,3–7] Under physiological conditions, IDPs exist in dynamic equilibrium as a broad ensemble of different conformations[5]

Biomaterials Science Series No. 10
Biomimetic Protein Based Elastomers: Emerging Materials for the Future
Edited by Namita Roy Choudhury, Julie C. Liu and Naba K. Dutta

Published by the Royal Society of Chemistry, www.rsc.org

and can adopt structures of increased order at elevated temperatures due to improved hydrophobic interactions.[8]

It was previously believed that IDRs functioned as passive linkers in globular protein molecules.[1] The advent of sophisticated biochemical and computational methods, however, has advanced the study of complementary functional aspects of IDRs in IDPs, and it has been observed that disordered regions in proteins act as interaction modules to regulate cellular processes by exposing short linear peptide motifs[9] and by undergoing posttranslational modifications to improve functionality.[1,10] Hence, there is an increasing interest in the disorder–function paradigm, which postulates that the absence of structural constraints allows IDPs to play a vital role in a variety of biological processes such as transcription, translation, signal transduction, and cell-cycle control,[1,11,12] and also in the development of human pathological conditions.[13] An increased tendency toward the aggregation of molecules such as amyloid β-peptide, α-synuclein, and huntingtin has been linked to several neurodegenerative diseases, in particular Alzheimer's, Parkinson's, and Huntington's diseases, respectively, in this case.[11,13–15]

The aggregation tendencies of IDRs and IDPs also play a significant role in the development of robust mechanical properties in both naturally derived and engineered protein-based materials. Structural proteins such as byssus fibers, abductin, spider silks, wheat gluten, collagen, elastin, and resilin have evolved to achieve distinct material properties to fulfill environmental requirements in a wide range of organisms.[16,17] Byssus fibers produced by mussels enable secure attachment to surfaces in turbulent intertidal zones.[16] The opening mechanism of mollusk shells is facilitated by abductin, a protein found in their inner hinge ligament.[16] Spiders manufacture different varieties of silk to support diverse functions such as web construction and prey capture. Web frames and safety lines are made up of rigid dragline silk, whereas flagelliform/viscid silk forms the sticky spiral used to entrap prey.[16] Gluten, an elastomer found in plants, acts as a nutrient store for developing seedlings.[18] Collagen and elastin are extracellular matrix proteins that form a substantial component of vertebrate connective tissues. Resilin, an outstanding elastomer, confers reversible and long-range elasticity to insect cuticle and acts as an efficient energy store.[16,17] These examples illustrate how structural proteins are essential for a miscellany of biological processes, wherein amino acid composition and organization define various functions and properties. Hence, the protein polymer research community continues to look to these sophisticated evolutionary designs to generate bioinspired and biomimetic macromolecules with tunable stimuli-responsive properties. Similar to natural IDPs, the intrinsically disordered protein polymers (IDPPs) undergo liquid–liquid phase separation (LLPS), which modulates their interactions, morphologies, and mechanical properties. Despite the different motivations for research in both fields, substantial synergy exists between both communities in harnessing the phase behavior of intrinsically disordered polypeptides.[19]

4.2 Structural Proteins: Naturally Occurring IDPs

Structural proteins were among the earliest studied IDPs,[20] although despite their abundant prevalence, only a limited number have been characterized in their native form in significant detail,[21–25] owing to the difficulty of isolation from their natural sources, their insolubility when crosslinked, and batch-to-batch variability. Although the advent of recombinant technology has alleviated some of these issues, early research was focused primarily on characterizing the mechanical properties and associated biological functions of this diverse family of proteins.[16] Collagen fibers are stiff and are excellent at storing energy. They are not particularly elastomeric, but their ability to deform reversibly allows them to be classified as elastic proteins.[16] Flagelliform silk and byssus threads are tough and have high extensibility, but their ability to return back to the original undeformed state is severely limited on account of low resilience.[16] Owing to their high resilience, reversible extensibility, and low stiffness, resilin and elastin are commonly classified as rubber-like proteins.[16] This makes elastin and resilin excellent candidates for storing elastic strain energy efficiently. Indeed, in very early studies, Weis-Fogh demonstrated that resilin from a dragonfly tendon had a perfectly elastic response to being strained twice its original length for two weeks.[26]

Structural proteins such as collagen, elastin, and silk fibroin have low or no levels of tryptophan or sulfur-containing residues.[27] Furthermore, their excellent mechanical properties can be attributed to the presence of large amounts of glycine in the polypeptide backbone, which provides conformational flexibility.[28] Resilin is unique in comparison with other structural proteins with respect to its amino acid composition. On account of having more acidic and basic groups (Glu, Asp, Arg, and Lys), resilin is more hydrophilic relative to elastin and collagen and these compositional differences may also explain its low isoelectric point and swelling behavior.[27] The proportion of hydrophobic residues such as Gly, Pro, Ala, Val, Leu, Ile, and Phe is significant at about 66%, similar to the proportion of hydrophobic residues in collagen but much lower when compared to silk fibroin and elastin. Hence, high amounts of Pro residues in resilin and elastin contribute to their inability to crystallize and to their propensity to remain disordered and act rubber-like when hydrated with appropriate solvents.

Rauscher *et al.* showed that above a threshold in proline and glycine composition, there is a pronounced increase in elastomeric properties.[28] Quiroz *et al.* analyzed the composition of Pro-Gly-rich IDPs and determined that P-X_n-G motifs (where X is any amino acid except Pro or Gly, and n varies from 0 to 4) are needed to encode disorder. Hence, P-X_n-G motifs separated by an indiscriminate selection of any combination of 3–15 amino acid residues could be used as a scaffold for designing IDPPs.[29]

It is theorized that elastomeric structural proteins such as elastin and resilin exhibit long-range rubber-like elasticity due to the tendency of the highly conserved Pro-Gly dyad to form β-turns.[18] These dynamic β-turns can increase chain entropy by sliding along the chain.[30,31] Upon application of a stress, this conformational entropy decreases due to reduced sliding; hence there is an entropy-driven elastic recoil upon removal of the applied force. The domain comprising the elastomeric repeat motif is accompanied by other nonelastic domains that facilitate crosslinking in most elastomeric proteins (except in the case of resilin and abductin, where crosslinking occurs within the elastic domain).[18] Their crosslinked nature facilitates the dispersion of stress throughout the network, and the elastic domain length and degree of crosslinking can tune the properties of the elastomeric proteins.[18]

The mechanical properties of elastin and resilin are highly dependent on strain rate, hydration level, and pH.[16] Reduced water content results in a corresponding decrease in the resilience,[16] and elastin has poor resilience for applications with higher strain rates. However, for resilin, the ratio of energy lost irreversibly as heat to the change in stored elastic energy (*i.e.*, the loss factor) does not increase proportionally at high frequencies,[32–34] and even at elevated frequencies around 200 Hz, resilin maintains its outstanding mechanical properties.[32] This makes resilin an attractive evolutionary choice for decreasing power expenditure in oscillating structures such as sound-producing organs of insects,[35,36] as well as an unusually responsive material for engineering applications, which has become an area of increasing research and development over the past decade.

In early studies of the function of resilin in fleas, resilin pads alone were proposed to function as efficient energy stores to power the fleas' jumping mechanism.[37] The elastic spring-like structures are able to store and release large amounts of energy very rapidly to facilitate short jumping times, enabling fast and powerful movements. However, more recent work by Burrows *et al.* shows that resilin alone cannot provide enough energy storage to satisfy the power demands of more accomplished jumpers like froghoppers and locusts.[38] In these insects, resilin, along with chitin, a hard polysaccharide material, forms a composite cuticle that is superior on account of its ability to prevent creep at high strains for long durations. The presence of resilin protects fracture of the hard chitinous material, and it enables quick restoration to the original undeformed shape, allowing high-frequency movements. Electron microscopy and X-ray diffraction experiments have shown that native resilin is highly amorphous and is devoid of regular structural patterns.[39] Researchers have postulated this lack of structure to be the underlying reason behind resilin's remarkable mechanical properties.[31] If no primary bonds are broken upon application of a stress, there is no subsequent effect on the structure, allowing the network to deform reversibly.

The compositional properties of resilin also affect the thermoresponsive properties of this IDPP. Resilin has more tyrosine residues compared to elastin and collagen, which is interesting in light of current knowledge about the propensity of recombinant rec1-resilin to undergo an upper critical solution temperature (UCST)-like phase separation upon cooling, especially considering that such UCST phase behavior is not observed in elastin, collagen, or silk.[40] The reaction of adjacent phenoxy radicals results in the conjugation of neighboring tyrosine residues, leading to the formation of dityrosine and trityrosine units, which serve as crosslinking sites for the elastomeric network.[40,41] The fluorescent property of these linkages facilitated the early noninvasive identification of resilin in the insect cuticle and provides handles to monitor changes in fluorescence that indicate variations in crosslink density and related mechanical properties. Studies by Quiroz *et al.* suggest that solvent-shielding interactions between aromatic residue Tyr and positively charged Arg could also be an origin of the UCST behavior observed in resilin.[29,42]

4.3 Recombinantly Derived Intrinsically Disordered Protein Polymers (IDPPs)

Given the highly repetitive nature of IDPs and structural proteins, recombinant methods have been widely applied to the production of polypeptides and protein polymers that are equipped with the advantageous mechanical and thermoresponsive behavior of the corresponding IDPs. The methods yield polymers with extremely low polydispersity, specific sequences and functional domains, and site-specific chemical functionality, which has led to the development of a wide library of building blocks for creating functional materials. Superior materials thus can be designed based on existing proteins by editing the wildtype sequence,[43–45] and engineered synthetic polypeptides of putative consensus repeat sequences can likewise provide simple yet efficient models of the behavior of the more complex natural structural proteins, obtaining insights into the fundamental origins of their behavior.[46] Derived from tropoelastin, elastin-like polypeptides (ELPs) have been widely applied models to study the lower critical solution temperature (LCST) phase behavior of artificial protein polymers.[21,29,47,48] Owing to their short, hydrophobic, and disordered consensus sequence, ELPs have emerged as a popular choice to model the biophysical and physicochemical properties of larger and more complex IDPs.[46]

Despite this demonstrated versatility of the elastins and the longstanding knowledge of the unique properties of the related resilin, the development of resilin-based materials was hindered due to the challenges associated with limited supply, batch-to-batch variability, and isolation from natural sources.[49–51] In 2001, Ardell and Andersen reported gene product CG15920, a 620-amino acid sequence with an amino acid composition similar to the sequence of resilin found in a variety of insect proteins.[52] The highly conserved repeat sequences were consistent with the encoding of a precursor for the resilin (or pro-resilin) found in *Drosophila melanogaster*,[52] which comprises the soluble, noncrosslinked polypeptide chains from the subcuticular space of insects; subsequent crosslinking between tyrosine residues in pro-resilin results in the formation of the insoluble, rubber-like elastomeric cuticular resilin network.[41] The gene product consists of three major regions and a signal peptide sequence. As shown in Figure 4.1, residues 26–322 form

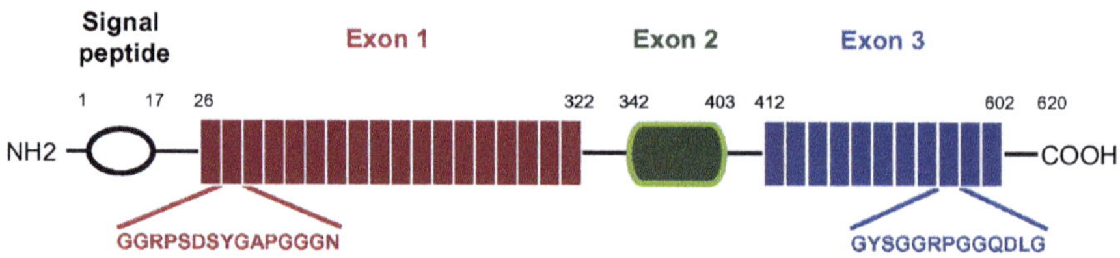

Figure 4.1 The putative resilin sequence from the *Drosophila melanogaster* CG 15290 gene product. The sequence consists of a signal peptide and three different exons (exons 1–3). The signal peptide is removed before secretion into the extracellular space. Exons 1 and 3 include 18 repeats of GGRPSDSYGAPGGGN and 11 copies of GYSGGRPGGQDLG, respectively. The sequence in exon 2 is involved in binding chitin. Reproduced from ref. 53 with permission from Elsevier, Copyright 2014.

exon 1, which is the N-terminal elastic domain. Similarly, residues 412–602 form exon 3, which is the C-terminal elastic domain. Residues 342–403 form exon 2, which is the chitin-binding domain.[52]

Putative pro-resilins from 12 *Drosophila* species were compared with pro-resilin-related proteins found in other insect species, namely honeybees, red flour beetles, parasitic wasps, body lice, and two mosquito species – *Anopheles gambiae* and *Culex quinquefasciatus*. GGRPSDSYGAPGQGN appeared as the conserved consensus in *Drosophila* pro-resilin. There were some remarkable differences between relative residue compositions of *Drosophila* resilin and the resilin sequences found in *A. gambiae*.[54] In particular, significant differences such as the absence of the chitin-binding domain and higher alanine and lower glycine content were reported in the resilin sequences found in *A. gambiae*.[54]

Based on the gene identification by Ardell and Andersen, Elvin *et al.* cloned and expressed the first exon of CG15920 as the first resilin-like protein (rec1-resilin) in *Escherichia coli*.[50] In order to validate the conservation of resilin and its properties in a wider range of insects, Lyons *et al.* amplified resilin gene transcripts found in fleas (*Ctenocephalides felis*), buffalo fly (*Haematobia irritans exigua*), and dragonfly (*Aeshna* sp.) pharate adults.[55] The recombinant crosslinked resilins derived from *D. melanogaster*, fleas, and buffalo flies were reported to have resilience values of 97%, 88%, and 87%, respectively.[55] The seminal report by Elvin *et al.*[50] laid the groundwork for developing a variety of recombinant resilin-like peptides (RLPs). Table 4.1 provides a compilation of the different resilin-like polypeptides that have been synthesized based on resilin found in different species. Moreover, the modular nature of recombinant protein synthesis allows for facile coupling of the putative elastomeric resilin sequence with different functional domains, which has provided opportunities for designing tunable RLPs for use as bioactive scaffolds in mechanically demanding regenerative medicine applications.[56–59] More recently, Quiroz *et al.* synthesized a rich library of RLPs to support heuristic predictions for encoding phase behavior *via* amino acid substitutions.[29]

Table 4.1 Summary of expression and purification conditions of recombinant resilin-like polypeptides. Adapted from ref. 53 with permission from Elsevier, Copyright 2014.

Name	Description	Molecular Weight (kDa)	Induction Method	Expression Temperature (°C)	Yield (mg L^{-1} of culture)	Purification Method	References
Species – Fruit fly (*Drosophila Melanogaster,* CG15920)							
rec1-resilin	Exon 1 (residues 19–321)	28	IPTG	37	15	Anion exchange and Ni–NTA affinity chromatography	50,60–65
			Autoinduction	37→27	30–35	Salting-out and heating	66,67
					60		
			IPTG + Lactose	27 or 37	300		67,68
			Lactose induction		120		
Dros16	16 repeats of a consensus sequence (GGRPSDSYGAPGGGN) from exon 1 Full-length	23	Autoinduction	Unknown	20	Salting-out and heating	66,68
Full-length resilin	Exon 1+ exon 2+ exon 3	60	IPTG	37	20	Salting-out and heating	69,70
6H-resChBD	Exon 1+exon 2 (residues 19–415) with His tag on N-terminus	41	IPTG	Unknown	100	Ni–NTA affinity chromatography followed by ion-exchange chromatography	69
Exon1, Exon3	Exon 1, exon 3	30,23	IPTG	30	25	Salting-out and heating	71
RLP12, RLP24, RLP36, RLP48	12, 24, 36, or 48 repeats of a consensus sequence (GGRPSDSYGAPGGGN) where Y is replaced with F. Sequence is flanked with K and interspersed with bioactive domains (cell-binding domain, MMP-sensitive degradation domain, and heparin-binding domain)	28 (RLP12) 49 (RLP24) 71 (RLP36) 93 (RLP48)	Autoinduction	37→27	70–80 (RLP12) 100 (RLP24) 79 (RLP36) 53 (RLP48)	Ni–NTA affinity chromatography	51,58,72

Table 4.1 (*Continued*)

Name	Description	Molecular Weight (kDa)	Induction Method	Expression Temperature (°C)	Yield (mg L^{-1} of culture)	Purification Method	References
RLP-X (X = RGD, RDG, or MMP)[a]	12 repeats of a consensus sequence (GGRPSDSYGAPGGGN) where Y is replaced with F and M. Sequence is flanked with K.	23–24	IPTG	37	20–30	Ni–NTA affinity chromatography	57
REC	Resilin–elastin–collagen-like polypeptide (resilin domain: 4 repeats of a consensus sequence [GGRPSDSYGAPGGGN])	11.5	IPTG	37	40–50	Ni–NTA affinity chromatography	73,74
RS R32	Resilin-*Bombyx mori* copolypeptide 5 repeats of [$(GGRPSDSYGAPGGGN)_4$-$(GAGAGS)_4$] or 32 repeats of consensus sequence (GGRPSDSYGAPGGGN)	38 (RS) 49 (R32)	IPTG	16	50–100	Ni–NTA affinity chromatography	75
R4C R8C	Resilin-spider silk block copolypeptide 4 or 8 repeats of a consensus sequence (GGRPSDSYGAPGGGN) coupled with dragline silk C-terminal domain of spider *Nephila clavipes*	21 (R4C) 27 (R8C)	IPTG	37→30	600–800	Ni–NTA affinity chromatography	76
R8 R32 R64	8, 32, or 64 repeats of a consensus sequence from exon 1 (GGRPSDSYGAPGGGN)	16 (R8) 49 (R32) 93 (R64)	IPTG	37→30	530 (R8) 770 (R32) 370 (R64)	Ni–NTA affinity chromatography	77

Species – African malaria mosquito (*Anopheles gambiae*)							
An16	16 repeats of a consensus sequence (AQTPSSQYGAP)	18.9	Autoinduction	Unknown	220	Salting-out and heating	31,66
			Lactose induction	Unknown	450		68
An4-, An8-, An16-, and An32-EGFP	4, 8, 16, or 32 repeats of the consensus sequence (AQTPSSQYGAP) with enhanced green fluorescence protein (EGFP)	Unknown	Autoinduction	37 → 28	70 (An4E) 110 (An8) 115 (An16) 14 (An32E)	Salting and cold-coacervation	78
RZ10-RGD RZ10-RDG	10 repeats of a consensus sequence (AQTPSSQYGAP) where Y is replaced with F and K is added	18	IPTG	37	22–58	Salting-out and heating	59,79–81
RZ30-BA	30 repeats of a consensus sequence (AQTPSSQYGAP) where Y is replaced with F, K is added, and bioactive (BA) domains are interspersed	50	Autoinduction	37 → 20–25	Unknown	Salting-out and heating	82
Species – Flea (*Ctenocephalides felis*)							
Cf-resB	Isoform B transcript of full-length resilin, which is devoid of chitin-binding domain	47	Autoinduction	Unknown	Unknown	Salting-out and heating	55
Species – Buffalo fly (*Haematobia irritans*)							
Hi-resB	Isoform B transcript of full-length resilin, which is devoid of chitin-binding domain	52	Autoinduction	Unknown	Unknown	Salting-out and heating	55
Sequence heuristics							
N/A	Protein scaffold described by P-X_n-G containing motifs (where n varies from 0 to 4, X and neighboring residues are Arg based, charged, or zwitterionic)	Unknown	IPTG	37	200–300	Salting and cold-coacervation	29

[a]RGD = Arg-Gly-Asp cell-binding domain; RDG = Arg-Asp-Gly scrambled negative control; MMP = matrix metalloproteinase.

RLPs mainly have been expressed in *E. coli via* the T5/T7 expression system with three different induction methods (Table 4.1): isopropyl β-D-1-thiogalactopyranoside (IPTG) induction, autoinduction, and lactose induction. The choice of a particular expression system is guided by interests in a simpler expression protocol and by the amount of RLP that may be required for a given study or application. Compared to IPTG induction, Studier's autoinduction method offers a simpler bacterial culture protocol that does not require monitoring of cell growth to identify an optimal time for induction. Lactose induction in conjunction with Studier's induction is ideal for high cell density fermentation, affording 20-fold increased yields in comparison with conventional IPTG induction.[67]

As shown in Table 4.1, a variety of methods have been used to purify recombinantly expressed resilins. The salting out and heating method takes advantage of resilin's high temperature stability to enrich resilin-based proteins or polypeptides at high salt concentrations by the precipitation of nontarget proteins at high temperatures.[66] Indeed, tunable thermoresponsive properties can be harnessed to purify a wider range of recombinant RLPs with fewer steps. The cold temperature coacervation method was developed based on UCST behavior observed at low temperatures for RLP solutions.[78] Furthermore, a polyhistidine tag with affinity to nickel–nitrilotriacetate (Ni–NTA) resin can be incorporated during the modular polypeptide design; affinity chromatography employed in conjunction with temperature cycling protocols permits more selective and efficient RLP purification.[57] More detailed comparisons between different induction and purification methods can be found in a recent review by Su *et al.*[53]

4.3.1 Mechanical Properties and Advantages

The mechanical properties of natural resilins vary from insect to insect depending on the specific function of the protein and its amino acid sequence.[83,84] In general, crosslinked materials from recombinant resilins, and in particular, those based on the original clones designed by Elvin[50] and Qin,[69] have lower resilience and elastic modulus than their natural counterparts.[85] The resilience of resilin from locust and dragonflies has been reported as 95%[24] and 99%,[83] respectively, whereas the resilience of the recombinant version is reported to be approximately 90%.[85] Regarding the elastic modulus, the natural materials from locust, dragonfly, and cockroach are in the range of 10^6 Pa,[24,83,84] whereas the synthetic analogs exhibit storage moduli that are 15-fold lower.[85]

As mentioned above, the mechanical properties of these elastomeric proteins are enabled by their distinct conformational flexibility. Recent computational studies, supported by circular dichroism experiments, show that the secondary structure of rec1-resilin is 85–95% random coil. This lack of structure is attributed to the presence of alternating glycine and proline residues that together avoid folding and promote the extension of the chain.[86] Recent molecular dynamics simulations show that the SYGAP section of the RLP motif (PSSSYGAPGGGNGGR), compared with other less abundant sections like SYSAP, TYGAP, or TYSAP, is slightly folded, which makes the total end-to-end length of a single motif to be 20.68 Å in comparison to the 40.04 Å, 38.33 Å, and 37.68 Å, respectively, to the other motifs lacking the SYGAP fragment. When several repeats of the RLP motif are present in the polypeptides, these variations in length are amplified, being apparently one of the reasons for the origin of the entropic rubber-like elastic behavior of rec1-resilin and related recombinant sequences.[87] Oscillatory rheology, small-strain microindentation, and large-strain microindentation have been used to characterize the mechanical and micromechanical performance of RLP-based hydrogels. The combination of the three techniques has helped to determine that the highest resilience values are obtained above the overlap polypeptide concentration (c*), where the more effective crosslinking reactions lead to a denser network with fewer defects.[88] The excellent resiliency (97%), extensibility (300%), and negligible stress relaxation of RLPs[51,72] make them interesting candidates for the regeneration of mechanically active tissues such as cardiovascular and vocal folds (VFs), compared to other elastomers (*e.g.*, elastin-like polypeptide-based materials) that can show significant hysteresis upon application and release of strain.[56]

In addition, biomaterials with hybrid mechanical properties, based on RLPs and poly(ethylene glycol) (PEG), conserve the inherent properties described for RLP-only systems,[89,90] with the added advantage of the LLPS of the solutions into PEG-rich and RLP-rich phases. Upon crosslinking, the evolving microstructure of the emulsion can be captured in the hydrogel, which allows the localization of mechanically distinct RLP-rich microdomains. Micromechanical characterization of the hydrogel domains *via* atomic force microscopy (AFM) indentation has shown that these materials follow the rule of mixtures.[91] The Young's moduli of the RLP-rich and PEG-rich domains are similar at early stages of phase separation (1.5 kPa and 1.4 kPa, respectively). In contrast, as the phases enrich over time, the mechanical properties begin to differentiate, resulting in Young's moduli of 2.2 kPa for the RLP-rich domains and 1.7 kPa for the PEG-rich domains when the emulsions are incubated 10 minutes before crosslinking. The limits for Young's moduli in these materials are marked by the bulk mechanical properties of the RLP-rich and PEG-rich phases crosslinked at equilibrium (2.7 kPa and 1.1 kPa, respectively). Thus, using the rule of mixtures, the localized mechanical properties of the microdomains in the hydrogel can be correlated with the incubation time of the emulsion before crosslinking.[91] Additionally, the dynamic nature of the phase separation phenomenon allows temporal control over the dimensions of the microdomains (Figure 4.2), and opportunities to guide cell localization around mechanically relevant regions of adequate dimensions, predefining the distribution of cells,[89,91,92] which may afford distinct advantages in guiding tissue regeneration.

The RLP extensibility and elasticity have also been transferred to collagen when crosslinked parallel to the surface of the triple helix bundles as a resilin-reinforced collagen fiber. Only 5% of RLP covering the collagen

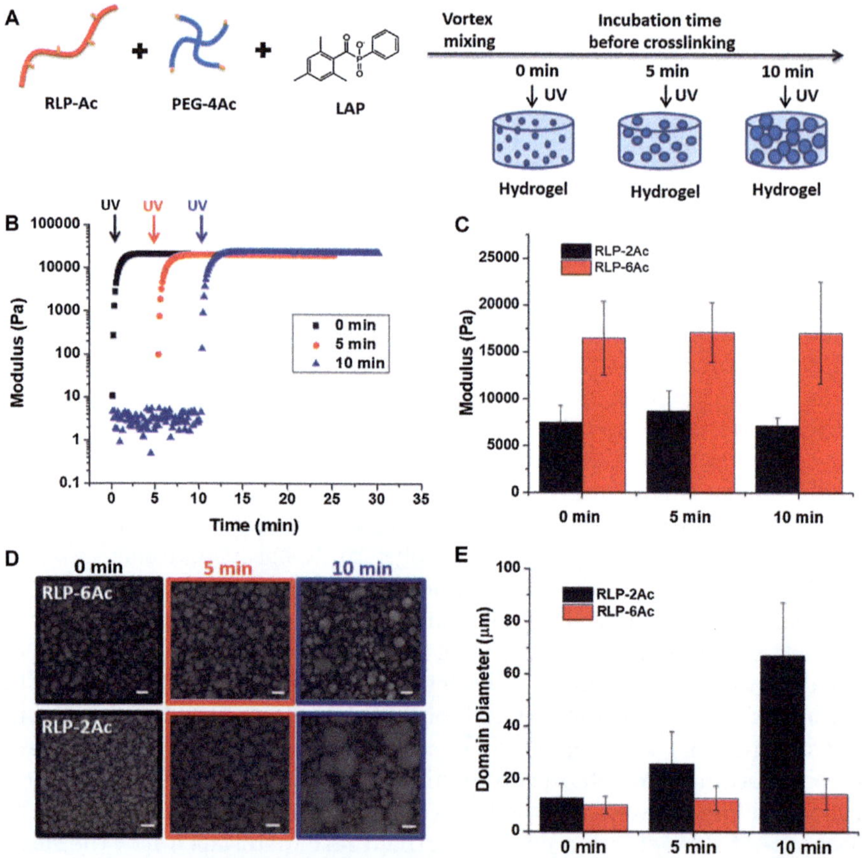

Figure 4.2 Temporally controlled microstructured hydrogels. (A) Schematic of hydrogel formation and microstructure development. (B) Time sweep of 10 wt% 50/50 RLP-6Ac/PEG-4Ac with UV irradiation at 0, 5, and 10 minutes after vortex mixing. (C) Storage moduli comparison for RLP-2Ac and RLP-6Ac with UV irradiation at 0, 5, and 10 minutes after mixing. (D) Autofluorescence images of photo-crosslinked 10 wt% 50/50 RLP-6Ac/PEG-4Ac and 10 wt% 50/50 RLP-2Ac/PEG-4Ac hydrogels; microscale RLP-rich domains grow in diameter when precursors were incubated at room temperature for 0, 5, and 10 minutes prior to photo-crosslinking. (E) Average domain diameters of the RLP-rich domains over time for RLP-2Ac and RLP-6Ac solutions with PEG-4Ac. Reproduced from ref. 91, https://doi.org/10.1002/advs.201701010, under the terms of the CC BY 4.0 license https://creativecommons.org/licenses/by/4.0/.

surface has been enough to increase the strain at break from 6% to 21% and the elastic modulus from 22 MPa to 49 MPa, making these reinforced fibers excellent candidates for the regeneration of the human Achilles tendon.[93]

Besides applications in tissue regeneration, RLP sequences have also been explored for the production of other high-performance materials, including in the production of stretchable self-adhesive wearable sensors[77] and elastic wood fibers.[94] In the former work, hydrogels are made from high-molecular-weight RLP (93.5 kDa), plasticized with glycerol to impart skin adhesion, and enriched with RLP-coated graphene oxide particles to impart conductivity. The resulting material was shown to detect conductivity variations from very small vibrations coming from phonation up to high stretching in finger knuckles.[77] In the latter, RLP sequences have been expressed *in vivo* in the cell wall of Eucalyptus. Although elastic wood-based fibers have not yet been produced, current reports indicate the increased stiffness of the cell wall due to interactions between resilin and lignin that result in an increased crosslinking density.[94] Other studies have tried to mimic the mechanical properties of resilin utilizing natural and synthetic polymers. In particular, cellulose crosslinked with isoprene in adequate ratios has been shown to reach up to 180% extensibility, an elastic modulus of 3.9 MPa, and resilience of 98%, matching the mechanical properties of resilin closely[85] and offering a synthetic alternative to the natural polymer.[95]

4.3.2 Thermally Responsive Properties and Advantages

The amino acid sequence not only has an impact on chain flexibility and the mechanical behavior of IDPs, as explained previously for the case of resilin, but it also controls phase separation behavior that is characteristic of the IDPs and that dictates the stimuli under which the IDPs form coacervates. In the case of RLPs and ELPs (and the ELPs in particular), variations in their sequences have been widely employed to manipulate the stimuli that trigger LLPS under specific conditions useful for applications such as drug delivery. In both cases, the disorders in the protein and polypeptide sequences are consistent and are a prerequisite for phase separation.[19,28]

Garcia-Quiroz and Chilkoti created an extensive library of synthetic polypeptides, demonstrating with turbidimetry assays that IDPs rich in hydrophobic amino acids (*i.e.*, ELPs) tend to have LCST-like transitions while those containing positively charged, negatively charged, and aromatic residues (*i.e.*, RLPs) tend to show UCST-like transitions.[29] These experiments were confirmed by Dignon *et al.* using a novel coarse-grained model that accounts for temperature-dependent solvent-mediated interactions, where the calculated Flory scaling exponent (ν) is the parameter that determines whether the peptide will collapse ($\nu < 0.5$) or will remain in solution ($\nu > 0.5$).[42] In total, this model has been able to match 36 of the 39 experimental results[42] and thus offers new opportunities to understand coacervation behavior in these sequences and leverage that understanding in materials design.

Resilin-like polypeptides, and rec1-resilin specifically, have a dual-phase transition behavior, with rec1-resilin exhibiting a UCST transition at approximately 6 °C and an LCST transition at approximately 70 °C.[60] These observations are completely consistent with the subsequent experiments by Quiroz and Chilkoti and computational studies by Dignon *et al.*, as well as consistent with the presence of both a charged/aromatic sequence (GGRPSDSY, UCST behavior) and a hydrophobic region (GAPGGGN, which supports LCST behavior). Modifications to the rec1-resilin sequence, such as the one developed by Li *et al.*,[96] in which the characteristic tyrosine in the consensus motif was replaced by phenylalanine and methionine residues, have been shown to abolish the UCST transition while maintaining the LCST transition at a temperature similar to that reported for rec1-resilin. This suggests the key role of this amino acid in the RLP motif and its importance when designing new sequences that can phase separate at different temperatures.

Coassemblies of rec1-resilin with other polypeptides such as polyproline and silk fibroin have also been shown to impact the LCST transition as well as the elastic modulus of swollen hydrogels, reaching up to 6 MPa, which is hundreds of times higher than other RLP hydrogels.[97] In both cases, the hydrophobic peptides (polyproline and silk fibroin) promote a reduction of the LCST transition of rec1-resilin down to physiologically relevant temperatures. The structures formed by the irreversible phase separation have been shown to be spherical micelles with the hydrophobic polypeptides at the core and the hydrophilic rec1-resilin at the outer layer,[98] and further formation of nanofibrils upon annealing at 37 °C for 12 hours.[75] RLPs have also been shown to form distinct assembled structures upon mixture and subsequent crosslinking with peptide amphiphiles.[99] These experiments show that acrylate-functional RLPs and thiol-functional amphiphiles form nanospheres and nanofibrous networks, respectively. A combination of both nanostructures, a beaded fibrous network, is achieved by combining both materials, and the density of beads along the fibers can be tuned with the initial concentrations of the precursor solutions (Figure 4.3).

The coassembled hydrogels with potential applications in tissue engineering scaffolds and drug release systems show self-healing properties and high extensibility, with the latter attributed to the RLP in the system.[99]

In functional constructs of rec1-resilin with cellulose-binding motifs (CBMs) and hydrophobin (HFBI), the formation of RLP coacervates can be triggered at room temperature by pH changes above or below the isoelectric point.[100] The former forms individual coacervates at pH above 6 or below 3 (the pI of the construct is 4.5), and transitions of the latter appear to be independent of the pH and result in clusters of coacervates, likely due to the hydrophobic attraction of the HFBI motif.[100] An advantage of adding functional domains to the RLP sequence, outside of any tuning of coacervation, is the enhancement of adhesion to different substrates. The CBM improves the coacervate adhesion to cellulose nanofibrils, acting as a crosslinker that improves the storage modulus of the fibril composites,[101] while the HFBI permits adhesion to graphene surfaces, opening the door to bifunctional RLPs as linkers in substrates with different hydrophobicity.

4.4 IDPPs in High-performance Materials

The multistimuli-responsive nature of IDPPs has been harnessed for creating several high-performance materials for a variety of applications. IDPs play a

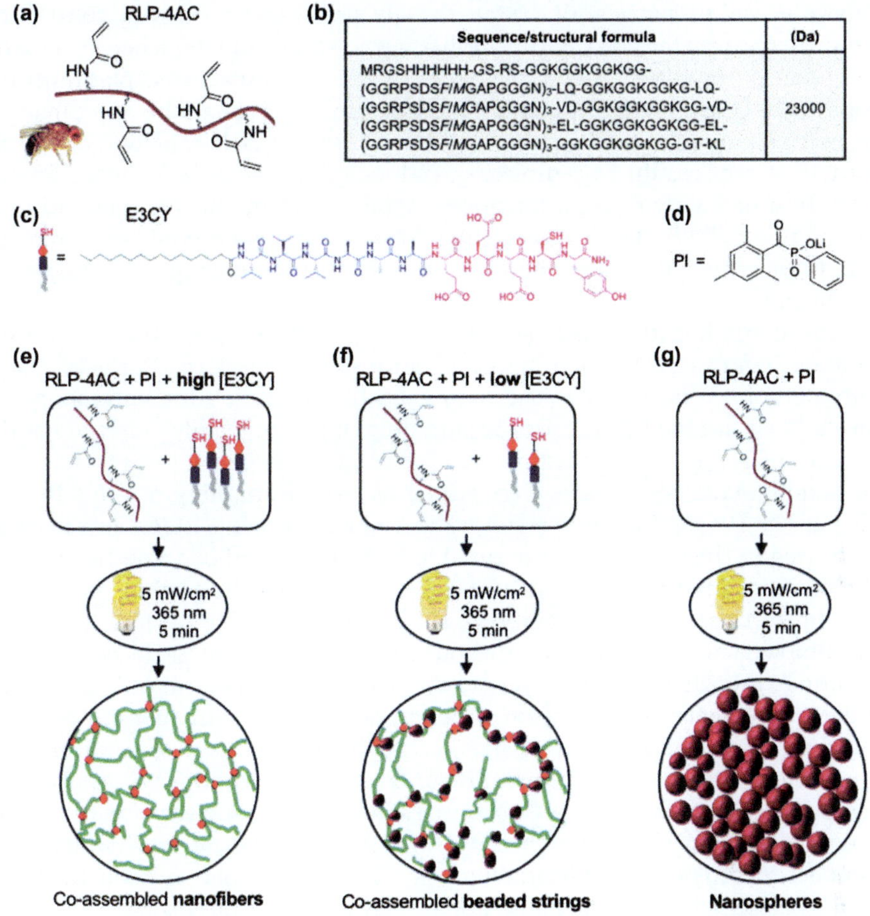

Figure 4.3 (a) Schematic representation of RLP structure functionalized with four acrylamide moieties, (b) amino acid sequence of RLP, (c) structure of E3CY peptide amphiphile, and (d) structure of the photoinitiator (PI) used. Schematic representation of photopolymerization of (e) RLP-4AC and high concentration of E3CY, (f) RLP-4AC and low concentration of E3CY, and (g) RLP-4AC to generate nanostructures coassembled hydrogels with nanofibers and beaded strings and single-component hydrogels with nanosphere nanostructures. Reproduced from ref. 99 with permission from the Royal Society of Chemistry.

pivotal role in biomineralization processes in nature.[102] This phenomenon has inspired sustainable strategies that use IDPPs as stabilizers, structure-directing agents, and multifunctional templates for nanomaterial synthesis.[62,103] Han *et al.* attached ELPs to the silica-precipitating silaffin R5 peptide to construct micellar templates for the synthesis of near-monodisperse silica nanoparticles *via* silicification.[104] RLP sequences enriched in tyrosine residues have been used in the synthesis of noble metal nanoparticles with tunable sizes and photophysical properties.[65,105] Nanoparticle sizes can be tuned by controlling the protein-to-metal ratio, with RLP being the soft template, whose conformation can be modified based on pH. Balu *et al.* manufactured platinum nanoparticles based on their patented method of using rec1-resilin as a directing agent for use as catalysts in fuel cells.[106] Furthermore, Balu *et al.* demonstrated the use of rec1-resilin as a directing, reducing, and stabilizing agent for the formation of highly fluorescent noble metal quantum clusters made of gold and platinum.[105,107] These nanoparticles and nanoclusters could be employed as stimuli-responsive fluorophores in biosensing and bioimaging theragnostic applications.

Surface modification with stimuli-responsive IDPPs provides an external handle to control the physicochemical features of biointerfaces. Surfaces coated with ELPs have been used to design reusable platforms for harvesting cell sheets.[108] Above the LCST, cell adhesion domains were readily presented at the surface to facilitate cell attachment. At temperatures below the LCST, a conformational change was triggered, which led to the masking of the adherent Arg-Gly-Asp (RGD) domains, resulting in cell detachment. RLPs have likewise been used in the development of tunable interfaces[64] and patterned surfaces.[63] The conformation of adsorbed RLP can be modulated by altering surface energy by tuning physical conditions at the substrate.[63] Entropically favored monomolecular structures were obtained on hydrophilic surfaces, whereas columnar morphologies were reported for hydrophobic surfaces.[63] Thin-film hydrogels, constructed based on differences in rec1-resilin morphology as a function of substrate characteristics, could provide strategies for fabricating drug delivery reservoirs or templates for sensing applications.[62] Truong *et al.* demonstrated the effect of pH on the orientation, packing density, thickness, and viscoelasticity of the adsorbed RLP layer.[64] RLPs can reversibly switch from compact to brush-like conformations as a function of pH and can hence be used to create pH-responsive functional interfaces for biosensors.

4.5 IDPPs in Structured Biomaterials

4.5.1 IDPPs as Building Blocks for Self-assembled Structures

The multistimuli-responsive and tunable physicochemical phase behavior of IDPPs makes them an attractive choice as building blocks for generating biomaterials with well-defined nano- and microstructures for biotherapeutic cargo delivery. The thermoresponsive properties of ELPs to undergo reversible coacervation above their LCST have been used in the creation of materials for elastomeric adhesives and drug delivery. Altering the chain length and guest residue in the VPGXG sequence can be used to modulate the LCST to physiologically relevant temperatures. This temperature-triggered coacervation has been used to fabricate self-assembled nanoconstructs as controlled-release vessels and tissue engineering scaffolds.[109,110] In addition to conventional stimuli such as temperature and pH ionic strength, next-generation ELPs have been engineered to respond to more sophisticated cues such as light and biomolecular recognition.[109] Functionalization with chromophores such as azobenzene[111–113] and spiropyran[114] results in light-triggered coacervation mediated by altering hydrophobic interactions within the ELPs, thereby affording spatiotemporal control.

Covalent linkage of ELPs to biomolecules that can induce isothermal phase transition upon target-specific binding can be used in the design of highly selective and tunable biopolymer switches.[115,116] The fusion of UCST motifs (RLPs) with LCST motifs (ELPs) could offer a strategy to program dual UCST and LCST phase behavior in IDPPs.[29] Block copolymers of recombinant ELPs and RLPs have been used to fabricate self-assembled spherical micelles and phase-inverted cylindrical micelles.[117] The control over the final structure is tuned by changes in the molecular weight of the copolymer as well as the hydrophobic to hydrophilic ratio between both peptide sections.[117]

A majority of the literature regarding IDPPs refers to recombinantly expressed sequences with over tens or hundreds of pentapeptide repeats. Short synthetic ELP constructs fabricated by means of solid-phase peptide synthesis (SPPS) have not been widely useful for physiologically relevant applications due to transition temperatures (LCST or T_t) much higher than 37 °C.[29,118,119] Luo *et al.* overcame this limitation by conjugating short ELP sequences with collagen-like peptides (CLPs). Similar to native collagen, CLPs adopt triple-helical conformations below their melting temperatures.[120] CLP triple helices dramatically lowered the ELP T_t by enriching the local concentration of the ELP domains, facilitating nanovesicle formation.[121] Subsequent work by Qin *et al.* (Figure 4.4) highlights the importance of the amino acid composition of the ELP sequence on the self-assembly morphology and T_t.[122] Increasing the amount of W residues in the ELP sequence over a specific threshold results in the formation of platelet morphologies.

4.5.2 IDPPs in Microstructured Hydrogels

As introduced above, IDPPs offer tremendous opportunities for the fabrication of stimuli-responsive hydrogels for mechanically demanding applications,[46,57,123,124] with added opportunities to control coacervation and thus microstructure. Physical or chemical encapsulation of microstructures within resulting hydrogels can be used for the mechanical reinforcement and controlled delivery of bioactive molecules from isolated domains.[125] Microstructural heterogeneity also provides opportunities to better capture some of the complex chemical, topographical, and mechanical properties of the native extracellular matrix (ECM).

Over the years, a variety of methods such as electrospinning, gas foaming, electrospinning, 3D printing, and solvent casting have been used to make microstructured materials for a variety of applications.[126–128] Photopatterning or two-photon laser lithography offers spatiotemporal control over the fabrication of UV-crosslinkable and UV-cleavable hydrogel systems.[129–132] Since sequential 2D patterning involving multiple steps and complicated laser systems is required to create high-fidelity 3D structures, the fabrication

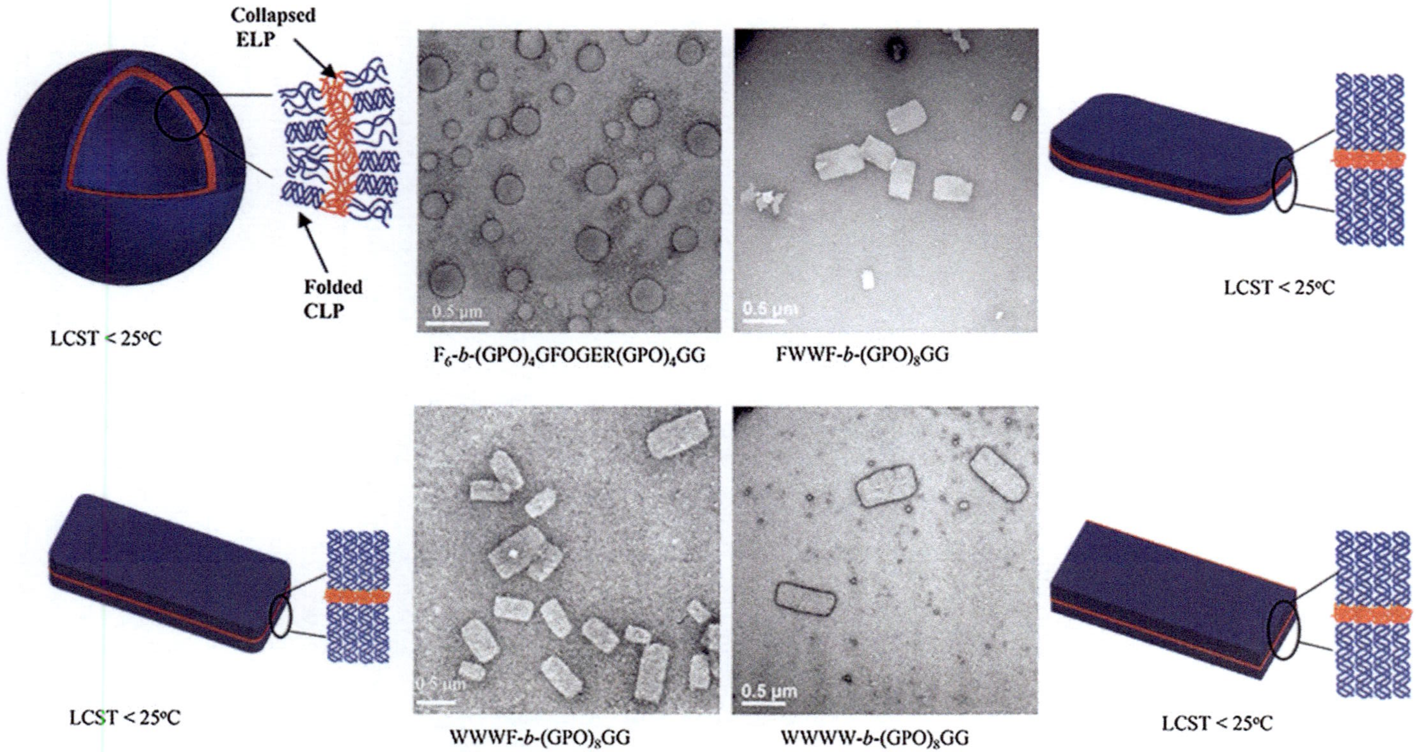

Figure 4.4 Nanostructure morphology can be modulated *via* amino acid substitution. Reproduced from ref. 122 with permission from American Chemical Society, Copyright 2019.

of bulk structures can prove to be costly with respect to time and energy.[126] Microfluidics approaches can be used to encapsulate a variety of preformed microstructures in hydrogel precursors to form gel-in-gel structures.[126,133,134] Although highly sophisticated morphologies can be obtained by modulating relative channel position, rate of injection, solution type, and viscosity, microfluidics approaches are currently limited to lower processing volumes.[126]

As mentioned above, LLPS has interested biological researchers due to its regulatory implication in cells.[135–138] Native proteins undergo LLPS under specific conditions from condensed liquid droplets (membraneless organelles), which perform various cellular functions.[126,135,139] LLPS outside of the cell permits the production of a variety of materials for tissue regeneration, drug delivery, and/or isolated chemical reactions, with significant design flexibility owing to the sensitivity of phase separation to variables such as temperature, molecular weight, and composition.[126,140–142] LLPS has also been widely used in the development of protein, polysaccharide, and DNA purification protocols.[143,144] Typically, proteins that undergo LLPS are hydrophilic, zwitterionic, charged, multivalent, and have intrinsically disordered regions,[135,145] offering additional sequence-specific design parameters that can be employed to create materials. In addition to the *de novo* design of proteins, factors such as temperature, concentration, molecular weight, ionic strength, and charge interaction can be used to modulate phase behavior. If phase separation can be induced at concentrations below the gel point, enriched local concentration upon phase separation can lead to hydrogel formation.[146] The underlying morphology of phase separation can also be captured *via* crosslinking, thereby generating a 3D microstructure in one step. Wang *et al.* reported a method for fabrication of ELP hydrogels with bead-string microstructures, which did not require amino acid substitution for controlling microstructure morphology[147] but instead relied on variations in the relative rate of chemical crosslinking *versus* thermally triggered self-assembly (Figure 4.5).

Roberts *et al.* demonstrated that the incorporation of ordered polyalanine helices within ELPs led to the formation of physically crosslinked viscoelastic and porous fractal networks, which could be harnessed as subcutaneous depots.[148]

Recent work from the Kiick group has shown that the manipulation of LLPS provides a facile one-pot strategy to create defined hydrogel microstructures in RLP/PEG hydrogels.[89,91,92,149] The LLPS of RLP/PEG solutions provides a simple platform for the creation of microstructured, two-component hydrogels with distinct micromechanical properties.[91,92,150] Phase separation behavior and subsequent scaffold morphology can be tuned by manipulating temperature and protein/polymer concentrations and compositions.[92] Control over the microstructure morphology can be obtained by exploiting specific regions of the phase diagram *via* a well-informed selection of initial concentration and composition of RLP and PEG.[92] The microstructured RLP/PEG materials can be produced *via* a range of crosslinking strategies. A Mannich-type reaction between the crosslinker

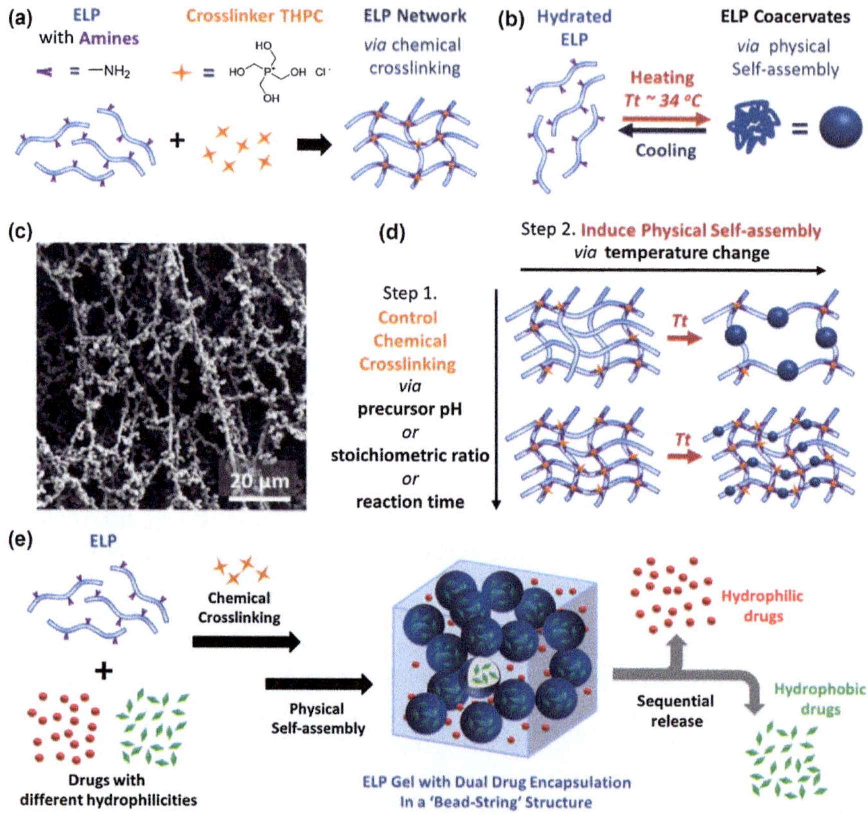

Figure 4.5 Schematic of ELP hydrogel formation with controlled "bead-string" structure. (a) Covalent crosslinking of ELP network using chemical crosslinker tetrakis(hydroxymethyl)phosphonium chloride (THPC). (b) Thermally triggered physical self-assembly of ELP. (c) SEM image of "bead-string" microstructured ELP gel (5 wt% ELP crosslinked at a 1 : 1 stoichiometric ratio for 2 minutes at room temperature [RT], followed by 10 minutes at 37 °C). (d) Proposed strategy to control hydrogel microstructure by modulating chemical crosslinking and physical self-assembly processes of ELP. (e) Simple, one-pot formation of microstructured ELP hydrogel for dual encapsulation and sequential release of hydrophilic and hydrophobic drugs. Reproduced from ref. 147 with permission from American Chemical Society, Copyright 2018.

tris(hydroxymethyl phosphine) (THP) and amine groups has been found to yield RLP-rich domains in a PEG continuous matrix, although the crosslinking requires approximately an hour to reach complete conversion and thus limits control over the final microstructure.[57,92,126] Thiol-mediated click chemistries have also been employed; emulsions of cysteine-functionalized RLPs mixed with vinyl sulfone-functionalized PEG have been used to fabricate microstructured hydrogels for cardiovascular applications.[58,89,149]

However, in the chemical approaches above, a lack of independent control over crosslink density and reaction kinetics provides challenges for the desired tuning of mechanical properties and microstructure.[91]

Photo-triggered crosslinking methodologies to capture microstructure offer spatiotemporal control over scaffold morphology and mechanical properties (Figure 4.2).[91] LLPS of an aqueous mixture of acrylamide-functionalized RLP and acrylate-funtionalized PEG was arrested *via* UV light to produce RLP-PEG hybrid hydrogels; the same was achieved with an aqueous mixture of norbornene-functionalized RLP and thiol-functionalized PEG.[91] Increasing the duration of UV exposure led to a higher degree of photopolymerization, offering control over mechanical properties.[91] Modulation of incubation time before crosslinking directly controlled the domain diameters of the RLP-rich phases.[91] Although to date, primarily, spherical morphologies of the RLP-rich phase are obtained *via* nucleation and growth mechanisms, it should also be possible to generate bicontinuous structures *via* spinodal decomposition, but it proves to be more challenging due to percolation-to-cluster transitions caused by fluctuations in local composition.[126,151,152] Biocompatible emulsion stabilization can be achieved by using block copolymer surfactants or Pickering stabilizers such as silica nanoparticles.[126,153–155] Spatial localization of a variety of cell types in the RLP-rich domains presents exciting opportunities for producing microstructured hydrogels for targeted biomedical applications such as drug delivery and tissue engineering.[58,91]

4.6 IDPPs for Tissue Engineering Applications

On account of their tunable viscoelasticity and structural similarities to the ECM, hydrogels are a popular choice for creating biomimetic tissue engineering scaffolds.[156] Their water content and swollen nature make them highly permeable, facilitating nutrient and metabolite transport.[157] Collagen, gelatin, chitosan, fibrin, alginate, hyaluronic acid, and agarose are some of the commonly used naturally derived polymers, while polycaprolactone, poly(vinyl alcohol), poly(*N*-isopropylacrylamide), PEG, and poly(2-hydroxyethyl methacrylate) are some of the popular synthetic polymers used to fabricate tissue engineering hydrogels. Advances in recombinant methodologies have expanded the use of biosynthetic protein polymers as scaffolds. *De novo* sequence design and the high fidelity of synthesis allow modular control over chemical composition, chain length, structure, and functionality.[158] These advantages have led to the adoption of recombinantly produced proteins such as ELPs, RLPs, CLPs, silk-like polypeptides (SLPs), and others in the design of ECM mimics.[159] Of specific interest are hydrogels derived from elastomeric proteins for applications in engineering mechanically active tissues.

4.6.1 Vocal Fold Tissue Engineering

Their outstanding elasticity and resilience at elevated frequencies corresponding to the functioning of human VFs make resilin-like polypeptides excellent candidates for VF tissue engineering. Sound production is a function of the vibration of the pliable VFs,[56] which is dependent on the viscoelasticity of the lamina propria.[160] Mechanical stresses, environmental factors, and disease pathology can have undesired effects on the mechanical properties of the VFs, resulting in hoarseness, fatigue, and a variety of associated voice disorders.[56,161] Treatment *via* surgical intervention can result in the formation of scar tissue, further complicating rehabilitation,[162] which has motivated the development of minimally invasive injectable materials-based approaches for VF tissue regeneration. A range of materials, including hyaluronic acid and elastin-based materials, have been employed,[56,163,164] although with challenges with long-term feasibility owing to mechanical mismatch. Appropriate ECM mimics would exhibit analogous mechanical properties at high frequencies, similar degradation rates *in vivo*, and appropriate biomolecule transport profiles.

Resilin's mechanical properties suggest that it may be a useful material for VF regeneration.[51,165] The modular nature of the recombinant polypeptide design allows the incorporation of cell-binding domains (RGDSP),[166] heparin immobilization domains (KAAKRPKAAKDKQTK),[167,168] and proteolytic degradation domains (GPQGIWGQ),[169,170] along with a domain comprising 12 repeat units of the putative resilin consensus sequence from exon 1 of *D. melanogaster*.[51] Phosphine-based crosslinking of the lysine residues generates mechanically robust hydrogels with shear storage moduli of approximately 10–50 kPa at strains lower than 15%,[51,72] with similarities to values reported for native VF tissue: 500 to 5000 Pa at low frequencies (1 to 10 Hz) and 200 to 2000 Pa at elevated frequencies (30 to 150 Hz).[56,72,171] Modular polypeptide design allows for the facile decoupling of mechanical properties from the biological composition as indicated in Figure 4.6.

Rapid gelation kinetics facilitated the encapsulation of human mesenchymal stem cells (hMSCs). Studies showed high cell viabilities, cell spreading, and mechanical integrity of the RLP hydrogels after 3 weeks, indicating a promising application as an injectable hydrogel and cell vehicle.[172] *In vitro* investigation of the inflammatory response to hydrogel was tested *via* the characterization of tumor necrosis factor-alpha (TNF-α) levels in murine-derived RAW 264.7 macrophage-like cells seeded on gel surfaces.[172] *The in vivo* biocompatibility of preformed RLP hydrogels (20 wt%, $10 \times 10 \times 1$ mm^3 films) was tested by subcutaneous transplantation in a rat model for 7 days.[172] Histology studies showed minimal inflammatory cell types (CD3, CD31) at the gel–tissue interface, providing preliminary validation for *in vivo* compatibility.[172] Recently, King *et al.* studied the performance of hyaluronic acid/RLP (HA/RLP) hybrid injectable hydrogels in rabbit VF models.[173] Chemically crosslinked RLP-acrylamide and HA-thiol networks provided opportunities to encapsulate and deliver human bone marrow mesenchymal stem cells, with only mild inflammation in the lamina propria up to 3 weeks.[173] Altogether, these studies highlight the potential utility of resilin-based hydrogels in the regeneration of VFs.

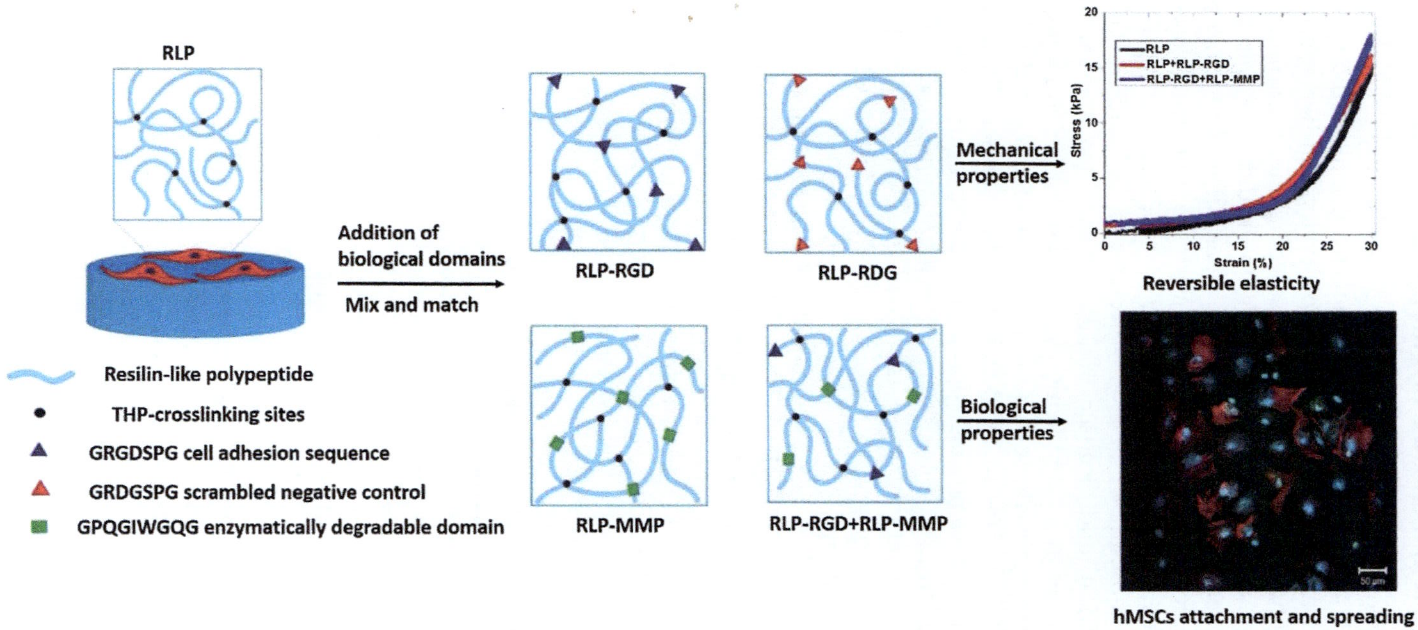

Figure 4.6 Schematic of approaches for independently tailoring mechanical and biological properties of RLPs with three repeats of stress–strain cyclic tensile testing mechanical properties at 20 wt% with different material compositions and 2D hMSC attachment biological responses with a 20 wt% RLP-RGD hydrogel. Original image reproduced from ref. 172 with permission from John Wiley & Sons, Copyright © 2015 WILEY-VCH Verlag GmbH & Co. KGaA, Weinheim; adapted image reproduced from ref. 56 with permission from Elsevier, Copyright 2016.

4.6.2 Cardiovascular Tissue Engineering

Myocardial infarction events lead to cell and cardiac muscle death due to obstructed blood flow.[174] The regeneration capacity of adult cardiomyocytes is severely limited, which causes scar tissue to develop, further affecting the mechanical and functional integrity of the diseased tissue.[175] Cardiomyocytes and biomolecules have been injected directly into the diseased heart tissue to promote tissue regeneration.[176,177] However, more efficient treatments need to be developed since cells need to integrate with the underlying vasculature and mend diseased tissue.[178,179] Engineering vascularized mimics of cardiovascular tissue is particularly challenging because of the complex biological, chemical, mechanical, and electrical requirements.[180] On account of their minimally invasive nature, there is a high level of interest in developing robust injectable hydrogel systems that can act as cell carriers for targeted delivery.[181] Hydrogels derived from functional biopolymers like collagen, fibrin, and alginate have been shown to provide appropriate biochemical cues to promote angiogenesis and correct fibrosis.[180,182,183] However, due to their poor mechanical and electrical properties, the use of these biopolymers has been limited to the delivery of growth factors and other biomolecules to assist cardiac tissue engineering.[180,184,185]

ELPs have been used to coat the luminal surfaces of stents and grafts to reduce thrombogenic complications.[186] Recently, Annabi *et al.* created a model myocardium by seeding neonatal cardiomyocytes in micropatterned thin-film patches of methacrylated tropoelastin (MeTro).[187] They showed that the attachment, spreading, and elongation of primary cardiomyocytes was elevated in the micropatterned MeTro gels. Electrical stimulation seemed to further enhance the functionality of the MeTro patterned patches, indicating potential success as synthetic mimics for cardiovascular tissue engineering.[179,187]

Elastomeric hydrogels are particularly suitable for cardiovascular tissue engineering on account of their ability to deform reversibly under cyclic loading arising from hemodynamic processes.[179] McGann *et al.* reported the application of multicomponent, cell-instructive RLP/PEG hydrogels for cardiac tissue engineering.[58] Recursive ligation was introduced to create longer RLP sequences to increase cysteine functionality, which would form macromolecular networks upon interacting with PEG–vinyl sulfone. Encapsulation of human aortic adventitial fibroblasts showed high cell viabilities and the development of spindle morphologies after 1 week. This shift in phenotype indicated that the cells were responding to the biochemical cues presented by the RLP hydrogels.[58] Subsequently, norbornene-functionalized RLP and thiol-functionalized PEG were used to create spatiotemporally controlled hybrid photopolymerized hydrogels *via* thiol–ene click chemistry.[149] Mechanical testing of these hybrid hydrogels revealed shear storage moduli within 5 to 10 kPa, validating their potential application for cardiovascular therapies. Kim *et al.* showed that using enzymatic methods for crosslinking RLP hydrogels showed improved long-term cell adhesion for cardiovascular applications.[81] Upon initially seeding the same number of cells, after 1 week, a drastically lower number of cells were observed on THP crosslinked matrices as opposed to matrices crosslinked with enzymatic crosslinker transglutaminase.[81]

4.6.3 Cartilage Tissue Engineering

Trauma, aging, and disease can cause cartilage to undergo deleterious alterations leading to osteoarthritis, which results in loss of function and diminished quality of life.[188] Potential for self-healing is limited due to limited lymphatic and vascular cues and load-bearing requirements, along with architectural anisotropy with respect to matrix composition.[189] Current surgical intervention strategies such as microfracture, abrasion, drilling, and chondrocyte implantation have resulted in the subpar regeneration of inferior fibrocartilage tissue.[188] Hence there is a need for improved biomaterial platforms to facilitate cartilage regeneration *via* tissue engineering strategies. Injectable hydrogels have emerged as popular choices for minimally invasive biologic delivery and *in situ* gelation in irregularly shaped defects.[188] Hydrogels made by using biopolymers like collagen and gelatin to manufacture mimics of the cartilage have been shown to promote chondrogenic differentiation of MSCs without the addition of external growth factors.[190] However, on account of their poor load-bearing capacity, it is necessary to use composites made by incorporating other materials such as hyaluronic acid, chondroitin sulfate, silk fibroin, and anegen.[188,191–195]

Above their LCST, ELPs form gel-like coacervates, which can provide biochemical cues to promote chondrogenesis *in vitro*.[196] However, the load-bearing capacity of these coacervates is severely lacking. The dynamic shear stiffness required for articular cartilage tissue mimics is around 100 to 500 kPa.[196,197] McHale *et al.* reported an enzymatic strategy to covalently crosslink ELPs *in situ via* transglutaminase to form hydrogels in irregularly shaped defects.[197] The reported maximum shear modulus of 310 kPa was within the suitable range necessary for cartilage tissue engineering. Upon incorporating chondrocytes into these gels, the elevation of glycosaminoglycan and type II collagen levels indicated the formation of hyaline cartilage *in vitro*.[123]

Expression of RLPs containing the *A. gambiae* consensus sequence reported higher yields compared to those derived from *D. melanogaster*. Hence Renner *et al.* characterized the application of these mosquito-derived RLP constructs for cartilage tissue engineering.[59]

Table 4.2 highlights that the unconfined compressive modulus of mosquito-derived RLPs is four times higher than that of natural resilin and is one order of magnitude higher than elastin while being around the required value for articular tissue. This could be attributed to increased hydration in RLP gels due to the hydrophilicity of the consensus sequence compared to that of the more hydrophobic ELPs.[61] Modular incorporation of the RGD cell-binding domain resulted in the formation of F-actin networks and spreading of MSCs, further validating the application of this stiffer mosquito resilin for cartilage regeneration.[59]

Table 4.2 Comparison of mechanical properties for fibrocartilage, articular cartilage, natural resilin, and elastin- and resilin-based proteins.

	Fibrocartilage	Articular cartilage	Natural resilin	Elastin-based proteins	Resilin-based proteins	Renner *et al.* 2012 study[a]
Complex modulus G^* (kPa)	100–600[198,199]	180–2500[200,201]		2–45.8[202,203]	0.5–10[51,72]	22 ± 1[59]
Yield strain (%)					450[51]	63[59]
Unconfined compressive modulus (MPa)	0.036–0.120[204]	2.25–15.3[205]	0.600[26]	0.00626–0.215[206]		2.4 ± 0.2[59]

[a]Data in Renner *et al.* 2012 study are represented as mean modulus ± the standard deviation ($n = 4$ for complex modulus and $n = 3$ for unconfined compressive modulus). Reproduced from ref. 59 with permission from American Chemical Society, Copyright 2012.

4.7 Conclusions and Perspectives

This review highlights how multistimuli-responsive properties of IDPPs, with a focus on resilin-based materials, can be utilized to make high-performance materials for biotherapeutic delivery, diagnostics, sensing, actuation, and tissue engineering. IDPPs that undergo aqueous phase separation behavior offer a tunable platform for generating sophisticated materials with distinct chemical, morphological, and mechanical properties. The phase behavior of these polypeptides can be encoded *de novo via* sequence determinants. Appropriate functional group placement could permit precise control over self-assembly facilitated by the adoption of thermodynamically favored conformations. Ternary mixtures incorporating polymeric cosolutes offer additional handles for facile manipulation of LLPS to create hybrid hydrogels with distinct morphologies.

Challenges remain with respect to understanding complex inter- and intra-peptide interactions driving the phase behavior in multicomponent systems. Going forward, computational simulations that can capture and predict experimental observations will provide insightful tools for harnessing the rich phase behavior of IDPPs and broadening the relevant sequence space. Selective and strategic incorporation of order within these disordered polypeptides could be employed in the fabrication of bioinspired hierarchical architectures. The synergistic interplay between ordered and disordered domains can be used to control mesoscale phase separation and to program reversible and hysteretic aggregation behavior. This nonequilibrium behavior of protein assemblies offers significant opportunities to encode tunable thermal hysteresis in phase separation, which could enable memory effects that guide self-assembly.

References

1. R. van der Lee, M. Buljan, B. Lang, R. J. Weatheritt, G. W. Daughdrill, A. K. Dunker, M. Fuxreiter, J. Gough, J. Gsponer, D. T. Jones, P. M. Kim, R. W. Kriwacki, C. J. Oldfield, R. V. Pappu, P. Tompa, V. N. Uversky, P. E. Wright and M. M. Babu, *Chem. Rev.*, 2014, **114**, 6589–6631.
2. C. P. Brangwynne, P. Tompa and R. V. Pappu, *Nat. Phys.*, 2015, **11**, 899–904.
3. F.-X. Theillet, L. Kalmar, P. Tompa, K.-H. Han, P. Selenko, A. K. Dunker, G. W. Daughdrill and V. N. Uversky, *Intrinsically Disord. Proteins*, 2013, **1**, e24360.
4. C. J. Oldfield and A. K. Dunker, *Annu. Rev. Biochem.*, 2014, **83**, 553–584.
5. V. N. Uversky, *Front. Phys.*, 2019, **7**, 10.
6. W. Zheng, G. Dignon, M. Brown, Y. C. Kim and J. Mittal, *J. Phys. Chem. Lett.*, 2020, 3408–3415.
7. G. H. Zerze, R. B. Best and J. Mittal, *J. Phys. Chem. B*, 2015, **119**, 14622–14630.
8. V. N. Uversky, *Protein J.*, 2009, **28**, 305–325.

9. N. E. Davey, K. Van Roey, R. J. Weatheritt, G. Toedt, B. Uyar, B. Altenberg, A. Budd, F. Diella, H. Dinkel and T. J. Gibson, *Mol. BioSyst.*, 2012, **8**, 268–281.
10. L. M. Iakoucheva, P. Radivojac, C. J. Brown, T. R. O'Connor, J. G. Sikes, Z. Obradovic and A. K. Dunker, *Nucleic Acids Res.*, 2004, **32**, 1037–1049.
11. M. M. Babu, *Biochem. Soc. Trans.*, 2016, **44**, 1185 LP–1181200.
12. H. J. Dyson and P. E. Wright, *Nat. Rev. Mol. Cell Biol.*, 2005, **6**, 197–208.
13. V. N. Uversky, C. J. Oldfield and A. K. Dunker, *Annu. Rev. Biophys.*, 2008, **37**, 215–246.
14. M. Madan Babu, R. Van Der Lee, N. Sanchez De Groot, J. Rg Gsponer, J. Gough and K. Dunker, *Curr. Opin. Struct. Biol.*, 2011, **21**, 1–9.
15. F. Chiti and C. M. Dobson, *Annu. Rev. Biochem.*, 2006, **75**, 333–366.
16. J. Gosline, M. Lillie, E. Carrington, P. Guerette, C. Ortlepp and K. Savage, *Philos. Trans. R. Soc., B*, 2002, **357**, 121–132.
17. L. Li, M. B. Charati and K. L. Kiick, *Polym. Chem.*, 2010, **1**, 1160–1170.
18. A. S. Tatham and P. R. Shewry, *Philos. Trans. R. Soc., B*, 2002, **357**, 229–234.
19. M. Dzuricky, S. Roberts and A. Chilkoti, *Biochemistry*, 2018, **57**, 2405–2414.
20. W. R. Gray, L. B. Sandberg and J. A. Foster, *Nature*, 1973, **246**, 461–466.
21. B. Vrhovski and A. S. Weiss, *Eur. J. Biochem.*, 1998, **258**, 1–18.
22. B. Vrhovski, S. Jensen and A. S. Weiss, *Eur. J. Biochem.*, 1997, **250**, 92–98.
23. S. O. Andersen, *Biochim. Biophys. Acta*, 1963, **69**, 249–262.
24. T. Weis-Fogh and S. O. Andersen, *Adv. Insect Physiol.*, 1964, **2**, 1–65. London Acad. Press.
25. R. E. Shadwick and J. M. Gosline, in *Metabolic Biochemistry and Molecular Biomechanics*, Elsevier, 1983, pp. 399–430.
26. T. Weis-Fogh, *J. Exp. Biol.*, 1960, **37**, 889–907.
27. K. Bailey and T. Weis-Fogh, *Biochim. Biophys. Acta*, 1961, **48**, 452–459.
28. S. Rauscher, S. Baud, M. Miao, F. W. W. Keeley and R. Pomès, *Structure*, 2006, **14**, 1667–1676.
29. F. G. Quiroz and A. Chilkoti, *Nat. Mater.*, 2015, **14**, 1164–1171.
30. A. M. Tamburro, V. Guantieri, L. Pandolfo and A. Scopa, *Biopolymers*, 1990, **29**, 855–870.
31. K. M. Nairn, R. E. Lyons, R. J. Mulder, S. T. Mudie, D. J. Cookson, E. Lesieur, M. Kim, D. Lau, F. H. Scholes and C. M. Elvin, *Biophys. J.*, 2008, **95**, 3358–3365.
32. J. Martin, T. Weis-Fogh and P. J. W. Sutton, *Philos. Trans. R. Soc., B*, 1962, **245**, 137–169.
33. M. B. Van Eldijk, C. L. McGann, K. L. Kiick and J. C. M. Van Hest, in *Peptide-Based Materials*, ed. T. Deming, Springer Berlin Heidelberg, Berlin, Heidelberg, 2011, vol. 310, pp. 71–116.
34. L. Li and K. L. Kiick, in *Polymer Science: A Comprehensive Reference*, ed. K. Matyjaszewski and M. Möller, Elsevier B.V., Amsterdam, 2012, vol. 9, pp. 105–116.
35. D. Young and H. Bennet-Clark, *J. Exp. Biol.*, 1995, **198**, 1001–1020.

36. N. Skals and A. Surlykke, *J. Exp. Biol.*, 1999, **202**, 2937–2949.
37. H. C. Bennet-Clark and E. C. A. Lucey, *J. Exp. Biol.*, 1967, **47**, 59–76.
38. M. Burrows and G. P. Sutton, *J. Exp. Biol.*, 2012, **215**, 3501 LP–3503512.
39. A. F. Huxley, *J. Mol. Biol.*, 1965, **13**, 791–795.
40. R. Aeschbach, R. Amadoò and H. Neukom, *Biochim. Biophys. Acta, Protein Struct.*, 1976, **439**, 292–301.
41. S. O. Andersen, *Biochim. Biophys. Acta, Gen. Subj.*, 1964, **93**, 213–215.
42. G. L. Dignon, W. Zheng, Y. C. Kim and J. Mittal, *ACS Cent. Sci.*, 2019, **5**, 821–830.
43. J. Brindha, M. M. Balamurali and K. Chanda, *RSC Adv.*, 2019, **9**, 34720–34734.
44. Y. Wang, P. Katyal and J. K. Montclare, *Adv. Healthcare Mater.*, 2019, **8**, 1801374.
45. H. Uludag, A. Ubeda and A. Ansari, *Front. Bioeng. Biotechnol.*, 2019, **7**, 131.
46. S. Roberts, M. Dzuricky and A. Chilkoti, *FEBS Lett.*, 2015, **589**, 2477–2486.
47. C. De las Heras Alarcón, S. Pennadam and C. Alexander, *Chem. Soc. Rev.*, 2005, **34**, 276–285.
48. D. E. Meyer and A. Chilkoti, *Nat. Biotechnol.*, 1999, **17**, 1112–1115.
49. J. Nehete, R. Bhambar, M. Narkhede and S. Gawali, *Pharmacogn. Rev.*, 2013, **7**, 107–116.
50. C. M. Elvin, A. G. Carr, M. G. Huson, J. M. Maxwell, R. D. Pearson, T. Vuocolo, N. E. Liyou, D. C. C. Wong, D. J. Merritt and N. E. Dixon, *Nature*, 2005, **437**, 999–1002.
51. M. B. Charati, J. L. Ifkovits, J. A. Burdick, J. G. Linhardt and K. L. Kiick, *Soft Matter*, 2009, **5**, 3412–3416.
52. D. H. Ardell and S. O. Andersen, *Insect Biochem. Mol. Biol.*, 2001, **31**, 965–970.
53. R. S. C. Su, Y. Kim and J. C. Liu, *Acta Biomater.*, 2014, **10**, 1601–1611.
54. S. O. Andersen, *Insect Biochem. Mol. Biol.*, 2010, **40**, 541–551.
55. R. E. Lyons, D. C. C. Wong, M. Kim, N. Lekieffre, M. G. Huson, T. Vuocolo, D. J. Merritt, K. M. Nairn, D. M. Dudek, M. L. Colgrave and C. M. Elvin, *Insect Biochem. Mol. Biol.*, 2011, **41**, 881–890.
56. L. Li, J. M. Stiadle, H. K. Lau, A. B. Zerdoum, X. Jia, S. L. Thibeault and K. L. Kiick, *Biomaterials*, 2016, **108**, 91–110.
57. L. Li, Z. Tong, X. Jia and K. L. Kiick, *Soft Matter*, 2013, **9**, 665–673.
58. C. L. McGann, E. A. Levenson and K. L. Kiick, *Macromol. Chem. Phys.*, 2013, **214**, 203–213.
59. J. N. Renner, K. M. Cherry, R. S. C. Su and J. C. Liu, *Biomacromolecules*, 2012, **13**, 3678–3685.
60. N. K. Dutta, M. Y. Truong, S. Mayavan, N. Roy Choudhury, C. M. Elvin, M. Kim, R. Knott, K. M. Nairn and A. J. Hill, *Angew. Chem., Int. Ed.*, 2011, **50**, 4428–4431.
61. M. Y. Truong, N. K. Dutta, N. R. Choudhury, M. Kim, C. M. Elvin, K. M. Nairn and A. J. Hill, *Biomaterials*, 2011, **32**, 8462–8473.

62. R. Balu, J. Whittaker, N. K. Dutta, C. M. Elvin and N. R. Choudhury, *J. Mater. Chem. B*, 2014, **2**, 5936–5947.
63. N. K. Dutta, N. R. Choudhury, M. Y. Truong, M. Kim, C. M. Elvin and A. J. Hill, *Biomaterials*, 2009, **30**, 4868–4876.
64. M. Y. Truong, N. K. Dutta, N. R. Choudhury, M. Kim, C. M. Elvin, A. J. Hill, B. Thierry and K. Vasilev, *Biomaterials*, 2010, **31**, 4434–4446.
65. S. Mayavan, N. K. Dutta, N. R. Choudhury, M. Kim, C. M. Elvin and A. J. Hill, *Biomaterials*, 2011, **32**, 2786–2796.
66. R. E. Lyons, E. Lesieur, M. Kim, D. C. C. Wong, M. G. Huson, K. M. Nairn, A. G. Brownlee, R. D. Pearson and C. M. Elvin, *Protein Eng., Des. Sel.*, 2007, **20**, 25–32.
67. M. Kim, C. Elvin, A. Brownlee and R. Lyons, *Protein Expression Purif.*, 2007, **52**, 230–236.
68. R. E. Lyons, K. M. Nairn, M. G. Huson, M. Kim, G. Dumsday and C. M. Elvin, *Biomacromolecules*, 2009, **10**, 3009–3014.
69. G. Qin, S. Lapidot, K. Numata, X. Hu, S. Meirovitch, M. Dekel, I. Podoler, O. Shoseyov and D. L. Kaplan, *Biomacromolecules*, 2009, **10**, 3227–3234.
70. G. Qin, X. Hu, P. Cebe and D. L. Kaplan, *Nat. Commun.*, 2012, **3**, 1003.
71. G. Qin, A. Rivkin, S. Lapidot, X. Hu, I. Preis, S. B. Arinus, O. Dgany, O. Shoseyov and D. L. Kaplan, *Biomaterials*, 2011, **32**, 9231–9243.
72. L. Li, S. Teller, R. J. Clifton, X. Jia and K. L. Kiick, *Biomacromolecules*, 2011, **12**, 2302–2310.
73. A. Bracalello, V. Santopietro, M. Vassalli, G. Marletta, R. Del Gaudio, B. Bochicchio and A. Pepe, *Biomacromolecules*, 2011, **12**, 2957–2965.
74. F. Sbrana, C. Fotia, A. Bracalello, N. Baldini, G. Marletta, G. Ciapetti, B. Bochicchio and M. Vassalli, *Bioinspiration Biomimetics*, 2012, **7**, 046007.
75. S. Huang, Z. Qian, A. Dan, X. Hu, M. Zhou and X. Xia, *ACS Biomater. Sci. Eng.*, 2017, **3**, 1576–1585.
76. F. Luo, Z.-G. G. Qian and X.-X. X. Xia, *Polymers*, 2018, **10**, 915.
77. X. Hu, X. X. Xia, S. C. Huang and Z. G. Qian, *Biomacromolecules*, 2019, **20**, 3283–3293.
78. R. E. Lyons, C. M. Elvin, K. Taylor, N. Lekieffre and J. A. M. Ramshaw, *Biotechnol. Bioeng.*, 2012, **109**, 2947–2954.
79. R. S.- C. Su, R. J. Galas, C. Y. Lin and J. C. Liu, *Macromol. Biosci.*, 2019, **19**, 1900122.
80. R. S.-C. Su, E. E. Gill, Y. Kim and J. C. Liu, *J. Mech. Behav. Biomed. Mater.*, 2019, **91**, 68–75.
81. Y. Kim, E. E. Gill and J. C. Liu, *Biomacromolecules*, 2016, **17**, 2530–2539.
82. J. N. Renner, Y. Kim, K. M. Cherry and J. C. Liu, *Protein Expression Purif.*, 2012, **82**, 90–96.
83. R. J. King, Degree Thesis, Virginia Polytechnic Institute and State University, 2010.
84. U. Choudhury, Degree Thesis, Virginia Polytechnic Institute and State University, 2012.

85. M. S. K. Khandaker, D. M. Dudek, E. P. Beers and D. A. Dillard, *J. Mech. Behav. Biomed. Mater.*, 2017, **69**, 385–394.
86. R. Balu, R. Knott, N. P. Cowieson, C. M. Elvin, A. J. Hill, N. R. Choudhury and N. K. Dutta, *Sci. Rep.*, 2015, **5**, 10896.
87. M. S. K. Khandaker, D. M. Dudek, E. P. Beers, D. A. Dillard and D. R. Bevan, *J. Mech. Behav. Biomed. Mater.*, 2016, **61**, 110–121.
88. S. Rattan, L. Li, H. K. Lau, A. J. Crosby and K. L. Kiick, *Soft Matter*, 2018, **14**, 3478–3489.
89. C. L. McGann, R. E. Akins and K. L. Kiick, *Biomacromolecules*, 2016, **17**, 128–140.
90. L. Li and K. L. Kiick, *Front. Chem.*, 2014, **2**, 1–13.
91. H. K. Lau, A. Paul, I. Sidhu, L. Li, C. R. Sabanayagam, S. H. Parekh and K. L. Kiick, *Adv. Sci.*, 2018, **5**, 1–13.
92. H. K. Lau, L. Li, A. K. Jurusik, C. R. Sabanayagam and K. L. Kiick, *ACS Biomater. Sci. Eng.*, 2016, **3**, 757–766.
93. M. Sanami, Z. Shtein, I. Sweeney, A. Sorushanova, A. Rivkin, M. Miraftab, O. Shoseyov, C. O'Dowd, A. M. Mullen, A. Pandit and D. I. Zeugolis, *Biomed. Mater.*, 2015, **10**, 065005.
94. I. Preis, M. Abramson and O. Shoseyov, *Mol. Biotechnol.*, 2018, **60**, 310–318.
95. Z. Wang, Z. Wang, L. Yuan, F. Jiang, Y. Zhang and C. Tang, *ACS Macro Lett.*, 2016, **5**, 220–223.
96. L. Li, T. Luo and K. L. Kiick, *Macromol. Rapid Commun.*, 2015, **36**, 90–95.
97. J. L. Whittaker, N. K. Dutta, C. M. Elvin and N. R. Choudhury, *J. Mater. Chem. B*, 2015, **3**, 6576–6579.
98. J. L. Whittaker, N. K. Dutta, R. Knott, G. McPhee, N. H. Voelcker, C. Elvin, A. Hill and N. R. Choudhury, *Langmuir*, 2015, **31**, 8882–8891.
99. B. O. Okesola, H. K. Lau, B. Derkus, D. K. Boccorh, Y. Wu, A. W. Wark, K. Kiick and A. Mata, *Biomater. Sci.*, 2020, **8**, 846–857.
100. W. Fang, Nonappa, M. Vitikainen, P. Mohammadi, S. Koskela, M. Soikkeli, A. Westerholm-Parvinen, C. P. Landowski, M. Penttilä, M. B. Linder and P. Laaksonen, *Colloids Surf., B*, 2018, **171**, 590–596.
101. W. Fang, A. Paananen, M. Vitikainen, S. Koskela, A. Westerholm-Parvinen, J. J. Joensuu, C. P. Landowski, M. Penttilä, M. B. Linder and P. Laaksonen, *Biomacromolecules*, 2017, **18**, 1866–1873.
102. A. L. Boskey and E. Villarreal-Ramirez, *Matrix Biol.*, 2016, **52–54**, 43–59.
103. A. Ibáñez-Fonseca, T. Flora, S. Acosta and J. C. Rodríguez-Cabello, *Matrix Biol.*, 2019, **84**, 111–126.
104. W. Han, S. R. MacEwan, A. Chilkoti and G. P. López, *Nanoscale*, 2015, **7**, 12038–12044.
105. R. Balu, L. Bourgeois, C. M. Elvin, A. J. Hill, N. R. Choudhury and N. K. Dutta, *J. Mater. Chem. B*, 2015, **3**, 6580–6586.
106. R. Balu, N. R. Choudhury, J. P. Mata, L. De Campo, C. Rehm, A. J. Hill and N. K. Dutta, *ACS Appl. Mater. Interfaces*, 2019, **11**, 9934–9946.
107. R. Balu, R. Knott, C. M. Elvin, A. J. Hill, N. R. Choudhury and N. K. Dutta, *Biosensors*, 2019, **9**, 128.

108. M. Pierna, M. Santos, F. J. Arias, M. Alonso and J. C. Rodríguez-Cabello, *Biomacromolecules*, 2013, **14**, 1893–1903.
109. S. R. MacEwan and A. Chilkoti, *Biopolymers*, 2010, **94**, 60–77.
110. F. Garcia Quiroz, N. K. Li, S. Roberts, P. Weber, M. Dzuricky, I. Weitzhandler, Y. G. Yingling and A. Chilkoti, *Sci. Adv.*, 2019, **5**, eaax5177.
111. M. Alonso, V. Reboto, L. Guiscardo, V. Maté and J. C. Rodríguez-Cabello, *Macromolecules*, 2001, **34**, 8072–8077.
112. J. C. Rodríguez-Cabello, M. Alonso, L. Guiscardo, V. Reboto and A. Girotti, *Adv. Mater.*, 2002, **14**, 1151–1154.
113. L. A. Strzegowski, M. B. Martinez, D. C. Gowda, D. W. Urry and D. A. Tirrell, *J. Am. Chem. Soc.*, 1994, **116**, 813–814.
114. M. Alonso, V. Reboto, L. Guiscardo, A. San Martín and J. C. Rodríguez-Cabello, *Macromolecules*, 2000, **33**, 9480–9482.
115. J. P. Dhandhukia, D. A. Brill, A. Kouhi, M. K. Pastuszka and J. A. MacKay, *Protein Sci.*, 2017, **26**, 1785–1795.
116. B. Kim and A. Chilkoti, *J. Am. Chem. Soc.*, 2008, **130**, 17867–17873.
117. I. Weitzhandler, M. Dzuricky, I. Hoffmann, F. Garcia Quiroz, M. Gradzielski and A. Chilkoti, *Biomacromolecules*, 2017, **18**, 2419–2426.
118. J. R. McDaniel, D. C. Radford and A. Chilkoti, *Biomacromolecules*, 2013, **14**, 2866–2872.
119. H. Nuhn and H. A. Klok, *Biomacromolecules*, 2008, **9**, 2755–2763.
120. T. Luo and K. L. Kiick, *J. Am. Chem. Soc.*, 2015, **137**, 15362–15365.
121. T. Luo, M. A. David, L. C. Dunshee, R. A. Scott, M. A. Urello, C. Price and K. L. Kiick, *Biomacromolecules*, 2017, **18**, 2539–2551.
122. J. Qin, T. Luo and K. L. Kiick, *Biomacromolecules*, 2019, **20**, 1514–1521.
123. A. Chilkoti, T. Christensen and J. A. MacKay, *Curr. Opin. Chem. Biol.*, 2006, **10**, 652–657.
124. F. Arias, M. Santos, A. Fernández-Colino, G. Pinedo and A. Girotti, *Curr. Top. Med. Chem.*, 2014, **14**, 819–836.
125. L. L. Palmese, R. K. Thapa, M. O. Sullivan and K. L. Kiick, *Curr. Opin. Chem. Eng.*, 2019, **24**, 143–157.
126. C. Garcia Garcia and K. L. Kiick, *Acta Biomater.*, 2019, **84**, 34–48.
127. N. Annabi, J. W. Nichol, X. Zhong, C. Ji, S. Koshy, A. Khademhosseini and F. Dehghani, *Tissue Eng., Part B*, 2010, **16**, 371–383.
128. K. J. De France, F. Xu and T. Hoare, *Adv. Healthcare Mater.*, 2018, **7**, 1700927.
129. Y. Li, B. H. San, J. L. Kessler, J. H. Kim, Q. Xu, J. Hanes and S. M. Yu, *Macromol. Biosci.*, 2015, **15**, 52–62.
130. K. Kobayashi, S. H. Oh, C. K. Yoon and D. H. Gracias, *Macromol. Rapid Commun.*, 2018, **39**, 1700692.
131. C. Loebel, N. Broguiere, M. Alini, M. Zenobi-Wong and D. Eglin, *Biomacromolecules*, 2015, **16**, 2624–2630.
132. J. Torgersen, X. H. Qin, Z. Li, A. Ovsianikov, R. Liska and J. Stampfl, *Adv. Funct. Mater.*, 2013, **23**, 4542–4554.
133. Y. Liu, N. O. Nambu and M. Taya, *Biomed. Microdevices*, 2017, **19**, 55.
134. J. Ma, Y. Wang and J. Liu, *Micromachines*, 2017, **8**, 255.

135. E. H. Reed and D. A. Hammer, *Soft Matter*, 2018, **14**, 6506–6513.
136. S. Alberti and A. A. Hyman, *BioEssays*, 2016, **38**, 959–968.
137. A. E. Conicella, G. L. Dignon, G. H. Zerze, H. B. Schmidt, A. M. D'Ordine, Y. C. Kim, R. Rohatgi, Y. M. Ayala, J. Mittal and N. L. Fawzi, *Proc. Natl. Acad. Sci. U. S. A.*, 2020, **117**, 5883–5894.
138. B. S. Schuster, G. L. Dignon, W. S. Tang, F. M. Kelley, A. K. Ranganath, C. N. Jahnke, A. G. Simpkins, R. M. Regy, D. A. Hammer, M. C. Good and J. Mittal, *Proc. Natl. Acad. Sci.*, 2020, **117**, 5883–5894.
139. A. A. Hyman, C. A. Weber and F. Jülicher, *Annu. Rev. Cell Dev. Biol.*, 2014, **30**, 39–58.
140. S. G. Lévesque, R. M. Lim and M. S. Shoichet, *Biomaterials*, 2005, **26**, 7436–7446.
141. B. S. Schuster, E. H. Reed, R. Parthasarathy, C. N. Jahnke, R. M. Caldwell, J. G. Bermudez, H. Ramage, M. C. Good and D. A. Hammer, *Nat. Commun.*, 2018, **9**, 2985.
142. L. Faltova, A. M. Küffner, M. Hondele, K. Weis and P. Arosio, *ACS Nano*, 2018, **12**, 9991–9999.
143. J. A. Asenjo and B. A. Andrews, *J. Chromatogr. A*, 2011, **1218**, 8826–8835.
144. C. D. Keating, *Acc. Chem. Res.*, 2012, **45**, 2114–2124.
145. S. Boeynaems, S. Alberti, N. L. Fawzi, T. Mittag, M. Polymenidou, F. Rousseau, J. Schymkowitz, J. Shorter, B. Wolozin, L. Van Den Bosch, P. Tompa and M. Fuxreiter, *Trends Cell Biol.*, 2018, **28**, 420–435.
146. E. A. Scott, M. D. Nichols, R. Kuntz-Willits and D. L. Elbert, *Acta Biomater.*, 2010, **6**, 29–38.
147. H. Wang, A. Paul, D. Nguyen, A. Enejder and S. C. Heilshorn, *ACS Appl. Mater. Interfaces*, 2018, **10**, 21808–21815.
148. S. Roberts, T. S. Harmon, J. L. Schaal, V. Miao, K. (Jonathan) Li, A. Hunt, Y. Wen, T. G. Oas, J. H. Collier, R. V. Pappu and A. Chilkoti, *Nat. Mater.*, 2018, **17**, 1154–1163.
149. C. L. McGann, R. E. Dumm, A. K. Jurusik, I. Sidhu and K. L. Kiick, *Macromol. Biosci.*, 2016, **16**, 129–138.
150. H. K. Lau, S. Rattan, H. Fu, C. G. Garcia, D. M. Barber, K. L. Kiick and A. J. Crosby, *Macromol. Biosci.*, 2020, **20**, 1900360.
151. J. Läuger, R. Lay and W. Gronski, *J. Chem. Phys.*, 1994, **101**, 7181–7184.
152. H. Takeno, M. Iwata, M. Takenaka and T. Hashimoto, *Macromolecules*, 2000, **33**, 9657–9665.
153. H. Chen, C. McFaul, I. Titushkin, M. Cho and R. Lee, *Regener. Eng. Transl. Med.*, 2018, **4**, 1–10.
154. S. Khoee and M. R. Karimi, *Polymer*, 2018, **142**, 80–98.
155. X. Zhao, G. Yu, J. Li, Y. Feng, L. Zhang, Y. Peng, Y. Tang and L. Wang, *ACS Sustainable Chem. Eng.*, 2018, **6**, 4105–4114.
156. S. Van Vlierberghe, P. Dubruel and E. Schacht, *Biomacromolecules*, 2011, **12**, 1387–1408.
157. K. Y. Lee and D. J. Mooney, *Chem. Rev.*, 2001, **101**, 1869–1879.
158. K. L. Kiick, *Polym. Rev.*, 2007, **47**, 1–7.
159. M. S. Tjin, P. Low and E. Fong, *Polym. J.*, 2014, **46**, 444–451.

160. X. Jia, J. A. Burdick, J. Kobler, R. J. Clifton, J. J. Rosowski, S. M. Zeitels and R. Langer, *Macromolecules*, 2004, **37**, 3239–3248.
161. S. L. Thibeault, S. D. Gray, D. M. Bless, R. W. Chan and C. N. Ford, *J. Voice*, 2002, **16**, 96–104.
162. M. S. Benninger, D. Alessi, S. Archer, R. Bastian, C. Ford, J. Koufman, R. T. Sataloff, J. R. Spiegel and P. Woo, *Otolaryngol.-Head Neck Surg.*, 1996, **115**, 474–482.
163. S. E. Grieshaber, A. J. E. Farran, S. Lin-Gibson, K. L. Kiick and X. Jia, *Macromolecules*, 2009, **42**, 2532–2541.
164. L. A. Hughes, J. Gaston, K. McAlindon, K. A. Woodhouse and S. L. Thibeault, *Acta Biomater.*, 2015, **13**, 111–120.
165. L. Li and K. L. Kiick, *ACS Macro Lett.*, 2013, **17**, 40.
166. U. Hersel, C. Dahmen and H. Kessler, *Biomaterials*, 2003, **24**, 4385–4415.
167. N. Yamaguchi, L. Zhang, B. S. Chae, C. S. Palla, E. M. Furst and K. L. Kiick, *J. Am. Chem. Soc.*, 2007, **129**, 3040–3041.
168. S. H. Kim and K. L. Kiick, *Peptides*, 2007, **28**, 2125–2136.
169. H. Nagase and G. B. Fields, *Biopolymers*, 1996, **40**, 399–416.
170. M. P. Lutolf, J. L. Lauer-Fields, H. G. Schmoekel, A. T. Metters, F. E. Weber, G. B. Fields and J. A. Hubbell, *Proc. Natl. Acad. Sci. U. S. A.*, 2003, **100**, 5413–5418.
171. T. Jiao, A. Farran, X. Jia and R. J. Clifton, *Exp. Mech.*, 2009, **49**, 235–246.
172. L. Li, A. Mahara, Z. Tong, E. A. Levenson, C. L. Mcgann, X. Jia, T. Yamaoka and K. L. Kiick, *Adv. Healthcare Mater.*, 2016, **5**, 266–275.
173. R. E. King, H. K. Lau, H. Zhang, I. Sidhu, M. B. Christensen, E. W. Fowler, L. Li, X. Jia, K. L. Kiick and S. L. Thibeault, *Regener. Eng. Transl. Med.*, 2019, **5**, 373–386.
174. L. S. Mehta, T. M. Beckie, H. A. DeVon, C. L. Grines, H. M. Krumholz, M. N. Johnson, K. J. Lindley, V. Vaccarino, T. Y. Wang, K. E. Watson and N. K. Wenger, *Circulation*, 2016, **133**, 916–947.
175. M. Sepantafar, R. Maheronnaghsh, H. Mohammadi, S. Rajabi-Zeleti, N. Annabi, N. Aghdami and H. Baharvand, *Biotechnol. Adv.*, 2016, **34**, 362–379.
176. A. Singh, A. Singh and D. Sen, *Stem Cell Res. Ther.*, 2016, **7**, 82.
177. M. Korf-Klingebiel, M. R. Reboll, S. Klede, T. Brod, A. Pich, F. Polten, L. C. Napp, J. Bauersachs, A. Ganser, E. Brinkmann, I. Reimann, T. Kempf, H. W. Niessen, J. Mizrahi, H. J. Schönfeld, A. Iglesias, M. Bobadilla, Y. Wang and K. C. Wollert, *Nat. Med.*, 2015, **21**, 140–149.
178. J. L. Olson, A. Atala and J. J. Yoo, *Chonnam Med. J.*, 2011, **47**, 1.
179. Z. Li and J. Guan, *Polymers*, 2011, **3**, 740–761.
180. B. Peña, M. Laughter, S. Jett, T. J. Rowland, M. R. G. Taylor, L. Mestroni and D. Park, *Macromol. Biosci.*, 2018, **18**, 1800079.
181. A. Moorthi, Y.-C. Tyan and T.-W. Chung, *Biomater. Sci.*, 2017, **5**, 1976–1987.
182. N. J. R. Blackburn, T. Sofrenovic, D. Kuraitis, A. Ahmadi, B. McNeill, C. Deng, K. J. Rayner, Z. Zhong, M. Ruel and E. J. Suuronen, *Biomaterials*, 2015, **39**, 182–192.

183. T. Hao, J. Li, F. Yao, D. Dong, Y. Wang, B. Yang and C. Wang, *ACS Nano*, 2017, **11**, 5474–5488.
184. C. M. Brougham, T. J. Levingstone, S. Jockenhoevel, T. C. Flanagan and F. J. O'Brien, *Acta Biomater.*, 2015, **26**, 205–214.
185. M. D. Brigham, A. Bick, E. Lo, A. Bendali, J. A. Burdick and A. Khademhosseini, *Tissue Eng., Part A*, 2009, **15**, 1645–1653.
186. S. W. Jordan, C. A. Haller, R. E. Sallach, R. P. Apkarian, S. R. Hanson and E. L. Chaikof, *Biomaterials*, 2007, **28**, 1191–1197.
187. N. Annabi, K. Tsang, S. M. Mithieux, M. Nikkhah, A. Ameri, A. Khademhosseini and A. S. Weiss, *Adv. Funct. Mater.*, 2013, **23**, 4950–4959.
188. C. Celik, V. T. Mogal, J. H. P. Hui, X. J. Loh and W. S. Toh, *Hydrogels*, Springer, Singapore, 2018, pp. 315–337.
189. W. Seong Toh, C. Bindzus Foldager, J. Hoi Po Hui, B. Reino Olsen and M. Spector, *Curr. Stem Cell Res. Ther.*, 2016, **11**, 618–625.
190. L. Zhang, T. Yuan, L. Guo and X. Zhang, *J. Biomed. Mater. Res., Part A*, 2012, **100 A**, 2717–2725.
191. L. S. Kontturi, E. Järvinen, V. Muhonen, E. C. Collin, A. S. Pandit, I. Kiviranta, M. Yliperttula and A. Urtti, *Drug Delivery Transl. Res.*, 2014, **4**, 149–158.
192. S. Dinescu, B. Gălăţeanu, M. Albu, A. Lungu, E. Radu, A. Hermenean and M. Costache, *BioMed Res. Int.*, 2013, **2013**, 1–11.
193. I. L. Kim, R. L. Mauck and J. A. Burdick, *Biomaterials*, 2011, **32**, 8771–8782.
194. Y. Liu, X. Z. Shu and G. D. Prestwich, *Tissue Eng.*, 2006, **12**, 3405–3416.
195. W. Shi, M. Sun, X. Hu, B. Ren, J. Cheng, C. Li, X. Duan, X. Fu, J. Zhang, H. Chen and Y. Ao, *Adv. Mater.*, 2017, **29**, 1701089.
196. H. Betre, L. A. Setton, D. E. Meyer and A. Chilkoti, *Biomacromolecules*, 2002, **3**, 910–916.
197. M. K. McHale, L. A. Setton and A. Chilkoti, *Tissue Eng.*, 2005, **11**, 1768–1779.
198. J. C. Iatridis, S. Kumar, R. J. Foster, M. Weidenbaum and V. C. Mow, *J. Orthop. Res.*, 1999, **17**, 732–737.
199. D. R. Anderson, S. L.-Y. Woo, M. K. Kwan and D. H. Gershuni, *J. Orthop. Res.*, 1991, **9**, 550–558.
200. L. A. Setton, V. C. Mow and D. S. Howell, *J. Orthop. Res.*, 1995, **13**, 473–482.
201. W. Zhu, V. C. Mow, T. J. Koob and D. R. Eyre, *J. Orthop. Res.*, 1993, **11**, 771–781.
202. D. W. Lim, D. L. Nettles, L. A. Setton and A. Chilkoti, *Biomacromolecules*, 2008, **9**, 222–230.
203. D. W. Lim, D. L. Nettles, L. A. Setton and A. Chilkoti, *Biomacromolecules*, 2007, **8**, 1463–1470.
204. A. Gabrion, P. Aimedieu, Z. Laya, E. Havet, P. Mertl, R. Grebe and M. Laude, *Surg. Radiol. Anat.*, 2005, **27**, 507–510.
205. J. R. Parsons, in *Handbook of Biomaterial Properties*, Springer US, Boston, MA, 1998, pp. 40–47.
206. E. M. Srokowski and K. A. Woodhouse, *J. Biomater. Sci., Polym. Ed.*, 2008, **19**, 785–799.

CHAPTER 5

Resilin-mimetic Polypeptides and Elastomeric Modular Protein Polymers: Amino Acid Sequence, Conformational Ensemble, and Stimuli Responsiveness

RAJKAMAL BALU, NABA K. DUTTA AND NAMITA ROY CHOUDHURY*

Chemical and Environmental Engineering, School of Engineering, RMIT University, Melbourne, Victoria 3000, Australia
*Email: namita.choudhury@rmit.edu.au

5.1 Introduction

Resilin is a rubber-like elastomeric insect protein. It is a structural protein involved in jumping, sound production, and flying actions of many insects and arthropods with a fatigue life of over 300 million cycles.[1] In insects, the apical surface of the epidermal cells secretes pro-resilins (uncrosslinked resilins) into the subcuticular space, which are then enzymatically crosslinked to form crosslinked resilin elastomer.[2] Resilin was first discovered by Weis-Fogh in the 1960s, in locust wing-hinges and dragonfly tendons (Figure 5.1A), and is postulated as the most energy-efficient (resilient) elastic protein known.[3,4] Early investigations on the composition of resilin revealed the protein to contain ~66% hydrophobic residues (much lower than elastin protein, which is ~95%)

Biomaterials Science Series No. 10
Biomimetic Protein Based Elastomers: Emerging Materials for the Future
Edited by Namita Roy Choudhury, Julie C. Liu and Naba K. Dutta

Published by the Royal Society of Chemistry, www.rsc.org

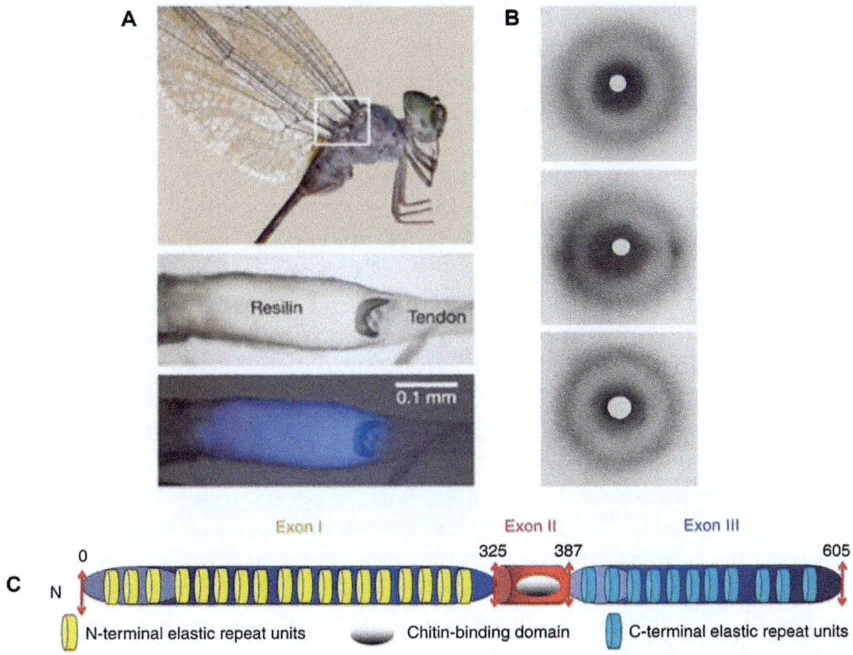

Figure 5.1 (A) Location of resilin in the wing tendon from adult dragonfly (top panel) and photomicrographs of the tendon in phosphate-buffered saline under white light (middle panel) and ultraviolet light (bottom panel). Reproduced from ref. 4 with permission from Springer Nature, Copyright 2005. (B) X-ray diagrams from dried dragonfly tendons taken at unstretched (top), 3.3 times stretched – showing an equatorial reflection (middle) and returned to rest length (bottom) state. The ring is about 4 Å. Reproduced from ref. 7 with permission from Elsevier, Copyright 1965. (C) Amino acid scheme of the fruit fly *Drosophila melanogaster* resilin. The red arrows with numbers represent the position or length of amino acid residues of the exons. Reproduced from ref. 13 with permission from Springer Nature, Copyright 2012.

with ~45% proline and glycine residues combined.[5] Moreover, the crosslinks in resilin were identified as di- and trityrosine,[6] and the protein exhibited a highly amorphous structure under X-ray diffraction and electron microscopy techniques (Figure 5.1B).[7] Since then, resilin has been identified as an integral part of many other insects and arthropods, including scorpions, honey bees, cockroaches, and snow fleas.[8–11] The amino acid sequence of resilin was first identified by Ardell and Anderson in 2001 from the product of the resilin CG15920 gene obtained from the fruit fly *Drosophila melanogaster*.[12] The CG15920 gene product comprises 605 amino acid residues and is arranged into three regions (Figure 5.1C): (1) exon-1 with the N-terminal domain consisting of 18 copies of the amino acid repeat sequence GGRPSDSYGAPGGGN, (2) exon-2 with the chitin-binding domain, and (3) exon-3 with the C-terminal domain consisting of 11 copies of the amino acid repeat sequence GYSGGRPGGQDLG.[12]

In 2005, the first biomimetic and genetically engineered resilin-like polypeptide (RLP), namely Rec1-resilin (obtained from the exon-1 of CG15920 gene), was expressed by Elvin *et al.*[4] as water-soluble polypeptides in the bacteria *Escherichia coli*. The authors also successfully demonstrated photochemical crosslinking of the synthesized RLP to form dityrosine crosslinked hydrogel, which exhibited resilience (~97%) greater than that of natural elastin protein hydrogel (~90%) and synthetic polybutadiene rubber (~80%).[4] Fascinatingly, Rec1-resilin exhibits intrinsically disordered structure,[14] and unusual multi-stimuli (including pH, temperature, ion, and other macromolecules) responsiveness.[15,16] Followed by Rec1-resilin, several other RLPs of different positions and lengths from CG15920 and other insect resilin genes, including BX619161 from the mosquito *Anopheles gambiae* have been synthesized (yield in the range of 15–450 $mg\,L^{-1}$ of culture) and studied over the years (Figure 5.2).[17–24] However, hydrogels fabricated from as-synthesized RLPs showed poor elastic storage modulus and no cell adhesion,[25,26] which persuaded the development of RLP-based modular protein polymers with improved cell attachment and mechanical properties.[27,28] Over the last decade, several RLP-based modular protein polymers with additional domains and improved physicochemical and biological properties have been synthesized.[29–37]

In general, the structure and properties of elastomeric proteins are dictated by their amino acid sequence and composition, where the amino acid sequence is characterized by specific amino acid repeat motifs and their composition is characterized by mean net charge, hydrophobicity, and specific amino acid content. Therefore, understanding the similarities and differences in amino acid sequence, composition, molecular structure, responsive behavior, molecular interaction, and self-assembly of different RLPs and RLP-based modular protein polymers, along with their synthesis and purification protocols, can provide in-depth information about the key requirements for rational design and development of resilin protein-based smart biomaterials.

5.2 Synthesis and Purification

Biomimetic RLPs are conventionally synthesized as water-soluble biopolymers using recombinant technology.[38] Briefly, the recombinant polypeptide synthesis typically involves the construction of genetic vector using intended amino acid repeat sequence, followed by the transfection of constructed genetic vector into the host organism, overexpression of the polypeptide in the host organism using isopropyl β-D-1-thiogalactopyranoside (IPTG) or autoinduction methods, host cell lysis, and extraction and purification of synthesized polypeptides.[39,40] The final yield of RLPs ($mg\,L^{-1}$ of culture) can also be significantly increased by lactose-induced fermentation.[41] The obtained water-soluble RLPs are commonly purified using salting-out and heating, affinity, and/or ion-exchange chromatography methods, with the help of sonication and centrifugation techniques.[39,40] In addition, recent developments in design strategies for advanced functional biomimetic

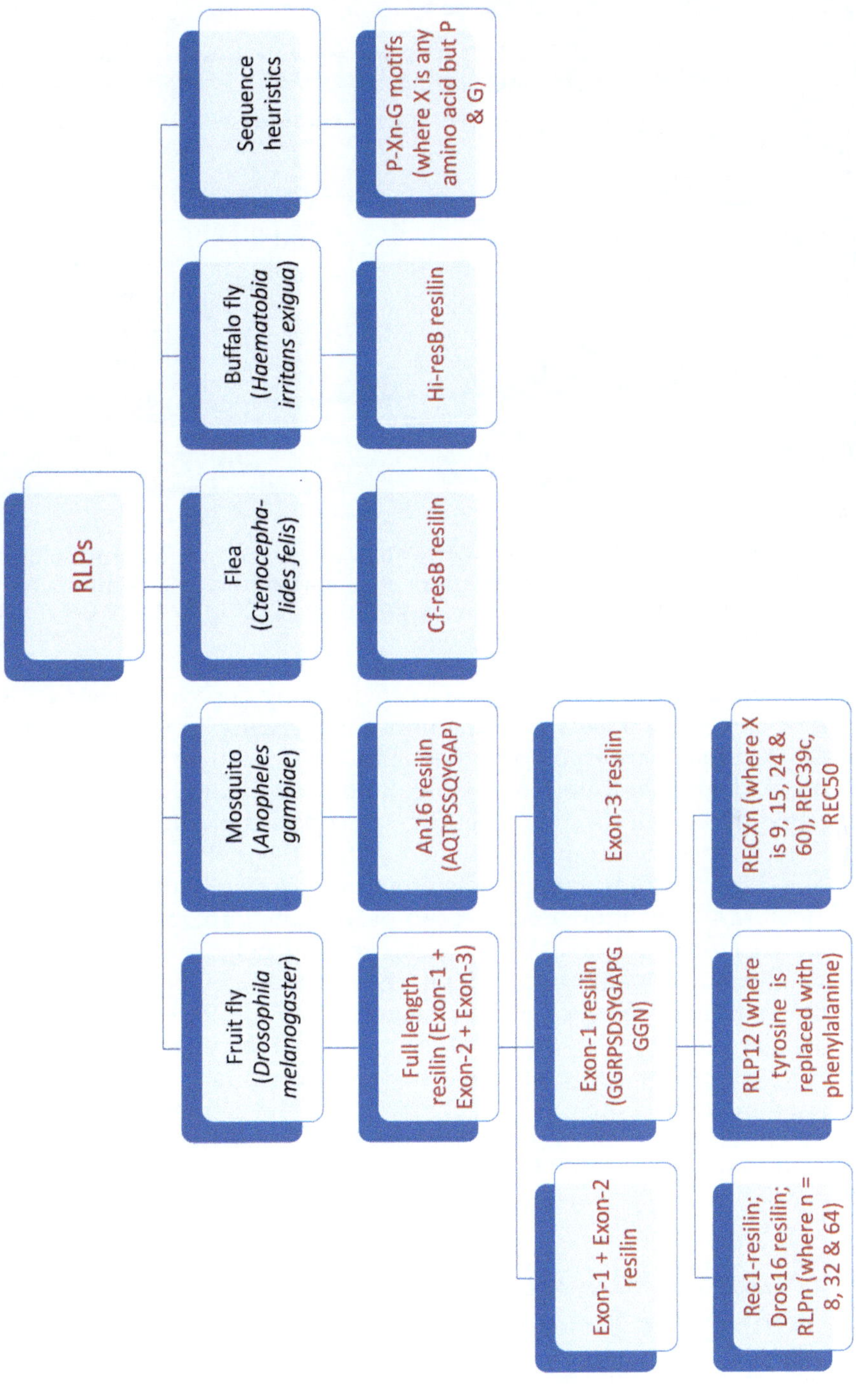

Figure 5.2 Scheme of resilin-like polypeptides (RLPs) reported in the literature.[4,17–24]

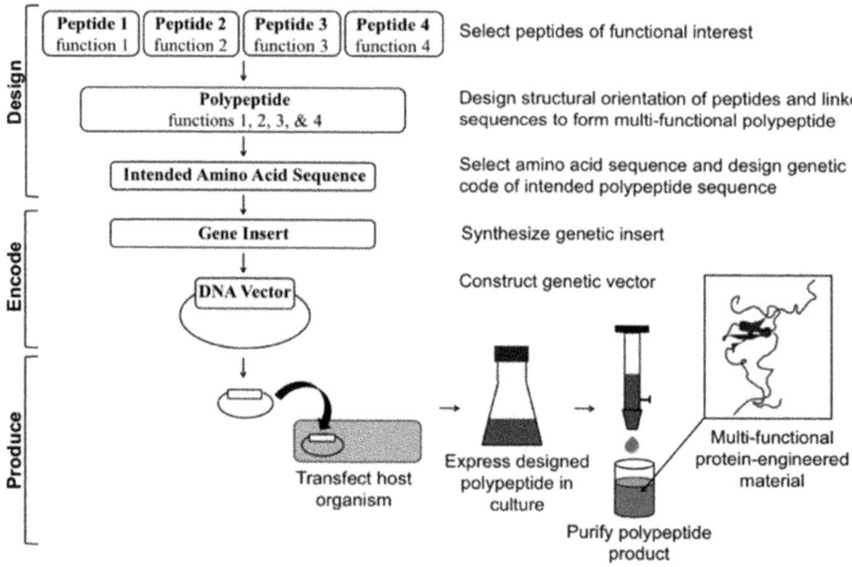

Figure 5.3 Schematic of steps involved in the design and synthesis of modular protein engineered polypeptides. Reproduced from ref. 42 with permission from John Wiley & Sons, Copyright © 2012 WILEY-VCH Verlag GmbH & Co. KGaA, Weinheim.

materials, such as the modular protein engineering approach (Figure 5.3), allows for the selection, design, and utilization of genetic sequence encoding peptide domains of repurposed functionalities (for example, structural, cell binding, biodegradable, molecular recognition, crosslinking, *etc.*) in addition to the traditional selection of repetitive structural motifs of natural functionalities to develop truly multifunctional RLP-based modular protein polymers.[42] Furthermore, modular protein engineering, along with an expanding library of peptide domains, can provide the unparalleled potential to create elastomeric RLP-based modular protein polymers with multiple and complex functionalities. In general, the methods applied for RLPs were successfully adapted for the synthesis and purification of RLP-based modular protein polymers. However, the obtained yield of RLP-based modular protein polymers was always less than their pristine counterparts. A summary of RLP-based modular protein polymers reported in the literature is presented in Table 5.1. Further understanding and development of the sequence-based high-yield expression of RLP-based modular protein polymers can certainly pave the way for bench-top to scale up production facility.

5.3 Amino Acid Sequence, Composition, and Molecular Structure Prediction

Anderson *et al.*[45] compared amino acid sequence and composition of resilins from 12 *Drosophila* and some other insect species to determine the

Table 5.1 Summary of RLP-based modular protein polymers synthesized.

Modular protein polymers[a]	Added domains	Property improvements	Reference
RLP_{12}–RGD, RLP_{12}–HBD, RLP_{12}–MMP, RLP_{12}–RGD_2–MMP–HBD	Cell-binding domain (RGD); Heparin-binding domain (HBD); Matrix metallopeptidase domain (MMP)	Cell attachment, biodegradation, and molecular recognition	22, 27
GB1–RLP_4, GB1–RLP–$(GB1)_5$–RLP–$(GB1)_4$–RLP	Artificial elastomeric protein (GB1)	Mechanical property	28
RLP_4–ELP_7–CLP_2	Elastin-like polypeptide (ELP); Collagen-like polypeptide (CLP)	Self-assembly	29
An_X-EGFP (where $X=4$, 8, 16, and 32)	Enhanced green fluorescence protein (EGFP)	Autofluorescence	30
RZ_{10}–RGD, RZ_{10}–BMP2, RZ_{10}–QK	Cell-binding domain (RGD), bone morphogenetic peptide (BMP2), vascular endothelial growth factor mimicking peptide (QK)	Cell attachment and differentiation	31, 35, 36
RLP_n–SLP_n (where $n=1$, 4, and 8)	Silk-like polypeptide (SLP)	Temperature responsiveness and self-assembly	32, 43
CBM–RLP–CBM; CBM_2–RLP–HFBI	Cellulose binding module (CBM); Hydrophobin protein domain (HFBI)	pH responsiveness, self-assembly, and selective substrate adhesion	33, 44
RLP_{12}–LCD_5	Lysine-rich domain (LCD)	Lysine crosslinking sites	34
$(RLP)_n$–$(X\text{-}ELP)_m$ (where $n=20$, 40, 60, 80, and 100; $m=40$, 80, and 160; X is the guest amino acid residue A/G, S, and V)	Elastin-like polypeptide (ELP)	Temperature responsiveness and self-assembly	37

[a]The RLPs, namely An and RZ, are derived from the BX619161 gene, whereas the others are from the CG15920 gene.

structural features important for the characterization of resilin properties. The gene products from 12 *Drosophila* species varied in length from 570 to 616 amino acid residues, whereas they showed only very little difference in the composition, where the content of proline varied between 8.8 and 11.2%, glycine between 33.3 and 35.4%, serine between 9.5 and 11.7%, alanine between 4.4 and 7.0%, and tyrosine between 5.8 and 6.5%.[45] The N-terminal repeat sequences (exon-1) of the protein from 12 *Drosophila* species showed a 6.1% difference in position with the number of repeats varying between

13 and 18, whereas C-terminal repeat sequences (exon-3) showed a 2.6% difference in position with the number of repeats varying between 9 and 13. The exon-2 showed a very minor difference of 0.2% of the positions.[45] Moreover, a common consensus sequence of GGRPSDSYGAPGQGN and GYSGGRPGGQDLG for exon-1 and exon-3, respectively, was observed to be conserved in all 12 *Drosophila* species.[45] However, a comparison between *Drosophila* and other insect RLPs reveals striking differences in their amino acid sequence and consensus regions, where some of the proteins contain only exon-1, while others contain both exon-1 and exon-3, and some do not contain the chitin-binding consensus sequence.[45] Table 5.2 shows variants of the N-terminal conserved region of RLPs from different insect species. The amino acid composition of all RLPs resembles that of pro-resilin obtained from *D. melanogaster* with over 30% glycine and 10% proline, and a common consensus sequence of EPPVN-SYLPPS observed to be conserved in the N-terminal region.[45]

The physical determinants of the structure and properties of proteins of elastomeric nature have been previously established with molecular simulations, where a fundamental requirement of the elastomeric domain is to remain disordered, which is directed by a minimum threshold of total glycine and proline amino acid content.[46] The amino acid proline interrupts the transmission of stable secondary structure in protein, whereas the amino acid glycine engenders flexibility. The proline and glycine composition threshold is not limited to the hydrophobic domains of resilin but is also observed for other elastomeric proteins, such as abductin, elastin, flagelliform silk, mussel byssus, and wheat gluten.[46] Figure 5.4 shows the proline *versus* glycine content plot of resilin compared to different elastomeric and amyloidogenic sequences, where resilin clearly falls into the elastomeric or coexistence region.

On the other hand, archetypical methods, such as Uversky analysis plots, mean hydrophobicity *versus* mean net charge and provide information about the native protein disorder with a solid line representing the border between

Table 5.2 RLPs comprising variants of the N-terminal conserved region. Adapted from ref. 45, with permission from Elsevier, Copyright 2010.

Species	Gene number	N-terminal region
Apis mellifera	AmelCPR15	RSEPPVNSYLPPS
Apis mellifera	XP_001120921.1	ADAPISGSYLPPS
Anopheles gambiae	AGAP002367	EAPLPPSGSYLPPS
Culex quinquefasciatus	XP_001843145.1	EAPLPGGSYLPPS
Drosophila melanogaster	CG15920	RPEPPVNSYLPPS
Drosophila melanogaster	CG7709	GAEPPVNNAYLPPS
Nasonia vitripennis	NvitCPR25	RPEPPVNSYLPPG
Nasonia vitripennis	XP_001605137.1	AEAPISGNYLPPS
Pediculus humanus	XP_002428637.1	LAEPPVKTSYLPPS
Tribolium castaneum	XP_001807028.1	RAEPPVNSYLPPS
Tribolium castaneum	XP_975482.1	PSGGPANSYLPPG

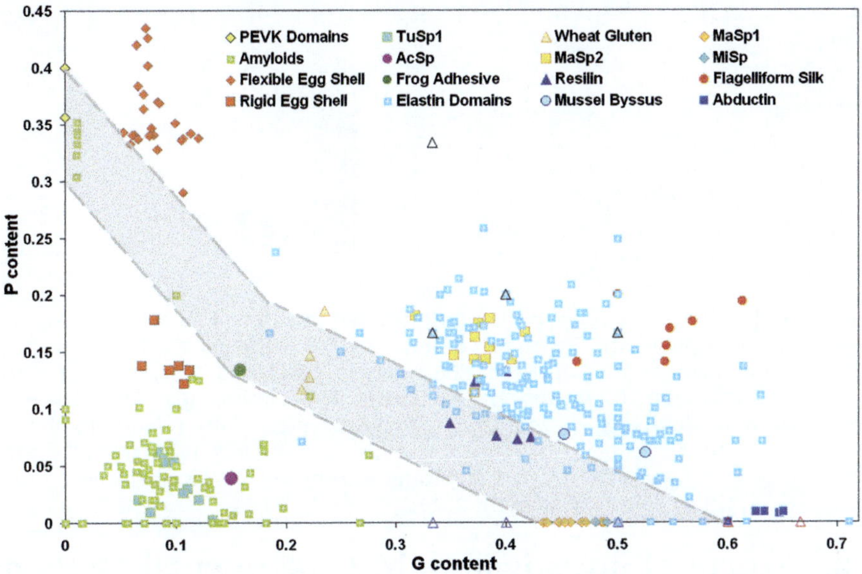

Figure 5.4 Elastomeric and amyloidogenic sequences plotted as a function of proline and glycine amino acid content. The grey region represents the coexistence of amyloidogenic and elastomeric properties, whereas the above and below regions represent elastomeric and amyloidogenic, respectively. Reproduced from ref. 46 with permission from Elsevier, Copyright 2006.

unfolded and folded proteins.[47] A combination of large net charge and low overall hydrophobicity represents natively unfolded or intrinsically disordered protein (IDP) structural feature, which lacks a fixed or ordered 3D structure. Figure 5.5A shows the Uversky plot of resilin compared to different elastomeric, natively unfolded, and folded proteins, where resilin clearly falls into the natively unfolded protein region. Therefore, the molecular structure of resilin was hypothesized to be that of natively unfolded or IDP. Moreover, protein secondary structure prediction methods, such as the discrimination of protein secondary structure class (DPSC), have suggested the structure of RLPs to be disordered with largely or completely random coil conformations.[15,48] Lately, Quiroz *et al.*[23] identified the proline and glycine patterns in repeat sequences that would provide a minimal and generic scaffold for IDPs. The authors mapped the proline (P) and glycine (G) residue pairs as P–X_n–G motifs across IDPs (Figure 5.5B), where X is other amino acids except P and G and n ranges from 0 to 4. The recurring P–X_4–G motifs were also identified to be more than 30% in resilin. Moreover, the unusual high glycine content (>30%) along with structure-breaking P–G pairs repeating every 4–9 residues defines a robust IDP-like scaffold supporting the structure for resilin to be that of IDP.[23] The predicted structure of resilins has also been analyzed and established through various experimental techniques, which are discussed in detail in the following section.

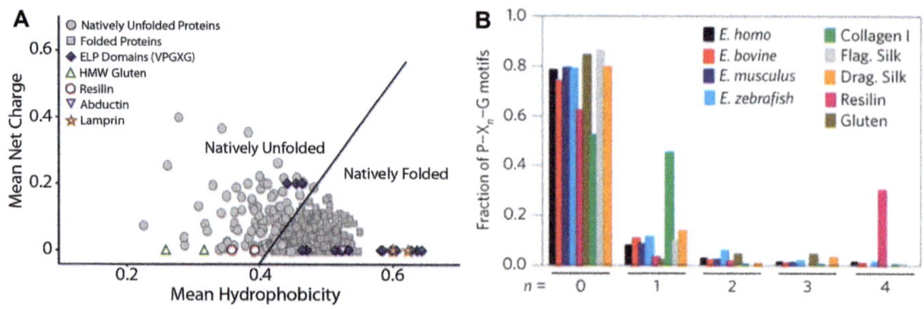

Figure 5.5 (A) Repeat sequence proteins plotted as a function of mean net charge and hydrophobicity. Reproduced from ref. 49 with permission from John Wiley & Sons, Copyright FEBS Letters 589 (2015) 1873–3468 © 2015 Federation of European Biochemical Societies. (B) Abundance of individual P–X_n–G motifs among prototypical IDPs relative to P–X_n–G motifs per protein in total. X is other amino acids apart from P and G, and n ranges from 0 to 4. Reproduced from ref. 23 with permission from Springer Nature, Copyright 2015.

5.4 Structural Investigation by Experimental Methods

Proteins exhibit different forms of local segments called secondary structures. The two 'regular' secondary structures, namely α-helices and β-sheets, are linked by tight turns and loose/flexible loops, whereas random coil exhibits irregular conformation. The 'regular' secondary structures exhibit a certain pattern of hydrogen bonds formed between the carboxyl oxygen atoms and amino hydrogen atoms in the peptide backbone. These secondary structures may interact (*via* ionic and hydrogen bonds, hydrophilic and hydrophobic interactions, and disulfide bridges) and bond in a number of ways to form tertiary structures.[50] The structure of IDPs is characterized by substantial conformational flexibility (lack of fixed or ordered 3D structure). Given the structure of resilin predicted to be intrinsically disordered (from amino acid sequence and composition), the experimental technique for precise protein structure analysis, namely X-ray crystallography, failed to identify secondary structures of resilin because it does not crystallize.[17] Therefore, experimental spectroscopic techniques more suitable for characterizing IDPs were applied for studying the structure and dynamics of RLPs.

5.4.1 Circular Dichroism (CD) Spectroscopy

Ultraviolet CD spectroscopy is a commonly used experimental technique to investigate the secondary structures of proteins.[51] In the CD spectroscopy experiment, a dilute solution of protein of interest is subjected to circularly polarized light. The basis for secondary structural characterization is provided by the measured difference between the extent to which the left-hand circular and right-hand circular polarization (two possible spin angular momentum states for a photon) are absorbed by the sample.[52] The obtained sample data are typically presented as degrees of ellipticity *versus* wavelength plots. Figure 5.6A

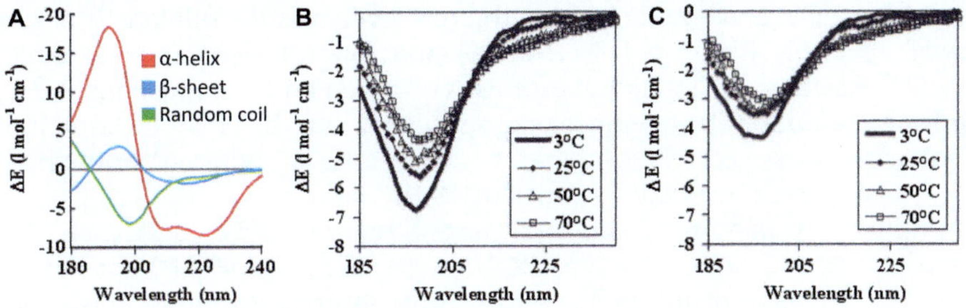

D Fit Parameters obtained Using ContinLL with the SP37A Basis Set

	α-helix	β-sheet	turns	PPII	unordered
An16	0.05	0.10	0.11	0.10	0.64
Rec1-resilin	0.02	0.18	0.11	0.11	0.58

Figure 5.6 (A) Typical CD spectra of α-helix (red), β-sheet (blue), and random coil (green) secondary structures. CD spectra of (B) An16 and (C) Rec1-resilin RLPs in 10 mM phosphate-buffered saline solution (at a range of temperature) along with (D) secondary structure of samples (3 °C equilibrated) estimated with the ContinLL algorithm using the SP37A basis set. Reproduced from ref. 53 with permission from American Chemical Society, Copyright 2009.

shows typical CD spectra of protein secondary structures, namely α-helix, β-sheet, and random coil. The α-helix exhibit positive bands around 193 nm and negative bands around 208 nm and 222 nm. Antiparallel β-pleated sheets exhibit positive and negative bands around 195 nm and 218 nm, respectively, whereas unordered random coils exhibit negative bands around 195 nm and weak ellipticity above 210 nm.[51] Lyons *et al.*[53] first investigated and compared the secondary structure of two RLPs, namely Rec1-resilin and An16 (Figure 5.6B and C), at different temperatures using CD spectroscopy.[53] An16 is synthesized from the mosquito *Anopheles gambiae* gene BX619161, which is significantly different from that of Rec1-resilin (synthesized from the fruit fly *D. melanogaster*), where the An16 consists of ∼9% glycine and ∼46% nonpolar residues, whereas Rec1-resilin contains ∼34% glycine and ∼56% nonpolar residues.[48] Both RLPs exhibited unordered secondary structure characteristics in 10 mM sodium phosphate buffer (pH 7) at all temperatures.[54] In addition, no significant spectral differences for the RLPs at different pH and in the presence of metal ions were observed.[48,55,56] Quantitative estimations of the secondary structures of RLPs were also obtained by using the CDPro fitting program interface with the ContinLL fitting algorithm and the SP37A basis reference data set (containing model proteins for poly-L-proline II, PPII structure).[53] The fit results (Figure 5.6D) imply that the structures of RLPs are largely unordered (≥58%) with some turns, PPII, and β-sheet conformations.[53] Simultaneously, Qin *et al.*[18,19] compared the secondary structure of exon-1 and exon-3 RLPs with respective full-length fruit fly resilin. The authors reported that the partial (exon-1 and exon-3) resilins exhibited largely unordered (turns and random coils; ∼60%) secondary structure, with ~5% α-helix and ~23% β-sheet structures, whereas the full-length resilin showed relatively higher α-helix (~10%) and lower β-sheet (~16%) structures.[18,19] However, the structural estimates obtained are nonindependent (*i.e.*, highly dependent on the reference data sets used), and the obtained structural information is less specific, compared to more complex and accurate methodologies like nuclear magnetic resonance (NMR).

Tamburro *et al.*[21] also reported unordered and PPII structures for N-terminal RLPs, namely RES9n, RES15n, RES24n, and RES60n in water, where the intensity of minima for RLPs was observed to increase with an increase in RLP length. The dynamic equilibrium between PPII and β-turn conformations (isoelliptic points near 208 nm, widely present in elastomeric proteins) is reported to be the origin of the relaxed-state high entropy of resilin.[57] Also, the resilins Cf-resB and Hi-resB, synthesized from flea and buffalo fly genes, respectively, showed spectra similar to that of Rec1 resilin, with an average disorder score of 0.477 and 0.493, respectively.[20] Moreover, Kiick and coworkers reported conformational properties (largely unordered structure with contribution from turns) of a suite of RLP-based modular protein polymers, namely RLP_{12}–RGD_2–MMP–HBD, RLP_{12}–RGD_2, RLP_{12}–MMP, and RLP_{12}–HBD to be similar to that of RLPs reported by Tamburro[21] and other research groups.[22,27] Similar results were also reported for other RLP-based modular protein polymers, namely RLP_4–ELP_7–CLP_2 and RZ_{10}–RGD, where the RLP-backbone, RLP_4, was derived from fruit fly gene, whereas the RZ_{10} was derived from mosquito gene.[29,31]

5.4.2 Fourier-transform Infrared Spectroscopy

Fourier-transform infrared (FTIR) spectroscopy is a versatile nondestructive experimental technique to investigate the secondary structures and local conformational changes of proteins.[58] In the FTIR spectroscopy experiment, a solution or freeze-dried powder of protein of interest is illuminated by an incident beam containing many combinations of frequencies of infrared radiation. The absorbed frequency of radiation is then Fourier transformed into absorption for each wavelength, which provides the basis for secondary structural characterization.[58] The obtained sample data are typically presented as absorbance *versus* wavenumber (the reciprocal of the wavelength of the wave) plots. A typical FTIR spectrum of protein often shows nine amide bands (A, B, and I–VII) related to protein secondary structural components, where the amide I band (1700–1600 cm^{-1}) associated with the C=O stretching vibration is the most sensitive and prominent vibrational band of the protein backbone, as shown in Figure 5.7A.[59] From the amide I band, secondary structures of protein in water, namely α-helix, β-sheet, turn and unordered conformations, can be identified from their vibrational bands observed in the ranges of 1648–1657, 1623–1641, 1642–1657 and 1662–1686 cm^{-1}, respectively.[59] The conformationally sensitive amide II band results from the C–N stretching vibrations and N–H bending; however, it is considered a less useful predictor for quantifying the secondary structure of proteins.[59] Qin *et al.*[18,19] first investigated and compared the

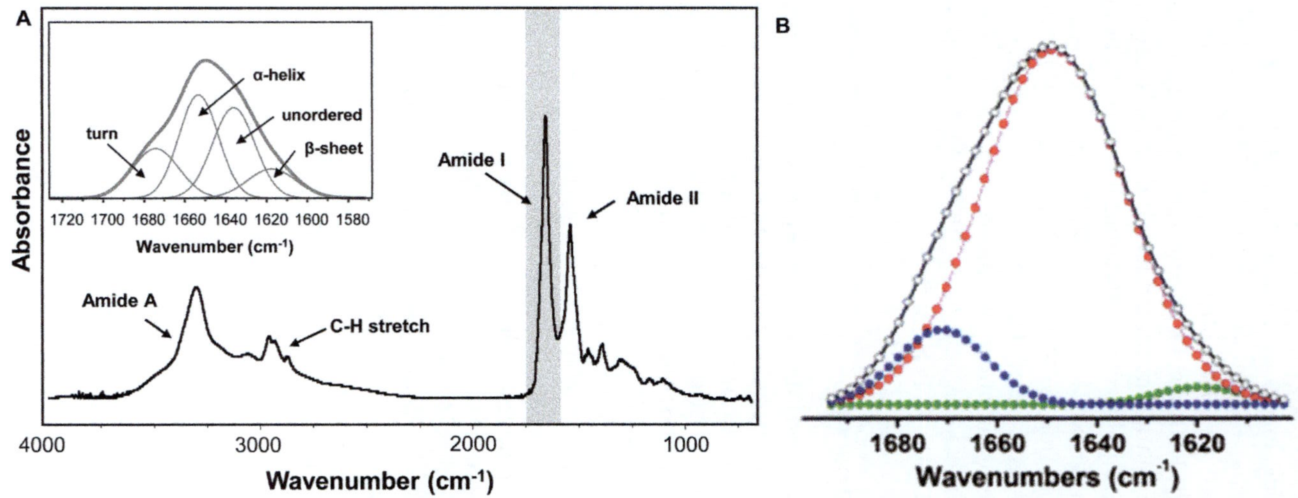

Figure 5.7 (A) Typical FTIR spectra of protein. Inset: expanded view of the amide I band. Reproduced from ref. 59 with permission from Elsevier, Copyright 2013. (B) Amide I region of the FTIR spectrum of the modular protein polymer, RLP_{12}–RGD_2–MMP–HBD. The experimental data (open circles) was fit with peaks corresponding to β-turn (blue), random coil (red), and β-sheet (green) structures. Reproduced from ref. 27 with permission from the Royal Society of Chemistry.

secondary structure of freeze-dried full-length fruit fly resilin with respective exon-1 and exon-3 RLPs using FTIR spectroscopy. The authors demonstrated (from FTIR amide I peak deconvolution using above conformational band peaks) that both partial (exon-1 and exon-3) and full-length resilins exhibit largely unordered (turns and random coils; 68–73%) secondary structure, with 12–13% α-helix and 15–20% β-sheet structures.[18,19] In comparison to CD spectroscopy results, the FTIR results of exon-1 and exon-2 resilins show higher α-helix content, which can be attributed to the data set's dependency on CD spectra fitting and overlapping conformational bands of FTIR spectra deconvolution.[18,19] On the other hand, Tamburro *et al.*[21] reported largely unordered with PPII conformations for RLPs, namely RES15n, RES24n, and RES39c in heavy water or deuterium oxide (D_2O). Similar results (largely unordered with no α-helix and β-sheet structures) have also been reported for RLP-based modular protein polymers, namely RLP_{12}–RGD_2–MMP–HBD (Figure 5.7B), CBM–RLP–CBM, and CBM_2–RLP–HFBI.[27,44]

5.4.3 NMR Spectroscopy

NMR is a very diverse experimental technique, which can provide atomic-resolution information on structure, dynamics, and interactions of IDPs in solution.[60] In the NMR spectroscopy experiment, a solution of protein of interest is placed in a very strong magnetic field and subjected to electromagnetic radiation (radiofrequency waves). This promotes transitions between nuclear energy levels (resonance) in atomic nuclei (consisting of an odd number of neutrons and/or protons, *e.g.*, ^1H, ^{13}C, ^{15}N, *etc.*), which provide the basis for the analysis of secondary structural elements.[61] In the case of IDPs, each atomic nucleus experiences many different local environments with time (due to their dynamic behavior). Therefore, the observed chemical shift is an average of chemical shifts of each possible structure. As a result, the range of chemical shifts in IDPs is exceptionally narrow, which makes the spectra of IDPs extremely crowded and hard to analyze. This is generally overcome by increasing the dimensionality of the NMR experiment.[62,63] Two-dimensional NMR spectroscopy is a set of NMR methods that provide additional molecular information compared to 1D NMR, where the data is plotted in two frequency axes rather than one. Different types of 2D NMR spectroscopy use different correlations, such as homonuclear through-bond correlation (magnetization transfer occurs between nuclei of the same type, *e.g.*, exclusive correlation spectroscopy, ECOSY), heteronuclear through-bond correlation (magnetization transfer occurs between nuclei of a different type, *e.g.*, heteronuclear single-quantum correlation spectroscopy, HSQC), and through-space correlation methods (magnetization transfer occurs between nuclei that are physically close to each other regardless of whether there is a bond between them, *e.g.*, nuclear Overhauser effect spectroscopy, NOESY), which are commonly used for structural analysis of IDPs.[61,63]

Nairn *et al.*[64] first studied the structure of An16 resilin dissolved in $H_2O:D_2O$ (9 : 1) buffer (containing phosphate and sodium chloride) using NMR spectroscopy in phase-sensitive mode, where both homonuclear and heteronuclear single-quantum correlation 2D experiments were recorded.[64] Figure 5.8A shows the experimental chemical shift of H_α and C_α as a function of An16 amino acid sequence and as compared to a range of chemical shifts for various secondary structures of protein reported by Zhang *et al.*[65] The H_α chemical shifts of An16 indicate no significant β-strand content, whereas C_α chemical shifts indicate minimum α-helix. This implies the majority of configurations in the protein is random coil.[64] Moreover, the measured H_N–H_α coupling constants fall between 6.5 Hz and 8.3 Hz, which exclude α-helix and β-sheet structures and indicate PPII or randomly fluctuating (coil) structures for An16.[64] Tamburro *et al.*[21] investigated the structure of N-terminal RLPs, namely RES9n, RES15n, RES24n, and RES60n; C-terminal RLP, namely RES39c, and resilin-inspired polypeptide $(PGGGN)_{10}$ dissolved in $H_2O:D_2O$ (9 : 1); and deuterated trifluoroethanol : H_2O (8 : 2), where 2D spectra were collected (using the stated method) in the phase-sensitive mode. Figure 5.8B shows the experimental chemical shifts of H_α and N_H protons of RES9n, RES15n, and RES24n plotted as a function of amino acid sequence, where chemical shifts of all H_α protons are in the range of random coil with no shifts observed toward α-helix and β-sheet. However, the RLPs along with RES60n and RES39c showed a certain propensity toward β-turn (both type I and type II) conformation and not α-helix.[21] The secondary structural investigation of the RLPs using the NMR technique revealed that the majority of backbone carbons and protons of RLPs exhibit chemical shifts characteristic of that of the random coil structural configuration, and the coexistence of quasiextended structures (*e.g.*, PPII conformation) and folded β-turns, with no apparent features of α-helix or β-sheet.[21,64] However, a number of structural features of RLPs related to their overall size, shape, and organization in solutions remain elusive to NMR, which were obtained by other complementary techniques such as small-angle scattering.

5.4.4 Small-angle Scattering

Small-angle scattering (SAS) provides low-resolution information about the size, shape, structure, and conformation of small molecules and their intermolecular assemblies (1–1000 nm) in solutions.[67] SAS techniques, such as small-angle X-ray scattering (SAXS) and small-angle neutron scattering (SANS), are complementary to each other. SAXS is sensitive to heavier elements in a sample, whereas SANS is sensitive to lighter elements, such as hydrogen.[67] In the SAS experiment, a solution of the protein of interest is placed in a collimated X-ray or neutron beam, which isotropically scatters (as proteins are randomly oriented in solution) the incident beam. The intensity of the scattered beam is then measured as a function of the angle of scattering using 2D detectors, which provides the basis for structural analysis.[67] The measured SAS data is typically presented as double logarithmic scattering intensity $[I(q)]$ *versus* scattering vector $[q]$ plot, where q is obtained as a function of angle of scattering and the wavelength of irradiation.[68] In general, the SAS profile of a monodispersed system has three distinct regions,

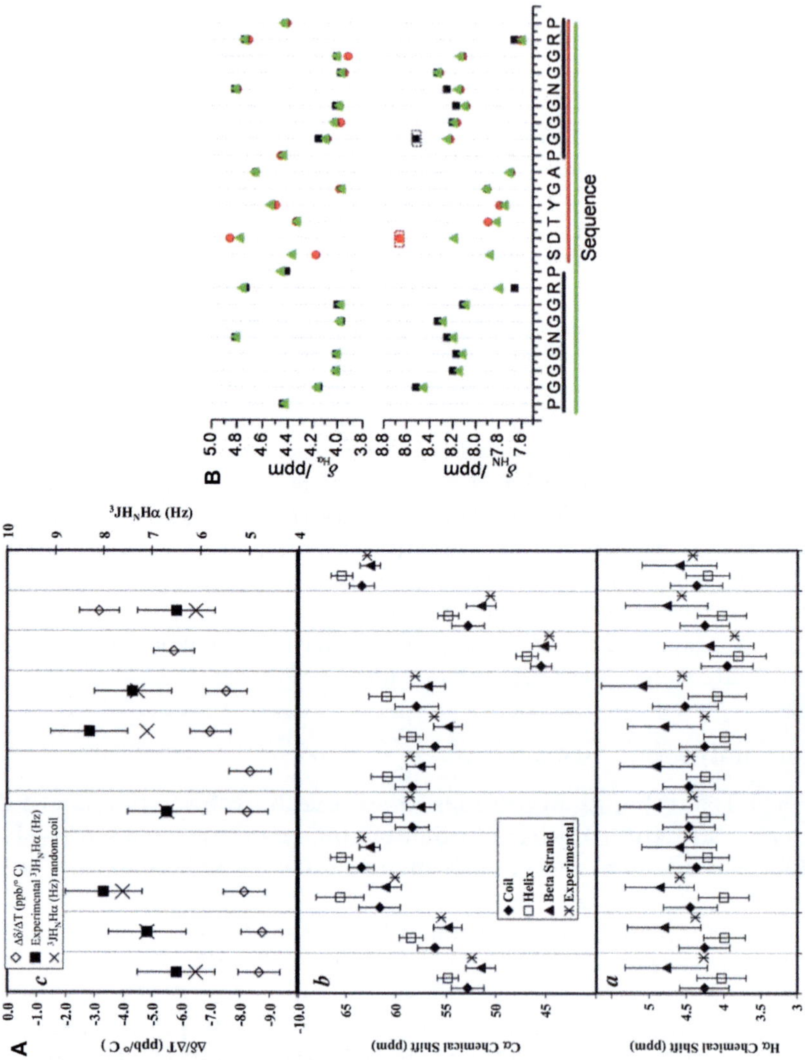
A
B
Δδ/ΔT (ppb/° C)
Experimental ³J$_N$Hα (Hz)
³J$_N$Hα (Hz) random coil
³JH$_N$Hα (Hz)
Cα Chemical Shift (ppm)
Hα Chemical Shift (ppm)
Coil
Helix
Beta Strand
Experimental
Ala1 Gln2 Thr3 Pro4 Ser5 Ser6 Gln7 Tyr8 Gly9 Ala10 Pro11
Residue
δ$_{Hα}$ /ppm
δ$_{HN}$ /ppm
P G G G N G G R P S D T Y G A P G G G N G G R P
Sequence

namely Porod, Fourier, and Guinier regions, as shown in Figure 5.9A. The Porod region (high-q) is characterized by scattering intensity slope and defines the local structure and/or fractal dimension, whereas the Fourier region (mid-q) defines the overall shape of the particles, and the Guinier region (low-q), characterized by scattering intensity plateau, defines the overall size of the particles in terms of radius-of-gyration (R_g).[69] Dutta and coworkers have extensively studied the intrinsic structure of RLPs, such as Rec1-resilin and An16, using SAXS and SANS techniques.[14,48,55] Figure 5.9B shows the typical SAXS profile of Rec1-resilin and An16 in 10 mM phosphate-buffered saline (PBS).[48,55] The measured slope value of 2 in the Porod region qualitatively represents Gaussian coil structure (*i.e.*, random coil secondary structure) for RLPs.[70] However, secondary structures of proteins cannot be quantitatively estimated using SAS, whereas IDPs can be qualitatively identified for their "unfoldedness" or "random coil" likeness using Kratky plot ($I(q)^{*}q^2$ *versus* q), which shows distinct curve shapes for fully unordered and partially unordered structures, as shown in Figure 5.9C.[71] The low-q region of the Kratky plot (Figure 5.9D) shows a typical monotonous increase followed by a plateau in the high-q region, which supports the structure of the proteins to be intrinsically disordered.

Among various models, the shape-independent pair-distance distribution $P(r)$ function uses the total scattering spectrum (rather than Guinier approximation, which extrapolates $I(q)$ to zero concentration at low-q) for the estimation of real space R_g, therefore being considered more appropriate for IDPs. The $P(r)$ curve is a histogram of all interatomic distances (r) in the scattering particle, which is obtained by inverse Fourier transformation of the complete scattering data and can provide molecular structures of scattering particles more intuitively.[68,72] A symmetric bell-shaped $P(r)$ curve is observed for globular proteins, whereas an extended tail curve for unfolded proteins (Figure 5.10A).[73] The $P(r)$ curve of Rec1-resilin and An16 (Figure 5.10B) show an asymmetric extended tail curve (which is characteristic of an elongated molecule) with an estimated maximum molecular dimension (Dmax) of ~20.0 nm and ~18.0 nm, respectively. Moreover, a real space R_g of ~50.0 nm and ~47.2 nm was obtained for Rec1-resilin and An16, respectively.[48,71] On the other hand, the $P(r)$ output can also be used as a guide for *ab initio*

Figure 5.8 (A) The NMR parameters plotted against the amino acid sequence of An16, where (a) and (b) are H_α and C_α chemical shifts, respectively, compared with individual secondary structure reference ranges of protein from Zhang *et al.*,[65] whereas (c) is the temperature coefficient (left axis) of H_N chemical shift and comparison of experimental $^3JH_NH_\alpha$ coupling constants (right axis) with those for a random coil predicted by Smith *et al.*[66] Reproduced from ref. 64 with permission from Elsevier, Copyright 2008. (B) Comparison of chemical shifts of H_α and N_H proton resonances of RES9n (black), RES15n (red), and RES24n (green), where the boxes refer to N-terminal adjacent N_H protons. Reproduced from ref. 21 with permission from John Wiley & Sons, Copyright © 2010 WILEY-VCH Verlag GmbH & Co. KGaA, Weinheim.

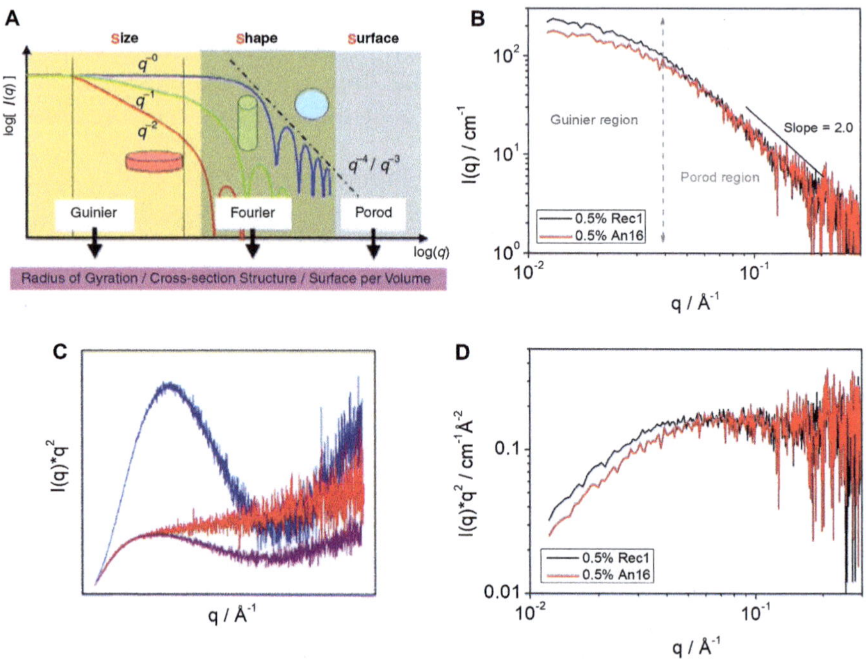

Figure 5.9 (A) Regions of SAS profile and data that may be extracted from each. Reproduced from ref. 69 with permission from Anton Paar, Copyright 2013. (B) SAXS profile of 0.5% Rec1-resilin and An16 in PBS. (C) Kratky plot of globular (blue), partially unordered (purple), and completely disordered (red) proteins. Reproduced from ref. 71 with permission from the Royal Society of Chemistry. (D) Kratky plot of 0.5% Rec1-resilin and An16 in PBS.

modeling of protein structure reconstruction, where one among infinite possible *ab initio* 3D structural ensembles of An16 in water are reconstructed using the GASBOR computer program, as shown in Figure 5.10C.[74]

In addition, the residual structure of IDPs can also be assessed from the ratio of R_g (estimated from SAS data) and hydrodynamic radius, R_h (measured using dynamic light scattering, dynamic light scattering [DLS] technique), *i.e.*, R_g/R_h. The R_g/R_h of globular proteins is approximately 0.77, whereas it is 1.40 for fully denatured proteins.[75] The R_h of Rec1-resilin and An16 measured using DLS is ~5.2 nm and ~4.6 nm, respectively, which gives an R_g/R_h value of ~0.96 and ~0.91, respectively, and suggests the existence of residual structure (molten globule and/or premolten globule) in the equilibrium structure of RLPs in solution.[75] Lately, based on data collected from several chemically unfolded and globular proteins, Kohn *et al.*[76] and Narang *et al.*[77] proposed power-law relationships between the number of monomers in the polymer chain and ensemble average R_g, which can be used to estimate the R_g of proteins at both unfolded and folded states. For Rec1-resilin (with 310 amino acid residues) and An16 (with 185 amino acids), the R_g value is expected to be ~59.5 Å and ~46.5 Å, respectively, if fully unfolded (based on chemically

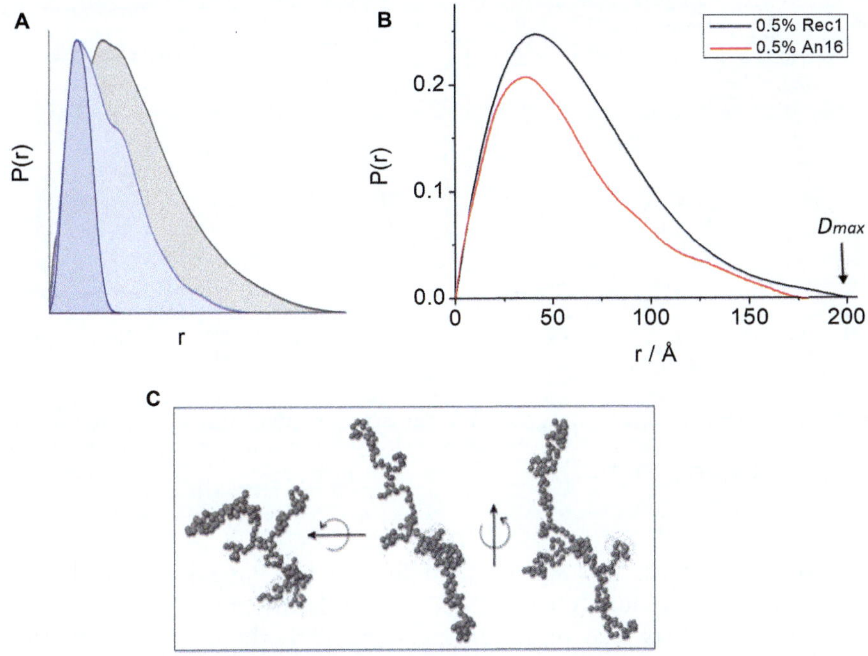

Figure 5.10 (A) $P(r)$ curve of globular (dark blue), partially unordered (light blue), and completely disordered (gray) proteins. Reproduced from ref. 73 with permission from John Wiley & Sons, Copyright FEBS Letters 589 (2015) 0014-5793 ©2015 The Authors. Published by FEBS Press and John Wiley & Sons Ltd. (B) P(r) curve of 0.5% Rec1-resilin and An16 in PBS. (C) *Ab initio* 3D structural ensemble (chain-like dummy residues) of An16 in water reconstructed using the GASBOR computer program. A snapshot from different angles for one among possible infinite ensembles is provided. Reproduced from ref. 48 with permission from Elsevier, Copyright 2014.

unfolded proteins data set) or ~19.6 Å and ~16.3 Å, respectively, if compact or globular (based on globular proteins data set).[76,77] Therefore, based on both theoretical predictions and SAXS experimental results, the RLPs are established as IDPs, where An16 is considered to be completely unfolded in aqueous solution, whereas Rec1-resilin to be largely unfolded but not completely disordered. In recent years, the ensemble optimization method (EOM) has been developed as a tool to express the properties of IDPs in solution in terms of their size and shape distributions (*i.e.*, ensembles of reliable conformations).[78] The EOM is a SAXS data-driven strategy for optimizing ensemble structures, where the experimental profile is considered to be derived from a number of coexisting conformational states of the proteins.[71] Balu *et al.*[55] applied the EOM strategy to describe the structural properties of Rec1-resilin, where the presented pool considers all the degrees of freedom of the protein sequence and provided unimodal distribution with R_g and Dmax values of ~5.1 nm and ~15.3 nm, respectively (Figure 5.11A and B).[71]

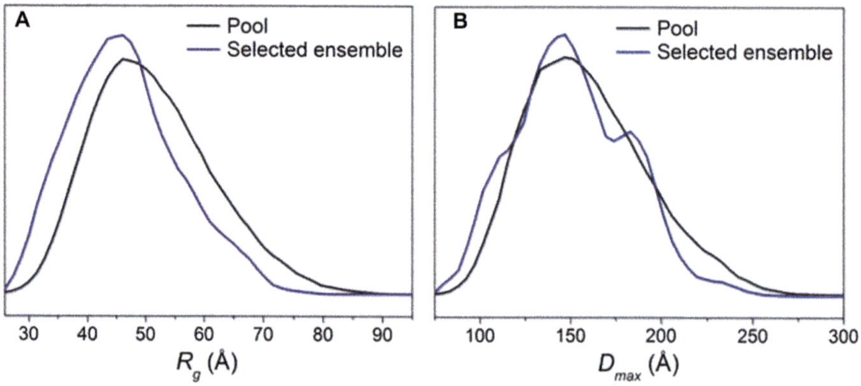

Figure 5.11 (A) R_g and (B) Dmax distribution curves of Rec1-resilin obtained using EOM. Reproduced from ref. 55 with permission from Springer Nature, Copyright 2015.

The observed shift in the R_g (4.7 nm) of selected ensemble distribution of Rec1-resilin suggested that <100% of the structure is free to move (unordered), which is also in good agreement with theoretical predictions based on the chemically unfolded proteins data set.[76] Therefore, the molecular structures of RLPs and RLP-based modular protein polymers were experimentally established to be that of IDPs. The potential of resilins with largely unordered or IDP structure to exhibit outstanding resilience encouraged detailed investigation on their stimuli-responsive properties and self-assembly. Their responsive properties, along with their structural change with respect to change in the environment, are discussed in detail in the following sections.

5.5 Multistimuli Responsiveness

Stimuli-responsive biomaterials, such as RLPs and ELPs, alter their physical and/or chemical properties upon exposure to different environmental conditions and have great potential for a wide range of biomedical and bionanotechnology applications, such as tissue engineering, drug delivery, biosensing, and smart coating. The responsiveness of RLPs and RLP-based modular protein polymers to external stimuli originates from different classes of amino acids in their primary structure and the types of peptide interactions (Figure 5.12).[79] The naturally occurring amino acids can undergo various types of interactions, such as covalent interactions (sulfur-containing amino acids), noncovalent interactions (electrostatic), hydrogen bonding (polar amino acids), π–π stacking (aromatic amino acids), and hydrophobic and steric interactions (strand directing amino acids). Moreover, these interactions also depend on external stimuli, such as temperature, ions, pH, light, and other macromolecules.[79]

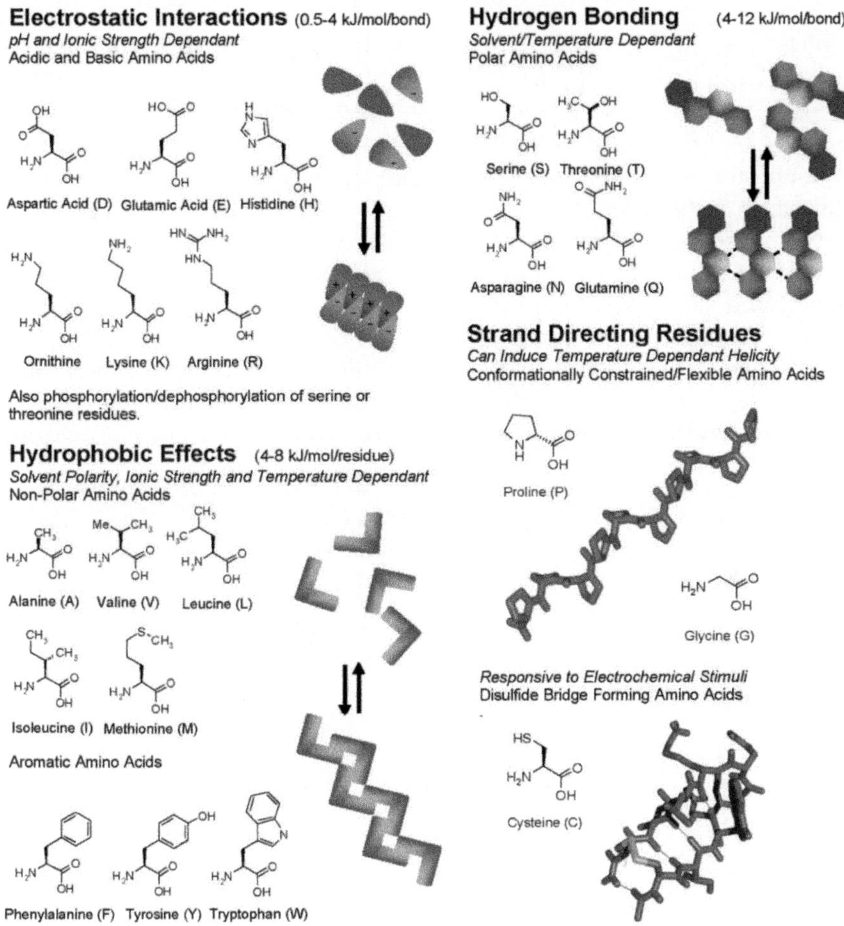

Figure 5.12 Schematic description of different classes of amino acids and their different types of peptide interactions. Reproduced from ref. 79 with permission from the Royal Society of Chemistry.

5.5.1 Temperature and Ion Responsiveness

IDPs can undergo four types of thermoresponsive phase transitions (Figure 5.13A), where above (upper) or below (lower) a critical solution temperature the protein molecules are miscible for all compositions. Among IDPs, ELPs were arguably the first reported to exhibit lower critical solution temperature (LCST). However, they did not exhibit upper critical solution temperature (UCST).[80] The thermoresponsive nature of RLPs was first observed by Elvin *et al.*[4] at low temperature (formation of coacervates) and was applied for their purification. The exhibition of both LCST and UCST by a single protein molecule solution is rare and has been predicted in the past.[81] It was in 2011 when Dutta *et al.*[15] first experimentally demonstrated Rec1-resilin to exhibit both UCST and LCST (dual-phase transition behavior, DPTB), with an increase in hydrodynamic diameter (measured by DLS) in aqueous solutions (Figure 5.13B). Rec1-resilin exhibited a UCST of around 6 °C and an LCST of around 70 °C. The LCST behavior of Rec1-resilin displayed slow reversal kinetics, whereas the UCST behavior was observed to be completely reversible without hysteresis. Moreover, the molecular structure and organization of Rec1-resilin at UCST (Figure 5.13B) has also been observed using cryo-transmission electron microscopy (TEM) and SANS, where cryo-TEM showed a morphology of interconnected gel particle network, whereas SANS showed an increase in low-q intensity and slope extending beyond the measurement range (>140 nm).[14,15]

Based on extensive molecular simulations in the past couple of decades, it has been widely accepted that IDPs undergoing phase transition behavior do not collapse into entropically disfavored β-turns rich structure but rather collapse into amorphous coacervates due to unfavorable solvent entropy.[80]

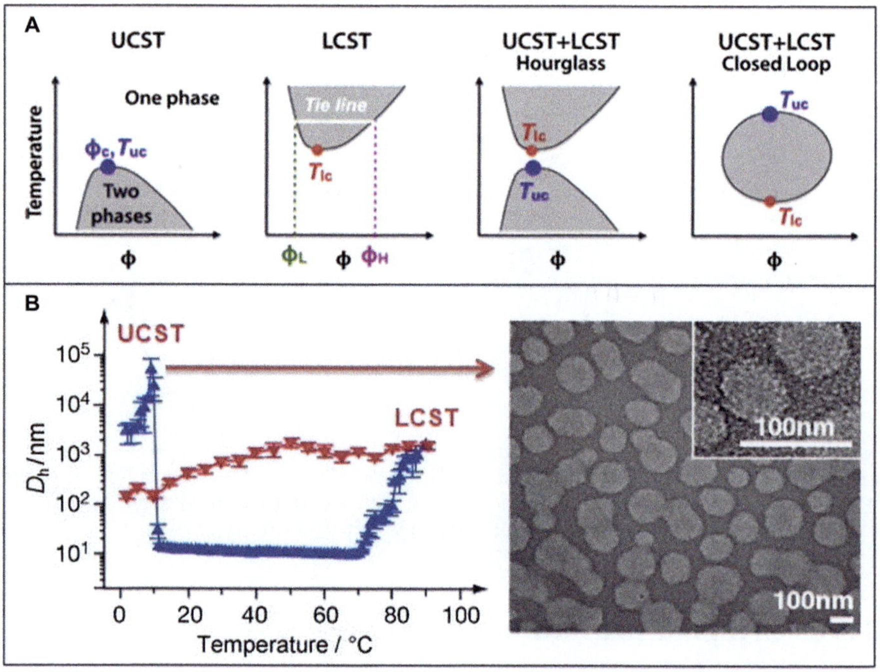

Figure 5.13 (A) Types of thermoresponsive phase transitions of IDPs. T_{uc} and T_{lc} denote UCST and LCST, respectively, whereas $Ø_C$ denotes the volume fraction of IDP at critical solution temperature. Conversely, $Ø_L$ and $Ø_H$ denote volume fractions of IDP coexisting at dilute and dense phases, respectively. Reproduced from ref. 80 with permission from Elsevier, Copyright 2018. (B) Hydrodynamic diameter, D_h of Rec1-resilin in phosphate-buffered saline solution (10 mM) as a function of temperature measured using DLS (thermal cycling; blue – heating and red – cooling cycles) and corresponding cryo-TEM micrograph at UCST. Reproduced from ref. 15 with permission from John Wiley & Sons, Copyright © 2011 WILEY-VCH Verlag GmbH & Co. KGaA, Weinheim.

Although high glycine and proline residues promote disorder with elastomeric properties, additional sequence features are required to modulate precise phase behavior control of IDPs. For example, resilin and tropoelastin have high glycine and proline content, where resilin exhibits both UCST and LCST, whereas tropoelastin exhibits only LCST.[23] Based on the primary sequence, resilins are enriched in polar and charged residues with a preference for arginine over lysine residues (Figure 5.14A), whereas tropoelastin is enriched in nonpolar hydrophobic residues with a preference for lysine over arginine residues. Lately, Quiroz *et al.*[23] tested and reported the underlying sequential feature differences between resilin and tropoelastin and suggested the following set of sequence features required for UCST behavior of IDPs at physiological temperatures: (1) Overall fraction of charged residues >0.2, (2) overall fraction of aromatic residues >0.1, (3) fraction of positively charged residue (arginine) >0.85, and (4) net charge per residue between −0.05 and 0.05. Moreover, the authors suggested that the dual-phase thermal transition behavior of RLPs with consensus repeat unit GRPSDSYGAPGN comes from the fusion of a UCST motif (GRPSDSYG) and a putative LCST motif (APGGGN).[23]

Conversely, the modular protein polymer, namely RLP_4–SLP_4, exhibited increased particle size with an increase in temperature, *i.e.*, from ~369 nm (4 °C) to 1149 nm (70 °C). Moreover, RLP_4–SLP_4 was observed to form two

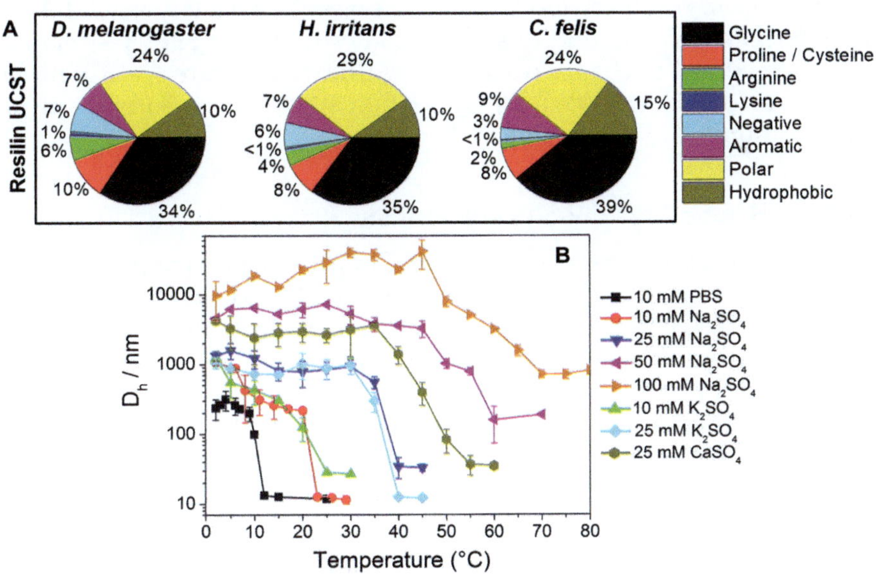

Figure 5.14 (A) Composition of resilin from various insect species showing UCST behavior. Reproduced from ref. 80 with permission from Elsevier, Copyright 2018. (B) Hydrodynamic diameter, D_h, of An16 measured over a wide range of temperature as a function of ion-type concentration (following the Hofmeister series) using DLS. Reproduced from ref. 48 with permission from Elsevier, Copyright 2014.

nanostructured populations with differing sizes at temperatures between 25 and 45 °C.[32] However, other variants, such as RLP_4–SLP and RLP_8–SLP, showed turbidity (UCST-like behavior) at 4 °C, whereas negligible or only a slight increase in solution turbidity at 70 °C.[43] Conversely, equilibrium co-mixing/blending of RLPs with regenerated silk fibroin protein molecules decreased the LCST of Rec1-resilin to physiologically relevant temperatures.[16] Weitzhandler *et al.*[37] investigated the thermoresponsive properties of a suite of RLP-based modular protein polymers, namely $(RLP)_n$–$(X\text{-}ELP)_m$, where $n = 20, 40, 60, 80$, and 100; $m = 40, 80$, and 160; and X is the guest amino acid residue A/G (equal amount of A and G), S, and V. Interestingly, the modular protein polymer $(RLP)_{40}$–$(S\text{-}ELP)_{80}$ showed both UCST (at 35 °C with micellar structure) and LCST (at 58 °C with coacervate structure) behavior with soluble polymer chains between them. Moreover, decreasing the RLP block length to 20 resulted in no UCST and LCST, whereas increasing the RLP block length (≥40) and/or substitution of S residue with A/G and V resulted in only UCST.[37]

Dutta and coworkers have extensively studied the ion responsiveness of RLPs in aqueous solutions.[15,48] At physiological pH and temperature, the hydrophilic amino acid residues in RLPs form a hydrogen-bonded network structure and hydration shell with the surrounding water. However, at UCST, RLPs (example An16) expel most of the associated water (collapsed hydration shell), forming a collapsed microgel structure.[48] The responsiveness of RLPs to ion (kosmotropes) arise from interactions between ions and amino acid residues, and ions and the water molecules, which follows the Hofmeister series and affects the solubility of RLPs.[82] The UCST of An16 was demonstrated to change (increase) significantly (Figure 5.14B) with the addition of both anion (SO_4^{2-}) and cation (Ca^{2+}) as a function of ion concentration.[48]

5.5.2 pH and Light Responsiveness

RLPs are responsive to pH, where a change in pH can alter the protonated or deprotonated state of acidic and basic amino acid residues, which dynamically changes the conformation, net surface charge, hydrodynamic diameter, and organization of the protein in aqueous solutions.[79] For example, at physiological pH, Rec1-resilin and An16 exhibit negative zeta-potentials, which become more negative upon changing the pH to basic and positive changing the pH to acidic (along with a systematic change in conformation and D_h). However, the random coil secondary structure of RLPs does not change with change in pH (studied using CD spectroscopy),[48,55] and the isoelectric point (IEP) at which the net surface charge becomes zero was observed around a pH of 4.9.[15,48] Moreover, the reversible conformational change of Rec1-resilin as a function of pH, i.e. compact to brush-like conformations (Figure 5.15A) has been reported using quartz crystal microbalance with dissipation monitoring (QCM-D).[83] The expansion and contraction of RLP chains, cycling between IEP and basic pH, is much larger relative to those between acidic pH and IEP, which is due to the strong interaction between

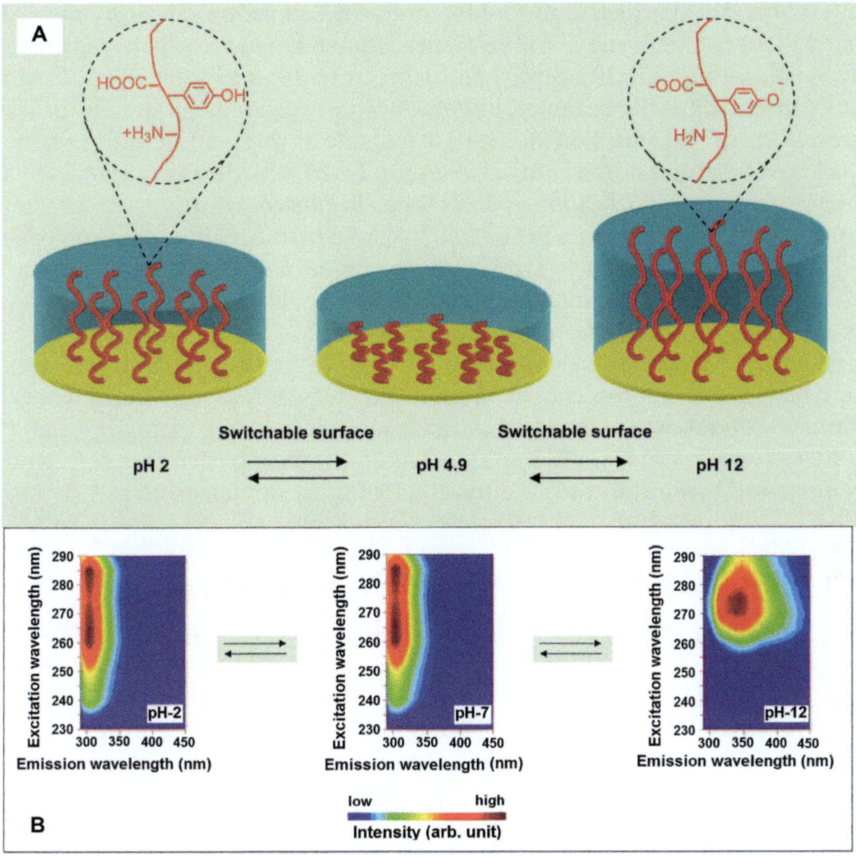

Figure 5.15 (A) Responsive behavior of Rec1-resilin as a function of pH measured using QCM-D. Reproduced from ref. 83 with permission from Elsevier, Copyright 2010. (B) Excitation–emission 3D contour plots of An16 measured using fluorescence spectrometer as a function of pH. Reproduced from ref. 48 with permission from Elsevier, Copyright 2014.

water at basic pH and polar amino acid residues (asparagine, serine, and tyrosine). The RLPs were also observed to exhibit different packing densities as a function of pH at the solid (gold)–liquid interface.[83] The pH responsiveness of RLPs is also associated with their thermal and light responsiveness. For example, the UCST of Rec1-resilin in aqueous solution was observed to increase with a decrease in pH and *vice versa*, whereas LCST does not change significantly.[15]

On the other hand, coacervates of the RLP, Rec1-resilin, and RLP-based modular protein polymers, namely $(CBM)_2$–RLP–HFBI and CBM–RLP–CBM, showed pH-dependent size variations in 0.5 M sodium phosphate buffers.[44] The RLP and modular RLPs showed an increasing trend in coacervate size with an increase in pH and solute concentration, whereas the $(CBM)_2$–RLP–HFBI showed bimodal size distribution. Moreover, the salt-induced coacervation can be attributed to the random coil to β-sheet transformation, which was characterized using FTIR spectroscopy (band shift from 1670 to 1634 cm^{-1}).[44] RLPs exhibit bright blue fluorescence under UV light, which originates from excitation (275 nm) and emission (303 nm) of tyrosine amino acid residues, whereas crosslinked resilin exhibits emission around 420 nm due to dityrosine.[15,48] Increase in pH significantly affects the fluorescence property of RLPs (Figure 5.15B), where above a pH of 10.5 (pK_a of tyrosine residue), the molecular transformation of tyrosine→tyrosinate occurs, which shifts its emission to ~350 nm.[84] The fluorescence property of RLPs is also affected by the surface plasmon resonance and fluorescence property of noble metal nanoparticles and nanoclusters,[85,86] respectively, in close vicinity, where the protein fluorescence is quenched by fluorescent metal nanoclusters above a certain concentration.[56,86] Notably, a suite of RLP-based modular protein polymers, namely An_X-EGFP (where $X = 4$, 8, 16, and 32), were synthesized by Lyons *et al.*,[30] which demonstrated green fluorescent emission (520 nm) when excited at 485 nm.

5.5.3 Responsiveness to Different Substrate Surfaces

RLPs are responsive to the physical condition of substrate surfaces (*i.e.*, surface energy, charge, and hydrophilicity), where their interfacial behavior on various substrate surfaces is dictated by the balance of entropic and enthalpic effects.[87] For example, Dutta *et al.*[88] have extensively studied and reported the various molecular architectures of Rec1-resilin in response to various substrates, using atomic force microscopy (AFM). Figure 5.16A–C shows the AFM images of the organization of dry Rec1-resilin on the surface of mica, graphite, and silicon wafer, which has water contact angles of 16°, 92°, and 63°, respectively.[88] While an energetically favorable side-on adsorption (monomolecular layer structure) was obtained on hydrophilic surfaces, a parallel end-on ordering (columnar structure) was obtained on hydrophobic surfaces, where the thickness of protein islands on substrates was observed in the following order: graphite > silicon wafer > mica.[88] The responsive behavior of RLPs on different surfaces can be explained by

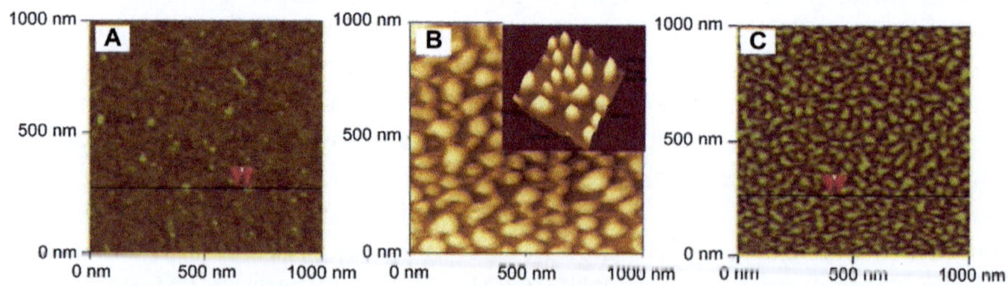

Figure 5.16 AFM images of Rec1-resilin morphologies formed on the surface of (A) mica, (B) graphite, and (C) Si–wafer. Reproduced from ref. 88 with permission from Elsevier, Copyright 2009.

thermodynamics, where the apparent free energy of adsorption is lower on hydrophobic surfaces (favoring protein–substrate interactions) than on hydrophilic surfaces.[89] Moreover, the pH of the aqueous medium greatly influences the adsorption characteristics of RLPs, where at higher pH, the solubility of protein increases, leading to denser surface coverage (protein–surface interaction is attractive), and *vice versa*.[88] Lately, responsiveness to different substrates has also been demonstrated for RLP-based modular protein polymers, namely $(CBM)_2$–RLP–HFBI and CBM–RLP–CBM.[44] The $(CBM)_2$–RLP–HFBI coacervate formed or adhered as a noncontinuous layer on both the cellulose and graphene surfaces, whereas the CBM–RLP–CBM coacervate formed isolated microsized particles only on cellulose surface.[44]

5.5.4 Mechanical Responsiveness

Using CD spectroscopy, which is sensitive to PPII conformation, Bochicchio *et al.*[57] initially proposed the mechanical stress-responsive behavior of resilin to be that of an entropic spring. At a relaxed state, resilin has an equilibrium conformation between that of an open and folded structure, *i.e.*, high entropic, which significantly decreases when mechanically stressed. Upon release, resilin returns to its initial high-entropy relaxed state.[57] Qin *et al.*[13] studied the elastic properties (Figure 5.17) and conformational transition of full-length exon-1 and exon-3 resilin (cast thin films) under mechanical energy input using AFM and FTIR. Interestingly, exon-1 resilin (segment with more hydrophilic blocks) exhibits resilience >90% (similar to full-length resilin), whereas exon-3 resilin (segment with a large hydrophobic block in the middle) exhibits resilience around 63%. On the other hand, an increase in β-turn structure was observed for both full-length and exon-3 resilin when stretched up to 1.5 times the original length, whereas no significant change was observed for exon-1 resilin.[13] Based on the observations, the authors proposed that the exon-1 segment absorbs the applied mechanical energy and remains unstructured, whereas the exon-3 segment receives energy from exon-1 and transforms into an ordered β-turn structure to store energy. Upon release, the exon-3 segment undergoes reversible structural transformation and transfers the absorbed energy to the exon-1 segment.[13] Moreover, the polar groups of resilin form strong hydrogen bonding with surrounding water molecules, which act as deformation energy-absorbing layer, thereby improving elasticity.[25,90]

Lately, Griffo *et al.*[33] investigated the effect of pH on the conformation (contour length) and mechanical property (work of rupture) of exon-1 resilin using single-molecule force spectroscopy. The study was performed using the exon-1 resilin-based modular protein polymer $(CBM)_2$–RLP–HFBI. The modular RLPs were adhered (through the CBM) on a thin layer of cellulose, while hydrophobic octadecyltrichlorosilane (OTS)-coated AFM tip acted as an attachment site for modular RLPs (through the HFBI).[33] The extension experiments were performed at pH 5 (using 10 mM sodium acetate buffer) and pH 11 (using 10 mM phosphate buffer). At pH 5, the RLP in $(CBM)_2$–RLP–HFBI exhibited random coil conformation and could not be fully

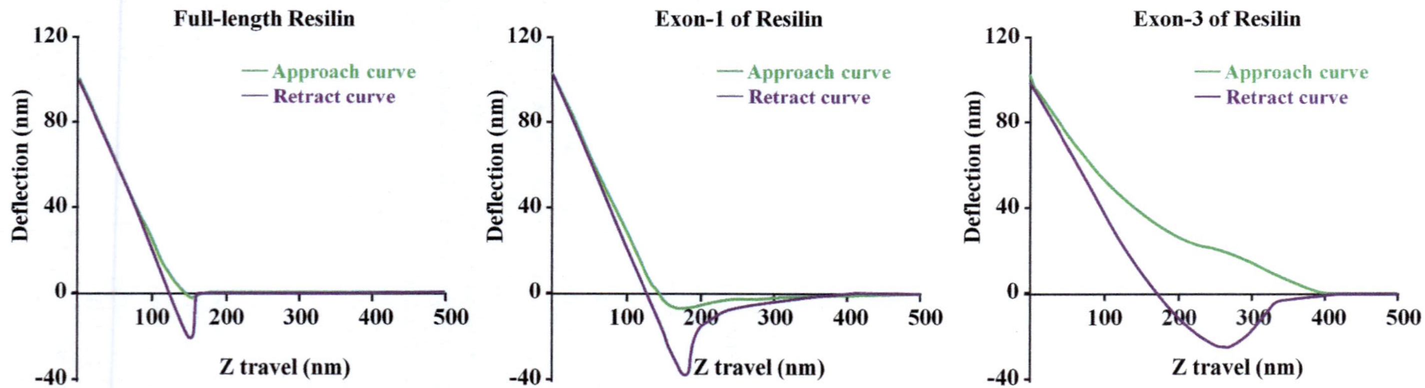

	Full-length Resilin	Exon-1	Exon-3
Elastic Modulus (MPa)	2.55 ± 1.05	2.35 ± 0.87	0.63 ± 0.21
Resilience/Elasticity (%)	94.5 ± 2.1	90.2 ± 3.1	63.5 ± 7.9

Figure 5.17 AFM force–distance curves and estimated mechanical properties of full-length, exon-1, and exon-3 resilin. Reproduced from ref. 13 with permission from Springer Nature, Copyright 2012.

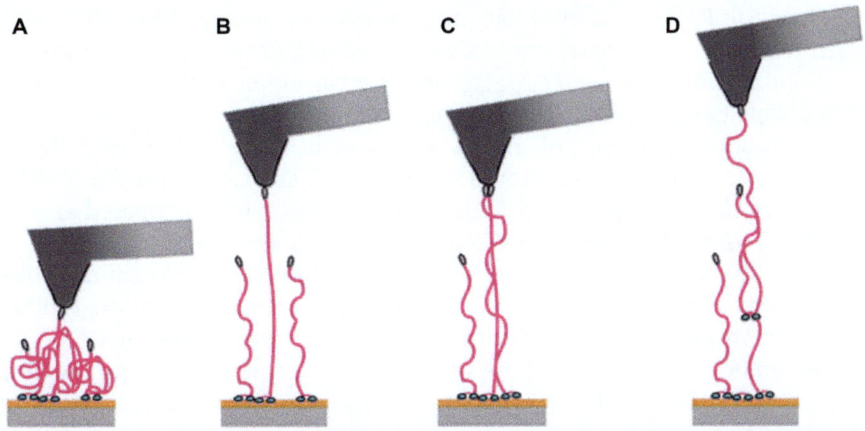

Figure 5.18 Schematic of different conformations of the RLP-based modular protein polymer $(CBM)_2$–RLP–HFBI at the detachment of the molecule. The modular RLP assumes (A) a random coil conformation at pH 5 and (B) an extended conformation at pH 11. Moreover, at pH 11, the modular RLP shows attachment of multiple entanglements (C) and (D). Reproduced from ref. 33 with permission from American Chemical Society, Copyright 2017.

extended by tensile stress as it appears that the uncoiling force of RLP was higher than the HFBI rupture force. Conversely, at pH 11, the RLP in $(CBM)_2$–RLP–HFBI exhibited swollen coil conformation, and full extension of the RLP was obtained (rupture force of HFBI exceeding uncoiling force of RLP).[33] Moreover, at pH 11, strong entanglement (tendency toward network-like structure) of RLP in $(CBM)_2$–RLP–HFBI was also observed. A schematic of different conformations of the modular RLPs at pH 5 and 11 are shown in Figure 5.18.[33] At pH 5, the work of rupture and contour length for $(CBM)_2$–RLP–HFBI were measured at around 485 E/kT and 29 nm, respectively, while the same were 6725 E/kT and 64 nm, respectively, at pH 11. The event of a single molecule stretching was observed around 6% at pH 5 for $(CBM)_2$–RLP–HFBI, whereas it was observed to be 18% at pH 11.[33] The stimuli-responsive properties of RLPs that originate from their amino acid design can be potentially harnessed to develop self-assembled nanostructures with modular properties, which are further discussed in the following section.

5.6 Modular Self-assembly

Self-assembling polypeptides offer an amazing platform for nanomaterials-based drug delivery applications.[91] Tamburro *et al.*[21] first reported the tendency of random coil RLPs, namely RES24n, RES60n, and RES50, to form fibrillar structures at room temperature, where RES50 showed fibers and networks of variegate morphologies with self-assembled micron-long fibers (50–80 nm in diameter) in water without any propensity for twisting, which was dissimilar to both amyloid- and elastin-like fibers.[21] Similar observations were also reported for a trimeric RLP-based modular protein polymer, namely RLP_4–ELP_7–CLP_2, which self-assembled into a network of aligned and flexible fibers in water with a Young's modulus between 0.1 MPa and 3 MPa, which is dissimilar to that of twisted rope-helical structures observed for some ELPs rich in glycine.[29] The trimeric modular protein polymer, when incubated at 37 °C, formed isolated fibrils at day 2 and bundles of fibrils (aligned fibrillar structure) at day 3 with no evidence of further aggregation with time, whereas, when incubated at 50 °C, the aggregation of bundles of fibrils into very large fibers of about 1 μm diameter was observed.[29] Conversely, Rec1-resilin exhibited self-assembled spherical morphology particles above and below critical solution temperatures,[15] whereas the RLP-based modular protein polymer, namely RLP_4–SLP_4, demonstrated self-assembled nanofibrillar structure at 37 °C (Figure 5.19A).[32] When equilibrated at 37 °C, the RLP_4–SLP_4 self-assembled into curly and relatively short nanofibers at 4 hours, which evolved into extended and longer nanofibers at 8 hours, and finally into uniform microfibers of ~1 μm length at 12 hours.[32] Conversely, equilibrium mixing of RLPs with regenerated silk fibroin protein molecules resulted in the hydrophobic interaction-driven self-assembly of rod-like coassembled structures at ambient temperature.[16]

Lately, Weitzhandler *et al.*[37] reported the self-assembly of a suite of RLP-based modular protein polymers, $(RLP)_n$–$(X\text{-}ELP)_m$, where $n = 20$, 40, 60, 80, and 100; $m = 40$, 80, and 160; and X is the guest amino acid residue A/G (equal amount of A and G), S, and V. The modular protein polymers demonstrated self-assembly behavior for RLP block length (n) from 40 to 100, with no self-assembly for $n = 20$. The self-assembly transformation of spherical micelle to cylindrical micelle morphology was observed with an increase in RLP block length, decrease in ELP block length, and increase in hydrophobicity (S<A/G<V) of ELP block (Figure 5.19B), where the spherical $[(RLP)_{40}$–$(A/G\text{-}ELP)_{80}]$ and cylindrical $[(RLP)_{80}$–$(A/G\text{-}ELP)_{80}]$ morphology of modular protein polymers have been confirmed by SANS results.[37] Such self-assembled modular structures of RLPs offer a versatile platform as drug delivery systems for the targeted and controlled release of therapeutic agents.

5.7 Conclusions

In summary, the largely unordered structure of RLPs originates from high glycine and proline amino acid content (>30%) similar to that of elastomeric IDPs, and the presence of a large fraction (30%) of recurring P–X_4–G motifs (where X is the other amino acid apart from P and G, and n varies from 0 to 4) repeating every 4 to 9 residues in the primary sequence. The intrinsically disordered structure of RLPs and RLP-based modular protein polymers has also been experimentally established using techniques, such as CD spectroscopy, FTIR, NMR, and SAS, which reveal an ensemble of random coil secondary structure with the coexistence of small fractions of turns and PPII conformation in equilibrium. The multistimuli responsiveness and self-assembly of RLPs originate from their unique amino acid sequence, hydrophilic composition, and dynamic architecture, where the dual-phase

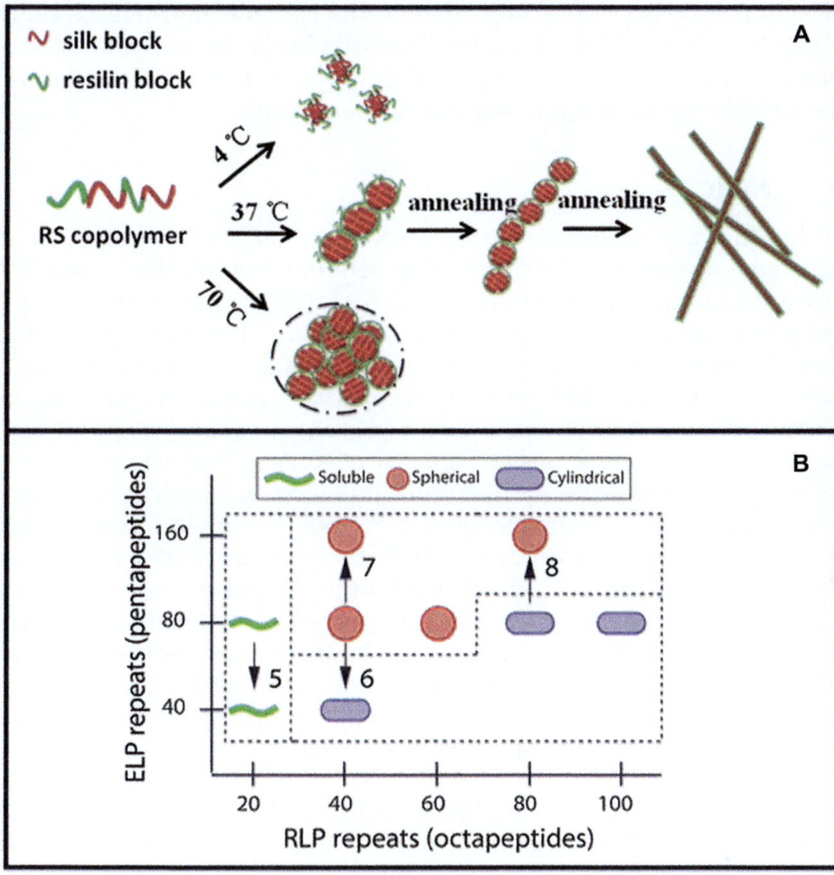

Figure 5.19 (A) Proposed model for hierarchical assembly of the RLP-based modular protein polymer RLP_4–SLP_4. Reproduced from ref. 32 with permission from American Chemical Society, Copyright 2017. (B) Trends in sets 1 and 3: RLP-based modular protein polymers, $(RLP)_n$–$(X\text{-}ELP)_m$, where n and X represent the number of repeat units and amino acid residue, respectively. Reproduced from ref. 37 with permission from American Chemical Society, Copyright 2017.

transition behavior comes from the fusion of a putative LCST motif (APGGGN) and a UCST motif (GRPSDSYG) in the primary sequence. The critical solution temperature behavior of RLPs is related to unfavorable solvent entropy, where the polypeptides collapse into amorphous coacervates and their LCST can be modulated by either altering the hydrophobicity of amino acids or their charge. Moreover, the UCST behavior of RLPs is considered to originate from a set of sequence-specific features, such as the overall fraction of charged residues, arginine and aromatic residues, and net charge per residue. The self-assembly (soluble, micellar, fibrillar, and coacervates) of RLPs and RLP-based modular protein polymers are substantially influenced by the RLP chain length, sequence, and hydrophobicity of copolypeptides, their overall composition, and solvent pH. The highlighted key information shall provide fundamental knowledge and understanding of RLPs for improved rational design and development strategies of resilin protein-based smart biomaterials.

5.8 Outlook

Peptide- and protein-based stimuli-responsive materials provide an unprecedented opportunity for the development of functional biomaterials for a variety of biomedical applications, such as tissue engineering, drug delivery, and biosensors. The intrinsically disordered structure of RLPs, along with their multistimuli-responsive properties, has challenged the traditional protein structure paradigm of fixed 3D structure-dependent function and has demonstrated their potential as a multifunctional biomaterial. Particularly, the dual-thermoresponsive behavior of RLPs exhibiting soluble structures around physiological temperature provides an excellent opportunity for local delivery of therapeutic agents. Moreover, modular protein engineering, along with the expanding library of peptide domains and sequence heuristics approaches, has provided the unparalleled potential to develop smart RLP-based materials with tunable properties and complex functionalities. Such advances present an incredible opportunity for the design and development of novel modular RLP-based biomaterials with new functionality, such as cell attachment, bioadhesion, biodegradation, and modular self-assembly required for specific biomedical applications and clinical reality, where the length and hydrophobicity of additional or secondary domains can influence the overall molecular property.[92] However, high-yield expression of RLP-based modular protein polymers needs further optimization to pave the way for bench-top to scale up production facility. Furthermore, the recent development in the expression of RLPs in transgenic plants to improve their stem's mechanical properties has also opened up new opportunities for the use of RLPs in plant-based materials.[93] However, several fundamental research questions, including those of how the responsive behavior of RLPs is encoded in their amino acid sequence and how tyrosine amino acid residue influences the thermal responsiveness of RLPs, need to be realized for a better understanding and the further development of novel RLP-based biomaterials.

Over the last decade and a half, several strategies have been developed to fabricate crosslinked RLP-based hydrogels. However, the development of RLP-based 3D constructs using advanced biofabrication techniques, such as extrusion-based 3D printing and electrospinning, is still in its infancy, which can be extended by addressing the current flow property limitations of RLP-based materials.[94] Such developments could potentially expand their applications in the field of tissue engineering with tunable morphological and mechanical properties. On the other hand, the unique amino acid sequence, molecular flexibility, and multistimuli responsiveness of RLPs have demonstrated the potential to act as size/structure-directing agent, metallic precursor reducing agent, and multifunctional soft templates to facilitate the synthesis or biomineralization of metallic nanostructures, which can be potentially applied for catalysis and bioimaging. In this context, developments of RLP-based fluorescent bionanoconjugates incorporating receptor binding and/or therapeutic components can further expand their targeted diagnostic and therapeutic applications. Conversely, by applying several design parameters, such as size, shape, hydrophobicity, stiffness, surface charge, and molecular recognition, one can also possibly regulate the kinetics of multiple steps in the immune response and interplay of RLP-based materials for a wide range of biomedical applications.[95] Such an endeavor can offer advancements in the clinical and therapeutic applications of RLP-based smart biomaterials.

Abbreviations

AFM	Atomic force microscopy
BMP2	Bone morphogenetic peptide
CMB	Cellulose-binding module
CD	Circular dichroism
CLP	Collagen-like polypeptide
DLS	Dynamic light scattering
DPTB	Dual-phase transition behavior
EGFP	Enhanced green fluorescence protein
ELP	Elastin-like polypeptide
EOM	Ensemble optimization method
FTIR	Fourier transform infrared
GB1	Artificial elastomeric protein
HBD	Heparin-binding domain
HFBI	Hydrophobin protein domain
IDP	Intrinsically disordered protein
IEP	Isoelectric point
LCD	Lysine-rich domain
LCST	Lower critical solution temperature
MMP	Matrix metallopeptidase domain
NMR	Nuclear magnetic resonance
PPII	Poly-L-proline II
QCM-D	Quartz crystal microbalance with dissipation monitoring
QK	Vascular endothelial growth factor mimicking peptide
RGD	Cell-binding domain
RLPs	Resilin-like polypeptides
SANS	Small-angle neutron scattering
SAS	Small-angle scattering
SAXS	Small-angle X-ray scattering
SLP	Silk-like polypeptide
TEM	Transmission electron microscopy
UCST	Upper critical solution temperature

Acknowledgements

This work was supported by the Australian Research Council (ARC) Discovery research grant (DP160101267).

References

1. P. R. Shewry, A. S. Tatham and A. J. Bailey, *Elastomeric Proteins: Structures, Biomechanical Properties, and Biological Roles*, Cambridge University Press, 2010.
2. G. C. Coles, *J. Insect Physiol.*, 1966, **12**, 679–691.
3. T. Weis-Fogh, *J. Exp. Biol.*, 1960, **37**, 889.
4. C. M. Elvin, A. G. Carr, M. G. Huson, J. M. Maxwell, R. D. Pearson, T. Vuocolo, N. E. Liyou, D. C. C. Wong, D. J. Merritt and N. E. Dixon, *Nature*, 2005, **437**, 999–1002.
5. K. Bailey and T. Weis-Fogh, *Biochim. Biophys. Acta*, 1961, **48**, 452–459.
6. S. O. Andersen, *Biochim. Biophys. Acta, Gen. Subj.*, 1964, **93**, 213–215.
7. G. F. Elliott, A. F. Huxley and T. Weis-Fogh, *J. Mol. Biol.*, 1965, **13**, 791–795.
8. S. Govindarajan and G. S. Rajulu, *Experientia*, 1974, **30**, 908–909.
9. H. R. Hermann and D. E. Willer, *Int. J. Insect Morphol. Embryol.*, 1986, **15**, 107–114.
10. S. F. Frazier, G. S. Larsen, D. Neff, L. Quimby, M. Carney, R. A. DiCaprio and S. N. Zill, *J. Comp. Physiol., A*, 1999, **185**, 157–172.
11. M. Burrows, *J. Exp. Biol.*, 2011, **214**, 2362.
12. D. H. Ardell and S. O. Andersen, *Insect Biochem. Mol. Biol.*, 2001, **31**, 965–970.
13. G. Qin, X. Hu, P. Cebe and D. L. Kaplan, *Nat. Commun.*, 2012, **3**, 1003.
14. R. Balu, J. P. Mata, R. Knott, C. M. Elvin, A. J. Hill, N. R. Choudhury and N. K. Dutta, *J. Phys. Chem. B*, 2016, **120**, 6490–6503.
15. N. K. Dutta, M. Y. Truong, S. Mayavan, N. Roy Choudhury, C. M. Elvin, M. Kim, R. Knott, K. M. Nairn and A. J. Hill, *Angew. Chem. Int. Ed.*, 2011, **50**, 4428–4431.
16. J. L. Whittaker, N. K. Dutta, R. Knott, G. McPhee, N. H. Voelcker, C. Elvin, A. Hill and N. R. Choudhury, *Langmuir*, 2015, **31**, 8882–8891.
17. R. E. Lyons, E. Lesieur, M. Kim, D. C. C. Wong, M. G. Huson, K. M. Nairn, A. G. Brownlee, R. D. Pearson and C. M. Elvin, *Protein Eng., Des. Sel.*, 2007, **20**, 25–32.
18. G. Qin, S. Lapidot, K. Numata, X. Hu, S. Meirovitch, M. Dekel, I. Podoler, O. Shoseyov and D. L. Kaplan, *Biomacromolecules*, 2009, **10**, 3227–3234.
19. G. Qin, A. Rivkin, S. Lapidot, X. Hu, I. Preis, S. B. Arinus, O. Dgany, O. Shoseyov and D. L. Kaplan, *Biomaterials*, 2011, **32**, 9231–9243.
20. R. E. Lyons, D. C. C. Wong, M. Kim, N. Lekieffre, M. G. Huson, T. Vuocolo, D. J. Merritt, K. M. Nairn, D. M. Dudek, M. L. Colgrave and C. M. Elvin, *Insect Biochem. Mol. Biol.*, 2011, **41**, 881–890.

21. A. M. Tamburro, S. Panariello, V. Santopietro, A. Bracalello, B. Bochicchio and A. Pepe, *ChemBioChem*, 2010, **11**, 83–93.
22. L. Li, Z. Tong, X. Jia and K. L. Kiick, *Soft Matter*, 2013, **9**, 665–673.
23. F. G. Quiroz and A. Chilkoti, *Nat. Mater.*, 2015, **14**, 1164–1171.
24. X. Hu, X.-X. Xia, S.-C. Huang and Z.-G. Qian, *Biomacromolecules*, 2019, **20**, 3283–3293.
25. M. Y. Truong, N. K. Dutta, N. R. Choudhury, M. Kim, C. M. Elvin, K. M. Nairn and A. J. Hill, *Biomaterials*, 2011, **32**, 8462–8473.
26. A. V. Vashi, J. A. M. Ramshaw, V. Glattauer, C. M. Elvin, R. E. Lyons and J. A. Werkmeister, *Biofabrication*, 2013, **5**, 035005.
27. M. B. Charati, J. L. Ifkovits, J. A. Burdick, J. G. Linhardt and K. L. Kiick, *Soft Matter*, 2009, **5**, 3412–3416.
28. S. Lv, D. M. Dudek, Y. Cao, M. M. Balamurali, J. Gosline and H. Li, *Nature*, 2010, **465**, 69–73.
29. A. Bracalello, V. Santopietro, M. Vassalli, G. Marletta, R. Del Gaudio, B. Bochicchio and A. Pepe, *Biomacromolecules*, 2011, **12**, 2957–2965.
30. R. E. Lyons, C. M. Elvin, K. Taylor, N. Lekieffre and J. A. M. Ramshaw, *Biotechnol. Bioeng.*, 2012, **109**, 2947–2954.
31. J. N. Renner, K. M. Cherry, R. S. C. Su and J. C. Liu, *Biomacromolecules*, 2012, **13**, 3678–3685.
32. S.-C. Huang, Z.-G. Qian, A.-H. Dan, X. Hu, M.-L. Zhou and X.-X. Xia, *ACS Biomater. Sci. Eng.*, 2017, **3**, 1576–1585.
33. A. Griffo, H. Hähl, S. Grandthyll, F. Müller, A. Paananen, M. Ilmén, G. R. Szilvay, C. P. Landowski, M. Penttilä, K. Jacobs and P. Laaksonen, *ACS Omega*, 2017, **2**, 6906–6915.
34. C. L. McGann, R. E. Dumm, A. K. Jurusik, I. Sidhu and K. L. Kiick, *Macromol. Biosci.*, 2016, **16**, 129–138.
35. Y. Kim, J. N. Renner and J. C. Liu, *Biomater. Sci.*, 2014, **2**, 1110–1119.
36. Y. Kim and J. C. Liu, *Biomater. Sci.*, 2016, **4**, 1761–1772.
37. I. Weitzhandler, M. Dzuricky, I. Hoffmann, F. Garcia Quiroz, M. Gradzielski and A. Chilkoti, *Biomacromolecules*, 2017, **18**, 2419–2426.
38. J. Cappello, J. Crissman, M. Dorman, M. Mikolajczak, G. Textor, M. Marquet and F. Ferrari, *Biotechnol. Prog.*, 1990, **6**, 198–202.
39. R. Balu, J. Whittaker, N. K. Dutta, C. M. Elvin and N. R. Choudhury, *J. Mater. Chem. B*, 2014, **2**, 5936–5947.
40. R. S. C. Su, Y. Kim and J. C. Liu, *Acta Biomater.*, 2014, **10**, 1601–1611.
41. M. Kim, C. Elvin, A. Brownlee and R. Lyons, *Protein Expression Purif.*, 2007, **52**, 230–236.
42. R. L. DiMarco and S. C. Heilshorn, *Adv. Mater.*, 2012, **24**, 3923–3940.
43. F. Luo, Z.-G. Qian and X.-X. Xia, *Polymers*, 2018, **10**, 915.
44. W. Fang, Nonappa, M. Vitikainen, P. Mohammadi, S. Koskela, M. Soikkeli, A. Westerholm-Parvinen, C. P. Landowski, M. Penttilä, M. B. Linder and P. Laaksonen, *Colloids Surf., B*, 2018, **171**, 590–596.
45. S. O. Andersen, *Insect Biochem. Mol. Biol.*, 2010, **40**, 541–551.
46. S. Rauscher, S. Baud, M. Miao, F. W. Keeley and R. Pomès, *Structure*, 2006, **14**, 1667–1676.

47. V. N. Uversky, J. R. Gillespie and A. L. Fink, *Proteins: Struct., Funct., Bioinf.*, 2000, **41**, 415–427.
48. R. Balu, N. K. Dutta, N. R. Choudhury, C. M. Elvin, R. E. Lyons, R. Knott and A. J. Hill, *Acta Biomater.*, 2014, **10**, 4768–4777.
49. S. Roberts, M. Dzuricky and A. Chilkoti, *FEBS Lett.*, 2015, **589**, 2477–2486.
50. C. I. Branden and J. Tooze, *Introduction to Protein Structure*, CRC Press, 2012.
51. N. J. Greenfield, *Nat. Protoc.*, 2006, **1**, 2876–2890.
52. K. Nakanishi, N. Berova and R. W. Woody, *Circular Dichroism: Principles and Applications*, Wiley, 1994.
53. R. E. Lyons, K. M. Nairn, M. G. Huson, M. Kim, G. Dumsday and C. M. Elvin, *Biomacromolecules*, 2009, **10**, 3009–3014.
54. L. B. Chemes, L. G. Alonso, M. G. Noval and G. de Prat-Gay, in *Intrinsically Disordered Protein Analysis: Volume 1, Methods and Experimental Tools*, ed. V. N. Uversky and A. K. Dunker, Humana Press, Totowa, NJ, 2012, pp. 387–404.
55. R. Balu, R. Knott, N. P. Cowieson, C. M. Elvin, A. J. Hill, N. R. Choudhury and N. K. Dutta, *Sci. Rep.*, 2015, **5**, 10896.
56. R. Balu, L. Bourgeois, C. M. Elvin, A. J. Hill, N. R. Choudhury and N. K. Dutta, *J. Mater. Chem. B*, 2015, **3**, 6580–6586.
57. B. Bochicchio, A. Pepe and A. M. Tamburro, *Chirality*, 2008, **20**, 985–994.
58. H. Yang, S. Yang, J. Kong, A. Dong and S. Yu, *Nat. Protoc.*, 2015, **10**, 382–396.
59. L. M. Miller, M. W. Bourassa and R. J. Smith, *Biochim. Biophys. Acta, Biomembr.*, 2013, **1828**, 2339–2346.
60. E. B. Gibbs, E. C. Cook and S. A. Showalter, *Arch. Biochem. Biophys.*, 2017, **628**, 57–70.
61. J. Cavanagh, N. J. Skelton, W. J. Fairbrother, M. Rance and A. G. Palmer, *Protein NMR Spectroscopy: Principles and Practice*, Elsevier Science, 2010.
62. K. Grudziąż, A. Zawadzka-Kazimierczuk and W. Koźmiński, *Methods*, 2018, **148**, 81–87.
63. R. Konrat, *J. Magn. Reson.*, 2014, **241**, 74–85.
64. K. M. Nairn, R. E. Lyons, R. J. Mulder, S. T. Mudie, D. J. Cookson, E. Lesieur, M. Kim, D. Lau, F. H. Scholes and C. M. Elvin, *Biophys. J.*, 2008, **95**, 3358–3365.
65. H. Zhang, S. Neal and D. S. Wishart, *J. Biomol. NMR*, 2003, **25**, 173–195.
66. L. J. Smith, K. A. Bolin, H. Schwalbe, M. W. MacArthur, J. M. Thornton and C. M. Dobson, *J. Mol. Biol.*, 1996, **255**, 494–506.
67. D. I. Svergun, M. H. J. Koch, P. A. Timmins and R. P. May, *Small Angle X-Ray and Neutron Scattering from Solutions of Biological Macromolecules*, OUP Oxford, 2013.
68. L. A. Feigin and D. I. Svergun, *Structure Analysis by Small-Angle X-Ray and Neutron Scattering*, Springer US, 2013.
69. H. Schnablegger and Y. Singh, *The SAXS Guide: Getting Acquainted with the Principles*, Anton Paar GmbH, Austria, 3rd edn, 2013.
70. O. Glatter and O. Kratky, *Small Angle X-ray Scattering*, Academic Press, 1982.

71. P. Bernadó and D. I. Svergun, *Mol. BioSyst.*, 2012, **8**, 151–167.
72. P. Bernadó, *Eur. Biophys. J.*, 2010, **39**, 769–780.
73. A. G. Kikhney and D. I. Svergun, *FEBS Lett.*, 2015, **589**, 2570–2577.
74. H. Liu and P. H. Zwart, *J. Struct. Biol.*, 2012, **180**, 226–234.
75. R.-B. Veronique and D. Dominique, *Curr. Protein Pept. Sci.*, 2012, **13**, 55–75.
76. J. E. Kohn, I. S. Millett, J. Jacob, B. Zagrovic, T. M. Dillon, N. Cingel, R. S. Dothager, S. Seifert, P. Thiyagarajan, T. R. Sosnick, M. Z. Hasan, V. S. Pande, I. Ruczinski, S. Doniach and K. W. Plaxco, *Proc. Natl. Acad. Sci. U. S. A.*, 2004, **101**, 12491–12496.
77. P. Narang, K. Bhushan, S. Bose and B. Jayaram, *Phys. Chem. Chem. Phys.*, 2005, **7**, 2364–2375.
78. G. Tria, H. D. T. Mertens, M. Kachala and D. I. Svergun, *IUCrJ*, 2015, **2**, 207–217.
79. R. J. Mart, R. D. Osborne, M. M. Stevens and R. V. Ulijn, *Soft Matter*, 2006, **2**, 822–835.
80. K. M. Ruff, S. Roberts, A. Chilkoti and R. V. Pappu, *J. Mol. Biol.*, 2018, **430**, 4619–4635.
81. J. Li, R. Rajagopalan and J. Jiang, *J. Chem. Phys.*, 2008, **128**, 235104.
82. Y. Zhang and P. S. Cremer, *Curr. Opin. Chem. Biol.*, 2006, **10**, 658–663.
83. M. Y. Truong, N. K. Dutta, N. R. Choudhury, M. Kim, C. M. Elvin, A. J. Hill, B. Thierry and K. Vasilev, *Biomaterials*, 2010, **31**, 4434–4446.
84. T. Truong, R. Bersohn, P. Brumer, C. K. Luk and T. Tao, *J. Biol. Chem.*, 1967, **242**, 2979–2985.
85. S. Mayavan, N. K. Dutta, N. R. Choudhury, M. Kim, C. M. Elvin and A. J. Hill, *Biomaterials*, 2011, **32**, 2786–2796.
86. R. Balu, R. Knott, C. M. Elvin, A. J. Hill, N. R. Choudhury and N. K. Dutta, *Biosensors*, 2019, **9**, 128.
87. A. Baszkin and W. Norde, *Physical Chemistry of Biological Interfaces*, Taylor & Francis, 1999.
88. N. K. Dutta, N. R. Choudhury, M. Y. Truong, M. Kim, C. M. Elvin and A. J. Hill, *Biomaterials*, 2009, **30**, 4868–4876.
89. E. A. Vogler, *Biomaterials*, 2012, **33**, 1201–1237.
90. R. Kappiyoor, G. Balasubramanian, D. M. Dudek and I. K. Puri, *Soft Matter*, 2011, **7**, 11006–11009.
91. N. Habibi, N. Kamaly, A. Memic and H. Shafiee, *Nano Today*, 2016, **11**, 41–60.
92. R. Balu, N. K. Dutta, A. K. Dutta and N. R. Choudhury, *Nat. Commun.*, 2021, **12**, 149.
93. I. Preis, M. Abramson and O. Shoseyov, *Mol. Biotechnol.*, 2018, **60**, 310–318.
94. R. D. Pedde, B. Mirani, A. Navaei, T. Styan, S. Wong, M. Mehrali, A. Thakur, N. K. Mohtaram, A. Bayati, A. Dolatshahi-Pirouz, M. Nikkhah, S. M. Willerth and M. Akbari, *Adv. Mater.*, 2017, **29**, 1606061.
95. J. I. Andorko and C. M. Jewell, *Bioeng. Transl. Med.*, 2017, **2**, 139–155.

CHAPTER 6

Advanced Biomedical Applications of Self-assembled Elastin-like Recombinamer Structures

J. C. RODRÍGUEZ-CABELLO,* CARMEN GARCÍA-ARÉVALO AND LUIS QUINTANILLA-SIERRA

BIOFORGE (Group for Advanced Materials and Nanobiotechnology), CIBER-BBN, University of Valladolid, 47011 Valladolid, Spain
*Email: roca@bioforge.uva.es

6.1 Introduction

Several years ago, the role and durability of elastin in the native extracellular matrix (ECM) of tissues subjected to repetitive deformations sparked interest in this unique behavior for exploitation in the field of materials science. Shortly thereafter, the development of genetic engineering techniques that yield high levels of elastin analogs, now known as elastin-like recombinamers (ELRs), transformed the field of biomaterials and their potential applications in biomedicine, where ELRs have generally shown both versatile and suitable properties for such tasks. The recent explosion of research into ELRs as intrinsically disordered proteins (IDPs) has generated a renewed interest in understanding the mechanism governing their self-assembly (SA), which may improve our ability to predict the tendency of a particular sequence to form multiple structures in

Biomaterials Science Series No. 10
Biomimetic Protein Based Elastomers: Emerging Materials for the Future
Edited by Namita Roy Choudhury, Julie C. Liu and Naba K. Dutta

Published by the Royal Society of Chemistry, www.rsc.org

response to biological signals. This chapter brings together some of the basic aspects that govern SA and translates them to this family of genetically engineered protein-based polymers (ELRs) and their most prominent biomedical applications.

6.2 Elastin-like Recombinamers (ELRs): Definition, Description, and Characteristics

ELRs are a family of genetically engineered protein-based polymers whose composition is the result of an engineered design step and the creation of a synthetic gene based on the sequence of natural elastin.[1,2] The most well-known members within the ELR family are based on the pentapeptide VPGVG sequence (or its permutations), which is present in the mammalian elastic protein elastin, with some of their main characteristics being derived from those of the natural protein.[1]

Natural elastin is a hydrophobic protein belonging to the fibrous component of ECM macromolecules embedded in the highly hydrated connective tissue that is critical to many vertebrate tissues.[3] It is involved both in providing specific mechanical properties and in cellular communications based on biochemical and physical signals that determine the cellular fate and define the biological properties of the tissue as a whole. In some tissues, such as the large arterial blood vessels, lung parenchyma, elastic ligaments, and skin, elastin is the major component and provides the extensibility and elastic recoil necessary for their function as they are subjected to constant and reversible physiological deformations.[4]

As is the case for many structural proteins, elastin is synthesized as a monomeric precursor, known as tropoelastin, which is encoded by a single gene in humans that translates into 786 amino acids over 35 exons, and it is secreted as a ~60 kDa nonglycosylated protein.[5] This precursor subsequently assembles and stabilizes into a stable polymeric structure in the ECM by covalent crosslinking. The principal function of this matrix is to extend and contract in repetitive motion when hydrated.[5,6] The functional parts of tropoelastin include hydrophobic regions that specialize in elasticity, association, assembly, and cell attachment, alternating with hydrophilic crosslinking domains containing predominantly alanine and lysine residues.[5,7]

Glycine, valine, proline, and alanine residues dominate (>80%) over hydrophobic domains and are frequently arranged into motifs based on combinations of the sequence elements PGV, GVA, GV, and GGV.[8] These arrangements are highly conserved in elastins from all species.[9] Urry's early work in this field showed that three repeating peptide sequences occur in tropoelastin: a tetrapeptide (VPGG), a pentapeptide (VPGVG), and a hexapeptide (VAPGVG),[10] the latter motif being found in domain 24, the most common and longest repeat element in human tropoelastin.[11] However, there is no obvious single protein sequence that confers elasticity and several studies have attempted to establish the minimal sequence requirements for elastin-like derivatives and how the sequence and structure of elastin contribute to these unusual properties.[4]

The modular nature of tropoelastin's sequence, and particularly the content and context of proline and glycine residues, undoubtedly seems to play a role in determining the properties and SA behavior of both the monomeric and polymeric forms of the protein.[8] In accordance with such requirements, the classical pentapeptide consensus sequence for elastin (VPGVG) has served as a model for many other elastin-based materials, such as the recombinantly produced ELRs.[9,12]

The fourth amino acid (valine [Val]) in the consensus repeat unit $(VPGVG)_n$ has been generalized as $(VPGXG)_n$, where the fourth residue "X" is known as the "guest residue". This position can host any amino acid except proline and determines the physicochemical properties of the recombinamer, thus making it a key parameter during ELR design.[13,14] Other substitutions, such as those involving glycine for alanine in the third position,[15,16] and those that change the amino acid order,[17] can influence their function, mechanical properties, and phase behaviors.

All functional ELRs undergo a liquid–liquid phase separation in response to different stimuli, which can be used as a pathway to material assembly. The phase separation of elastin and ELRs exhibits a temperature dependence that results in the self-association of disordered hydrophobic domains, with a concomitant increase in solvent entropy, in a process known as "coacervation", inverse temperature transition (ITT), or lower critical solution temperature (LCST) phase transition behavior.[1,18,19] In aqueous solution and below a critical temperature (transition temperature, T_t), the polymer chains remain fully hydrated and shielded by stable hydrogen-bonded clathrate water structures packing the elastin-like molecules, thus preventing their self-aggregation. Above T_t, however, the clathrate water structure breaks down, thus resulting in exposure of the hydrophobic domains, which assemble to form a phase-separated state containing 63% water and 37% polymer by weight (Figure 6.1). The apparent increase in order upon raising the temperature is a thermodynamically favorable and entropically driven process because the increased order of the ELR aggregates is more than offset by the decreased order of the surrounding water.[2,20]

The highly cooperative and intermolecular nature of the coacervation process is modulated by the sequence of the ELR, which also modulates the growth into different assemblies, stability, interactions, reversibility of the process, and the mechanical properties of the resulting elastin-based material, which in turn provides a unique approach for material design.[19] However, the amino acid composition is not the only factor that can affect the LCST behavior: other extrinsic factors, such as ionic strength,[21,22] pH of the solution,[23] the presence of cosolvents,[24] the concentration of the polypeptide,[25] the number and arrangement of hydrophobic domains,[2,26,27] *etc.* may also have an influence.

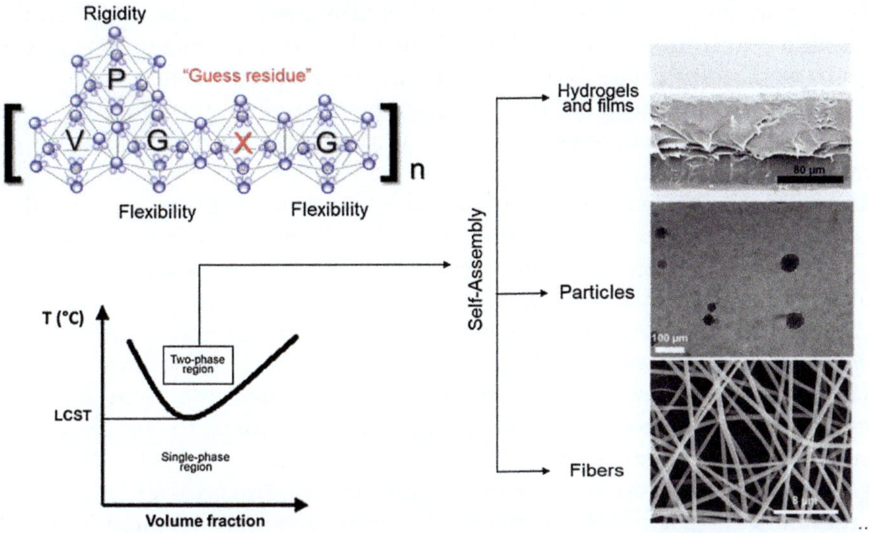

Figure 6.1 Representation of the consensus repeat pentapeptide $(VPGXG)_n$ that acts as an ELR building block, surrounded by a hydrophobic hydration cover, where X represents the guest residue and n is the number of tandem repeats. The chemical composition, chain length, and distribution of the building blocks in the ELR condition the thermally triggered lower critical solution temperature (LCST) phase transition in aqueous solution and secondary structure formation after self-assembly.

6.3 Self-assembly in ELRs

6.3.1 Concept of Self-assembly

According to Whitesides and Grzybowski, SA is defined as the "autonomous organization of components into patterns or structures",[28] with a lower order being found in the isolated components than in the self-assembled system.[29]

Although SA processes are observed on all scales, herein, we will focus our attention on SA at the nanoscale. In this case, the size of self-assembled structures usually ranges from a few tens to hundreds of nanometers, which is of noticeable interest in the field of nanotechnology. More specifically, we will concentrate on SA in the field of chemistry and materials science. Beyond the limits of molecular chemistry (based on the covalent bond), in the field of supramolecular chemistry, *i.e.*, "chemistry beyond the molecule", Lehn defined molecular SA as "pattern-formation processes tending toward equilibrium".[30–33]

Bilayers, micelles, vesicles, tubular structures, liquid crystal phases, and Langmuir monolayers are different morphologies that have been reported to be suitable for the formation of supramolecular assemblies[34,35] using atoms and molecules, together with different structures with diverse chemical compositions, shapes, and functionalities, as building blocks or "bricks".

Several noncovalent interactions (such as capillarity; π–π stacking; van der Waals forces; electrostatic, dipolar, hydrophobic, and hydrogen bonds) lead to self-assembled structures, with these interactions acting on a strictly local level. Thus, the structure builds itself and the features of the self-assembled structure (*e.g.*, size, geometry, *etc.*) are conditioned by those of the individual components and their interaction.[36,37]

The strength of covalent interactions is typically more energetic than that of noncovalent ones, with covalent bonds typically having a strength of 40–100 kcal mol^{-1} and van der Waals and hydrogen bonds 0.1–5 kcal mol^{-1}.[38] Several dynamic characteristics of self-assembled systems, including their inherent ability to self-heal and error correction, reversibility, and significant sensitivity to perturbations coming from the external environment, are associated with the magnitude of these interactions. Thus, these weak interactions are able to rearrange and adapt to the landscape around them.[39] Moreover, they must overcome interactions that could break the system up in order to achieve stability *via* synergism over large areas of the molecular surface of the building blocks and cooperativity.[38,40] According to this idea, thermodynamic analysis provides some hints about the desirable features that the individual units should provide. Thus, since the SA process occurs in the absence of external forces, the Gibbs free energy must be minimized to ensure that the self-assembled structures are thermodynamically more stable than the isolated components. Consequently, energy minimization drives the process and a thermodynamic equilibrium between the self-assembled system and its components is reached.[41]

The subtle balance between enthalpy (ΔH) and entropy (ΔS) becomes a key detail in the discussion.[42] Indeed, although a wide range of enthalpies is found for the different molecular interactions taking place in SA structures, a magnitude of around 2–20 kcal nm^{-2} is typical.[42] From the viewpoint of the formation entropy, two contributions are considered. First, when two particles come together at dilute concentrations, the contribution to the Gibbs free energy of the loss in translational entropy is around +5.5 kcal mol^{-2}. Secondly, the integration of many contributions to the loss in conformational entropy is about +0.7 kcal mol^{-1} as a result of freezing conformations around bonds in a large and flexible molecule. As such, rigidity and multiple contacts are considered to be interesting features of the "bricks" in the self-assembled structure.[43] Whitesides and Grzybowski[28] also proposed that two types of SA can be considered depending on whether or not energy is dissipated for the SA structure. The former is considered to be a "dynamic" SA, whereas the latter corresponds to a "static" SA. Oscillating chemical reactions and most folded, globular proteins correspond, respectively, to examples of these two types of SA.[44,45] However, Halley and Winkler suggest that the dynamic category actually corresponds to self-organization in the biology domain.[29,46]

Two additional variants of SA, namely "template" SA, where the formed structures follow the boundaries or template in the environment of the components, and "biological" SA, were also proposed by Whitesides and Grzybowski.[28]

In this chapter, we will focus our interest exclusively on self-assembled structures formed from materials based on ELRs and intended for biomedical applications.

6.3.2 Morphologies of Self-assembled ELR Structures

Four kinds of morphologies, namely nanoparticles, fibers, hydrogels, and other structures, are considered.

6.3.2.1 Nanoparticles

Two categories can be distinguished depending on whether the self-assembled nanostructure is based on only ELRs (single component) or the combination of ELRs with other materials (multicomponent or hybrid nanoparticles).

3.2.1.1 Single-component Nanoparticles. Elastin-like corecombinamers are obtained when combining two or more different ELR blocks. An amphiphilic diblock is the simplest design to give rise to self-assembled nanoparticles. Above the hydrophobic block transition temperature, hydrophobic and hydrophilic blocks arrange at the core and corona of the micelle, respectively. In addition, phase aggregates appear as a result of micellar collapse above the hydrophilic block transition temperature. The critical micelle temperature and micelle aggregation temperature are determined by the hydrophobic block and the hydrophilic corona, respectively.[47]

The first self-assembled ELR nanostructure was reported by Conticello *et al.*[48] Specifically, this group designed an amphiphilic diblock, including a cationic block and a phenylalanine-containing block. Investigation of aqueous ELR solutions by dynamic light scattering indicated the presence of micelles (diameter of around 50 nm), which formed cylindrical micellar aggregates as the temperature increased.

The impact of some design parameters at the molecular level (*e.g.*, identity, sequence, molecular weight, and length of each individual block comprising the ELR corecombinamer and the hydrophilic-to-hydrophobic ratio) is summarized by describing the nanostructures obtained.

The effect of a macromolecular architecture was studied by Wright and Conticello by varying the identity and proportion of blocks forming a corecombinamer,[49] specifically, amphiphilic di- (AB) and triblocks (ABA or BAB) (A and B are hydrophilic and hydrophobic blocks, respectively). The effect of concentration was analyzed at ambient temperature under basic conditions. These authors found that, whereas dilute aqueous solutions of the diblock polymer generate spherical nanoparticles (hydrodynamic diameter of 50–90 nm) due to collapse of the hydrophobic block, aggregation of particles, which leads to different micellar phases (worm-like micelles and micellar clusters), predominates at higher concentrations.

The effect of individual block molecular weight and the hydrophilic-to-hydrophobic ratio was studied by Dreher *et al.*,[50] who found that, when this ratio is varied between 1 : 2 and 2 : 1 in several linear amphiphilic diblock ELRs, the polymer self-assembles into monodisperse spherical micelles (hydrodynamic radius of around 30–40 nm) triggered by a mild hyperthermia (37–42 °C).

Gradients between the hydrophilic and hydrophobic blocks have also been studied recently by MacEwan *et al.*[51] Thus, for a diblock corecombinamer, the size and shape of micellar nanostructures are modulated by the incorporation of an interface block, and the micellar hydrodynamic radius is proportional to the molecular weight of the corecombinamer blocks.

Several block arrangements and lengths have been designed by Rodríguez-Cabello *et al.*[26] In particular, three architectures (E50A40, E100A40, and E50A40E50) with a hydrophobic-to-hydrophilic length ratio of more than 1 : 2, in which the E-block and A-block are a glutamic acid-containing block and a fixed-size block containing an alanine residue, respectively, were designed. Above the transition temperature of the hydrophobic block, spherical micelles (for E50A40) and vesicles (for E50A40E50) with a hydrodynamic radius ranging from 75 to 100 nm were observed. The incorporation of different salts (NaCl, KCl, and Na_2HPO_4) at different concentrations (for NaCl)[22] showed that, although the morphologies were maintained, a marked size decrease was observed in all cases, with hydrodynamic diameters ranging from 40 to 80 nm. This experimental result suggests that the arrangement of hydrophobic blocks is more compact.

Finally, a quantitative approach has been used[47,52] to systematize the relationship between the transition temperature and key design parameters at a molecular level. Thus, Janib *et al.*[52] synthesized an exhaustive library of mono- and diblock ELRs with a hydrophobic-to-hydrophilic ratio of 1 : 1. Their mathematical model paid special attention to the impact of the molecular weight of the individual blocks in the co-ELR on the self-assembled structures. Several conclusions were reached based on this model, as highlighted below:

- a minimum molecular weight of the individual ELRs of around 30 kDa is required to obtain stable structures; and
- the length of the hydrophobic block plays an essential role in stabilizing the nanoparticle, with a minimum length of 48 pentamers being required.

3.2.1.2 Hybrid Nanoparticles. The SA of different blocks is required in order to get closer to the complexity level typically found in nature. In this regard, Okesola and Mata[38] have reviewed multicomponent self-assembled systems using proteins and peptides as individual blocks. With regard to ELRs only, covalent conjugation and a solution at the gene level are two strategies for preparing these hybrid nanostructures.[38] An additional approach based on physical interactions will be considered in Section 6.5.2.

- ***Chemical Conjugation***

  Covalent bonding has been achieved by click chemistry using Cu and an ELR and a collagen-like peptide modified with alkynes and azides, respectively (Luo and Kiick[53]). A significant decrease in the ELR transition temperature was observed and found to be modulated by helical peptide formation. The nanostructures obtained are vesicles with an average diameter of about 80–100 nm.
- ***Genetically Encoded Hybrid ELRs***

  One of the smallest ELR-based nanostructures reported in the literature is a three-armed star-shaped micelle with a radius of 10–15 nm, which was obtained under low-salt conditions.[54] A negatively charged peptide including 27 amino acids was attached to the C-terminal of an ELR and, at temperatures above the transition temperature, the foldon domain folds and the aforementioned stable nanostructures were observed.

  Asymmetric amphiphilic ELRs consisting of an ELR genetically fused to $(XGG)_8$ and $(XG_y)_8$, where X and *y* represent the hydrophobic amino acid and the number of glycine spacers, respectively, have been designed.[55] Cylindrical micelles with a core radius of 21–23 nm and a length of around 170 nm (from small-angle neutron scattering measurements) were observed. Two conclusions were reached: first, the design parameter *y* strongly controls the SA process, and second, π–π stacking may participate in the hydrophobic interactions in the core of the nanostructure as the amino acids used (X = Tyr, Phe, and Trp) contain aromatic groups.

  As a final example, Xia *et al.*[56] have designed a monomer based on the combination of silk-like motifs alternating with an ELR segment. Micelle formation is observed in dilute aqueous solution, and both the SA process and the micelle radius are modulated by the silk-to-elastin ratio. Above the ELR transition temperature, micelles (with a radius of a few tens of nanometers) arrange hierarchically in spherical nanostructures with a hydrodynamic radius of about 220 nm.

6.3.2.2 Fibers

A similar behavior to elastin is found in ELRs, given that both contain the same supramolecular structures.[57] For example, fiber-like structures (with a radius of about 2–2.5 nm) are found in the solution for the pentapeptide (VPGVG). These fibers are modeled following the β-spiral model[58] and the combination thereof results in twisted filaments. The Gly-Gly amino acid sequence in ELRs such as (VGGVG)[59] or (VGGLG)[60] also gives rise to fibrils organized into filaments.

SA into amyloid-like fibers has also been reported in the literature,[61] with some characteristics at the amino acid level being found to induce the formation of this kind of fiber. These include the specific amino acid sequence, where the motif (XGGZG) was identified (X, Z = V, A, or L),[62] a length of at least 15 amino acids,[63] and the combination of proline and glycine.[8] Trifluoroethanol has also been suggested to be a suitable solvent for forming these types of fibers.[64]

The conjugation of diblock copolymers has also been used to form fibers. Thus, at 45 °C, aqueous solutions of proline-rich and glycine-rich polypeptides self-assemble into nanoparticles that eventually generate beaded fibers (diameter of about 20–70 nm, length of more than 10 μm) after 7 days.[65]

6.3.2.3 Hydrogels

Collaborative physical interactions are the foundation of ELR hydrogels despite the weakness of these interactions when considered individually. As such, physical crosslinking can be obtained in ELRs following different approaches. Firstly, in the guest position of the pentamer ELR structure itself, by way of ionic interactions between amino acids with an opposite electrical charge, for instance, glutamic acid and lysine side chains.[66] Hydrogels can also be obtained in the presence of periodic cysteine residues as a result of intermolecular disulfide crosslinking.[67] Secondly, in the case of diblock corecombinamers, the repulsion between amphiphilic (hydrophilic and hydrophobic) blocks with opposite polarity results in local segregation, thus resulting in periodic nanostructures.[68,69] With regard to triblock (hydrophobic–hydrophilic–hydrophobic) copolymers,[49,70] the hydrogels reported are based on micellar aggregates that operate as effective crosslinks when concentrated aqueous solutions are used.

Polymer concentration has also been used to transform the morphology of self-assembled structures. Thus, spherical micelles observed in dilute aqueous solution (with an average hydrodynamic radius of 50–100 nm) form a lyotropic gel at higher concentrations.[71]

Finally, silk-like[72] or leucine zipper[73] motifs have been used to implement physical crosslinking. For the silk-like motif, a complex gelation process has been suggested to occur as the temperature increases, with an interplay between the ELR and the silk domain and the silk blocks undergoing annealing and forming organized structures (β-sheets).[74] With regard to zipper motifs, coiled-coil dimers[75,76] are found to be related to both ionic and hydrophobic interactions within the zipper block. The usual arrangement of blocks consists of a triblock, with the leucine zippers being disposed at both ends.[73] Given the reported biocompatibility of these hydrogels, they may find a role in biomedical applications.[77]

6.3.2.4 Other Structures

A broad range of sophisticated structures, including ELR–pDNA polyplexes,[78] hollow spheres with diameter sizes from 100 to 1000 nm,[79] multilayer assemblies obtained by using the layer-by-layer approach,[80] a plasmonic matrix consisting of polypeptides interfaced with gold nanorods,[81] an elastin-mimetic dendrimer where a peptide was attached to polyamidoamine dendrimers,[82] and graphene–ELR nanocomposites,[83] have been produced taking advantage of the conformational changes in ELRs induced by environmental stimuli.

6.4 ELRs as Intrinsically Disordered Proteins and Their Impact on Self-assembled Structures: Concept and Characteristics

6.4.1 Concept of Intrinsic Protein Disorder

IDPs are defined as those that possess no well-defined 3D structure, either entirely or in part, but are still functional.[84]

Although intrinsic disorder is abundant in nature, the amount of disordered proteins in multicelled organisms differs significantly between organisms and genomes, with the level of disorder increasing with their complexity.[85] For example, half of the eukaryotic proteome contains small sequences predicted to contain disordered regions of 40 consecutive residues or more under physiological conditions.[86]

IDPs are important components of the cell signaling machinery and are involved in regulating a wide range of cellular processes, such as transduction signaling and regulation, transcription, differentiation, protein transport, mitosis, and meiosis.[87,88]

The terms "natively denatured", "natively unfolded", "intrinsically unstructured", or "intrinsically disordered" were gradually introduced in the literature in the 1990s. These almost synonymous terms rapidly expanded to emphasize the existence of structural differences compared with natively folded proteins, the existence of which was one of the most fruitful paradigms of protein science (protein structure determines function), now being questioned.[89]

Although the molecular 3D structure of a protein has been fundamentally tied to its function, the structure is important not only in terms of what it is but also in how it changes.[90] IDPs exist as dynamic ensembles of conformations in solution, which undergo temporal fluctuations that lack specific equilibrium values. Thus, IDPs are not fully disordered, and the intrinsic disorder exhibits a number of function-related transient short- and long-range structural organization effects covering a spectrum of states from fully unstructured or random-coil-like to partially structured in the form of (pre-)molten globules.[91] A single IDP can display several ensembles and one of the main aspirations in biology is to interpret how they function without a well-defined structure and to determine how changes in structure or binding affinities translate into function.[92,93]

Despite numerous databases and the availability of over 60 disorder prediction algorithms, the absence of standards and references for comparisons that predict intrinsic disorder makes it difficult to establish homogeneous guidelines for interpreting these predictions.[94] Furthermore, the plasticity of IDPs (in other words, their ability to bind to several targets and adopt different conformations on different targets) makes it even more difficult to understand the intrinsic disorder recognition process as there are no general mechanisms for the binding-induced folding of IDPs.[95] Moreover, folding coupled to the binding of IDPs is not always complete in the presence of a partner. This "fuzziness", in which the protein remains partially disordered in the bound state, makes it even more difficult to resolve IDP structure predictions.[96]

Several determinants strongly correlate with intrinsic disorder, with the primary structure or amino acid sequence being the main determinant for the absence of compaction in IDPs.[97] IDPs tend to have a biased composition characterized by a low mean hydrophobicity and a high net charge. As such, IDPs are often enriched in charged residues (such as Lys, Glu, Ser, and Gln) and structure-breaking amino acids (such as Pro and Gly), as well as Ala. They are also usually depleted in hydrophobic (such as aromatics: Trp, Tyr, Phe) and Cys, Ile, Leu, and Asn residues. These "order-promoting amino acids", which are key to folded proteins, appear to play a secondary role in intrinsic disorder.[84,86,97,98]

IDPs that function *via* molecular recognition usually use a large proportion of their surface to interact with their partners (around 50%) *via* short amino acid segments that have been denoted in different ways depending on whether the reference is made to their primary sequence or their structure, with "short linear motifs" (SLiMs, LMs, or MiniMotifs)[99] and "molecular recognition features" (MoRFs) or "preformed structural elements" (PSEs)[100] being the most commonly used terms. These key residues are generally shared, with some conservative substitutions, across all peptides recognized by an interacting surface on a peptide-binding domain, and they supply the majority of the binding specificity and affinity.[101–103] These critical residues for interaction are usually flanked by a highly disordered sequence with signs of residual structure that became structured when complexed to their partners.[84,104,105]

The high recurrence of SLiMs/PSEs/MoRFs along extensive functional regions in IDPs counterbalances the high entropic cost/loss upon binding to an interaction partner. The native state of IDRs is not a well-defined minimum energy state with a favored conformation as they exhibit a high conformational entropy in a relatively flat but rugged energy landscape, with numerous local energy minima separated by low energetic barriers. As such, they can exist in conformationally different but energetically similar states.[106,107] Several approaches have been suggested to counteract the apparently energetically disfavored binding of IDPs to their partners, with a comparable gain in binding enthalpy being one of these proposals. However, linear motifs less than 10 amino acids in length lack extensive interfaces for maximizing binding enthalpy. As such, instead of maximizing binding enthalpy, the special composition of IDRs allows these interactions to take place over their extensive interface *via* a large number of weak noncovalent interactions (multivalent IDRs with independent binding sites), thus limiting the entropic cost of the interaction. Nonetheless, other extrinsic mechanisms are also required to justify those situations in which a more limited surface area is involved. Thus, ligands, posttranslational modifications (PTMs), alternative splicing, pH changes, and other biological, physical, and chemical factors can shift the energy landscape and conformational ensemble of IDPs, thus allowing induced (un)folding, allosteric coupling, and cooperative binding phenomena.[107]

6.4.2 ELRs as Intrinsically Disordered Proteins

Elastic proteins such as tropoelastin (the soluble precursor of elastin), insect resilin, wheat gluten, or spider silks, among others, have provided inspiration for the design of protein-based polymers since the sequence origins of disorder were readily identified during early work with these proteins.[108] These rubber-like elastomeric properties require a high level of structural disorder for entropy-driven elastic recoil.[109]

ELRs with the consensus repeat and low-complexity basic sequence (VPGXG) can be considered to be a building block for IDPs.[110] Hydration and conformational disorder are fundamental requirements for their elastomeric function and for the aqueous liquid–liquid phase separation behavior of ELRs, which, in turn, rely on polymer chemistry and solvent quality in a given state, in a complicated balance between entropic and enthalpic forces. The phase separation of ELRs and other stimulus-responsive protein polymers can be explained by three pairwise interactions that involve polymer–solvent, polymer–polymer, and solvent–solvent contacts, with the average amino acid composition and distribution affecting the transition temperature, critical concentration, and the architecture of the resulting coacervates.[97,108]

Elastomeric chains remain hydrated and disordered even after aggregation because their backbone is unable to form extensive self-interactions. This is mainly due to the existence of a combined proline (P) and glycine (G) composition threshold in elastomeric proteins, where approximately two glycines are equivalent to one proline, and the percentage of prolines is a stronger determinant than the number of glycines.[6,8,111] Both amino acids promote disorder in the polypeptide backbone of elastomers, although for opposite reasons. Thus, the cyclic structure of proline renders a more rigid conformation than other amino acids as it does not contain backbone amide hydrogen atoms at physiological pH. This explains why proline is too stiff and is unable to form stabilizing hydrogen bonds in secondary structures, thereby negatively modulating their propagation when present.[112] In contrast, glycines are too flexible to do so in the presence of water, thus determining the extent of backbone hydration in monomers.[8,109] This can be rationalized in terms of polypeptide chain entropy. In multivalent IDPs, such as the case of ELRs, there is a synergy between the relative gain in configurational entropy as a result of both (i) minimization of conformational entropic cost due to the presence of flexible regions that contribute to polypeptide chain entropy and oppose hydrophobic packing and (ii) the limited number of configurations or accessible conformations since the backbone is inherently unable to form extensive self-interactions.[107] These findings are consistent with the fact that proline and glycine residues are frequent in loops, most of which are solvent exposed, rather than in the secondary structure of globular proteins.[8,113]

Solid-state NMR spectroscopy has shown that ELRs are in a well-hydrated and hence soluble state at temperatures below T_t or T_{LCST}, with relatively well-defined type II β-turns centered at the Pro-Gly position in the polypeptide repeat sequence.[114,115] Other contemporary computational approaches and molecular dynamics (MD) simulations that are not hindered by conformational disorder have therefore been useful in developing an atomic-level description of elastin's structure, although the scope of these studies has been limited to short time scales or small oligopeptides.[8] The latter shows that a single ELR molecule, such as $(VPGVG)_{18}$, exhibits a gradual decrease in the average value of the radius of gyration (R_g) as the temperature increases in the range between 290 and 350 K, with this decrease being related to the change in interactions between chemical groups in the polypeptide and solvent–water molecules. As the temperature increases, the ELR moves toward a more ordered secondary structure in the form of β-turns and β-strands, with a significant structural heterogeneity, in which Pro and Gly residues are involved in turn formation with highly dynamic and rapid interconversions. Type II β-turn conformations appear to be present at all temperatures.[18]

These observations are inconsistent with models requiring conformationally restricted structures, such as the β-spiral proposed by Urry,[116–118] and seem to be in line with other studies showing significant amounts of polyproline II structures and β-turns emerging from circular dichroism and Raman spectroscopy.[119,120]

The ability to recombinantly produce individual hydrophobic domains, or higher elastin-derived recombinamers (ELRs), has allowed predictions of disorder and conformational flexibility to be confirmed, with the formation of extended secondary structures being restricted in favor of transient and fluctuating local motifs, including β-turns, β-strands, and polyproline II (PPII) structures. In the absence of intramolecular and intermolecular hydrogen bonds, the latter can therefore interconvert between conformations.[6,11]

6.4.3 Example

Accurate repetition of units at the amino acid level has been related to certain degrees of disorder in ELRs.[121] As such, the relationship of the ELR transition temperature to parameters such as amino acid composition, molecular weight, and concentration has been quantitatively studied by McDaniel *et al.*[122] and Meyer *et al.*[25] Focusing our interest on the composition, both alanine and valine have been selected and mixed as guest residues in order to tune the transition temperature within a physiological range (30–45 °C). For simplicity, no pH sensitivity is obtained with these guest residues. In contrast to the spherical micelles reported by Wright and Conticello in certain corecombinamers,[49] no self-assembled structure was found in the alanine/valine-based ELR, thus indicating the significant impact of the trade-off between order and disorder at the amino acid level on the SA process.

6.5 Biomedical Applications of Self-assembled ELR Structures

The inherent features that govern the functional mechanisms of IDPs can be exploited in other unstructured protein-based polymers. ELRs offer the possibility of harnessing the vast array of functions that these biopolymers display in nature in biomaterials and bioinspired materials with many different applications. ELR changes in both the primary sequence and block arrangement may influence hydration and peptide–peptide hydrogen bonding and therefore aggregation tendencies. These arrangements have been exploited to obtain drug reservoirs, cell scaffolds for tissue engineering applications, and other systems that are particularly useful in the field of biomedicine.

6.5.1 Drug Delivery

The current high prevalence of cancer has resulted in an exponential growth in research into chemotherapeutic agents for the treatment of this disease.[123] Thus, the rational design of drug-loaded nanoparticles has been studied in an effort to minimize issues related to their injection into the bloodstream, increase their efficacy and the quantity located in the desired tissue, and at the same time limit the potential damage to healthy tissue. Since cancerous tissues are characterized by local hyperthermia (37–42 °C) and anomalous extracellular pH values,[124] these two parameters are often used to trigger SA in anticancer drugs. Multifunctional nanomedicines based on micelles have also been designed.[125,126]

As an example of a single component carrier, a histidine-rich, diblock ELR copolymer proposed by Callahan *et al.*[127] was shown to self-assemble in response to temperature, pH, and cosolutes (*e.g.*, Zn^{2+} at physiological concentrations), thus giving rise to spherical micelles with a hydrodynamic radius of around 30 nm.

Several strategies have been proposed in order to obtain (hybrid) nanocarriers based on several components, one of which is an ELR: first, covalent conjugation of the drug to the ELR, thus allowing several drug molecules to be linked to one ELR chain and improving the local drug concentration and its effectiveness. Doxorubicin is commonly employed as an anticancer drug and, as such, a large number of studies concerning the treatment of solid tumors can be found in the literature (see, for instance, ref. 128 and 129).

The aim of tumor-targeting nanomedicines is the selective elimination of the cancer cells or tumor microenvironments. As a specific example, doxorubicin has been conjugated to an ELR containing a tumor-homing peptide (F3) and cysteine-rich segment at its N- and C-termini, respectively, *via* a linker. In this case, nanocarriers with a spherical shape (hydrodynamic diameter of around 50–60 nm) were observed.[130]

Drug-loaded nanostructures, including sequences that allow their incorporation into cells, have also been proposed.[131,132] In particular, short peptides have been used to achieve internalization due to physical interactions with the cellular membrane.[133] Again, doxorubicin has been conjugated to the C-terminus of an ELR, along with several different peptides at the N-terminus.[134]

A high local concentration of arginine has also been observed to give rise to internalization.[135] Using this idea, five Arg residues were inserted into a cell-penetrating peptide included in an amphiphilic corecombinamer[136] that is thermoresponsive in the mild hyperthermia temperature range characteristic of tumors in order to self-assemble particles only inside the tumor. As can be seen in Figure 6.2, the resulting ELRs are soluble unimers at 37 °C with a low hydrodynamic radius and a low Arg density. However, at 42 °C, they self-assemble into micellar nanostructures (hydrodynamic diameter of about 40 nm) and the interfacial density of Arg residues in the corona

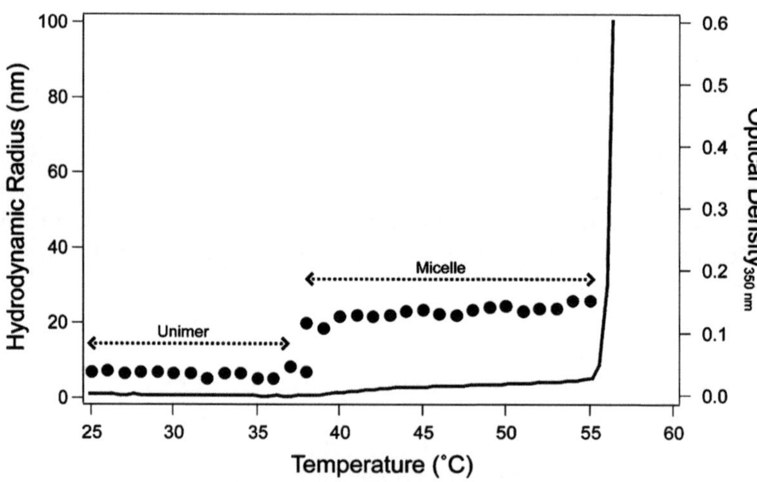

Figure 6.2 Evolution of the optical density (obtained by turbidimetry) and hydrodynamic radius (obtained by dynamic light scattering) as a function of temperature. Soluble unimers with a hydrodynamic radius of less than 10 nm are found at temperatures lower than 39 °C, whereas micelles with a radius of about 20 nm self-assemble above this temperature. Reproduced from ref. 136 with permission from American Chemical Society, Copyright 2012.

increases, thus resulting in cellular uptake of the micelles. Micelles form at temperatures of up to 56 °C; beyond this temperature, micron-size aggregates are formed. This dynamic system therefore allows smart digital control of the internalization process.

A second strategy for obtaining hybrid nanostructures involves genetic engineering. In this case, the ELR is genetically fused to another material, for example, a viral capsid protein proceeding from the cowpea chlorotic mottle virus,[137] a single-chain antibody[138] (anti-CD20 scFv: worm-like nanostructures) or an amphipathic peptide[139] (L4F: self-assembled vesicles). All the resulting structures show synergic effects between the ELR and the additional material, such as a reduction in the transition temperature of the free ELR. In the case of the virus, along with ELR-driven self-assembled particles (with a radius of about 9 nm), a pH-driven self-assembled nanostructure (with a radius of around 14 nm) was found using experimental conditions under which the ELR is unable to self-assemble. These nanocapsules could be used to control enzyme activity by switching their encapsulation or in a system intended for handling therapeutic agents that need to be protected from their surroundings.

6.5.2 Tissue Engineering

The simplest methodology for manufacturing an ELR-based tissue engineering scaffold involves raising the temperature of a concentrated solution of an ELR above its T_t, thus resulting in the formation of an insoluble coacervate phase.[140] This approach has been exploited in tissue engineering to obtain thermoresponsive surfaces that can be used to harvest cell sheets from different substrates for selective cell adhesion, cell shortening, and cell harvesting.[141,142] It can also be used to obtain ELR-based thin coatings to modify surfaces and control cell behavior, which is a key aspect in many applications in biology (substrates for cell tests and biological diagnosis) and biotechnology (cell expansion, biomaterials, and substrates for regenerative medicine).[143] However, ELR-based coacervates usually lack the structural stability and stiffness found in native tissues needed to be effective; therefore, many different strategies based on the formation of hydrogels by physical or chemical crosslinking have been developed in order to increase their mechanical resistance.

Hydrogels intended for tissue replacement or regeneration are more effective when their stress relaxation behavior matches that of native tissue as this behavior affects load transfer and nutrient transport.[144,145] Recombinant DNA techniques allow the inclusion of domains other than elastin-like domains in order to increase the moduli of ELR-based hydrogels *via* physical crosslinking. For example, in the study by Cipriani *et al.*[145] using an amphiphilic silk-elastin-like corecombinamer (SELR) in which the silk motif GAGAGS is a hexapeptide from *Bombyx mori* silk fibroin, a dual physical interaction, namely the ELR thermal transition of the hydrophobic block and the silk-based sequence responsible for the supramolecular rearrangement into β-sheets, triggers gel formation to obtain a rapid and stable gel (Figure 6.3). This study set up and optimized a thermal treatment (preannealing treatment), which accelerates the β-sheet formation into the SELR without losing the injectability of the material, based on the evolution of silk motifs into β-sheet structures and the phenomenon of thermal memory. The gel showed fast gelation, forming an ECM fibrillary structure directly after injection, and prompted *de novo* ECM formation, the absence of fibrocartilage, and the production of hyaline cartilage.

In another study, Fernandez-Colino *et al.*[77] achieved ELR physical cross-linking by incorporating leucine zipper motifs throughout the ELR sequence. These structural motifs function as eukaryotic transcription factors as they are able to heterodimerize and change their secondary structure during gene regulation, and they can also be exploited to drive the formation of hydrogels with appropriate mechanical properties to serve as scaffolds for tissue engineering applications.[146] The synergistic effect of the thermo-responsive process in these ELRs and the leucine zippers results in the formation of a hydrogel with stable mechanical properties above the T_t. These hydrogels have the attractive property of being injectable, thus allowing them to adapt to the defects of the lesion to be repaired.[77]

ELRs can also be chemically modified to bear reactant residues that stabilize hydrogel formation. For example, in a study by Gonzalez de Torre *et al.*,[144] two ELRs were modified to bear the reactive groups (azide and cyclooctyne) needed to form hydrogels *via* a Cu-free click chemistry reaction that takes place under physiological conditions. The porosity and intrinsic

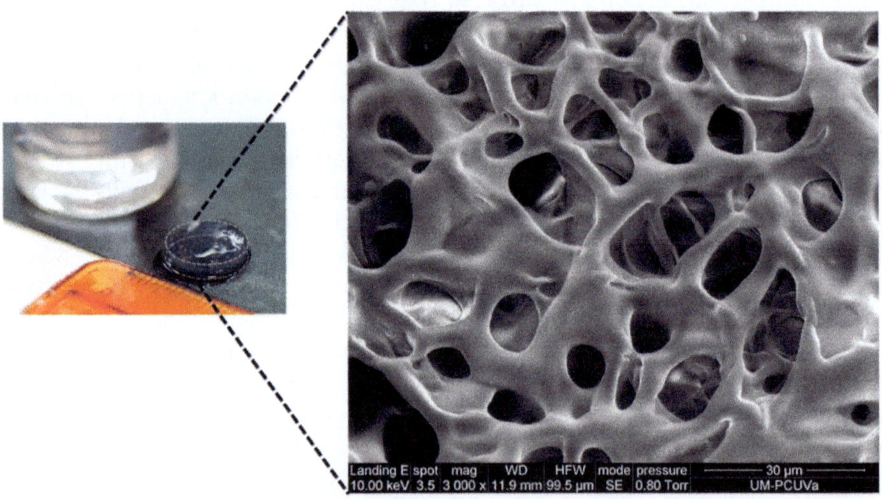

Figure 6.3 Macroscopic (left) and SEM image (right) of the fully hydrated hydrogel developed to repair joint cartilage based on an amphiphilic silk-elastin-like corecombinamer (SELR) and functionalized with the RGD adhesion domain. The micrograph shows the hydrogel at 120 mg mL^{-1} with a pore size of 10.23 ± 2.87 µm and wall thickness of 0.71 ± 0.12 µm.

viscoelastic properties of these hydrogels were characterized and they were found to be able to adapt to the most appropriate physiological environment, with storage and loss moduli in the range 1–10 kPa and 100–1000 Pa, respectively. In a subsequent study by Fernández-Colino *et al.*,[147] the interconnected porosity was improved by applying the salt-leaching gas-foaming technique under conditions that guarantee the cytocompatibility of the reaction to obtain scaffolds for use in tissue engineering.

Recombinant DNA techniques also allow all the above ELR-based materials to have different functionalities in their backbones. The most frequent functional motifs are adhesion sequences (RGD, REDV) or proteolytic sites (DRIR and GTAR),[148] although others that trigger vascularization processes (QK peptide)[149] or serve as a basis for other chemical modifications with noncanonical amino acids at specific points in the ELR chain[150] have been developed.

Although the SA of these biomaterials commonly results in hydrogels or sheets, other different structures, ranging from particles and hollow vesicles to hierarchical supraparticles and fibers, have been designed, as discussed in Section 6.3.2. For example, the same principle of a click chemistry reaction to stabilize the resulting material was applied to a pair of ELRs to obtain fibers with different diameters *via* wet spinning or electrospinning.[151]

As anticipated above, the ability of ELRs to assemble into different coatings means that the properties of surfaces can be modified by way of a protein adsorption phenomenon and their stiffness can be tuned depending on the amino acid sequence and substrate chemistry.[152] However, this interplay between disorder and order in ELRs can be used to program organic–inorganic interactions into hierarchically ordered mineralized structures by template SA.[153]

Several IDPs acquire a secondary structure when they interact with their binding partners, thus working as key elements for biomineralization. These partners usually include the component ions of the mineral being deposited, the crystals themselves, or the template on which the initial crystals form, as well as other IDPs and intrinsically disordered peptides.[154] ELRs have been used as boundary or template SA for biomineralization.[153,155]

For example, in a study by Li *et al.*,[156] two ELR-based hydrogels (HSS3 and REDV) were used for the template mineralization of hydroxyapatite nanocrystals using a 50 $\mu g\,mL^{-1}$ polyaspartate solution as a biomimetic polymer-induced liquid precursor (PILP). These anionic polyaspartic acids mimic the role of acidic proteins in biominerals, thus forming a liquid-like amorphous calcium phosphate (ACP) precursor complex. Mineralization using this system was able to induce precipitation within the framework of the cross-linked ELR-based hydrogels instead of only at the surface, as occurs in most conventional mineralization processes, thereby preserving the microporous structure of the ELRs.

Disorder at the aggregation level was found to be related to contributions from proline and glycine to the ELR backbone. In addition, these amino acids boost the formation of amyloid fibers.[8] In light of this, Elsharkawy *et al.*[153]

reported that the conformational ratio of disordered random coils to ordered β-sheets depends on the amount of crosslinker (hexamethyl diisocyanate, HDI) used with an amphiphilic co-ELR. When this ratio is higher than 0.26, spherulites are formed and subsequently act as a template, thus resulting in the formation of mineralized structures when the ELR membrane is incubated in a suitable solution.

In addition to these previous studies, in which the ELR sequence can recombinantly incorporate nonelastomeric sequences along the polymer backbone, which contribute to the ELR thermal transition assembly, other cellular studies with ELRs have taken advantage of "nonspecific supramolecular interactions".

Taking the physical interactions as a starting point, such as hydrophobic and electrostatic interactions, π–π stacking, hydrogen bonds, and van der Waals forces, grouped into the category of "non-specific supramolecular interactions" by Okesola and Mata,[38] Inostroza-Brito *et al.*[157] reported that, beyond the ELR transition temperature, mixing aqueous solutions of a peptide and an ELR results in coassembly *via* a diffusion–reaction mechanism, with a membrane with a multilayer structure being formed. The conformational transition of the ELR is tuned by the synergy between the ELR and the peptide. Specifically, the peptide self-assembles into fibers whose presence drives the conformational change of the ELR into an open conformation. Fluctuations trigger a morphogenesis in this membrane, thus resulting in a tubular 3D network with dynamic properties, such as self-healing. This is therefore an example of a dynamic self-assembled system according to the classification of Whitesides and Grzybowski.[28] Cellular cultures of primary mouse adipose-derived stem cells and human umbilical vein endothelial cells have been carried out using these tubular structures for 21 days. The seeding, growth, and metabolic activity of cells were observed, thus indicating their viability for tissue engineering and biocompatibility. Genipin was used as a biocompatible crosslinking agent to enhance the mechanical properties of these membranes.[158]

6.5.3 Biomedical Devices

The flexibility and binding specificity of IDPs allow them to be used to modulate the input/output response of their biorecognition systems upon binding to their target molecules.[159] Various types of PTMs, or other chemical modifications, are known to modulate the functionalities of many IDPs after biosynthesis. These modifications may involve the addition of functional groups, attachment of other proteins and peptides, changes to the chemical nature of the amino acids, and dissection of the backbone by proteolytic cleavage.[160]

The intrinsic disorder of ELRs and their outstanding environmental sensitivity, which is not restricted to the intrinsic triggers of temperature or ionic strength, mean that these elastomeric polymers have a great potential in the interconversion of energy as these elastic molecules can also stretch or contract in response to different stimuli, as is the case for most IDPs in different organisms. This sensitivity has led to ELRs being considered "elastic biomolecular machines", with the inverse temperature transition mechanism being at the heart of most biological energy interconversions.[161] These energy interconversions may find utility in a variety of applications, including sensors, bioanalytical instruments, and other biomedical devices, and some examples are provided below. For example, photomodulation of the inverse temperature transition of a modified ELR was first demonstrated by Strzegowski *et al.*,[162] with the incorporation of an azobenzene chromophore being sufficient to render the inverse temperature transition of an ELR copolymer photosensitive. Since then, additional photoresponsive ELRs have been synthesized and their performance improved, for example, by including cyclodextrins in order to increase the range of working temperatures in which the ELR may respond to light, thus validating their use for physiological applications.[163]

Intracellular calcium signal networks can also trigger the interconversion of ELRs, as demonstrated by Kim *et al.*,[164] who used a modular design that allosterically transduces calcium binding into an aqueous demixing phase transition with an ELR that is fused to a ligand-binding protein domain, namely calmodulin (CaM). The binding of calcium to CaM spontaneously triggers the ELR phase transition, thus leading to the formation of meso–microscale particles depending on the chain length of the ELR and with a reversible behavior. This allosteric behavior is probably facilitated by the dynamic and intrinsically disordered nature of ELRs and is considered to be the new paradigm for understanding allosteric mechanisms.[165]

Ibañez-Fonseca *et al.*[166] have explored Förster resonance energy transfer (FRET) of two different fluorescent proteins (FPs) fused to an amphiphilic ELR also containing protein sequences from silk-elastin-like corecombinamer (SELR) by spectroscopy (donor quenching) and confocal microscopy (acceptor photobleaching) in an effort to use ELRs as biological sensors. The absorption spectrum of SELR fused to one of the FPs (eqFP650) showed a peak greatly overlapping the emission spectrum of the SELR fused to the other FP (AcEGFP), hence enabling FRET upon the interaction of the two SELR molecules. Due to the stacking of silk domains at any temperature and to the aggregation of elastin-like blocks above the transition temperature, the resulting structure was proposed to be a novel system for enhancing the traceability of SELR-based hydrogels intended for tissue engineering and for the development of biosensors.

Recent studies have shown how redox reactions can modulate the temperature-responsive behavior of ELRs in an aqueous solution. For example, Petitdemange *et al.*[167] demonstrated how oxidation of the thioether groups from methionine residues present in the ELR backbone into both sulfoxide and sulfone derivatives is a useful means of precisely adjusting the residue polarity, and hence transition temperature, of the ELR. Given that most of the key sensors built into redox-regulated proteins undergo reversible thiol oxidation in response to changes in the oxidation status of the cellular environment, these findings could lead to the development of new ELR redox-based sensors.

Other recent studies have opened up new possibilities for the development of advanced biomaterials that allow the introduction of suitable functionalities to make ELRs sensitive to other stimuli that cannot be accessed by genetic engineering and could result in other types of responsiveness, such as to electromagnetic radiation or sugar concentration in the medium. The enzymatic modification of an ELR containing free γ-carboxylic groups by amidation with a commercial immobilized microbial lipase (Novozym 435 CAL-B) as biocatalyst has allowed the functionality of naturally produced and inspired materials to be combined with other synthetic materials, thus further expanding the current potential of these versatile polymers.[168]

6.6 Conclusions

The modular nature of the sequence of natural elastin, and hence of ELRs, which mimic both its structural and functional properties, has led to the development of a vast array of ELR-based self-assembled structures with interesting applications in the fields of chemistry and materials and biomedical sciences.

The phase behavior of ELRs seems to be similar to the performance of other IDPs, which interconvert between a very wide range of conformations, with proline and glycine residues playing an important role in these interconversions. However, other intrinsic and extrinsic factors also contribute to ensuring a thermodynamically favorable and entropically driven transition that tends toward equilibrium, with the presence of water being mandatory. The magnitude of these dynamic interactions can endow ELRs with reversibility, self-healing, and significant sensitivity to perturbations from the external environment, thus allowing them to rearrange and adapt to the landscape.

The rules that govern these interconversions, diverse chemical compositions, shapes, and functionalities have been reported, with ELRs typically being used as single-component building blocks or in combination with other materials. In this chapter, we have focused our attention on the most common ELR-based self-assembled structures intended for biomedical applications (nanoparticles, fibers, and hydrogels), although other more exotic structures have also been considered. All these structures can be obtained from these materials depending on their composition, how the different ELR blocks are combined, and their processing conditions.

ELRs offer the possibility of harnessing the vast array of functions exhibited by these biopolymers in nature, and their recombinant production allows sufficient quantities to be produced so that their application in the fields of tissue engineering, controlled drug delivery, biosensors, or medical devices can be scaled to meet the needs of the market.

References

1. J. C. Rodríguez-Cabello, L. Martín, M. Alonso, F. J. Arias and A. M. Testera, *Polymer*, 2009, **50**, 5159–5169.
2. J. C. Rodríguez-Cabello, I. G. de Torre, S. Acosta, S. Salinas and M. Herrero, in *Self-assembling Biomaterials*, ed. H. S. Azevedo and R. M. P. da Silva, Woodhead Publishing, 2018, pp. 49–78.
3. S. M. Mithieux and A. S. Weiss, *Adv. Protein Chem.*, 2005, **70**, 437–461.
4. F. W. Keeley, C. M. Bellingham and K. A. Woodhouse, *Philos. Trans. R. Soc., B*, 2002, **357**, 185–189.
5. S. M. Mithieux, S. G. Wise and A. S. Weiss, *Adv. Drug Delivery Rev.*, 2013, **65**, 421–428.
6. L. D. Muiznieks, A. S. Weiss and F. W. Keeley, *Biochem. Cell Biol.*, 2010, **88**, 239–250.
7. A. M. J. Coenen, K. V. Bernaerts, J. A. W. Harings, S. Jockenhoevel and S. Ghazanfari, *Acta Biomater.*, 2018, **79**, 60–82.
8. S. Rauscher, S. Baud, M. Miao, F. W. Keeley and R. Pomes, *Structure*, 2006, **14**, 1667–1676.
9. M. Miao, E. Sitarz, C. M. Bellingham, E. Won, L. D. Muiznieks and F. W. Keeley, *Biopolymers*, 2013, **99**, 392–407.
10. D. W. Urry, *Faraday Discuss. Chem. Soc.*, 1976, **61**, 205–212.
11. S. G. Wise, G. C. Yeo, M. A. Hiob, J. Rnjak-Kovacina, D. L. Kaplan, M. K. C. Ng and A. S. Weiss, *Acta Biomater.*, 2014, **10**, 1532–1541.
12. O. Selig, A. V. Cunha, M. B. van Eldijk and J. C. M. van Hest, *J. Phys. Chem.*, 2018, **122**, 8243–8254.
13. D. W. Urry, C. H. Luan, T. M. Parker, D. C. Gowda, K. U. Prasad, M. C. Reid and A. Safavy, *J. Am. Chem. Soc.*, 1991, **113**, 4346–4348.
14. D. W. Urry, *J. Phys. Chem. B*, 1997, **101**, 11007–11028.
15. J. Reguera, J. M. Lagarón, M. Alonso, V. Reboto, B. Calvo and J. C. Rodríguez-Cabello, *Macromolecules*, 2003, **36**, 8470–8476.
16. R. Herrero-Vanrell, A. C. Rincon, M. Alonso, V. Reboto, I. T. Molina-Martinez and J. C. Rodriguez-Cabello, *J. Controlled Release*, 2005, **102**, 113–122.
17. N. K. Li, S. Roberts, F. G. Quiroz, A. Chilkoti and Y. G. Yingling, *Biomacromolecules*, 2018, **19**, 2496–2505.
18. N. K. Li, F. G. Quiroz, C. K. Hall, A. Chilkoti and Y. G. Yingling, *Biomacromolecules*, 2014, **15**, 3522–3530.
19. L. D. Muiznieks and F. W. Keeley, *ACS Biomater. Sci. Eng.*, 2017, **3**, 661–679.
20. D. W. Urry, *What Sustains Life? Consilient Mechanisms for Protein-Based Machines and Materials*, Birkhäuser Boston, Boston (USA), 2006.
21. Y. Cho, Y. Zhang, T. Christensen, L. B. Sagle, A. Chilkoti and P. S. Cremer, *J. Phys. Chem. B*, 2008, **112**, 13765–13771.
22. G. Pinedo-Martín, M. Santos, A. M. Testera, M. Alonso and J. C. Rodríguez-Cabello, *Polymer*, 2014, **55**, 5314–5321.

23. A. Ribeiro, F. J. Arias, J. Reguera, M. Alonso and J. C. Rodriguez-Cabello, *Biophys. J.*, 2009, **97**, 312–320.
24. F. Dehghani, N. Annabi, P. Valtchev, S. M. Mithieux, A. S. Weiss, S. G. Kazarian and F. H. Tay, *Biomacromolecules*, 2008, **9**, 1100–1105.
25. D. E. Meyer and A. Chilkoti, *Biomacromolecules*, 2004, **5**, 846–851.
26. L. Martín, E. Castro, A. Ribeiro, M. Alonso and J. C. Rodríguez-Cabello, *Biomacromolecules*, 2012, **13**, 293–298.
27. J. Carlos Rodríguez-Cabello, J. Reguera, A. Girotti, M. Alonso and A. M. Testera, *Prog. Polym. Sci.*, 2005, **30**, 1119–1145.
28. G. M. Whitesides and B. Grzybowski, *Science*, 2002, **295**, 2418–2421.
29. J. D. Halley and D. A. Winkler, *Complexity*, 2008, **14**, 10–17.
30. J.-M. Lehn, *Angew. Chem., Int. Ed. Engl.*, 1988, **27**, 89–112.
31. J.-M. Lehn, *Angew. Chem., Int. Ed. Engl.*, 1990, **29**, 1304–1319.
32. J.-M. Lehn, *Science*, 2002, **295**, 2400–2403.
33. T. Aida, E. W. Meijer and S. I. Stupp, *Science*, 2012, **335**, 813–817.
34. B. M. Discher, Y.-Y. Won, D. S. Ege, J. C.-M. Lee, F. S. Bates, D. E. Discher and D. A. Hammer, *Science*, 1999, **284**, 1143–1146.
35. A. Katsuhiko, P. H. Jonathan, V. L. Michael, V. Ajayan, C. Richard and A. Somobrata, *Sci. Technol. Adv. Mater.*, 2008, **9**, 014109.
36. R. M. P. da Silva, D. van der Zwaag, L. Albertazzi, S. S. Lee, E. W. Meijer and S. I. Stupp, *Nat. Commun.*, 2016, **7**, 11561.
37. F. Tantakitti, J. Boekhoven, X. Wang, R. V. Kazantsev, T. Yu, J. Li, E. Zhuang, R. Zandi, J. H. Ortony, C. J. Newcomb, L. C. Palmer, G. S. Shekhawat, M. O. de la Cruz, G. C. Schatz and S. I. Stupp, *Nat. Mater.*, 2016, **15**, 469.
38. B. O. Okesola and A. Mata, *Chem. Soc. Rev.*, 2018, **47**, 3721–3736.
39. S. S. Lee, T. Fyrner, F. Chen, Z. Álvarez, E. Sleep, D. S. Chun, J. A. Weiner, R. W. Cook, R. D. Freshman, M. S. Schallmo, K. M. Katchko, A. D. Schneider, J. T. Smith, C. Yun, G. Singh, S. Z. Hashmi, M. T. McClendon, Z. Yu, S. R. Stock, W. K. Hsu, E. L. Hsu and S. I. Stupp, *Nat. Nanotechnol.*, 2017, **12**, 821.
40. G. Whitesides, J. Mathias and C. Seto, *Science*, 1991, **254**, 1312–1319.
41. L. F. L. Ian and M. Atkinson, *Self Assembly in Supramolecular Systems*, 2000.
42. K. Sharp, A. Nicholls, R. Fine and B. Honig, *Science*, 1991, **252**, 106–109.
43. D. B. Smithrud, T. B. Wyman and F. Diederich, *J. Am. Chem. Soc.*, 1991, **113**, 5420–5426.
44. S. Jakubith, H. H. Rotermund, W. Engel, A. von Oertzen and G. Ertl, *Phys. Rev. Lett.*, 1990, **65**, 3013–3016.
45. B. Hess, *Naturwissenschaften*, 2000, **87**, 199–211.
46. N. Goldenfeld and L. P. Kadanoff, *Science*, 1999, **284**, 87–89.
47. W. Hassouneh, E. B. Zhulina, A. Chilkoti and M. Rubinstein, *Macromolecules*, 2015, **48**, 4183–4195.
48. T. A. T. Lee, A. Cooper, R. P. Apkarian and V. P. Conticello, *Adv. Mater.*, 2000, **12**, 1105–1110.

49. E. R. Wright and V. P. Conticello, *Adv. Drug Delivery Rev.*, 2002, **54**, 1057–1073.
50. M. R. Dreher, A. J. Simnick, K. Fischer, R. J. Smith, A. Patel, M. Schmidt and A. Chilkoti, *J. Am. Chem. Soc.*, 2008, **130**, 687–694.
51. S. R. MacEwan, I. Weitzhandler, I. Hoffmann, J. Genzer, M. Gradzielski and A. Chilkoti, *Biomacromolecules*, 2017, **18**, 599–609.
52. S. M. Janib, M. F. Pastuszka, S. Aluri, Z. Folchman-Wagner, P. Y. Hsueh, P. Shi, Y. A. Lin, H. Cui and J. A. MacKay, *Polym. Chem.*, 2014, **5**, 1614–1625.
53. T. Luo and K. L. Kiick, *J. Am. Chem. Soc.*, 2015, **137**, 15362–15365.
54. A. Ghoorchian, J. T. Cole and N. B. Holland, *Macromolecules*, 2010, **43**, 4340–4345.
55. J. R. McDaniel, I. Weitzhandler, S. Prevost, K. B. Vargo, M.-S. Appavou, D. A. Hammer, M. Gradzielski and A. Chilkoti, *Nano Lett.*, 2014, **14**, 6590–6598.
56. X.-X. Xia, Q. Xu, X. Hu, G. Qin and D. L. Kaplan, *Biomacromolecules*, 2011, **12**, 3844–3850.
57. Y. Navon and R. Bitton, *Isr. J. Chem.*, 2016, **56**, 581–589.
58. D. W. Urry and T. M. Parker, in *Mechanics of Elastic Biomolecules*, ed. W. A. Linke, H. Granzier and M. S. Z. Kellermayer, Springer Netherlands, Dordrecht, 2003, pp. 543–559.
59. M. A. Castiglione Morelli, M. DeBiasi, A. DeStradis and A. M. Tamburro, *J. Biomol. Struct. Dyn.*, 1993, **11**, 181–190.
60. A. M. Tamburro, V. Guantieri and D. D. Gordini, *J. Biomol. Struct. Dyn.*, 1992, **10**, 441–454.
61. B. Bochicchio, M. Lorusso, A. Pepe and A. M. Tamburro, *Nanomedicine*, 2009, **4**, 31–46.
62. A. M. Tamburro, A. Pepe, B. Bochicchio, D. Quaglino and I. P. Ronchetti, *J. Biol. Chem.*, 2005, **280**, 2682–2690.
63. P. Moscarelli, F. Boraldi, B. Bochicchio, A. Pepe, A. M. Salvi and D. Quaglino, *Matrix Biol.*, 2014, **36**, 15–27.
64. F. Chiti, P. Webster, N. Taddei, A. Clark, M. Stefani, G. Ramponi and C. M. Dobson, *Proc. Natl. Acad. Sci. U. S. A.*, 1999, **96**, 3590–3594.
65. D. H. T. Le, R. Hanamura, D.-H. Pham, M. Kato, D. A. Tirrell, T. Okubo and A. Sugawara-Narutaki, *Biomacromolecules*, 2013, **14**, 1028–1034.
66. S. Zhang, T. Holmes, C. Lockshin and A. Rich, *Proc. Natl. Acad. Sci. U. S. A.*, 1993, **90**, 3334–3338.
67. D. Asai, D. Xu, W. Liu, F. Garcia Quiroz, D. J. Callahan, M. R. Zalutsky, S. L. Craig and A. Chilkoti, *Biomaterials*, 2012, **33**, 5451–5458.
68. M. Li and C. K. Ober, *Mater. Today*, 2006, **9**, 30–39.
69. J. Reguera, A. Fahmi, P. Moriarty, A. Girotti and J. C. Rodríguez-Cabello, *J. Am. Chem. Soc.*, 2004, **126**, 13212–13213.
70. X. Wu, R. Sallach, C. A. Haller, J. A. Caves, K. Nagapudi, V. P. Conticello, M. E. Levenston and E. L. Chaikof, *Biomacromolecules*, 2005, **6**, 3037–3044.

71. M. H. Misbah, L. Quintanilla, M. Alonso and J. C. Rodríguez-Cabello, *Polymer*, 2015, **81**, 37–44.
72. K. Tanaka, S. Inoue and S. Mizuno, *Insect Biochem. Mol. Biol.*, 1999, **29**, 269–276.
73. W. A. Petka, J. L. Harden, K. P. McGrath, D. Wirtz and D. A. Tirrell, *Science*, 1998, **281**, 389–392.
74. A. Fernández-Colino, F. J. Arias, M. Alonso and J. C. Rodríguez-Cabello, *Biomacromolecules*, 2014, **15**, 3781–3793.
75. J. Yang, C. Xu, P. Kopečková and J. Kopeček, *Macromol. Biosci.*, 2006, **6**, 201–209.
76. B. Tripet, K. Wagschal, P. Lavigne, C. T. Mant and R. S. Hodges, *J. Mol. Biol.*, 2000, **300**, 377–402.
77. A. Fernández-Colino, F. J. Arias, M. Alonso and J. C. Rodríguez-Cabello, *Biomacromolecules*, 2015, **16**, 3389–3398.
78. M. J. Piña, A. Girotti, M. Santos, J. C. Rodríguez-Cabello and F. J. Arias, *Mol. Pharmaceutics*, 2016, **13**, 795–808.
79. B. C. Dash, S. Mahor, O. Carroll, A. Mathew, W. Wang, K. A. Woodhouse and A. Pandit, *J. Controlled Release*, 2011, **152**, 382–392.
80. M. Swierczewska, C. S. Hajicharalambous, A. V. Janorkar, Z. Megeed, M. L. Yarmush and P. Rajagopalan, *Acta Biomater.*, 2008, **4**, 827–837.
81. H.-C. Huang, Y. Yang, A. Nanda, P. Koria and K. Rege, *Nanomedicine*, 2011, **6**, 459–473.
82. C. Kojima and K. Irie, *Pept. Sci.*, 2013, **100**, 714–721.
83. E. Wang, M. S. Desai, K. Heo and S.-W. Lee, *Langmuir*, 2014, **30**, 2223–2229.
84. J. Habchi, P. Tompa, S. Longhi and V. N. Uversky, *Chem. Rev.*, 2014, **114**, 6561–6588.
85. A. K. Dunker and Z. Obradovic, *Nat. Biotechnol.*, 2001, **19**, 805.
86. P. E. Wright and H. J. Dyson, *Nat. Rev. Mol. Cell Biol.*, 2014, **16**, 18.
87. H. J. Dyson and P. E. Wright, *Nat. Rev. Mol. Cell Biol.*, 2005, **6**, 197–208.
88. J. Gsponer, M. E. Futschik, S. A. Teichmann and M. M. Babu, *Science*, 2008, **322**, 1365–1368.
89. V. N. Uversky, *Protein Sci.*, 2002, **11**, 739–756.
90. Y. Zhang, B. Stec and A. Godzik, *Structure*, 2007, **15**, 1141–1147.
91. L. Staby, C. Shea, M. Willemoës, F. Theisen, B. B. Kragelund and K. Skriver, *Biochem. J.*, 2017, **474**, 2509.
92. P. Tompa, *Curr. Opin. Struct. Biol.*, 2011, **21**, 419–425.
93. R. B. Berlow, H. J. Dyson and P. E. Wright, *FEBS Lett.*, 2015, **589**, 2433–2440.
94. M. Vincent, V. N. Uversky and S. Schnell, *Proteomics*, 2019, **19**, 1800415.
95. P. Tompa, *FEBS Lett.*, 2005, **579**, 3346–3354.
96. P. Tompa and M. Fuxreiter, *Trends Biochem. Sci.*, 2008, **33**, 2–8.
97. J. A. Marsh and J. D. Forman-Kay, *Biophys. J.*, 2010, **98**, 2383–2390.
98. P. Tompa, *Trends Biochem. Sci.*, 2002, **27**, 527–533.
99. V. Neduva and R. B. Russell, *FEBS Lett.*, 2005, **579**, 3342–3345.

100. C. J. Oldfield, Y. Cheng, M. S. Cortese, P. Romero, V. N. Uversky and A. K. Dunker, *Biochemistry*, 2005, **44**, 12454–12470.
101. K. Van Roey, B. Uyar, R. J. Weatheritt, H. Dinkel, M. Seiler, A. Budd, T. J. Gibson and N. E. Davey, *Chem. Rev.*, 2014, **114**, 6733–6778.
102. R. van der Lee, M. Buljan, B. Lang, R. J. Weatheritt, G. W. Daughdrill, A. K. Dunker, M. Fuxreiter, J. Gough, J. Gsponer, D. T. Jones, P. M. Kim, R. W. Kriwacki, C. J. Oldfield, R. V. Pappu, P. Tompa, V. N. Uversky, P. E. Wright and M. M. Babu, *Chem. Rev.*, 2014, **114**, 6589–6631.
103. P. Tompa, *Trends Biochem. Sci.*, 2012, **37**, 509–516.
104. A. Mohan, C. J. Oldfield, P. Radivojac, V. Vacic, M. S. Cortese, A. K. Dunker and V. N. Uversky, *J. Mol. Biol.*, 2006, **362**, 1043–1059.
105. M. Fuxreiter, I. Simon, P. Friedrich and P. Tompa, *J. Mol. Biol.*, 2004, **338**, 1015–1026.
106. P. C. Whitford, K. Y. Sanbonmatsu and J. N. Onuchic, *Rep. Prog. Phys.*, 2012, 75, 076601.
107. T. Flock, R. J. Weatheritt, N. S. Latysheva and M. M. Babu, *Curr. Opin. Struct. Biol.*, 2014, **26**, 62–72.
108. M. Dzuricky, S. Roberts and A. Chilkoti, *Biochemistry*, 2018, **57**, 2405–2414.
109. S. Rauscher and R. Pomes, *Adv. Exp. Med. Biol.*, 2012, **725**, 159–183.
110. N. Srinivasan and S. Kumar, *Wiley Interdiscip. Rev.: Nanomed. Nanobiotechnol.*, 2012, **4**, 204–218.
111. L. D. Muiznieks and F. W. Keeley, *J. Biol. Chem.*, 2010, **285**, 39779–39789.
112. F. X. Theillet, L. Kalmar, P. Tompa, K. H. Han, P. Selenko, A. K. Dunker, G. W. Daughdrill and V. N. Uversky, *Intrinsically Disord. Proteins*, 2013, **1**, e24360.
113. F. Krieger, A. Möglich and T. Kiefhaber, *J. Am. Chem. Soc.*, 2005, **127**, 3346–3352.
114. M. Hong, D. Isailovic, R. A. McMillan and V. P. Conticello, *Biopolymers*, 2003, **70**, 158–168.
115. K. Ohgo, J. Ashida, K. K. Kumashiro and T. Asakura, *Macromolecules*, 2005, **38**, 6038–6047.
116. C. M. Venkatachalam and D. W. Urry, *Macromolecules*, 1981, **14**, 1225–1229.
117. D. W. Urry and M. M. Long, *CRC Crit. Rev. Biochem.*, 1976, **4**, 1–45.
118. D. T. McPherson, C. Morrow, D. S. Minehan, J. Wu, E. Hunter and D. W. Urry, *Biotechnol. Prog.*, 1992, **8**, 347–352.
119. B. Bochicchio and A. M. Tamburro, *Chirality*, 2002, **14**, 782–792.
120. B. Bochicchio, A. Ait-Ali, A. M. Tamburro and A. J. Alix, *Biopolymers*, 2004, **73**, 484–493.
121. R. K. Das and R. V. Pappu, *Proc. Natl. Acad. Sci. U. S. A.*, 2013, **110**, 13392–13397.
122. J. R. McDaniel, D. C. Radford and A. Chilkoti, *Biomacromolecules*, 2013, **14**, 2866–2872.
123. B. Sumer and J. Gao, *Nanomedicine*, 2008, **3**, 137–140.

124. R. B. Roemer, *Annu. Rev. Biomed. Eng.*, 1999, **1**, 347–376.
125. E. Blanco, C. W. Kessinger, B. D. Sumer and J. Gao, *Exp. Biol. Med.*, 2009, **234**, 123–131.
126. N. Nasongkla, E. Bey, J. Ren, H. Ai, C. Khemtong, J. S. Guthi, S.-F. Chin, A. D. Sherry, D. A. Boothman and J. Gao, *Nano Lett.*, 2006, **6**, 2427–2430.
127. D. J. Callahan, W. Liu, X. Li, M. R. Dreher, W. Hassouneh, M. Kim, P. Marszalek and A. Chilkoti, *Nano Lett.*, 2012, **12**, 2165–2170.
128. J. R. McDaniel, S. R. MacEwan, M. Dewhirst and A. Chilkoti, *J. Controlled Release*, 2012, **159**, 362–367.
129. J. Andrew MacKay, M. Chen, J. R. McDaniel, W. Liu, A. J. Simnick and A. Chilkoti, *Nat. Mater.*, 2009, **8**, 993.
130. J. Hu, L. Xie, W. Zhao, M. Sun, X. Liu and W. Gao, *Chem. Commun.*, 2015, **51**, 11405–11408.
131. D. Derossi, S. Calvet, A. Trembleau, A. Brunissen, G. Chassaing and A. Prochiantz, *J. Biol. Chem.*, 1996, **271**, 18188–18193.
132. E. S. Olson, T. A. Aguilera, T. Jiang, L. G. Ellies, Q. T. Nguyen, E. H. Wong, L. A. Gross and R. Y. Tsien, *Integr. Biol.*, 2009, **1**, 382–393.
133. J. B. Rothbard, T. C. Jessop, R. S. Lewis, B. A. Murray and P. A. Wender, *J. Am. Chem. Soc.*, 2004, **126**, 9506–9507.
134. L. Walker, E. Perkins, F. Kratz and D. Raucher, *Int. J. Pharm.*, 2012, **436**, 825–832.
135. P. A. Wender, D. J. Mitchell, K. Pattabiraman, E. T. Pelkey, L. Steinman and J. B. Rothbard, *Proc. Natl. Acad. Sci. U. S. A.*, 2000, **97**, 13003–13008.
136. S. R. MacEwan and A. Chilkoti, *Nano Lett.*, 2012, **12**, 3322–3328.
137. M. B. van Eldijk, J. C. Y. Wang, I. J. Minten, C. Li, A. Zlotnick, R. J. M. Nolte, J. J. L. M. Cornelissen and J. C. M. van Hest, *J. Am. Chem. Soc.*, 2012, **134**, 18506–18509.
138. S. R. Aluri, P. Shi, J. A. Gustafson, W. Wang, Y.-A. Lin, H. Cui, S. Liu, P. S. Conti, Z. Li, P. Hu, A. L. Epstein and J. A. MacKay, *ACS Nano*, 2014, **8**, 2064–2076.
139. M. K. Pastuszka, X. Wang, L. L. Lock, S. M. Janib, H. Cui, L. D. DeLeve and J. A. MacKay, *J. Controlled Release*, 2014, **191**, 15–23.
140. S. R. MacEwan and A. Chilkoti, *Biopolymers*, 2010, **94**, 60–77.
141. M. Mie, Y. Mizushima and E. Kobatake, *J. Biomed. Mater. Res., Part B*, 2008, **86**, 283–290.
142. M. Pierna, M. Santos, F. J. Arias, M. Alonso and J. C. Rodriguez-Cabello, *Biomacromolecules*, 2013, **14**, 1893–1903.
143. R. R. Costa, C. A. Custódio, A. M. Testera, F. J. Arias, J. C. Rodríguez-Cabello, N. M. Alves and J. F. Mano, *Adv. Funct. Mater.*, 2009, **19**, 3210–3218.
144. I. Gonzalez de Torre, M. Santos, L. Quintanilla, A. Testera, M. Alonso and J. C. Rodriguez Cabello, *Acta Biomater.*, 2014, **10**, 2495–2505.
145. F. Cipriani, M. Krüger, I. G. de Torre, L. Q. Sierra, M. A. Rodrigo, L. Kock and J. C. Rodriguez-Cabello, *Biomacromolecules*, 2018, **19**, 4333–4347.

146. M. K. Pastuszka and J. A. MacKay, *Wiley Interdiscip. Rev.: Nanomed. Nanobiotechnol.*, 2016, **8**, 123–138.
147. A. Fernández-Colino, F. Wolf, H. Keijdener, S. Rütten, T. Schmitz-Rode, S. Jockenhoevel, J. C. Rodríguez-Cabello and P. Mela, *Mater. Sci. Eng.: C*, 2018, **88**, 140–147.
148. T. Flora, I. G. D. Torre, M. Alonso and J. C. Rodríguez-Cabello, *Biofabrication*, 2019, **11**(3), 035008.
149. T. Flora, I. G. de Torre, M. Alonso and J. C. Rodriguez-Cabello, *J. Mater. Sci.: Mater. Med.*, 2019, **30**, 30.
150. F. Cipriani, D. Bernhagen, C. Garcia-Arevalo, I. G. de Torre, P. Timmerman and J. C. Rodriguez-Cabello, *Biomed. Mater.*, 2019, **14**, 035009.
151. I. Gonzalez de Torre, A. Ibanez-Fonseca, L. Quintanilla, M. Alonso and J. C. Rodriguez-Cabello, *Acta Biomater.*, 2018, **72**, 137–149.
152. R. R. Costa, M. González-Pérez, M. Herrero-Gutiérrez, R. A. Pires, M. Alonso, J. C. Rodriguez-Cabello, R. L. Reis and I. Pashkuleva, *Biomacromolecules*, 2018, **19**, 3401–3411.
153. S. Elsharkawy, M. Al-Jawad, M. F. Pantano, E. Tejeda-Montes, K. Mehta, H. Jamal, S. Agarwal, K. Shuturminska, A. Rice, N. V. Tarakina, R. M. Wilson, A. J. Bushby, M. Alonso, J. C. Rodriguez-Cabello, E. Barbieri, A. del Río Hernández, M. M. Stevens, N. M. Pugno, P. Anderson and A. Mata, *Nat. Commun.*, 2018, **9**, 2145.
154. A. L. Boskey and E. Villarreal-Ramirez, *Matrix Biol.*, 2016, **52–54**, 43–59.
155. Y. Li, J. C. Rodriguez-Cabello and C. Aparicio, *ACS Appl. Mater. Interfaces*, 2017, **9**, 5838–5846.
156. Y. Li, X. Chen, A. Fok, J. C. Rodriguez-Cabello and C. Aparicio, *ACS Appl. Mater. Interfaces*, 2015, **7**, 25784–25792.
157. K. E. Inostroza-Brito, E. Collin, O. Siton-Mendelson, K. H. Smith, A. Monge-Marcet, D. S. Ferreira, R. P. Rodríguez, M. Alonso, J. C. Rodríguez-Cabello, R. L. Reis, F. Sagués, L. Botto, R. Bitton, H. S. Azevedo and A. Mata, *Nat. Chem.*, 2015, **7**, 897.
158. K. E. Inostroza-Brito, E. C. Collin, A. Majkowska, S. Elsharkawy, A. Rice, A. E. del Río Hernández, X. Xiao, J. Rodríguez-Cabello and A. Mata, *Acta Biomater.*, 2017, **58**, 80–89.
159. K. A. Cissell, S. Shrestha, J. Purdie, D. Kroodsma and S. K. Deo, *Anal. Bioanal. Chem.*, 2008, **391**, 1721–1729.
160. H. Xie, S. Vucetic, L. M. Iakoucheva, C. J. Oldfield, A. K. Dunker, Z. Obradovic and V. N. Uversky, *J. Proteome Res.*, 2007, **6**, 1917–1932.
161. D. W. Urry, *Sci. Am.*, 1995, **272**, 64–69.
162. L. A. Strzegowski, M. B. Martinez, D. C. Gowda, D. W. Urry and D. A. Tirrell, *J. Am. Chem. Soc.*, 1994, **116**, 813–814.
163. M. Alonso, V. Reboto, L. Guiscardo, V. Maté and J. C. Rodríguez-Cabello, *Macromolecules*, 2001, **34**, 8072–8077.
164. B. Kim and A. Chilkoti, *J. Am. Chem. Soc.*, 2008, **130**, 17867–17873.
165. H. N. Motlagh, J. O. Wrabl, J. Li and V. J. Hilser, *Nature*, 2014, **508**, 331.

166. A. Ibanez-Fonseca, M. Alonso, F. J. Arias and J. C. Rodriguez-Cabello, *Bioconjugate Chem.*, 2017, **28**, 828–835.
167. R. Petitdemange, E. Garanger, L. Bataille, W. Dieryck, K. Bathany, B. Garbay, T. J. Deming and S. Lecommandoux, *Biomacromolecules*, 2017, **18**, 544–550.
168. A. M. Testera, M. Santos, A. Girotti, F. J. Arias, J. M. Báñez, M. Alonso and J. C. Rodríguez-Cabello, *Int. J. Biol. Macromol.*, 2019, **121**, 752–759.

CHAPTER 7

Smart Nanofibrous Materials for Tissue Engineering

KAMIA PUNIA,[a] APARAJITA BHATTACHARYA[a] AND JIN KIM MONTCLARE*[a,b,c]

[a] Department of Chemical and Biomolecular Engineering, New York University Tandon School of Engineering, Brooklyn, New York 11201, USA; [b] Department of Chemistry, New York University, New York, New York 10012, USA; [c] Department of Biomaterials, New York University College of Dentistry, New York, New York 10010, USA
*Email: montclare@nyu.edu

7.1 Introduction

Tissue engineering is a biomedical field of research aimed at fabricating temporary scaffolds for tissue repair or regeneration.[1] The composition and morphology of tissue engineering scaffolds play a significant role in the final fate of the desired engineered tissue/organ. One of the main prerequisites of a tissue engineering scaffold is to closely mimic the extracellular matrix (ECM), thereby providing a 3D porous scaffold for cell adhesion, signaling, migration, and differentiation.[1] The ECM is composed of a protein fiber network of collagen and elastin fibrils covered with cell-adhesive proteins such as laminin and fibronectin and polysaccharides such as heparan sulfate and hyaluronic acid occupying cavities in the fiber mesh.[2] An ideal tissue engineering scaffold should possess specific characteristics that include: a) biocompatibility of the scaffold material to provide a conducive environment for the cells to interact with surroundings, grow, and differentiate; b) biodegradability to allow for removal or adsorption of the scaffold

Biomaterials Science Series No. 10
Biomimetic Protein Based Elastomers: Emerging Materials for the Future
Edited by Namita Roy Choudhury, Julie C. Liu and Naba K. Dutta

Published by the Royal Society of Chemistry, www.rsc.org

after it induces tissue regeneration; and c) adaptability to the target application, thereby acclimatizing according to morphology, mechanical strength, and modulus of tissue/organ of interest.[3,4] There are a range of natural polymers that have been explored for their application in tissue engineering, such as collagen,[5] silk,[6] elastin,[7] fibrin,[8] chitosan,[9] and their derivatives. Synthetic polymer-based materials for tissue engineering have been mainly dominated by homo- or copolymers of lactic acid, glycolic acid, and caprolactone.[10]

Researchers have given considerable attention to nanofibrous scaffolds for tissue engineering owing to their characteristic resemblance to fibrous networks in the ECM, including nanoscale structure, high surface-area-to-volume ratio, and high porosity essential for cell migration, adhesion, and differentiation.[1] The porous structure of nanofibers allows unrestricted transport of nutrients and waste materials through the scaffolds.

Stimuli-responsive materials have tremendous potential in biomedical applications because of their responsiveness to external stimuli present in the system, such as pH, temperature, mechanical stress, magnetic field, and electrical stimulations.[11] Various techniques have been used to confer stimuli-responsiveness to nanofibrous scaffolds, including the incorporation of moieties for cellular recognition. The design of smart materials is dependent on the desired tissue/organ because of variations in morphology and function of the ECM for different tissues in the body. The consistent supply of growth factors and other biomolecules is essential for the growth and function of tissues. Smart nanofibers with extensive control over the release of protein and growth factors encapsulated within the fibers can be employed to obtain enhanced cell proliferation for tissue engineering applications.[12,13] Along with added functionalities, smart nanofibrous scaffolds have a 3D structure with high porosity and surface area critical for tissue engineering applications.

Researchers have shown great interest in developing nanofibrous scaffolds for tissue engineering with the ability to mimic endogenous inter- and intracellular responses. Therefore, it is essential to classify and evaluate various tissue engineering scaffolds for understanding cellular interactions. In this chapter, our main focus will be on protein and synthetic polymer-based smart nanofibers for tissue engineering. We will review the recent advances and potential applications of smart protein nanofibers with an emphasis on α-helical coiled-coil assemblies, elastin or elastin-like polypeptides (ELPs), silk or silk-like proteins, and collagen or collagen-mimetic proteins. Smart synthetic polymeric nanofibers will be discussed based on their responsiveness to various stimuli such as pH, temperature, mechanical stress, electric field, and magnetic field.

7.2 Self-assembling Peptides and Protein Nanofibers

A wide range of novel biological materials have been designed based on the understanding of supramolecular architecture in nature.[14–16] Biomaterials can self-organize into higher-order assemblies through the association of molecules governed by noncovalent interactions, including electrostatic forces, H-bonding, and van der Waals interactions.[17,18] Peptide and protein-based materials that can self-assemble into nano- to mesoscale fibers have tremendous potential in biomedical research including tissue engineering,[19] 3D cell culture[20] and drug delivery.[21] The assembly of proteins can be tuned by manipulating their structure using bioengineering techniques to optimize specific intra- or intermolecular interactions to create 3D nanofibrous scaffolds. Furthermore, self-assembling proteins/peptides can be bestowed with functionalities by integrating them with various molecules such as therapeutic drugs, peptides, and proteins. Proteins/peptides can be generated in a bottom-up manner either by chemical synthesis[22] or recombinant DNA technology.[23] Chemical synthesis is an efficient and cost-effective technique for the production of short peptides (<35 amino acids) in small quantities.[24] Proteins and peptides longer than 35 amino acids can be produced easily and in a cost-efficient manner by using recombinant biosynthesis using bacteria, fungi, animals, or plants as a host.[25] In this section, the focus will be on self-assembling smart nanofibers based on α-helical coiled-coil, elastin, silk, and collagen-based sequences (Table 7.1).

7.2.1 α-Helical Coiled-coil Nanofibers

α-Helical coiled-coil motif-based biomaterials have gained considerable interest in biomedical research due to their occurrence in a diverse range of natural proteins such as extracellular matrix, cytoskeleton, ligaments, and tendons.[26] A number of strategies have been designed to precisely create functional materials composed of coiled-coil domains for applications in tissue engineering, drug delivery, and 3D cell culture.[27] Coiled-coil proteins have two or more α-helices that wind around each other.[27] The α-helices are comprised of heptad repeats $(abcdefg)_n$ with hydrophobic residues at the *a* and *d* positions of the helix and polar residues at the *b*, *c*, *e*, *f*, and *g* positions.[28–31] The hydrophobic amino acids in the core of α-helix motif are responsible for its stability *via* hydrophobic and van der Waals interactions, enabling the self-assembly and oligomerization state of the coiled-coil. The Woolfson group has developed self-assembling fibers (SAFs) based on coiled-coil dimers.[32–34] They have designed two 28-residue peptides named SAF-p1 (Figure 7.1(II), sequence 1) and SAF-p2 (Figure 7.1(II), sequence 2) to form staggered heterodimers with overhanging ends that facilitate the assembly of coiled-coil fibers (Figure 7.1(II), Table 7.1). The sticky end-directed assembly of peptides leads to the formation of fibers 20 nm in diameter and hundreds of microns in length. Blunt-ended heterodimers formed by the association of SAF-p2 with a third peptide SAF-p3 (Figure 7.1(II), sequence 3) designed by permutation of SAF-p1 do not form any fibers (Figure 7.1(II), Table 7.1).[34]

However, by altering the length of the coiled-coil, Hartgerlink and coworkers demonstrate that sticky-ended monomers are not essential for

Table 7.1 Sequences of α-helical coiled-coil, elastin-, silk-, and collagen-based protein nanofibers.

Peptide/ Protein	Sequence	Reference
SAF-p1	KIAALKQKIASLKQEIDALEYENDALEQ	26
SAF-p2	KIRALKAKNAHLKQEIAALEQEIAALEQ	26
SAF-p3	EIDALEYENDALEQKIAALKQKIASLKQ	26
Peptide 1	EIKQLESEISKLEQEIQSLEK	27
Peptide 2	EISQLESEISQLEQEIQSLES	27
Peptide 3	EIAQLEYEISQLEQEIQALES	27
Peptide 4	EIAQLEYEISQLEYEIQALES	27
7HSAP1	KLAQAVE KLARAVE KLAYANE KLARAVE KLAQAVE	28
GCN4-pAA	MKVKQLADAV EELASANYHL ANAVARLAKA VGER	28
Q	MRGSHHHHHHGSIEGRVKEITFLKNTAPQMLREL QETNAALQDVRELLRQQSKL	29
COMPcc^S	MRGSHHHHHHGSGDLAPQMLRELQETNAALQDVREL LRQQVKEITFLKNTVMESDASGKLN	30
TriByp1	MKQIEDKIEEILKKIYHIENEIARIKKLIGE	31
TriByp2	MKQIEDKIEEILKKIYHIEKEIARIKKLIGE	31
TriByp3	MKQIEKKIEEILKKIYHIEKEIARIKKLIGE	31
ELPAdpA3	$(C_{16}H_{32}O_2)_2$-K(VPGAG)$_3$, where C, H, O = carbon, hydrogen, oxygen	32
A192	G(VPGAG)$_{192}$Y	32
ELPAdpV3	$(C_{16}H_{32}O_2)_2$-K(VPGVG)$_3$, where C, H, O = carbon, hydrogen, oxygen	32
ELPAdpI3	$(C_{16}H_{32}O_2)_2$-K(VPGIG)$_3$, where C, H, O = carbon, hydrogen, oxygen	32
	Elastin-PLGA TGPG[VGRGD(VGVPG)$_6$]$_{20}$WPC-PLGA	33
ELP-PCL	LATQVGQKTA-PCL	34
Silk fibroin	repeated motifs [GAGSGA]$_n$ and [GAGXGA]$_n$ X = tyrosine or valine	35
SELP-815K	MDPVVLQRRDWENPGVTQLNRLAAHPPFASDPM [GAGS(GAGGS)$_2$(GVGVP)$_4$GKGVP(GVGVP)$_{11}$ (GAGAGS)$_5$(AGVP)$_4$GKGVP(GVGVP)$_{11}$ (GAGAGS)5AGA]$_6$MDPGRYQDLRSHHHHHH	36
FS$_5$	[TGRGDSPAGG(GAGAGS)$_3$AS]$_5$	37
FES$_8$	[TGRGDSPA-(GVPGV)$_2$GG(GAGAGS)$_3$AS]$_8$	37
H4	(POG)$_3$POGLIGLIG(POG)$_4$	38
H6	(POG)$_3$LIGLIGLIG(POG)$_4$	38
NdQ	NYFYSLFdQG, dQ = D-glutamine	38
CPA1	C16-AAAAAKKKKG(GPO)$_3$GFOGER(GPO)$_3$G	39
CPA2	C16-AAAAAKKKKG(GPO)$_8$G	39
CPA3	C16-AAAAAKKKKG(GPO)$_2$GFOGERGPOG	39

α-helical coiled-coil fiber assembly.[35] They have synthesized four peptides comprised of 21 residues, varying in the amino acids at the peripheral region (*b*, *c*, and *f* positions) (Figure 7.1(III), Table 7.1). The assembly of blunt-ended peptides occurs above a minimum concentration (0.1%), resulting in the formation of nanofibers 4 nm in diameter and hundreds of nanometers in length. Hydrophobic leucine and isoleucine in positions *a* and *d*, respectively, allow the coiled-coil dimeric assembly and positively charged glutamic acid in positions *e* and *g* to impart pH responsiveness to the nanofiber assembly. The length and diameter of the fibers can be controlled by the selection of amino acids at positions *b*, *c*, and *f* (Figure 7.1(III)).

Xu *et al.* have employed the crystal structure information of the coiled-coil GCN4-pAA to redesign it and generate 7HSAP1 peptide (Figure 7.2(I), Table 7.1). The seven-helix bundle of GCN4-pAA, with A to G individual helices, resembles a supramolecular lock washer that facilitates noncovalent interactions between the coiled-coil protomers due to the high surface area of the displaced edge at the interface between A and G helices.[36] Arginine at the *d* position in GCN4-pAA is mutated into leucine based on their hypothesis that the presence of a positively charged arginine residue at the *d* position breaks the hydrophobic periodicity of the heptad, preventing the higher-order assembly. Therefore, the single mutant leads to the formation of high-aspect-ratio fibers with a diameter of 3 nm (Figure 7.2(I), Table 7.1).

The Montclare group has explored the coiled-coil domain of cartilage oligomeric matrix protein (COMPcc) as a candidate for creating higher-order nanofibers.[28,29,37–41] In one study, the protein Q is generated by re-engineering COMPcc to better distribute the surface charge by eliminating the extraneous C-terminal heptad repeat and then swapping the C-terminal portion after the glutamine residue at the *d* position to the N-terminus, allowing the lateral assembly of nanofibers (Figure 7.2(II), Table 7.1).[39] The homopentameric α-helical coiled-coil nanoscale assemblies of fibers with diameter 20–560 nm further assemble into mesoscale fibers in the presence of the small molecule curcumin (Figure 7.2(II)). These engineered protein mesofibers with a diameter of 16 μm, comparable to the dimensions of natural collagen, α-keratin, and spider silk fibers, have potential applications in tissue engineering and drug delivery.

In the current work, they employ a COMPcc coiled-coil bearing two serine mutations denoted as $COMPcc^{S}$ that assembles into nanofibers for the encapsulation and delivery of BMS493, a pan-retinoic acid receptor inverse antagonist for the potential treatment of osteoarthritis (Table 7.1).[37] The self-assembled fibers of 10–15 nm in diameter successfully protect BMS493 from degradation and deliver it *in vitro* to downregulate MMP13 mRNA levels in human articular chondrocytes.

Nambiar *et al.* have recently reported the pH-controlled self-assembly of bipyridine functionalized coiled-coil peptides into nano- and mesoscale structures.[42] The peptides are synthesized using Fmoc (fluorenylmethyloxy-carbonyl)-based solid-phase synthesis and modified with bipyridine groups through on-resin functionalization. Studies conducted on three peptide sequences, TriByp1, TriByp2, and TriByp3, with one, two, and three bipyridine groups in each monomer, demonstrate change in morphology from mesoscale rectangular blocks to nanoscale squares with an increase in bipyridine moieties (Figure 7.2(III), Table 7.1). The constructs also demonstrate reversible control of assembly from micro- to nanostructure when the pH changes from neutral to acidic (Figure 7.2(III)), making them attractive candidates for pH-controlled therapeutic delivery.

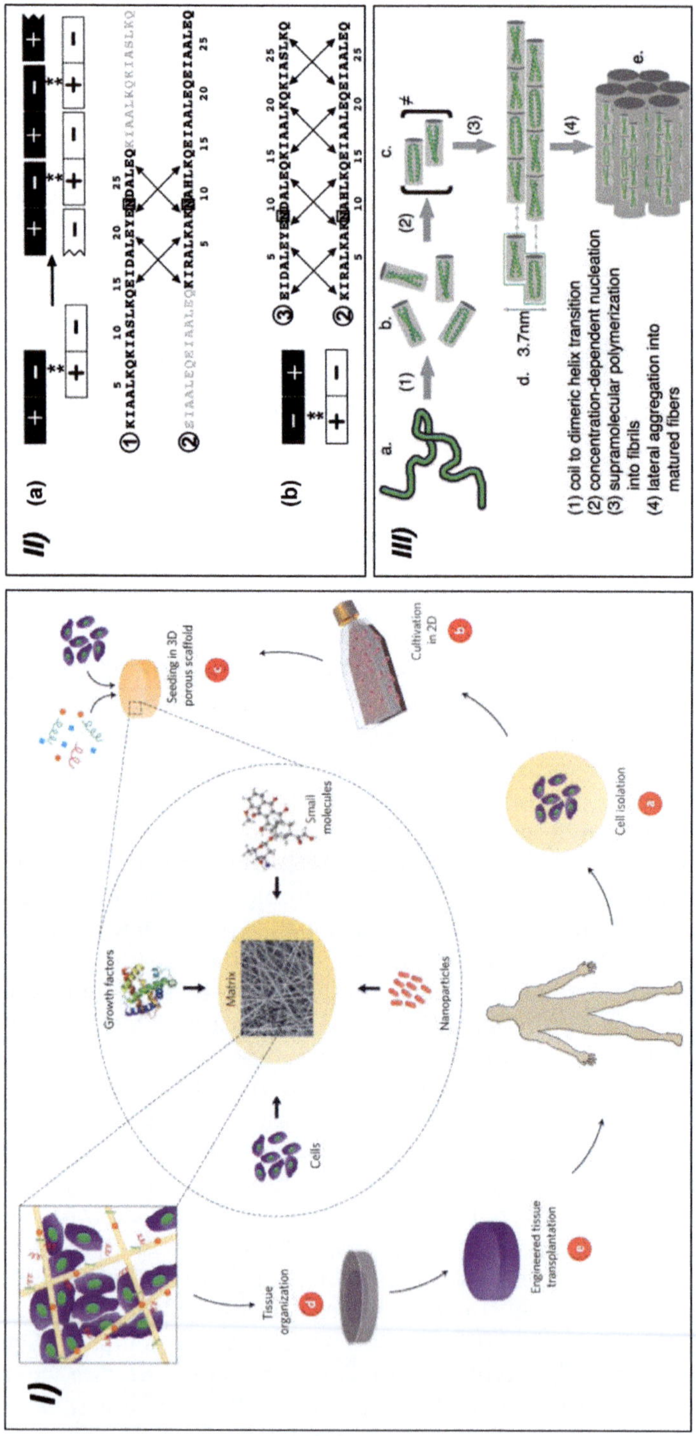
I)
Tissue organization
d
Engineered tissue transplantation
e
Cells
Growth factors
Matrix
Nanoparticles
Small molecules
Seeding in 3D porous scaffold
c
Cultivation in 2D
b
Cell isolation
a
II)
(a)
(b)
III)
a.
b.
c.
d. 3.7nm
e.
(1)
(2)
(3)
(4)
(1) coil to dimeric helix transition
(2) concentration-dependent nucleation
(3) supramolecular polymerization into fibrils
(4) lateral aggregation into matured fibers

Immense efforts have been made in recent years to explore α-helical coiled-coil protein-based nanofibers. A range of coiled-coil proteins/peptides have been shown to form supramolecular assemblies spanning from the nanoscale to the mesoscale. Furthermore, these α-helical assemblies can be modified with functional moieties for enabling a controlled switch to trigger reversible self-assembly. The incorporation of therapeutic molecules in the fibrous structures imparts them with another level of sophistication for biomedical applications.

7.2.2 Elastin and Elastin-like Polypeptide Nanofibers

Elastin is a major constituent in the ECM of mammalian connective tissues and organs, such as blood vessel linings, skin, and lungs. Elastin is responsible for imparting them with structural elasticity of varying degrees.[43] The precursor protein of elastin is called tropoelastin, a smaller, 67 kDa soluble molecule that crosslinks to form a larger, insoluble elastin product. Tropoelastin is a repetitive protein with alternating hydrophobic domains (such as tetra-, penta-, and hexapeptide sequences of VPGG, VPGVG, and VAPGVG) and hydrophilic domains consisting of mostly lysine and alanine residues. The hydrophobic domains are responsible for the two unique properties of elastin, namely its elasticity and thermal responsiveness. Tropoelastin makes use of its thermoresponsiveness during its maturation into elastin *via* its localization that occurs due to phase separation above its reversible inverse transition temperature (T_t).[43,44]

Artificial recombinant derivatives of elastin that reflect its unique properties are ELPs, consisting of the consensus sequence VPGXG, where X

Figure 7.1 (I) An example of a tissue engineering concept that involves seeding cells within porous, nanofibrous biomaterial scaffolds. (a) Cells are isolated from the patient. (b) Isolated cells are cultivated *in vitro* on 2D surfaces. (c) The cultivated cells are seeded in a 3D porous scaffold along with growth factors, small molecules, and micro- and/or nanoparticles. (d) Cell constructs are further cultivated in bioreactors to provide optimal conditions for organization into a functioning tissue. (e) Once a functioning tissue has been successfully engineered, the construct is transplanted on the defect to restore function. Reproduced from ref. 12 with permission from Macmillan Publishers Limited, Copyright 2011. (II) Design and sequences of the self-assembling fiber (SAF) peptides. (a) Proposed mechanism and amino acid sequences for a sticky-end assembly process. (b) Amino acid sequences of blunt-ended heterodimers. Reproduced from ref. 34 with permission from American Chemical Society, Copyright 2000. (III) Blunt-ended assembly of dimeric coiled-coil proteins. Proposed mechanism of self-assembly into nanofibers where (a) coiled-coil peptides assemble into (b) blunt-ended dimers, (c) antiparallel association of coiled-coils above a minimum concentration that serves as nucleation for fiber formation, (d) axial propagation of nanofibers into fibrils, and (e) lateral assembly of fibrils to form matured fibers. Reproduced from ref. 35 with permission from American Chemical Society, Copyright 2008.

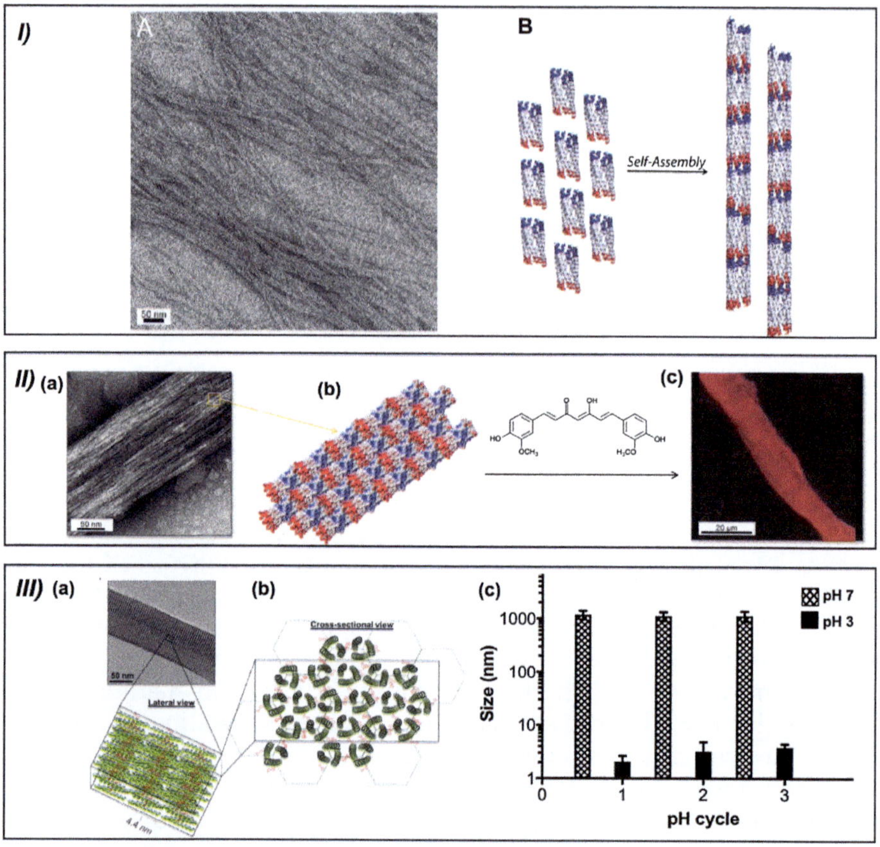

Figure 7.2 (I) Self-assembly of coiled-coil lock washer nanofibers. (A) Transmission electron micrograph (TEM) of 7HSAP1 nanotubes and (B) schematic representation of the assembly of the lock washer derived from the seven-helix bundle of the 7HSAP1 peptide. Blue surfaces represent the positively charged (N-terminal) heptads, and red represents negatively charged (C-terminal) heptads. Reproduced from ref. 36 with permission from American Chemical Society, Copyright 2013. (II) Homopentameric α-helical coiled-coil assemblies of Q fibers. (a) TEM of engineered protein fiber (b) Schematic representation of protein fiber assembly with staggered positive (red) and negative (blue) regions of the pentamer. (c) Reconstruction of 3D confocal XYZ data of protein in the presence of curcumin. Reproduced from ref. 39 with permission from American Chemical Society, Copyright 2014. (III) Self-assembly of bipyridine functionalized coiled-coil peptides. (a) TEM image of TriByp1 with inset depicting the coiled-coil trimer packing within the microstructure, (b) cross-sectional view depicting hexagonal close-packing of coiled-coils, and (c) reversible assembly of TriByp3 demonstrated by repeated shuttling of pH from 7.0 to 3.0 over three cycles. Reproduced from ref. 42 with permission from American Chemical Society, Copyright 2018.

is a guest residue that can be any amino acid other than proline.[45] ELPs are special because of their tendency to coacervate in solution above a certain inverse T_t that is unique to the particular pentapeptide. For various other smart polymers, the T_t has frequently been termed as the lower critical solution temperature (LCST).[46] The T_t depends on various factors such as the length of the polypeptide, the guest residue X, and environmental conditions such as ionic strength and pH. The tunability of T_t is useful in imparting thermoresponsiveness to the designed ELP biomaterial. This thermosensitive property can be harnessed to perform stimuli-responsive activities *in vivo,* with applications such as targeted drug delivery.[43] Essentially, protein nanofibers based on elastin and ELP-recombinant derivatives are not only useful in improving targeted drug delivery, but they have great promise for application in tissue engineering and gene therapy. This is due to their tunable self-assembling properties and thermoresponsive behavior.

Recent work by Aluri *et al.* has made use of this characteristic of ELPs to design peptide amphiphiles (PAs) that undergo self-assembly into nanofibers with the potential to deliver biological molecules incorporated in the scaffolds because of their T_t behavior.[47] These human tropoelastin-inspired ELPAs (elastin-like peptide amphiphiles) are the first ones to be reported.

They describe a new ELPA, dpX3, derived from small, non-phase-separating ELPs (Figure 7.3(II), Table 7.1). Three ELP-based peptide variants are synthesized using Fmoc solid-phase peptide synthesis. Each peptide consists of a short ELP sequence with Ala, Val, or Ile as a guest molecule, termed A3, V3, and I3, respectively. The N-termini of the peptides are dipalmitoylated to form the following ELPAs, dpA3, dpV3, and dpI3, respectively (Figure 7.3(III), Table 7.1). The addition of two saturated 16-carbon chains to these ELPs allows them to phase separate above a certain inverse T_t comparable to higher-molecular-weight ELPs such as A192 (Figure 7.3(II), Table 7.1). The ELPAs exhibit unique properties of being neutral in charge and exhibiting inverse temperature-dependent phase-separating behavior, making them useful for *in vivo* applications.[47]

Among these ELPAs, dpA3 forms nanofiber-like micelles of 1.7 ± 0.2 μm in length and 16.0 ± 2.0 nm in width above their critical micellar concentration (CMC), as confirmed using transmission electron microscopy (TEM). Atomic force microscopy (AFM) confirms that the nanofiber assembly is comprised of bundles of cylindrical micelles, with each individual strand bearing an average width of 13.9 ± 1.4 nm (Figure 7.3(III) a, b). The lengths can also be tuned by the addition of end-capping phospholipid, 1,2-dioleoyl-*sn*-glycero-3-phosphoethanolamine (DOPE) (Figure 7.3(III) c). The ratio of dpA3 to DOPE is varied to demonstrate that increasing the phospholipid leads to a decrease in the length of the assembly and increased cellular uptake. Furthermore, upon reaching equimolar ratios of dpA3 to DOPE, the length of the resultant nanofiber decreases 100-fold without any change in its radius, demonstrating for the first time a high degree of PA nanofiber length control.[47] More importantly, the high encapsulation efficiency of the small chemotherapeutic

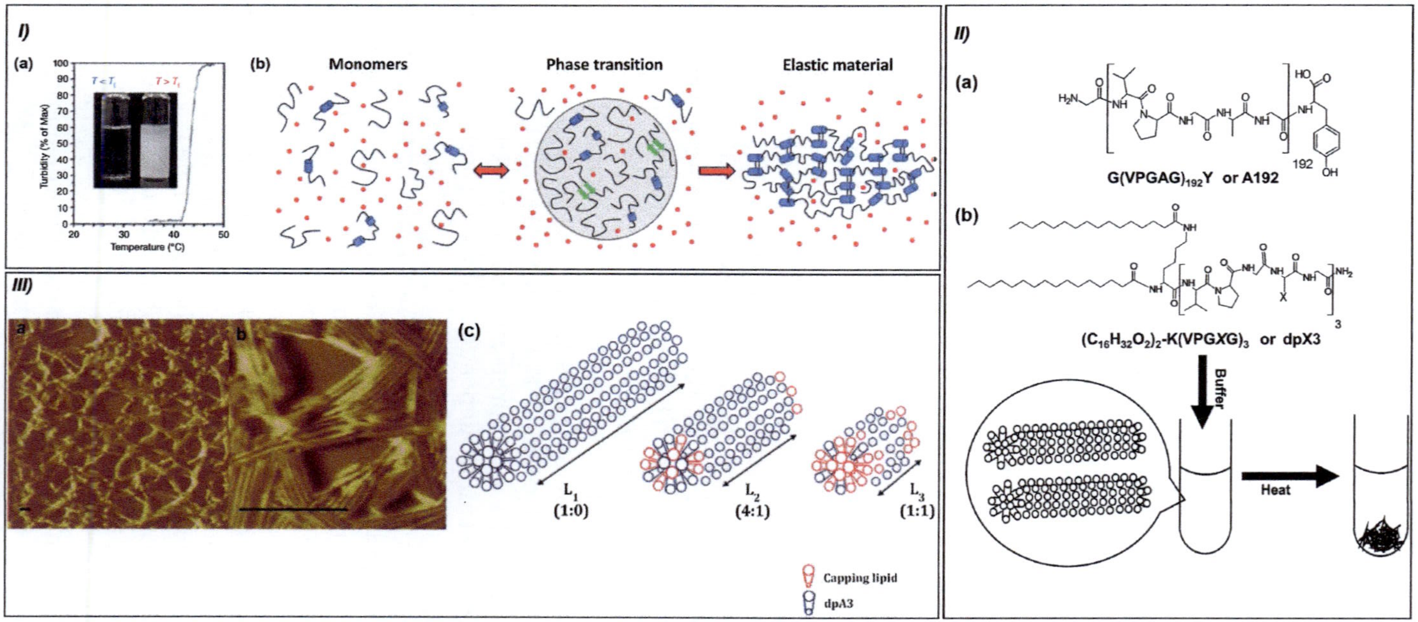

Figure 7.3 (I) A model for the self-assembly of ELPs undergoing phase separative coacervation with an increase in temperature and salt concentration. Reproduced from ref. 130 with permission from the National Academy of Sciences. (II) Design of ELPAs with T_t behavior. (a) A high-molecular-weight ELP, A192. (b) A low-molecular-weight ELPA construct modified with a dipalmitoyl lysine, dpA3. Reproduced from ref. 35 with permission from American Chemical Society, Copyright 2012. (III) ELPA nanofiber assembly. (a), (b) Bundles of cylindrical micelles visualized by AFM. (Scale bar = 250 nm). (c) Illustration of how increased DOPE:dpA3 ratios can modify nanofiber aspect ratio by reducing their length, making their cellular entry easier. Adapted from ref. 47 with permission from American Chemical Society, Copyright 2012.

hydrophobic molecule paclitaxel (PAX) into the hydrophobic pore of ELPA nanofibers is reported. Drug-loaded nanofibers also inhibit HeLa cell growth to the same degree as soluble PAX in dimethyl sulfoxide (DMSO), suggesting the potential use of dpA3 nanofibers as a delivery vehicle for chemotherapeutic cargo such as PAX.[47] This study demonstrates ELP-based nanofiber PAs with tunable length and thermoresponsiveness for applications in drug delivery and tissue engineering.

Foraida *et al.* have developed elastin-based hybrid nanofibers that display a significant advantage over purely synthetic nanofibers because of their increased hydrophilicity, elasticity, and cytocompatibility.[48] Hybrid elastin/poly(D-lactide-*co*-glycolide) (PLGA) electrospun nanofibers (Figure 7.4(I), Table 7.1) are generated either by blend electrospinning (EP) termed as EP-blend or by covalent surface modification (EP-covalent). The EP-blend nanofibers promote enhanced proliferation and polarization of modulated submandibular immortalized mouse salivary gland epithelial cells (SIMS) along the apical-basal cell axis, increased cell dampening (necessary for cell–scaffold interaction), greater elasticity, and biomimetic chemical structure as compared to purely synthetic PGLA nanofibers that have been previously developed by the group.[48,49] EP-covalent scaffolds show a significant increase in wettability due to higher elastin incorporation and a slight increase in the average height of the cell monolayer; however, it exhibits higher stiffness (Young's modulus: 20.72 MPa) as compared to control PLGA nanofibers (4.29 MPa).[48] An increase in the organization of epithelial cells is demonstrated on EP-blend scaffolds when compared to EP-covalent and control scaffolds, indicating that the elasticity of nanofibers rather than their chemical composition governs the cellular growth and organization.

ELP-based smart nanofibers can be used to deliver therapeutic genes to diseased tissues with the aim to restore the functionality of the dysfunctional gene(s) and correct the diseased phenotype.[50] Viruses are useful tools for gene therapy as they are known for infecting cells and delivering the genes packaged within the viral genome to the target cell.[51] Particularly for tissue engineering applications, viral vectors are used to induce genetic modification in cells that are, in turn, already supported by natural or hybrid biomaterial polymer scaffolds. The resultant applications of this form of viral gene delivery include bone and cartilage tissue regeneration, wound healing, and angiogenesis within diseased cardiac tissue.[52] A significant drawback with using viruses is their potential immunogenicity as well as possible interactions with the biomaterial scaffolds. To overcome these challenges, new and better viral vectors need to be used in combination with the most compatible scaffold materials. AAVs (adeno-associated viruses) hold great promise in this area as they exhibit no toxicity and immunogenicity in addition to the high efficiency of gene delivery.[50,51] Lee *et al.* have investigated a novel AAV r3.45 that previously exhibited high transduction efficiency in fibroblast infections. ELP/poly(ε-caprolactone) (PCL) electrospun nanofibers encapsulating the viral vector are employed to transduce NIH-3T3 fibroblasts seeded onto the scaffold. PCL, a

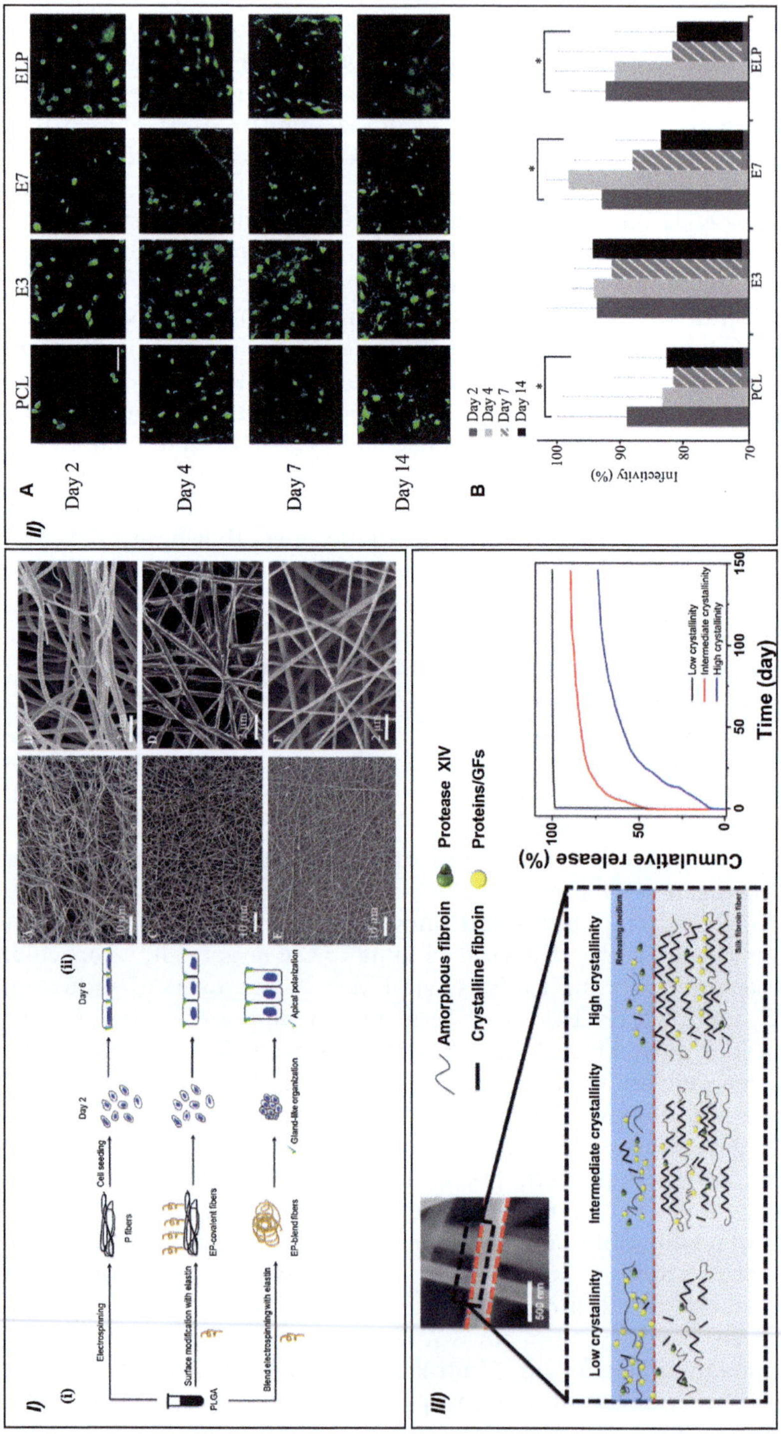
I)
(i)
(ii)
II)
A
Day 2
Day 4
Day 7
Day 14
PCL
E3
E7
ELP
B
Infectivity (%)
III)
Low crystallinity
Intermediate crystallinity
High crystallinity
Amorphous fibroin
Crystalline fibroin
Protease XIV
Proteins/GFs
Cumulative release (%)
Time (day)

biodegradable and cytocompatible synthetic polymer, is blended with the ELP domain composed of 128 repeating units of [VPGVG] to improve the mechanical properties of elastin-based scaffolds (Figure 7.4(II), Table 7.1). Gene delivery is effectively controlled by varying temperatures (*i.e.*, 4, 25, and 37 °C) and scaffold composition (ELP/PCL 0/100 (E0), 30/70 (E3), 50/50 (E5), 70/30 (E7), and 100/0 (E10)). Transduction efficiency is measured by observing differences in GFP expression under different conditions (Figure 7.4(II)). All the scaffolds show >80% GFP expression on the first day; E3 demonstrates constant release of AAV while maintaining high cellular transduction over a period of 14 days (Figure 7.4(II)). The ability of ELP-based nanofibers to modulate gene delivery and cellular transduction displays potential for smart gene therapy-induced tissue regeneration.

Elastin-like proteins, as a result of their elasticity and thermal responsiveness, have been explored to generate smart nanofibrous proteins. As demonstrated above, elastin-based nanofibers have been employed for therapeutic delivery, tissue engineering, and viral gene therapy-induced tissue regeneration. Thus, ELP-based nanofibers are promising biomaterials for a variety of tissue engineering applications.

7.2.3 Silk and Silk-like Protein Nanofibers

Silk is a natural protein produced by insects, spiders, and worms. Sericin and fibroin are the two main protein components of silk.[53] Sericin is the sticky substance present on the outside of silk fibers and needs to be removed in order to reach the second component, fibroin. Fibroin is more sought after because of its many unique properties, such as its high tensile strength and elastic modulus.[43,54] Among the different organisms that produce silk for structural support, the most studied are the silkworm *Bombyx mori* and spiders. Dragline spider silk is produced by the *Nephila clavipes* spider and has been known to be the toughest silk. Despite showing variations in its amino acid sequence based on the species that produces it, silk is made up of hydrophobic blocks composed of glycine, alanine, and

Figure 7.4 (I) Elastin/PLGA hybrid nanofibers. (i) Illustration of differential epithelial cell polarization as a result of differential methods of EP incorporation in PGLA-based nanofiber production.[48] (ii) SEM characterization of the nanofiber scaffolds. A, B: EP-blend nanofibers; C, D: EP-covalent nanofibers; and E, F: PLGA nanofibers. Reproduced from ref. 48 with permission from Elsevier, Copyright 2017. (II) Transduction of NIH-3T3 cells on various ELP/PCL nanofiber composites. (A) Fluorescence images of GFP-expressing cells at different time points. The scale bar is 50 μm. (B) Percentage of GFP-expressing cells on each scaffold. Reproduced from ref. 52 with permission from Elsevier, Copyright 2011. (III) Silk-based smart nanofibers. Proposed mechanism illustrating the effect of change in crystallinity of silk fibroin fibers on the hPL release from the nanofibers, demonstrating the 'smartness' of the material. Reproduced from ref. 58 with permission from Elsevier, Copyright 2018.

serine residues and hydrophilic blocks composed of charged amino acids.[43,54] The former imparts silk its high tensile strength by allowing it to form β-sheets, whereas the latter is responsible for silk fibroin's high elasticity. Silk fibroin produced by silkworm is not only cheap and easily available but is known for its use in biomaterial production due to its high biocompatibility, bioavailability, tunability, biodegradability, and low immunogenicity.[55]

A recent example of silk fibroin electrospun nanofiber usage in damaged tissue regeneration employs the incorporation of human platelet lysate (hPL) into the fibers during initial fiber production (Figure 7.4(III), Table 7.1). To obtain silk fibroin-based nanofibers, 25% poly(ethylene oxide) (PEO) (w/w_{silk}) is added to 60 mg mL^{-1} fibroin and allowed to stir overnight to obtain a homogenous solution. Fluorescein isothiocyanate (FITC)/albumin (2% w/w_{silk}) and hPL (7% w/w_{silk}) are added to the silk fibroin solution and nanofibers are spun *via* the electrospinning process. hPL is a highly nutritious conglomerate of multiple growth factors such as platelet-derived growth factor B (PDGF-BB), fibroblast growth factor (FGF), vascular endothelial growth factor (VEGF), epidermal growth factor (EGF), and transforming growth factor-beta 1 (TGF-β1), as well as cytokines that offer enormous potential for therapeutic recovery of damaged tissues.[56] Studies have shown hPL to be useful in many aspects of wound healing, such as cell growth, angiogenesis, regenerating bone tissue, and the recovery of wounds that are difficult to heal.[57] However, the limiting factors in employing hPL as a therapeutic tool are its low shelf-life at room temperature and difficulty in handling hPL gels that are currently used in the medical setting. Silk nanofibers can therefore be exploited for the controlled release of hPL by prolonging the biological activity of the lysate and improving its shelf-life *via* the mechanism of modulating the crystallinity of the fibroin protein itself (Figure 7.4(III)). It is useful for regulating the therapeutic concentration of the drug to avoid toxicity issues. This strategy also reduces the need for repeated administration over weeks, as there is a sustained release over time, under compatible conditions in terms of pH and temperature.[58]

Guzman-Puyol *et al.* have fabricated electrospun nanofibrous mats by blending fibroin with high-purity microcrystalline cellulose (MCC) or cellulose-rich parsley waste in equal proportions (Table 7.1).[59,60] Mechanical characterization reveals that the addition of MCC improves the stiffness of the mats and the use of cellulose-rich parsley produces mats with similar elastic modulus to fibroin-based mats, but with a greater extension capacity when loaded maximally. Cellulose-rich parsley also serves as a bioactive molecule with antioxidant properties for use in biomedicine. All of the prepared nanofibers with various compositions (fibroin, MCC/fibroin, and parsley/fibroin) demonstrate excellent cytocompatibility as well as promote proliferation of adhered fibroblasts, as shown by a colorimetric MTT assay performed after 24 hours of cell culture in Figure 7.5(I).[60]

Since naturally occurring silk is not pure and can cause an immunogenic response in live tissues, silk-like polypeptides (SLPs) are engineered to maintain the core properties of silk.[61] Cell adhesion domains like RGD can also be added to SLP for tissue engineering applications, and they can be designed to be free from any potential immunogenicity. Yang and coworkers have biosynthesized silk-like proteins FS_5 and FES_8, which both include the crystalline domain of silk fibroin $(GGAGS)_3$ and the TGRGDSPA sequence from fibronectin containing the cell-adhesive RGD domain. FES_8 is designed to also include the elastin domain $(GVPGV)_2$ between the silk and RGD domains. The incorporation of the elastin domain allows the processability of FES_8 into nonwoven fibers *via* electrospinning, whereas FS_5 could not be processed into fibers. Their studies also confirm greater cell adhesion for normal human dermal fibroblasts (NHDF) and African green monkey kidney (VERO) cells on both the SLPs when compared to silk fibroin only.[62]

Silk-elastin-like proteins (SELPs) are recombinant polypeptides made from genetically engineered hybrids that are a combination of silk and elastin.[63] This yields a polymer with a combination of physical and mechanical properties from silk as well as elastin. This unique combination of silk and elastin leads to the formation of a smart or stimuli-responsive nanobiomaterial due to the elastin domain's temperature sensitivity, as discussed in the previous section.[63] Furthermore, the smart characteristics of SELP proteins can be altered by varying the protein sequence, the number of amino acids, and the identity of the X residue in the elastin sequence (GXGVP).[43]

Mechanical force has been shown to induce chemical changes and trigger targeted delivery of molecules.[64] Chang *et al.* have used AFM to study the effect of mechanical force on the self-assembly of peptides.[44] Mechanical force can break the weak nanofibers that ultimately served as nucleation sites for the growth of new nanofibers. Preadsorbed silk-elastin-like (SELP-815k) polymers, composed of 8 silk units, 15 elastin units, and 1 lysine unit, are reported to undergo spontaneous local self-assembly into functional nanofibers on a mica substrate (Figure 7.5(II), Table 7.1). Upon application of nanomechanical stimulus using AFM in the tapping mode, the rate of self-assembly rises significantly, leading to the formation of nanofibrous structures. The assembly of nanofibers has been visualized by AFM over the time course of 28 minutes (Figure 7.5(II)). Although the impact of mechanical forces on nucleation rates of existing nanofibers had been reported before,[65] the above study implies that the growth of new nanofiber nuclei can be induced and enhanced by modulating nanomechanical force.

Overall, silk fibroin and recombinant variations of silk such as SLP and SELP-based protein nanofibers have been employed in tissue engineering for wound repair as well as scaffold formation and stimuli-induced targeted drug delivery. Recombinant silk-based proteins have been combined with elastin to impart unique mechanical and physical properties to the nanofibers. Bioengineering techniques facilitate fine-tuning of silk-based nanofibers for their applications in tissue engineering and regeneration and wound healing.[58,60,63] Thus, SELP-based nanofibers are another promising class of proteins with an exciting future in biomedical therapeutics research.

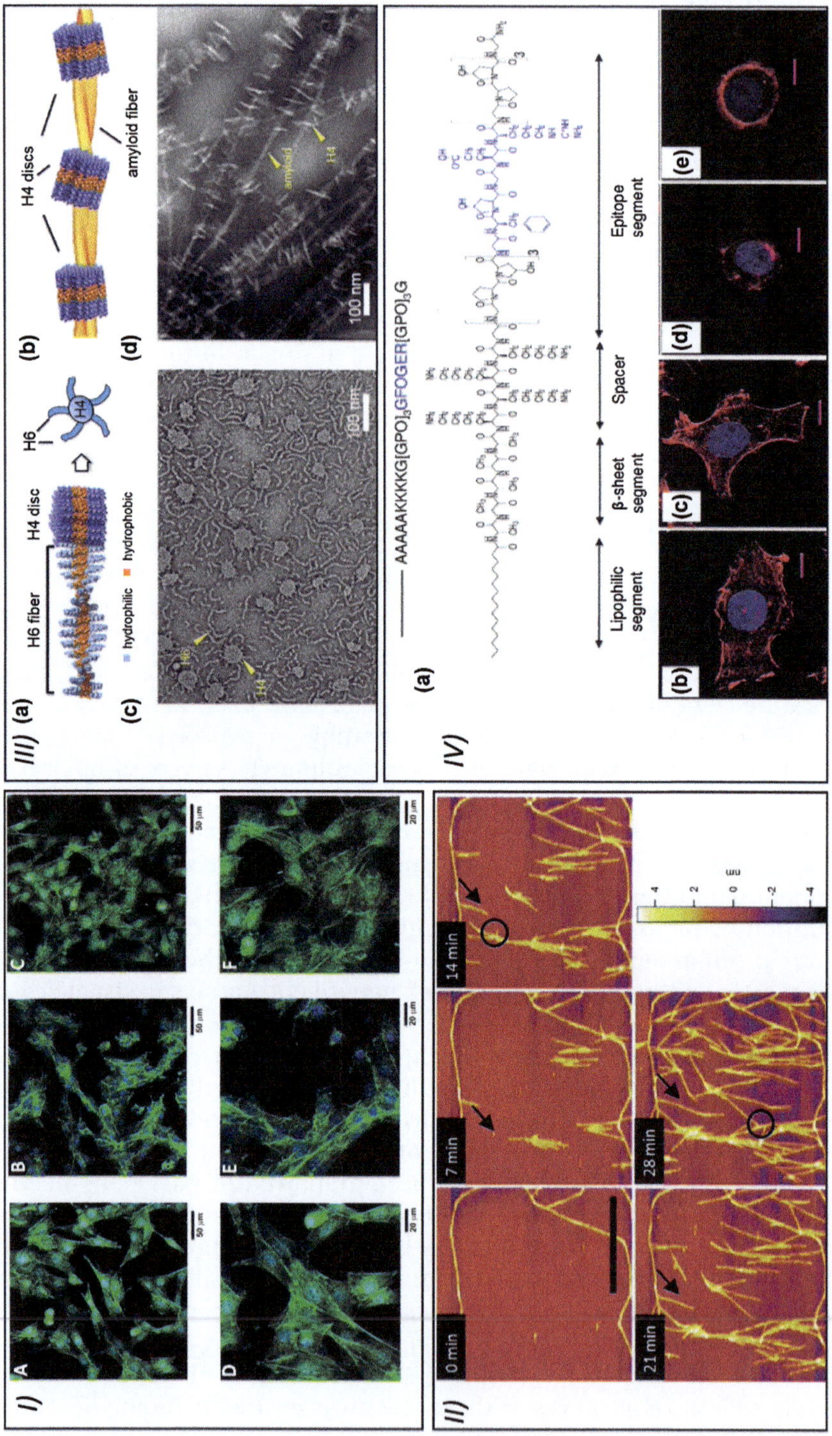

I)
II)
0 min
7 min
14 min
21 min
28 min
nm
III) (a)
H6 fiber
H4 disc
H6
H4
hydrophilic
hydrophobic
(b)
H4 discs
amyloid fiber
(c)
(d)
100 nm
IV)
(a)
AAAAAKKKKG[GPO]3GFOGER[GPO]3G
Lipophilic segment
β-sheet segment
Spacer
Epitope segment
(b)
(c)
(d)
(e)

7.2.4 Collagen and Collagen-mimetic Peptides

Collagen is the most abundant structural protein found in the ECM of mammalian connective tissues.[66] The primary amino acid sequence of native collagen is G-X-Y repeated, where X and Y are guest amino acids, most commonly proline and posttranslationally modified hydroxyproline, respectively.[67] The structure of collagen is a triple helix composed of three left-handed polyproline II-type helices.[68] This secondary structure is made possible due to the presence of the G residue, which is capable of forming intermolecular hydrogen bonds with hydroxyproline. Up to 29 different types (type I to XXIX) of collagen have been described. The quaternary structural arrangement of these triple helices is determined by the type of collagen. Types I–III, V, and XI are fibrous, formed by crosslinking as a result of lysine oxidase activity.[43,54,69] Among them, collagen type I is the most abundant and, therefore, most commonly used in designing artificial scaffolds for tissue engineering applications.

Various biomimetic tissue-engineered scaffolds closely approximating the natural ECM have been made *via* engineered collagen fibers.[70] They have been employed for regeneration of connective tissues such as bone, cartilage, skin, muscle, and nerve and for wound healing. The collagen-inspired biomaterials are especially useful in these applications because of their properties of cell adherence and the ability to stimulate cell proliferation due to their interaction with integrins at cell surfaces.[43]

Since animal-derived collagen presents many limitations such as batch-to-batch variations, lack of precise tunability of physical properties, and a limited quantity of collagens other than type I being harvested, the synthesis of collagen by chemical or biological methods is performed to address all the aforementioned problems. Collagen-mimetic peptides/proteins (CMPs)

Figure 7.5 (I) Biocompatibility assay of various fibroin-based nanofibers. Fluorescence imaging showing fibroblast growth on (A, D) fibroin; (B, E) MCC/fibroin; and (C, F) parsley/fibroin nanofibers. Reproduced from ref. 60 with permission from American Chemical Society, Copyright 2016. (II) Nanomechanical stimulus-induced SELP nanofiber assembly. AFM images of SELP nanofibrous assembly as a function of time obtained over time. Reproduced from ref. 44 with permission from American Chemical Society, Copyright 2011. (III) Hydrophobicity-driven self-assembly of CMPs. (a) Models suggesting the formation of nanostar structures due to hydrophobic interactions between H6 fiber and H4 disk upon mixing. (b) Model showing hydrophobic amyloid fiber of NdQ peptide interacting with the hydrophobic edges of H4 disks. (c) TEM of a 2 : 1 ratio of H6:H4 peptides displays H6 fibers attached to the disk edges. (d) TEM of a 1:1 ratio of H4 disks mixed with amyloids fibers. Reproduced from ref. 73 with permission from American Chemical Society, Copyright 2014. Further permissions related to the material excerpted should be directed to the ACS. (IV) Self-assembly and properties of CPAs. (a) CPA1 containing lipophilic, β-sheet, spacer, and epitope segments. Cell adhesion and spreading of HepG2 cells on (b) collagen, (c) CPA1, (d) CPA2, and (e) CPA3. Reproduced from ref. 74 with permission from American Chemical Society, Copyright 2011.

consist of collagen-like domains or short assemblies of the collagen consensus sequence GXY. Recombinant CMPs have been produced in numerous organisms such as bacteria, yeast, mammalian cells, and mice.[54] Cell adhesion domains such as RGD built within the CMPs contribute to adhesive as well as cell signaling properties *via* interacting with integrin receptors on cell surfaces.[71] Collagen has also been blended with other artificial polymers to improve and customize its mechanical properties and bioactivity.[72] Its overall biocompatibility, strength, hydrophilicity, lack of immunogenicity, and role in tissue regeneration render collagen an exemplary polymer for tissue engineering applications.

Nanda and coworkers explored hydrophobicity-driven collagen assembly by designing CMPs bearing a group of adjoining hydrophobic residues.[73] Minute changes in the position of hydrophobic residues led to drastic variations in the morphology. Diffusion-limited aggregation (DLA) simulations were used to study the hydrophobic self-assembly of peptides. The H4 peptide displayed disk-like morphology as observed from TEM images, consistent with the DLA prediction. The curved H4 disks possessed diameters ranging from 50 nm to 1 μm, and the thickness (~10 nm) of the disks was comparable to the length of a triple-helix. Additional hydrophobic sequences were added in the H6 peptide to produce disk-like morphologies as predicted by DLA. Nevertheless, fibers of H6 varying from 50 nm to microns in length and 6–11 nm in width were observed. H6 fibers assembled into bundles of micron-scale long fibers upon heating at 30 °C and annealing at 4 °C. As suggested by DLA, the mixture of H4 disks and H6 fibers formed nanostar-like structures due to hydrophobic interactions (Figure 7.5(III) a, c). H4 peptides were further shown to interact with a hydrophobic amyloidogenic peptide (NdQ) to produce micron-scale long fibers with multiple disks attached perpendicularly (Figure 7.5(III) b, d).

A novel collagen-mimetic peptide amphiphile (CPA) has been designed by Tong and coworkers by conjugating CMPs to a single hydrophobic carbon tail.[74] CPA self-assembles into nanofibers with the collagen triple-helix structure. CPAs, synthesized by solid-phase peptide synthesis, are composed of four segments: an N-terminal palmityl segment that spurs the aggregation of amphiphilic molecules into micelles, a β-sheet segment for the conversion of micelles to nanofibers by H-bonding, an epitome segment comprising CMP, and a lysine spacer to connect the epitome with the β-sheet and palmityl segments (Figure 7.5(IV) a). Three CPAs have been designed: CPA1, which consisted of a CMP with $(GPO)_3$ GFOGER$(GPO)_3$ in the epitome segment; CPA2 with no GFOGER motif; and CPA3 with the shorter $(GPO)_2$GFOGERGP motif. All the CPAs self-assemble into nanofibers with lengths reaching microns as observed from TEM micrographs of CPA1. The formation of a nanofibrous gel is also observed due to the entanglement of nanofibers upon an increase in the concentration of CPAs. The bioactivity of CPAs is examined with HepG2 cells. Both collagen and CPA1 exhibit high cell adhesion and spreading (Figure 7.5(IV) b, c), whereas decreased HepG2 cell spreading is observed in the case of CPA2 and CPA3 due to the absence of GFOGER or the triple-helical structure (Figure 7.5(IV) d, e). The results demonstrate that CPA nanofibers resemble the structure and biological properties of collagen fibers. The applications of collagen in tissue engineering are limited due to its immunogenicity and complexity in modification.[74] Efforts have been made to mimic the structure and biofunctions of collagen. The aforementioned properties of CMPs demonstrate their ability to not only self-assemble into triple-helix nanofibers but also illustrate similar biological functions as collagen and have garnered significant interest for tissue engineering applications.

7.3 Polymer-based Smart Nanofibers

Several polymer-based nanofibrous scaffolds closely resemble the structure and properties of native ECM. The characteristics of ECM are well demonstrated by the use of polymeric scaffolds for various tissue engineering applications, such as cartilage,[75] neuron,[76] bone,[77] and muscle.[78] Although the carbon-rich backbone of polymers renders them closer to biological tissues than inorganic materials, the presence of certain degradation products and modifiers such as plasticizers, coupling agents, and nucleating agents can interfere with biological systems.[79] In conjunction with physical properties, biodegradability and cytocompatibility are key parameters in polymer material selection for scaffold design. Biodegradable synthetic polymers such as PCL, poly(D,L-lactide) (PLA), poly(ethylene glycol) (PEG), and PLGA generally undergo hydrolytic degradation.[80] Smart polymers with responsiveness provide additional merits by encapsulating therapeutic molecules, such as drugs, growth factors, protein, DNA, and RNA, that are beneficial for tissue regeneration; they can enable controlled delivery of such therapeutic molecules in a biologically active state.[81,82] Furthermore, stimuli-responsive polymers allow site-specific targeting of molecules, enabling the administration of drugs at low dosage, leading to low systemic toxicity.[83]

Small physicochemical variations in the surrounding environment of polymer nanofibers can induce structural changes that lead to the release of bioactive molecules.[84] Smart polymers undergo reversible physical or chemical transitions based on the type of stimuli. These include pH-dependent electrostatic interactions of acidic or basic groups in the polymeric system, temperature-driven changes in hydrophilicity and lipophilicity, electroresponsive swelling and shrinking of conductive polymers, mechanically induced stress on piezoelectric polymers to generate electricity, and magnetically induced dipoles that cause interaction among particles of magnetoresponsive polymers.[85,86] Most changes are reversible, while in certain circumstances, the changes in polymer conformation can be irreversible.[82]

The most common and cost-effective technique to fabricate polymeric nanofibrous scaffolds is electrospinning.[87] Fibers are produced by applying a high-voltage electric field onto a polymer solution to extrude from a needle/spinneret in the form of a Taylor cone (Figure 7.6(I)). The solution is then dried along the path and the solid nanofibers are collected on a target. There

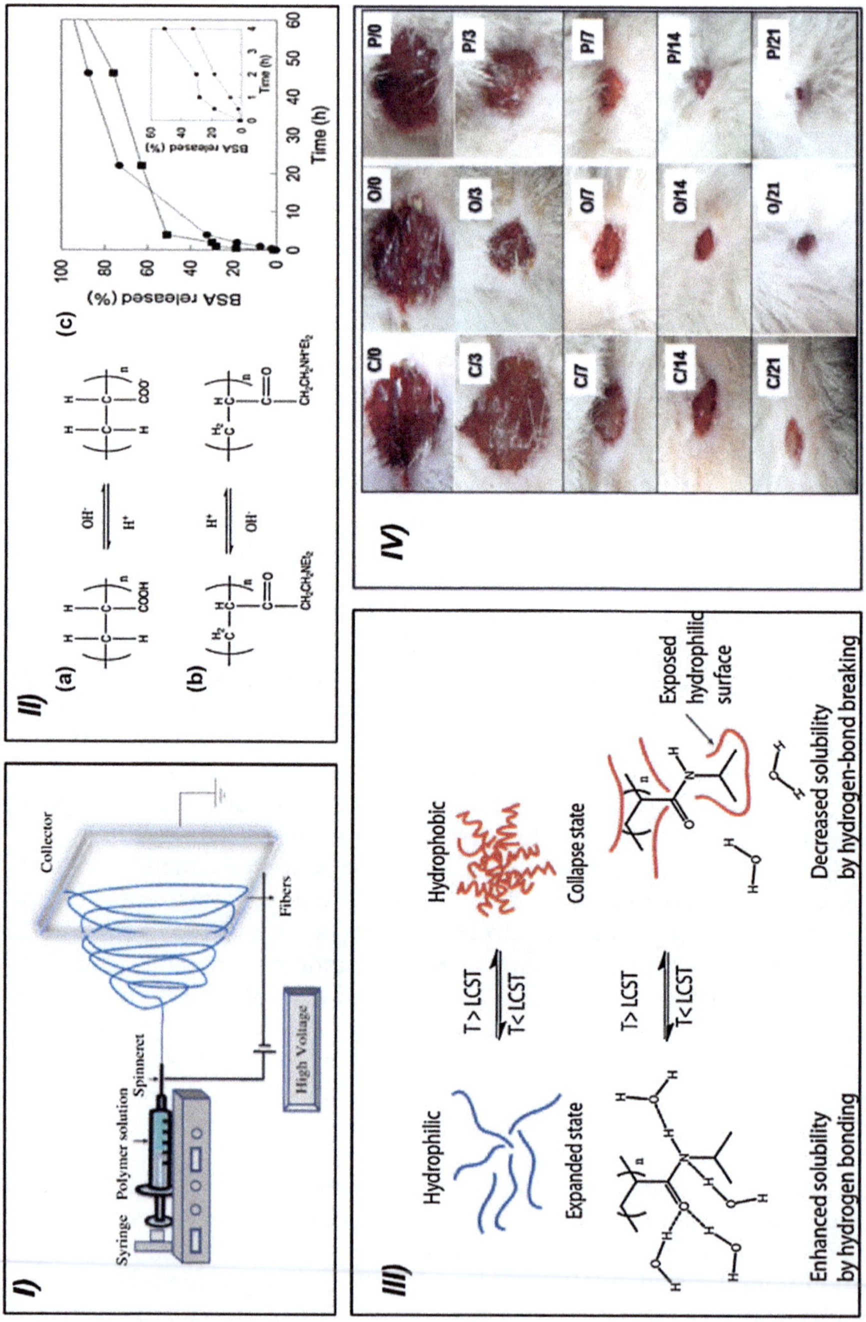
I)
Syringe
Polymer solution
Spinneret
Collector
Fibers
High Voltage
II)
(a)
(b)
(c)
BSA released (%)
Time (h)
III)
Hydrophilic
Expanded state
T > LCST
T < LCST
Hydrophobic
Collapse state
Exposed hydrophilic surface
Enhanced solubility by hydrogen bonding
Decreased solubility by hydrogen-bond breaking
IV)
C/0
O/0
P/0
C/3
O/3
P/3
C/7
O/7
P/7
C/14
O/14
P/14
C/21
O/21
P/21

Table 7.2 Various stimuli and responsive polymers.

Stimulus	Responsive polymer	References
pH	poly(acrylic acid), poly(methacrylic acid), poly(D,L-lactide)-poly(ethylene glycol), poly(*N*,*N*-dimethylaminoethyl-methacrylate), poly(*N*,*N*-diethylaminoethyl-methacrylate), chitosan, poly(ε-caprolactone)	88, 89, 90, 91, 92
Temperature	poly(*N*-isopropylacrylamide), poly(vinyl caprolactam), poly(ε-caprolactone)/poly(*N*-isopropylacrylamide)/egg albumin	89, 93, 94
Electric field	polypyrrole (PPy-AF, PPy-RF), polyaniline (PACL-2, PACL-3)	95, 96
Mechanical stress	polyvinylidene fluoride, polyvinylidene fluoride-trifluoroethylene polyvinylidene fluoride/polyurethane	77, 97, 98
Magnetic field	Superparamagnetic nanoparticles–incorporated polymers	99, 100, 101

are various parameters that govern the size and morphology of the fibers, including the choice of solvent, concentration and conductivity of the polymer solution, the voltage applied, and the distance between the spinneret and collector. The main characteristic features of electrospun nanofibers are high surface-area-to-volume ratio, high porosity, and ease of functionalization. Polymer-based smart nanofibers are promising candidates for tissue engineering scaffolds due to their cost-effective and facile synthesis, versatility, and ability to respond to various stimuli (Table 7.2).[82] In this section, we focus on various smart polymeric nanofibers: pH-responsive, thermosensitive, piezoelectric, and electro- and magneto-responsive.

7.3.1 pH-responsive Nanofibers

The human body has considerable variation in pH among various organs, with acidic pH (1–3) in the stomach to alkaline pH (7.9–8.5) in the colon.[88]

Figure 7.6 (I) Polymer-based nanofibers. Representation of electrospinning setup for fabrication of nanofibers. Reproduced from ref. 131 with permission from Elsevier, Copyright 2010. (II) pH-responsive nanofibers. Structural changes in (a) poly(acrylic acid) and (b) poly(*N*,*N*-diethylaminoethyl-methacrylate) with variation in pH. (c) Release of BSA from chitosan nanofiber (CNF) PEG hydrogels at pH 5.6 (■) and 7.4 (●). Reproduced from ref. 92 with permission from Elsevier, Copyright 2015. (III) LCST behavior of PNIPAAm. Reproduced from ref. 132 with permission from the Royal Society of Chemistry. (IV) Thermoresponsive PCL/PNIPAAm/egg albumin polyblend scaffolds. *In vivo* wound healing in a rat model with nanofiber formulations O, P (10%, 15% Gati) and control (C, no Gati) at 0, 7, 14, and 21 days. Reproduced from ref. 102 with permission from Elsevier, Copyright 2015.

The pH of the tumor's extracellular environment,[89] as well as that of inflamed tissues,[90] is slightly acidic. These attributes make pH-sensitive polymers a good candidate for the delivery of active biomolecules. In pH-sensitive polymers, acidic or basic moieties are present in the polymer architecture, resulting in stimuli responsiveness by either protonation or deprotonation with pH change in the surrounding environment. The polymers bearing weak basic groups such as poly(*N*,*N*-dimethylaminoethyl-methacrylate), poly(*N*,*N*-diethylaminoethyl-methacrylate), and chitosan are categorized under polybases (Table 7.2).[91] The amino group protonates under acidic conditions and the electrostatic effect causes swelling of fibers to trigger the drug release (Figure 7.6(II) b). Nitta and coworkers have reported chitosan and PEG-based nanofibrous hydrogels incorporating bovine serum albumin (BSA) (Table 7.2).[92] The nanofibrous hydrogel exhibits enhanced release of BSA with variation in pH from neutral to acidic due to the higher swelling rate of the fibers at lower pH. Approximately 32.4% and 51.5% release of BSA is observed in 4 hours when the pH of the buffer is 7.4 and 5.6, respectively (Figure 7.6(II) c). These results imply the applicability of these nanofibrous hydrogel scaffolds to carry bioactive substances and demonstrate pH-responsive drug release behavior.

pH-responsive polymers with acidic pendant groups are termed polyacids, for example, poly(acrylic acid) and poly(methacrylic acid) (Table 7.2).[93] The polyacids, depending on their pKa, become ionized with a change in pH that leads to repulsion in the negatively charged groups and subsequently causes structural changes in the polymer chain. Cui *et al.* have incorporated hydrolyzable acetal groups into the poly(D,L-lactide)–poly(ethylene glycol) (PLA–PEG) backbone to show pH-triggered drug release from the polymeric nanofibers.[94] PEG is reacted with benzaldehyde to introduce acetal groups in the resulting prepolymer termed PBE, which is further copolymerized with lactic acid. They have synthesized triblock copolymers PLA–PBE–PLA with PBE contents of 5% (PBELA-5) and 10% (w/w) (PBELA-10) to study the effect of different amounts of acid-labile groups in the polymers and PLA–PEG–PLA with 10% (w/w) of PEG (PELA-10) as a control. Electrospinning is used to fabricate polymer nanofibers containing 2% (w/w) of paracetamol as a model drug. Paracetamol release from various copolymer nanofibers is explored after incubation in buffer solutions with pH 4.0, 5.5, and 7.4. PELA-10 nanofibers with no acid-labile moiety exhibit similar release profiles at all the pH values tested. However, PBELA-5 and PBELA-10 nanofibers demonstrate enhanced release rates at pH 4.0 and 5.5 as compared to 7.4 due to pH-induced structural changes in the polymeric nanofibers. The nanofibers with the highest content of acid-labile units (PBELA-10) exhibit a significant difference in release profiles at acidic and neutral pH. Hence, the rate of release from the pH-responsive nanofibers can be controlled *via* different numbers of acid-labile units in the polymer matrix and the local pH environment.

The variation in pH of healthy and tumor tissues has been explored to assess the controlled drug-delivery properties of pH-responsive nanofibers. A continuous flow system to represent the environment exposed to the pH-responsive nanofibrous scaffolds can be designed to mimic the *in vivo* environment of the tissues.[95] Essentially, buffer at pH 2.5, 6.0, and 7.2 is allowed to flow (0.3 mL min^{-1}) through the syringe pump to represent the pH of the stomach, extracellular tumor environment, and healthy tissues, respectively. The doxorubicin (DOX)-loaded electrospun PCL scaffolds (Table 7.2) are placed in the Transwell inserts to study the *in vitro* drug release at different pH. A pH-dependent release profile is observed from the scaffolds with higher release at acidic pH. The change in pH results in the disruption of ionic interactions between the drug and PCL, causing the drug to release from these scaffolds at an acidic pH. The cell-viability assay performed with human embryonic kidney (HEK) cells maintained at pH 6.0 and 7.2 indicates a smaller number of viable cells at lower pH due to higher DOX release in a slightly acidic environment. *In vivo* applicability of DOX-loaded scaffolds is tested by implanting in zebra fish. The results indicate approximately 55% of DOX release within 10 hours after implantation. The aforementioned study supports the suitability of PCL scaffolds *in vivo* and pH-responsive drug release under simulated *in vivo* environment.

Variation in pH of the system causes changes in the morphology of the polymer to trigger the release of drugs. pH-triggered drug release from polymeric nanofibers can be explored for various biomedical applications.

7.3.2 Thermoresponsive Nanofibers

There has been tremendous interest in thermoresponsive polymers due to their applications in drug delivery and tissue engineering.[96] These polymers exhibit abrupt changes in conformation in response to variation in the temperature of specific organs due to certain ailments or *via* deliberate external heating procedures.[97] Poly(*N*-isopropylacrylamide) (PNIPAAm) is one of the widely explored thermosensitive polymers with LCST in the range of 31–35 °C (Table 7.2).[11] The intermolecular hydrogen bonding interactions between polymer chains and surrounding aqueous medium prevails below the LCST that contributes to the hydrophilicity of polymer chains. A small increase in temperature above the LCST prompts volume phase transition, *i.e.,* the deswelling/collapsing of polymer chains due to entropy-driven hydrophobic interactions (Figure 7.6(III)). This deswelling/collapsing of the polymer at higher temperatures allows the drug to be expelled out of the fibers. PNIPAAm bears an inherent limitation of processability; its fibers cannot be electrospun easily. Efforts have been made to improve its fiber formation by either blending it with other polymers or by fabricating fibers with PNIPAAm as the core and other polymers as the sheath.[98,99]

Poly(vinyl caprolactam) (PVCL) is another polymer that undergoes a phase transition at a LCST and is a better candidate for biomedical applications than PNIPAAm due to its higher biocompatibility (Table 7.2).[100] González and coworkers have recently studied thermoresponsive chemical hydrogel nanofibers synthesized by copolymerization of PVCL with thermally induced self-crosslinkable *N*-hydroxymethylacrylamide (NMA) monomer (Table 7.2).[101] They have synthesized poly(VCL-*co*-NMA) polymers with various compositions of PVCL to NMA: 95 : 5 (PN1), 90 : 10 (PN2), and 80 : 20 (PN3). The poly(VCL-*co*-NMA) copolymer nanofibers were produced *via* electrospinning and thermally cured at 210 °C for an hour to form chemically crosslinked hydrogel nanofibers. The temperature-dependent responsiveness of the nanofibers is demonstrated by their swelling/shrinking ability in water, below and above the LCST. The nanofibers with the lowest content of NMA exhibit the least change in volume that is attributed to the partial dissolution of fibers because of low crosslinking density. The temperature-dependent release of rhodamine B dye from PN2 nanofibrous scaffolds is assessed in water at 23 ± 2 and 52 ± 2 °C; a higher release rate is observed upon shrinking of hydrogel nanofibers above the LCST.

Bioactive thermoresponsive nanofiber mats have been synthesized for wound healing.[102] Various polyblend nanofiber formulations are generated comprising PNIPAAm, PCL, and egg albumin with different concentrations (0–20%) of drug, gatifloxacin hydrochloride (Gati). The nanofibers demonstrate rapid initial drug release for the first 10 hours and sustained release for 696 hours. Antibacterial properties of Gati-loaded nanofibers are tested against gram-positive *Staphylococcus aureus, a* commonly found pathogen in wounds. Significant inhibition of bacterial growth is observed around the Gati-loaded scaffolds, where antibacterial properties increase with the increasing concentration of Gati. The *in vivo* wound healing activity of nanofibers is tested in a rat model. The wounds treated with Gati-loaded nanofibers demonstrate significantly faster healing than the control gauge: 95% wound healing after 21 days with 15% Gati-loaded nanofibers as compared to 45% healing with gauge (Figure 7.6(IV)). The results validate the application of these thermoresponsive scaffolds for tissue regeneration.

Various thermoresponsive polymeric systems have been explored based on the chemistry of the functional groups, such as PVCL[101] and PNIPAAm.[11] The results discussed above demonstrate the potential of thermoresponsive nanofibers to achieve temperature-dependent release of biomolecules for tissue engineering.

7.3.3 Electroresponsive Nanofibers

Living tissues are shown to have an inherent electric field up to 500 mV mm^{-1}.[103] The transmembrane potential present in cells due to the difference in ionic concentrations between intracellular and extracellular environments is known to promote the proliferation of cells.[104] External electric stimulations can be applied to cells by using conductive materials in close proximity to tissues. An early report by Basset *et al.* has shown the positive effects of intramedullary implanting insulated batteries on bone formation.[105] Conductive polymers such as polyaniline (PANI) and polypyrrole (PPy) have been identified as potential materials to electrically stimulate tissues for biomedical applications (Table 7.2). The mechanism of conductivity of the conductive polymers lies in the delocalization of π-electrons over the π-conjugated polymeric system. The conductivity of conductive polymers can be enhanced by incorporating ionic dopants that add or remove electrons from the polymer to form polarons and bipolarons, facilitating additional charge flow (Figure 7.7(I)). The electroresponsive polymeric nanofibers can also be used for the delivery of bioactive molecules due to their fast responsiveness and precise control over release.[82]

Schmidt and coworkers have fabricated 3D conductive polymer scaffolds by coating the electrospun PLGA nanofibers with PPy for neuronal tissue scaffolds.[106] They have used PLGA as a nanofiber template for PPy coating because of its cytocompatibility and biodegradability. Highly aligned (AF) and random nanofibers (RF) of PLGA are electrospun on a stationary collector and drum rotator, respectively. A nanoscale-thick coating of the conductive polymer is deposited on the surface of PLGA nanofibers by polymerizing pyrrole and aggregating onto fibers to form PPy-AF and PPy-RF nanofibers. Ppy-coated PLGA fibers demonstrate electrical properties while maintaining the features of nanofibers, such as high porosity and surface-to-volume ratio. The cytocompatibility of PPy-AF and PPy-RF nanofibers is assessed by culturing PC12 cells and embryonic hippocampal neurons on the polymeric scaffolds. Ppy-coated PLGA nanofibers display compatible cellular interactions with no significant difference in cellular growth and proliferation as compared to control PLGA nanofibers (Figure 7.7(II) a–d). The application of electrical stimulation with a potential of 10 mV cm^{-1} through the polymer scaffolds exhibits noteworthy improvement in neurite formation as compared to nonstimulated cells.

Blended polymeric nanofiber scaffolds of electrically conductive PANI doped with camphorsulfonic acid (CPSA) and elastic poly(L-lactide-*co*-ε-caprolactone) (PLCL) have been synthesized by Jeong *et al.* in two volume ratios of PANI to PLCL of 85 : 15 and 70 : 30 (PACL-2 and PACL-3, respectively) along with pure PLCL fibers as a control (PACL-1) (Table 7.2).[107] The diameters of fibers for PACL-1, PACL-2, and PACL-3 are 430 ± 116 nm, 423 ± 100 nm, and 382 ± 102 nm, respectively. *In vitro* cellular response to CPSA-PANI/PLCL nanofiber scaffolds is assessed by culturing human dermal fibroblasts, NIH-3T3 fibroblasts, and mouse skeletal muscle cells (C2C12) on the nanofibers. An increase in the number of viable cells is observed with an increase in PANI content in the nanofibrous scaffolds. NIH-3T3 cells cultured on PACL-3 scaffolds reveal enhanced mitochondria metabolic activity upon the application of 20 mA of electric current as compared to 0 and 2 mA current. However, increasing the applied electric current to 200 mA leads to a detrimental effect on the growth of NIH-3T3 cells. The electrically conductive CPSA-PANI/PLCL nanofiber scaffolds demonstrate significantly higher adhesion and growth upon direct electrical stimulation.

Electrical stimulus can be applied to the tissues by conducting electrical signals through conductive polymers.[108] Electrical stimulation has been

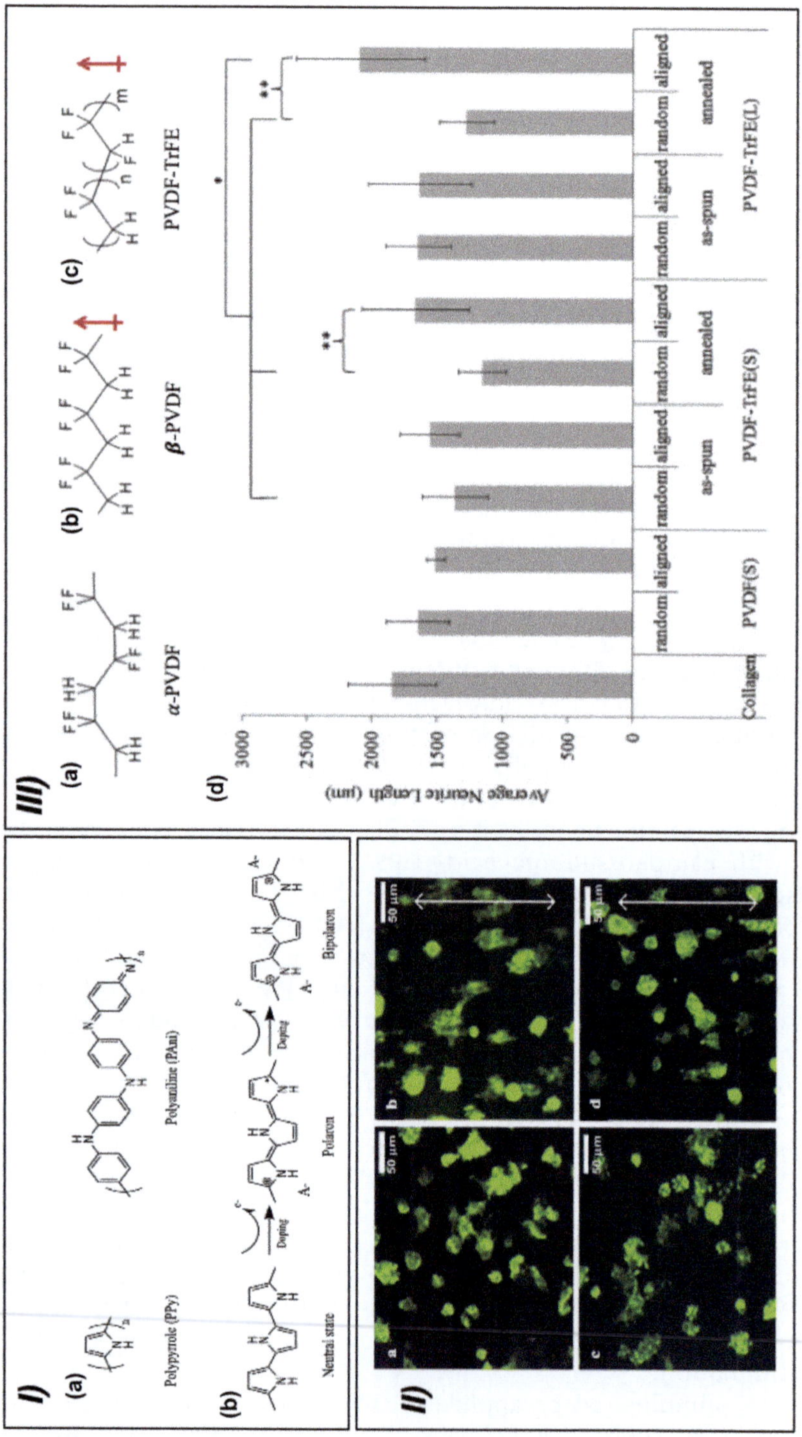
I)
(a)
Polypyrrole (PPy)
Polyaniline (PAni)
(b)
Neutral state
Doping
Polaron
Doping
Bipolaron
II)
50 μm
III)
(a)
α-PVDF
(b)
β-PVDF
(c)
PVDF-TrFE
(d)
Average Neurite Length (μm)
3000
2500
2000
1500
1000
500
0
Collagen
random
aligned
PVDF(S)
as-spun
annealed
PVDF-TrFE(S)
PVDF-TrFE(L)

shown to substantially enhance the growth of neurites, serving as a promising scaffold for nerve regeneration.[106,109,110] Considering the biocompatibility and excellent control over electrical stimulation of conductive polymeric nanofibers, these scaffolds can be used for neural tissue engineering applications.

7.3.4 Piezoelectric Nanofibers

Piezoelectric materials can generate electricity *via* minute mechanical forces such as compression and stretching while eliminating the need for the application of external electric sources.[111] Bones and cartilage have been reported to exhibit piezoelectricity. Collagen and hydroxyapatite present in bones are found to be responsible for the piezoelectric behavior.[112] There is a tremendous amount of interest in creating piezoelectric materials that can control the electric stimuli to enhance cellular development for tissue engineering applications. In one of the approaches, piezoelectric materials can be embedded into electrospun polymeric nanofibers.[113] Significant improvement has been observed in the adhesion and proliferation of mouse fibroblasts on ZnO-doped poly(urethane) (PU) nanofibers scaffolds as compared to only PU scaffolds. However, scaffolds comprising piezoceramics can be detrimental due to their cytotoxicity.[114,115] Kawai *et al.* have designed polyvinylidene fluoride (PVDF),[116] a nontoxic alternative to ceramics for tissue engineering applications (Table 7.2). PVDF does not show piezoelectricity in α-phase due to lack of net dipole moment in *trans-gauche* conformation; mechanical stretching is required to orient dipoles of α-PVDF into piezoelectric β-phase with net dipole moment in all-*trans* confirmation (Figure 7.7(III) a, b).[76] Alternatively, PVDF can be copolymerized with trifluoroethylene (TrFE) to achieve piezoelectric properties without the requirement of mechanical stretching due to the inherent net dipole moment of PVDF–TrFE (Figure 7.7(III) c).

Lee and coworkers have investigated the piezoelectric properties of PVDF–TrFE-based fibers and their cytocompatibility with dorsal root ganglion (DRG) neurons.[76] They have studied the effect of random and aligned PVDF–TrFE scaffolds with nano- to microscale dimensions. The

Figure 7.7 (I) Conductive polymers and their mechanism of conductivity. (a) Chemical structures of conductive polymers such as polypyrrole (PPy) and polyaniline (PANI). (b) Mechanism of conduction of polymers with PPy as an example. A^- : counterion/dopant. (II) *In vitro* cellular response to electroresponsive nanofibers. Fluorescence images demonstrating PC12 cells cultured on (a) RF, (b) AF, (c) PPy-RF, and (d) PPy-AF for 2 days. Reproduced from ref. 106 with permission from Elsevier, Copyright 2009. (III) Piezoelectric polymeric nanofibers. Chemical structure of (a) α-PVDF, (b) β-PVDF, and (c) PVDF–TrFE. Reproduced from ref. 111 with permission from Elsevier, Copyright 2015. SEM images of electrospun PVDF–TrFE nanosized random nanofibers (d) Effect of piezoelectric PVDF–TrFE nanofibers on the average neurite length of DRG neurons. Reproduced from ref. 76 with permission from Elsevier, Copyright 2011.

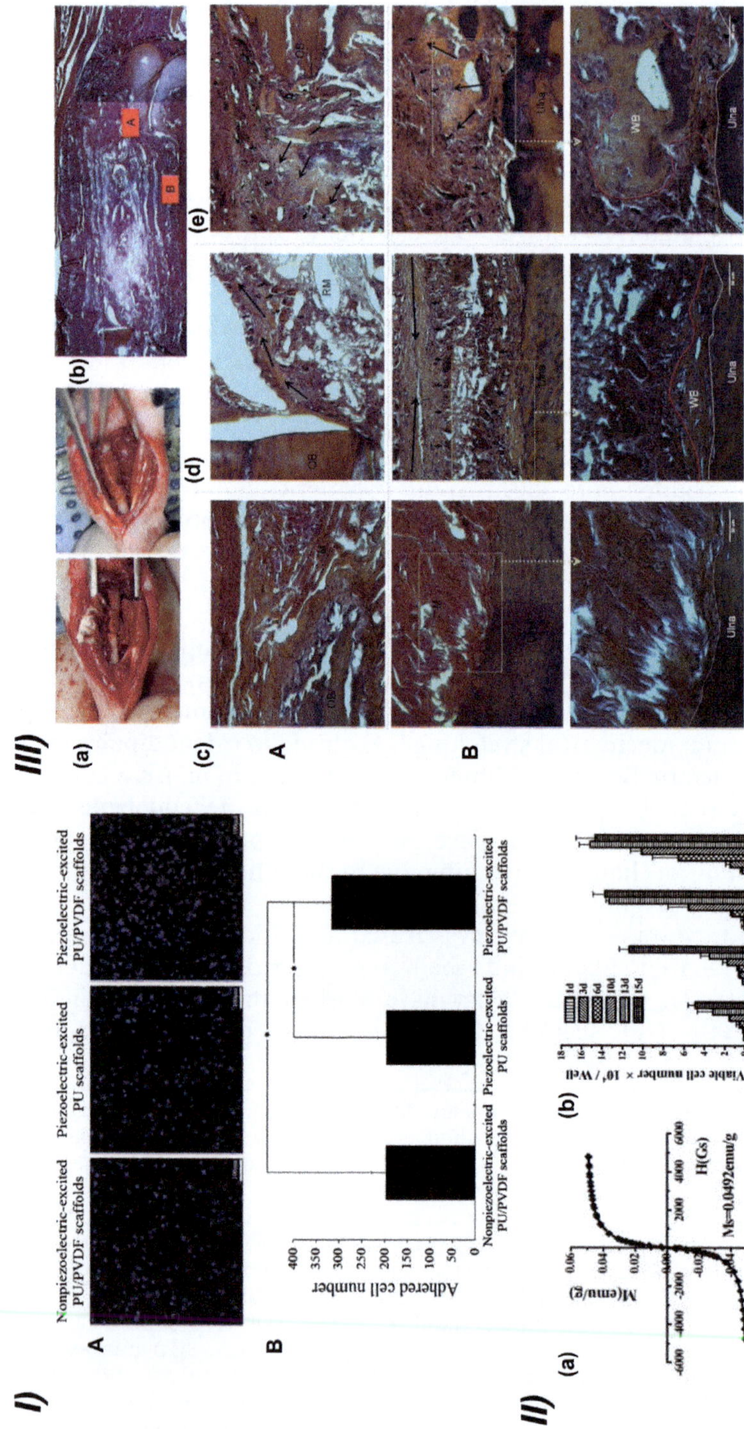
I)
A
Nonpiezoelectric-excited PU/PVDF scaffolds
Piezoelectric-excited PU scaffolds
Piezoelectric-excited PU/PVDF scaffolds
B
Adhered cell number
II)
(a)
M(emu/g)
H(Gs)
Ms=0.0492emu/g
(b)
Viable cell number × 10⁴ / Well
III)
(a)
(b)
(c)
(d)
(e)
A
B
WB
Ulna

piezoelectric properties of PVDF–TrFE scaffolds are enhanced by annealing and positively affect neuronal growth. DRG adhesion and neurite extension are tested on various fibrous surfaces: i) collagen-treated surface, ii) PVDF (random and aligned), iii) as-spun or annealed nanoscale PVDF–TrFE (random and aligned), and iv) as-spun or annealed micron-scale PVDF–TrFE (random and aligned). Neurite growth and extension are also assessed by using PVDF and PVDF-TrFE films. The poor adhesion and extension of DRG are observed on the films, indicating the significance of 3D fibrous structures for cellular interactions. The neurite extension is highest on micron-sized annealed aligned PVDF–TrFE scaffolds in comparison with random scaffolds (Figure 7.7(III) d).

Minute deformations in the polymeric fibers due to cellular growth and migration have been shown to induce piezoelectric nature in the scaffolds.[117] Guo *et al.* have investigated the activity of fibroblasts both *in vitro* and *in vivo* in the presence of piezoelectric composite scaffolds formed by the blending of PVDF with PU (Table 7.2).[118] PVDF is cospun with the thermoplastic polymer PU to enhance the elasticity of the fibrous scaffolds (PVDF/PU). It has been shown that α-PVDF transforms to piezoelectric β-phase upon the electrospinning of PVDF/PU scaffolds, imparting piezoelectric properties to the structure. A cell adhesion assay is performed to analyze the piezoelectric effect of PVDF/PU scaffolds on the adhesion of NIH-3T3 cells and compared to control PU scaffolds (Figure 7.8(I)). Enhanced fibroblast attachment is observed upon the application of deformation on PVDF/PU scaffolds relative to piezoelectric excited PU scaffolds and non-piezoelectric excited PVDF/PU scaffolds as negative controls.[118] Piezoelectric effects of PVDF/PU scaffolds on fibroblasts are tested *in vivo* by subcutaneously implanting the scaffolds in various regions of Sprague Dawley rats. Along with the PVDF/PU scaffolds, control PU scaffolds are also implanted

Figure 7.8 (I) *In vitro* piezoelectric effect of the PU/PVDF scaffolds on fibroblast adhesion. (A) Hoechst 33342-stained NIH-3T3 cells attached to the NIH-3T3 cell layers on the piezoelectric-excited PU/PVDF scaffolds and the control scaffolds (scale bar = 100 μm). (B) Number of the cells adhered to the cell layers on the piezoelectric-excited PU/PVDF scaffolds and the control scaffolds. Reproduced from ref. 118 with permission from Elsevier, Copyright 2012. (II) SPNs-incorporated magnetoresponsive nanofibers. (a) Magnetization curve of γ-Fe_2O_3/nHA/PLA nanofibers, (b) growth of preosteoblast cells on nHA/PLA and γ-Fe_2O_3/nHA/PLA scaffolds; + MF means the application of 0.9–1.0 mT magnetic field to the cells. Reproduced from ref. 126 with permission from the Royal Society of Chemistry. (III) (a) Image showing the application of magnetic scaffolds on rat segmental defects; (b) H&E-stained histologic appearance made after 8 weeks of the surgery. Histologic analysis was made in the regions 'A' and 'B' and presented in (c) control group free of scaffolds, (d) pure PCL scaffolding group, and (e) scaffolding group with magnetic nanoparticles.[128] Bracket regions in 'B' show the new bone formation. Reproduced from ref. 128, https://doi.org/10.1371/journal.pone.0091584, under the terms of the CC BY 4.0 license https://creativecommons.org/licenses/by/4.0/.

symmetrically and subjected to the same level of mechanical deformations. Piezoelectric stimulations on PVDF/PU scaffolds lead to an enhanced level of fibrosis (wound healing) as compared to control PU nanofibers.[118] These studies demonstrate the potential of piezoelectric materials in supporting cellular growth.

Numerous works of research have been performed to study the effect of piezoelectric materials for tissue engineering and regeneration.[111] Deformation of piezoelectric materials due to physiological stress results in neurite extension, enhanced cellular attachment, and proliferation. Piezoelectric nanofibers are promising candidates for tissue engineering scaffolds.

7.3.5 Magnetoresponsive Nanofibers

Magnetic fields can penetrate deep into the living tissue and are well tolerated by the human body (up to 7 Tesla)[119] as compared to other stimuli such as temperature and electric fields that have lower penetration and can be detrimental to cells when applied with high strength. Nanofibers with responsiveness to the magnetic field can be achieved by incorporating superparamagnetic nanoparticles (SPNs) into the polymer matrix during electrospinning (Table 7.2).[120] Magnetic nanoparticles can be biofunctionalized with growth factors and act as a carrier to transport the biomolecules under an externally applied magnetic field.[121] The biocompatibility and significant benefits of magnetoresponsive nanofibrous scaffolds in tissue regeneration have aroused tremendous interest in tissue engineering and drug delivery applications.[122–125] Meng *et al.* have fabricated magnetic field responsive nanofibrous films by blending γ-Fe_2O_3 nanoparticles coated with *meso*-2,3-dimercaptosuccinic acid, hydroxyapatite nanoparticles (nHA), and PLA using an electrospinning technique (Table 2).[126] γ-Fe_2O_3/nHA/PLA nanofibrous films demonstrate super paramagnetic behavior due to the incorporation of γ-Fe_2O_3 nanoparticles (Figure 7.8(II) a). Adhesion and proliferation of osteoblast cells have been tested on nHA/PLA and γ-Fe_2O_3/nHA/PLA films in the presence and absence of a magnetic field. A static magnetic field of 0.9–1.0 mT results in significantly improved growth and differentiation of osteoblast cells on the γ-Fe_2O_3/nHA/PLA scaffolds (Figure 7.8(II) b).[126] The increase in growth and proliferation of bone cells on scaffolds incorporating magnetic nanoparticles is attributed to the cellular response to magnetism-induced mechanical stress.[127] Singh and coworkers have demonstrated significantly enhanced attachment and penetration of osteoblasts on the magnetic nanofibrous scaffolds of PCL comprising SPNs (Table 7.2).[128] The magnetic scaffolds exhibit excellent tissue compatibility when subcutaneously implanted in rats while showing a higher rate of degradation of the scaffolds and considerable formation of new blood vessels as compared to control PCL nanofibers. The magnetic nanofiber scaffolds demonstrate the significant formation of new bones in the rat segmental defect sites (Figure 7.8(III) a–e). Magnetism-induced voltage/mechanical stress has been shown to increase in proliferation and differentiation of bone cells.[127] The results provided above suggest that SPNs-incorporated nanofibrous polymer scaffolds exhibit attractive biological and physicochemical characteristics for their applications in magnetism-induced tissue engineering.

Great advancements in polymeric nanofibers with tailor-made properties have been achieved to cater to a wide range of biomedical applications.[82] Polymer scaffolds provide a benign microenvironment for uninterrupted intra- and intercellular interactions while functioning as a carrier for a number of signaling molecules. The controlled and target-specific delivery of therapeutics can be successfully achieved through on/off switching of stimuli-responsive smart polymers. Smart polymeric nanofibers display exciting opportunities in the field of tissue engineering.

7.4 Conclusion

Nanofibrous scaffolds hold great promise in tissue engineering owing to their morphology and functional properties akin to ECM. A diverse range of nanofibrous biomaterials have been explored, ranging from natural proteins such as α-helical coiled-coil,[26] silk,[58] collagen,[66] and elastin[47] to recombinantly or chemically synthesized *de novo* proteins[27] to synthetic polymers[85] that are biocompatible and biodegradable and are mechanically strong for tissue engineering applications. The high surface area of nanofibrous scaffolds allows greater contact with cells to facilitate intracellular signaling. Smart materials can be triggered to control physical or chemical cues in the scaffold microenvironment to promote cellular migration and proliferation and overall enhancement in tissue regeneration. Tissue engineering scaffolds show promising future applications in a clinical setting for the treatment of various medical conditions involving the regeneration of bones, cartilage, liver, skin, ligaments, and tendons.[129] Moving forward, it is essential to explore the factors that are responsible for creating specific microenvironments conducive to different types of tissues.

Acknowledgements

This work was supported by NSF-DMREF under Award Number DMR 1728858, NSF-MRSEC Program under Award Number DMR 1420073, DoD under Award Number W81XWH-16-1-0483, and ARO under Award Number W911NF-19-2-0160.

References

1. L. E. Freed, F. Guilak, X. E. Guo, M. L. Gray, R. Tranquillo, J. W. Holmes, M. Radisic, M. V. Sefton, D. Kaplan and G. Vunjak-Novakovic, *Tissue Eng.*, 2006, **12**, 3285–3305.
2. R. O. Hynes, *Science*, 2009, **326**, 1216.

3. S. Kidoaki, I. K. Kwon and T. Matsuda, *Biomaterials*, 2005, **26**, 37–46.
4. K. Jayaraman, M. Kotaki, Y. Zhang, X. Mo and S. Ramakrishna, *J. Nanosci. Nanotechnol.*, 2004, **4**, 52–65.
5. J. X. Law, L. L. Liau, A. Saim, Y. Yang and R. Idrus, *Tissue Eng. Regener. Med.*, 2017, **14**, 699–718.
6. X. Zhang, M. R. Reagan and D. L. Kaplan, *Adv. Drug Delivery Rev.*, 2009, **61**, 988–1006.
7. D. B. Khadka and D. T. Haynie, *Nanomed.: Nanotechnol. Biol. Med.*, 2012, **8**, 1242–1262.
8. M. Dietrich, J. Heselhaus, J. Wozniak, S. Weinandy, P. Mela, B. Tschoeke, T. Schmitz-Rode and S. Jockenhoevel, *Tissue Eng., Part C*, 2012, **19**, 216–226.
9. K. Balagangadharan, S. Dhivya and N. Selvamurugan, *Int. J. Biol. Macromol.*, 2017, **104**, 1372–1382.
10. F. Khan and M. Tanaka, *Int. J. Mol. Sci.*, 2018, **19**, 17.
11. C. Huang, S. J. Soenen, J. Rejman, B. Lucas, K. Braeckmans, J. Demeester and S. C. De Smedt, *Chem. Soc. Rev.*, 2011, **40**, 2417–2434.
12. T. Dvir, B. P. Timko, D. S. Kohane and R. Langer, *Nat. Nanotechnol.*, 2010, **6**, 13.
13. M. von dem Bussche and M. H. Tuszynski, in *Encyclopedia of Neuroscience*, ed. L. R. Squire, Academic Press, Oxford, 2009, pp. 987–992.
14. M. Sarikaya, C. Tamerler, A. K. Y. Jen, K. Schulten and F. Baneyx, *Nat. Mater.*, 2003, **2**, 577–585.
15. R. V. Ulijn and A. M. Smith, *Chem. Soc. Rev.*, 2008, **37**, 664–675.
16. T. Aida, E. W. Meijer and S. I. Stupp, *Science*, 2012, **335**, 813.
17. M. Fujita, M. Tominaga, A. Hori and B. Therrien, *Acc. Chem. Res.*, 2005, **38**, 369–378.
18. S. I. Stupp, *Nano Lett.*, 2010, **10**, 4783–4786.
19. D. Li, T. Liu, X. Yu, D. Wu and Z. Su, *Polym. Chem.*, 2017, **8**, 4309–4321.
20. N. L. Francis, N. K. Bennett, A. Halikere, Z. P. Pang and P. V. Moghe, *ACS Biomater. Sci. Eng.*, 2016, **2**, 1030–1038.
21. M. Eriksson, S. Hassan, R. Larsson, S. Linder, T. Ramqvist, H. Lövborg, T. Vikinge, E. Figgemeier, J. Mueller, J. Stetefeld, T. Dalianis and S. Özbek, *Utilization of a Right-handed Coiled-coil Protein from Archaebacterium Staphylothermus marinus as a Carrier for Cisplatin*, 2009.
22. R. B. Merrifield, *J. Am. Chem. Soc.*, 1963, **85**, 2149–2154.
23. G. Morreale, E. Gyo Lee, D. B. Jones and A. P. J. Middelberg, *Biotechnol. Bioeng.*, 2004, **87**, 912–923.
24. A. K. Sato, M. Viswanathan, R. B. Kent and C. R. Wood, *Curr. Opin. Biotechnol*, 2006, **17**, 638–642.
25. S. Kyle, A. Aggeli, E. Ingham and M. J. McPherson, *Trends Biotechnol.*, 2009, **27**, 423–433.
26. K. Beck and B. Brodsky, *J. Struct. Biol.*, 1998, **122**, 17–29.

27. Y. Wu and J. H. Collier, *Wiley Interdiscip. Rev.: Nanomed. Nanobiotechnol.*, 2017, **9**, e1424.
28. S. K. Gunasekar, J. S. Haghpanah and J. K. Montclare, *Polym. Adv. Technol.*, 2008, **19**, 454–468.
29. W. Zhang, X. Yu, Y. Li, Z. Su, K. D. Jandt and G. Wei, *Prog. Polym. Sci.*, 2018, **80**, 94–124.
30. P. Katyal and J. K. Montclare, in *Protein Scaffolds: Design,* Synthesis, and Applications, ed. A. K. Udit, Springer New York, New York, NY, 2018, pp. 239–263.
31. J. A. Frezzo and J. K. Montclare, *Ther. Delivery*, 2015, **6**, 643–646.
32. M. G. Ryadnov and D. N. Woolfson, *Nat. Mater.*, 2003, **2**, 329–332.
33. M. G. Ryadnov and D. N. Woolfson, *J. Am. Chem. Soc.*, 2005, **127**, 12407–12415.
34. M. J. Pandya, G. M. Spooner, M. Sunde, J. R. Thorpe, A. Rodger and D. N. Woolfson, *Biochemistry*, 2000, **39**, 8728–8734.
35. H. Dong, S. E. Paramonov and J. D. Hartgerink, *J. Am. Chem. Soc.*, 2008, **130**, 13691–13695.
36. C. Xu, R. Liu, A. K. Mehta, R. C. Guerrero-Ferreira, E. R. Wright, S. Dunin-Horkawicz, K. Morris, L. C. Serpell, X. Zuo, J. S. Wall and V. P. Conticello, *J. Am. Chem. Soc.*, 2013, **135**, 15565–15578.
37. L. Yin, A. S. Agustinus, C. Yuvienco, T. Minashima, N. L. Schnabel, T. Kirsch and J. K. Montclare, *Biomacromolecules*, 2018, **19**, 1614–1624.
38. H. T. More, K. S. Zhang, N. Srivastava, J. A. Frezzo and J. K. Montclare, *Biomacromolecules*, 2015, **16**, 1210–1217.
39. J. Hume, J. Sun, R. Jacquet, P. D. Renfrew, J. A. Martin, R. Bonneau, M. L. Gilchrist and J. K. Montclare, *Biomacromolecules*, 2014, **15**, 3503–3510.
40. S. K. Gunasekar, M. Asnani, C. Limbad, J. S. Haghpanah, W. Hom, H. Barra, S. Nanda, M. Lu and J. K. Montclare, *Biochemistry*, 2009, **48**, 8559–8567.
41. S. K. Gunasekar, L. Anjia, H. Matsui and J. K. Montclare, *Adv. Funct. Mater.*, 2012, **22**, 2154–2159.
42. M. Nambiar, L.-S. Wang, V. Rotello and J. Chmielewski, *J. Am. Chem. Soc.*, 2018, **140**, 13028–13033.
43. M. S. Desai and S.-W. Lee, *Wiley Interdiscip. Rev.: Nanomed. Nanobiotechnol.*, 2015, **7**, 69–97.
44. J. Chang, X.-F. Peng, K. Hijji, J. Cappello, H. Ghandehari, S. D. Solares and J. Seog, *J. Am. Chem. Soc.*, 2011, **133**, 1745–1747.
45. D. L. Nettles, A. Chilkoti and L. A. Setton, *Adv. Drug Delivery Rev.*, 2010, **62**, 1479–1485.
46. A. Ribeiro, F. J. Arias, J. Reguera, M. Alonso and J. C. Rodríguez-Cabello, *Biophys. J.*, 2009, **97**, 312–320.
47. S. Aluri, M. K. Pastuszka, A. S. Moses and J. A. MacKay, *Biomacromolecules*, 2012, **13**, 2645–2654.

48. Z. I. Foraida, T. Kamaldinov, D. A. Nelson, M. Larsen and J. Castracane, *Acta Biomater.*, 2017, **62**, 116–127.
49. R. Jean-Gilles, D. Soscia, S. Sequeira, M. Melfi, A. Gadre, J. Castracane and M. Larsen, *J. Nanotechnol. Eng. Med.*, 2010, **1**, 31008.
50. Y. K. Sung and S. W. Kim, *Biomater. Res.*, 2019, **23**, 8.
51. K. Kamimura, T. Suda, G. Zhang and D. Liu, *Pharm. Med.*, 2011, **25**, 293–306.
52. S. Lee, J. S. Kim, H. S. Chu, G. W. Kim, J. I. Won and J. H. Jang, *Acta Biomater.*, 2011, **7**, 3868–3876.
53. B. Kundu, N. E. Kurland, S. Bano, C. Patra, F. B. Engel, V. K. Yadavalli and S. C. Kundu, *Prog. Polym. Sci.*, 2014, **39**, 251–267.
54. W. Shen, in *Biomaterials for Tissue Engineering Applications: A Review of the Past and Future Trends*, ed. J. A. Burdick and R. L. Mauck, Springer Vienna, Vienna, 2011, pp. 243–275.
55. N. Shehata, I. Kandas, I. Hassounah, P. Sobolčiak, I. Krupa, M. Mrlik, A. Popelka, J. Steadman and R. Lewis, *Nanomaterials*, 2018, 8, 585.
56. D. R. Knighton, K. F. Ciresi, V. D. Fiegel, L. L. Austin and E. L. Butler, *Ann. Surg.*, 1986, **204**, 322–330.
57. R. Spanò, A. Muraglia, M. R. Todeschi, M. Nardini, P. Strada, R. Cancedda and M. Mastrogiacomo, *J. Tissue Eng. Regener. Med.*, 2018, **12**, e82–e96.
58. C. Pignatelli, G. Perotto, M. Nardini, R. Cancedda, M. Mastrogiacomo and A. Athanassiou, *Acta Biomater.*, 2018, **73**, 365–376.
59. I. S. Bayer, S. Guzman-Puyol, J. A. Heredia-Guerrero, L. Ceseracciu, F. Pignatelli, R. Ruffilli, R. Cingolani and A. Athanassiou, *Macromolecules*, 2014, **47**, 5135–5143.
60. S. Guzman-Puyol, J. A. Heredia-Guerrero, L. Ceseracciu, H. Hajiali, C. Canale, A. Scarpellini, R. Cingolani, I. S. Bayer, A. Athanassiou and E. Mele, *ACS Biomater. Sci. Eng.*, 2016, **2**, 526–534.
61. K. Numata and D. L. Kaplan, *Adv. Drug Delivery Rev.*, 2010, **62**, 1497–1508.
62. M. Yang, C. Tanaka, K. Yamauchi, K. Ohgo, M. Kurokawa and T. Asakura, *J. Biomed. Mater. Res., Part A*, 2008, **84**, 353–363.
63. N. Dinjaski and D. L. Kaplan, *Curr. Opin. Biotechnol*, 2016, **39**, 1–7.
64. A.-S. Duwez, S. Cuenot, C. Jérôme, S. Gabriel, R. Jérôme, S. Rapino and F. Zerbetto, *Nat. Nanotechnol.*, 2006, **1**, 122–125.
65. H. Yang, S. Y. Fung, M. Pritzker and P. Chen, *Angew. Chem., Int. Ed. Engl.*, 2008, **47**, 4397–4400.
66. S. Ricard-Blum, *Cold Spring Harbor Perspect. Biol.*, 2011, **3**, a004978.
67. M. D. Shoulders and R. T. Raines, *Annu. Rev. Biochem.*, 2009, **78**, 929–958.
68. A. L. Fidler, S. P. Boudko, A. Rokas and B. G. Hudson, *J. Cell Sci.*, 2018, **131**, 7 jcs203950.
69. K. G. DeFrates, R. Moore, J. Borgesi, G. Lin, T. Mulderig, V. Beachley and X. Hu, *Nanomaterials*, 2018, 8, 457.
70. C. Dong and Y. Lv, *Polymers*, 2016, **8**, 42.

71. Z. Bazrafshan and G. K. Stylios, *Int. J. Biol. Macromol.*, 2019, **129**, 693–705.
72. P. J. Stahl, N. H. Romano, D. Wirtz and S. M. Yu, *Biomacromolecules*, 2010, **11**, 2336–2344.
73. K. McGuinness, I. J. Khan and V. Nanda, *ACS Nano*, 2014, **8**, 12514–12523.
74. J. Luo and Y. W. Tong, *ACS Nano*, 2011, **5**, 7739–7747.
75. S. C. Neves, L. S. Moreira Teixeira, L. Moroni, R. L. Reis, C. A. Van Blitterswijk, N. M. Alves, M. Karperien and J. F. Mano, *Biomaterials*, 2011, **32**, 1068–1079.
76. Y.-S. Lee, G. Collins and T. Livingston Arinzeh, *Acta Biomater.*, 2011, **7**, 3877–3886.
77. R. Khajavi, M. Abbasipour and A. Bahador, *J. Appl. Polym. Sci.*, 2016, **133**, 42883.
78. P. M. Martins, S. Ribeiro, C. Ribeiro, V. Sencadas, A. C. Gomes, F. M. Gama and S. Lanceros-Méndez, *RSC Adv.*, 2013, **3**, 17938–17944.
79. M. F. Maitz, *Biosurf. Biotribol.*, 2015, **1**, 161–176.
80. J. Kundu, F. Pati, Y. Hun Jeong and D.-W. Cho, in *Biofabrication*, ed. G. Forgacs and W. Sun, William Andrew Publishing, Boston, 2013, pp. 23–46.
81. J. Jiang, J. Xie, B. Ma, D. E. Bartlett, A. Xu and C. H. Wang, *Acta Biomater.*, 2014, **10**, 1324–1332.
82. H. Priya James, R. John, A. Alex and K. R. Anoop, *Acta Pharm. Sin. B*, 2014, **4**, 120–127.
83. A.-T. Khaled and S. Jagdish, *Recent Pat. Drug Delivery Formulation*, 2007, **1**, 65–71.
84. A. Aghabegi Moghanjoughi, D. Khoshnevis and A. Zarrabi, *Drug Delivery Transl. Res.*, 2016, **6**, 333–340.
85. E. Cabane, X. Zhang, K. Langowska, C. G. Palivan and W. Meier, *Biointerphases*, 2012, **7**, 9.
86. D. Roy, J. N. Cambre and B. S. Sumerlin, *Prog. Polym. Sci.*, 2010, **35**, 278–301.
87. J. Lannutti, D. Reneker, T. Ma, D. Tomasko and D. Farson, *Mater. Sci. Eng.: C*, 2007, **27**, 504–509.
88. C. Alvarez-Lorenzo and A. Concheiro, *Chem. Commun.*, 2014, **50**, 7743–7765.
89. P. Bawa, V. Pillay, Y. E. Choonara and L. C. du Toit, *Biomed. Mater.*, 2009, **4**, 022001.
90. J. Zhang, J. Koh, J. Lu, S. Thiel, B. S. H. Leong, S. Sethi, C. Y. X. He, B. Ho and J. L. Ding, *PLoS Pathog.*, 2009, **5**, e1000282.
91. E. Cabane, X. Zhang, K. Langowska, C. G. Palivan and W. Meier, *Biointerphases*, 2012, **7**, 9.
92. S. Nitta, S. Kaketani and H. Iwamoto, *Eur. Polym. J.*, 2015, **67**, 50–56.
93. S. J. Grainger and M. E. H. El-Sayed, in *Biologically-Responsive Hybrid Biomaterials*, World Scientific, 2010, pp. 171–190.

94. W. Cui, M. Qi, X. Li, S. Huang, S. Zhou and J. Weng, *Int. J. Pharm.*, 2008, **361**, 47–55.
95. M. Jassal, V. P. Boominathan, T. Ferreira, S. Sengupta and S. Bhowmick, *J. Biomater. Sci., Polym. Ed.*, 2016, **27**, 1380–1395.
96. M. A. Ward and T. K. Georgiou, *Polymers*, 2011, **3**, 1215–1242.
97. S. Mura, J. Nicolas and P. Couvreur, *Nat. Mater.*, 2013, **12**, 991.
98. Y. Lv, Q. Pan, S. W. A. Bligh, H. Li, H. Wu, Q. Sang and L.-M. Zhu, *J. Pharm. Sci.*, 2017, **106**, 1258–1265.
99. M. Chen, M. Dong, R. Havelund, V. R. Regina, R. L. Meyer, F. Besenbacher and P. Kingshott, *Chem. Mater.*, 2010, **22**, 4214–4221.
100. N. A. Cortez-Lemus and A. Licea-Claverie, *Prog. Polym. Sci.*, 2016, **53**, 1–51.
101. E. González and M. W. Frey, *Polymer*, 2017, **108**, 154–162.
102. M. D. Pawar, G. V. N. Rathna, S. Agrawal and B. S. Kuchekar, *Mater. Sci. Eng.: C*, 2015, **48**, 126–137.
103. M. E. Mycielska and M. B. A. Djamgoz, *J. Cell Sci.*, 2004, **117**, 1631.
104. S. Sundelacruz, M. Levin and D. L. Kaplan, *Stem Cell Rev. Rep.*, 2009, **5**, 231–246.
105. C. A. L. Bassett, R. J. Pawluk and R. O. Becker, *Nature*, 1964, **204**, 652–654.
106. J. Y. Lee, C. A. Bashur, A. S. Goldstein and C. E. Schmidt, *Biomaterials*, 2009, **30**, 4325–4335.
107. S. I. Jeong, I. D. Jun, M. J. Choi, Y. C. Nho, Y. M. Lee and H. Shin, *Macromol. Biosci.*, 2008, **8**, 627–637.
108. M. R. Gandhi, P. Murray, G. M. Spinks and G. G. Wallace, *Synth. Met.*, 1995, **73**, 247–256.
109. X. Liu, J. Qin, M. Chen and J.-H. Park, *Proc. Am. Soc. Inf. Sci. Technol.*, 2008, **45**, 1–10.
110. Z. Zhang, M. Rouabhia, Z. Wang, C. Roberge, G. Shi, P. Roche, J. Li and L. H. Dao, *Artif. Organs*, 2007, **31**, 13–22.
111. A. H. Rajabi, M. Jaffe and T. L. Arinzeh, *Acta Biomater.*, 2015, **24**, 12–23.
112. E. Fukada and I. Yasuda, *J. Phys. Soc. Jpn.*, 1957, **12**, 1158–1162.
113. T. Amna, M. S. Hassan, F. A. Sheikh, H. K. Lee, K.-S. Seo, D. Yoon and I. H. Hwang, *Appl. Microbiol. Biotechnol.*, 2013, **97**, 1725–1734.
114. W. P. Chen, H. L. W. Chan, F. C. H. Yiu, K. M. W. Ng and P. C. K. Liu, *Appl. Phys. Lett.*, 2002, **80**, 3587–3589.
115. J. Zhou, N. S. Xu and Z. L. Wang, *Adv. Mater.*, 2006, **18**, 2432–2435.
116. H. Kawai, *Jpn. J. Appl. Phys.*, 1969, **8**, 975–976.
117. J. Friedrichs, A. Taubenberger, C. M. Franz and D. J. Muller, *J. Mol. Biol.*, 2007, **372**, 594–607.
118. H.-F. Guo, Z.-S. Li, S.-W. Dong, W.-J. Chen, L. Deng, Y.-F. Wang and D.-J. Ying, *Colloids Surf., B*, 2012, **96**, 29–36.
119. J. M. Theysohn, S. Maderwald, O. Kraff, C. Moenninghoff, M. E. Ladd and S. C. Ladd, *Magn. Reson. Mater. Phys., Biol. Med.*, 2007, **21**, 63.

120. M. Wang, H. Singh, T. A. Hatton and G. C. Rutledge, *Polymer*, 2004, **45**, 5505–5514.
121. A. Tampieri, E. Landi, F. Valentini, M. Sandri, T. D'Alessandro, V. Dediu and M. Marcacci, *Nanotechnology*, 2010, **22**, 015104.
122. N. Bock, A. Riminucci, C. Dionigi, A. Russo, A. Tampieri, E. Landi, V. A. Goranov, M. Marcacci and V. Dediu, *Acta Biomater.*, 2010, **6**, 786–796.
123. A. Tampieri, S. Sprio, M. Sandri and F. Valentini, *Trends Biotechnol.*, 2011, **29**, 526–535.
124. R. De Santis, A. Gloria, T. Russo, U. D'Amora, S. Zeppetelli, C. Dionigi, A. Sytcheva, T. Herrmannsdörfer, V. Dediu and L. Ambrosio, *J. Appl. Polym. Sci.*, 2011, **122**, 3599–3605.
125. M. Bañobre-López, Y. Piñeiro-Redondo, R. De Santis, A. Gloria, L. Ambrosio, A. Tampieri, V. Dediu and J. Rivas, *J. Appl. Phys.*, 2011, **109**, 07B313.
126. J. Meng, Y. Zhang, X. Qi, H. Kong, C. Wang, Z. Xu, S. Xie, N. Gu and H. Xu, *Nanoscale*, 2010, **2**, 2565–2569.
127. S. Nomura and T. Takano-Yamamoto, *Matrix Biol.*, 2000, **19**, 91–96.
128. R. K. Singh, K. D. Patel, J. H. Lee, E.-J. Lee, J.-H. Kim, T.-H. Kim and H.-W. Kim, *PLoS One*, 2014, **9**, e91584.
129. T. A. E. Ahmed, E. V. Dare and M. Hincke, *Tissue Eng., Part B*, 2008, **14**, 199–215.
130. S. E. Reichheld, L. D. Muiznieks, F. W. Keeley and S. Sharpe, *Proc. Natl. Acad. Sci. U. S. A.*, 2017, **114**, E4408.
131. N. Bhardwaj and S. C. Kundu, *Biotechnol. Adv.*, 2010, **28**, 325–347.
132. J. Zhuang, M. R. Gordon, J. Ventura, L. Li and S. Thayumanavan, *Chem. Soc. Rev.*, 2013, **42**, 7421–7435.

CHAPTER 8

Biomimetic Adhesives for Clinical Applications

JESSICA E. TORRES,[a] SYDNEY HOLLINGSHEAD,[b] DANA BOUCHER[c] AND JULIE C. LIU*[a,c]

[a] Davidson School of Chemical Engineering, Purdue University, West Lafayette, IN 47907, USA; [b] Cook Biotech Inc., West Lafayette, IN 47906, USA; [c] Weldon School of Biomedical Engineering, Purdue University, West Lafayette, IN 47907, USA

*Email: julieliu@purdue.edu

8.1 Clinical Applications of Surgical Adhesives

8.1.1 Clinical Need for Surgical Adhesives

Millions of surgical procedures, the majority of which require surgical wound closure, are performed each year.[1] Sutures and surgical staples are the current standard of care for closing wounds but have many disadvantages compared to an ideal surgical adhesive. Sutures and staples inherently cause damage to tissue surrounding the wound, create tension at the wound site that can negatively affect healing,[2] and require a level of technical skill that takes many years of training to accomplish and maintain,[3,4] especially in cosmetic or microsurgery settings.

Surgical adhesives have great potential to address the large clinical need for wound closure. When used properly, surgical adhesives can reduce or prevent liquid or gas seepage from wounds. This reduction is especially critical in areas such as vascular, pleural, or gastrointestinal repair, where leakage can result in catastrophic morbidities or mortalities. If the adhesive is sufficiently

Biomaterials Science Series No. 10
Biomimetic Protein Based Elastomers: Emerging Materials for the Future
Edited by Namita Roy Choudhury, Julie C. Liu and Naba K. Dutta

Published by the Royal Society of Chemistry, www.rsc.org

strong, its use may also allow for a reduced number of sutures or staples in the wound closure. Currently, the only adhesives capable of use without mechanical support are limited to topical application. In general, surgical adhesives have a number of advantages compared to sutures and staples. Adhesives require less technical skill, do not inherently cause tissue damage, can reduce or redirect local tissue tension, and can reduce postoperative pain.[5] Because adhesives can be faster to apply, they can reduce surgery time and costs related to surgical suite use. Liquid adhesives can also be applied laparoscopically[6] and can be used in deep or narrow operative corridors where applying sutures or staples is difficult or impossible. Despite all these advantages and a great need, there are currently no FDA-approved adhesives for the internal closure of wounds in the absence of mechanical support.

8.1.2 Key Design Requirements – Performance and Safety

The adhesive strength of the material should be tuned to the tissue and area of use. Adhesives used in areas of high motion such as joints, skeletal muscle, or the face and neck may require higher adhesive strength than those used in areas of low motion, such as the intraabdominal region. In a sealant system, the adhesive burst strength should be able to withstand the maximum fluid or muscular pressures of the tissue. For example, the bladder pressure of normal body mass index (BMI) in women ranges from 13 mm Hg at rest to 97 mm Hg during coughing.[7] Intracranial pressure in healthy adults lying horizontally ranges from 7 to 15 mm Hg.[8]

The stiffness of the adhesive should, in general, match the stiffness of the surrounding tissue. An excessively stiff adhesive applied onto soft tissue may cause pain or excessive stress or strain under motion. Conversely, excessively soft adhesives on hard tissues may deform and weaken the wound closure site. Thus, adhesives applied to hard tissues, such as dental adhesives, traditionally have high stiffness. When adhesives are used in conjunction with mechanical closures such as sutures or staples, the mechanical properties of the closure will dominate, so the stiffness of the adhesive is less critical. The required elasticity of the adhesive also depends on the application. For tissues under repetitive pressure or flexion, such as the heart, skeletal muscle, or gastrointestinal tract, an elastic adhesive that moves and flexes with the tissue without failure is desired. Adhesives used in these applications must also be resilient in order to withstand thousands to millions of cycles of compression or tension before healing is finished. In contrast, adhesives used in wound closure may need to be more brittle and less elastic. This reduced pliability may prevent the wound faces from moving apart during normal motion since the opening of the wound may disrupt or halt the healing process. The mechanical properties of the adhesive should remain above the critical thresholds throughout the healing process to prevent delayed failure.

The adhesive should also be easily applied in its intended environment. Internal adhesives are often applied in a moist or wet environment. Surgeons can dry tissue surfaces, but excessive drying during surgery can lead to unwanted outcomes, and drying is not always achievable in some surgeries.[9] Thus, internal adhesives must be able to adhere in a moist environment. In general, moisture reduces adhesion by physically or chemically interfering with the adhesion mechanism of the material. Some adhesives overcome this issue by using a water-triggered adhesion mechanism. For example, cyanoacrylates rely on water to trigger polymerization and adhesion. On the other hand, there are some tissue systems, such as dural punctures, where liquid flows freely and a dry surface is impossible. In these instances, a liquid adhesive may not adhere well to the tissue surface due to a barrier layer of the free-flowing bodily fluid, and an adhesive patch that absorbs liquid as part of its adhesion mechanism may be preferred.

Another consideration is the need to tailor the viscosity or fluid cohesion of the adhesive to the intended method of application. Adhesives that are applied with a spraying device require a lower viscosity, whereas adhesives applied in small or tight areas, on tissue surfaces that are vertical or upside down, or in areas with sensitive tissue nearby may benefit from higher viscosity to prevent spreading or seeping. Furthermore, adhesives applied laparoscopically need to be shear-thinning or low enough in viscosity to travel through extensive tubing and narrow needles.

The adhesive should also be easy to use in a surgical setting. Adhesives that need to be shipped and stored refrigerated or frozen require more equipment, incur higher costs, and require additional preparation time to thaw. Adhesives that are complex to prepare by surgical staff also increase preparation time and the risk of incorrect preparation. Finally, the curing speed of the adhesive should be carefully tailored. A too-short cure time increases the risk of misapplication, inadvertent binding of tissue, and clogging of the delivery device, whereas a too-long cure time increases the time the patient is under surgery and may require additional mechanical support during curing. In general, a cure time longer than 60 seconds can feel excessively long to surgeons. Longer cure times can also contribute to surgery cost; the average cost of operating room time is $62 per minute.[10] All of these use and shipping characteristics should be carefully considered as an excessive cost in even a single area may be sufficient to make the product economically unfeasible, regardless of its performance in wound closure.

Beyond the performance and safety of the adhesive, the regulation of the materials must be considered. All currently marketed medical adhesive materials are regulated as medical devices or biologic products and fall under the regulatory requirements of the Center for Devices and Radiological Health and the United States Code of Federal Regulations.[11] All medical adhesives must undergo a suite of *in vitro* and *in vivo* biocompatibility testing[12] and show that the material is safe and does not cause toxic or extreme inflammatory effects.

Overall, researchers and engineers should consider the adhesive and mechanical requirements of their intended application, the restrictions of the application environment, the practicalities of the surgical suite, and the requirements of regulation when designing effective surgical adhesives.

8.1.3 FDA Regulation of Surgical Adhesives

Surgical adhesives are currently regulated as either medical devices or biologic products. Per the FDA, medical devices achieve their main function through physical action on the body and are not dependent on being metabolized to function.[13] Biologic products are considered a subset of drug products and are derived from a range of biological sources; however, in the context of surgical adhesives, they are overwhelmingly blood component derived[14] and tend to achieve their function through biological interactions.

Medical devices are grouped by product code (see Table 8.1), which are used by the FDA to identify and track medical devices by category of

Table 8.1 FDA categorization of surgical adhesives.

Device	Product code	Class	Category[a]
Bandage, liquid, skin protectant	NEC	I	Skin protectant/ topical burn dressing
Strip, adhesive, closure, skin	FPX	I	Topical wound closure
Tape and bandage, adhesive	KGX	I	Topical wound closure
Bandage, liquid	KMF	I	Topical wound closure
Adhesive, bracket and tooth conditioner, resin	DYH	II	Device fixative
Cement, dental	EMA	II	Device fixative
Cutaneous tissue adhesive with mesh	OMD	II	Topical wound closure
Tissue adhesive for the topical approximation of skin	MPN	II	Topical wound closure
Tissue adhesive for use in embolization of brain arteriovenous malformations	KGG	III	Embolic agent
Tissue adhesive for internal use	PJK	III	Plane fixative
Glue, surgical, arteries	MUQ	III	Sealant
Sealant, dural	NQR	III	Sealant
Ophthalmic sealant	PFZ	III	Sealant, topical wound closure
Sealant, polymerizing	NBE	III	Sealant
Tissue adhesive for ophthalmic use	LZQ	III	N/A[c]
Tissue adhesive, aortic aneurysmorrhaphy	ODT	HDE[b]	N/A[c]
Photochemical curing and adhesive	RGM	Unclassified	N/A[c]

[a]Categories defined by authors.
[b]Humanitarian Device Exemption.
[c]As of this review, these product codes do not have any approved devices and thus have no information on application that would allow for categorization.

application.[15] Biologic products do not currently have product codes. Product codes that currently have approved adhesive devices can be categorized as device and tissue fixatives, embolic agents, sealants, topical wound closure adhesives, and wound/burn dressing.[†] The FDA has also created several additional product codes for adhesives that do not currently have any approved devices. Codes without approved devices often exist when a company is pursuing a product that does not fit any existing codes, and the company requests a designation from the FDA. These codes include adhesives for ophthalmic use, adhesives for aortic aneurysmorrhaphy, and photochemically cured adhesives.

A review of existing product codes gives insight into which application areas have been more thoroughly explored and which areas are likely to be in development by manufacturers. Most existing product codes can be categorized as topical adhesives or sealants. This dominance indicates that adhesives for internal use in mechanically challenging applications may be an unmet need.

In general, medical devices are also classified as class I, II, or III depending on the risk level of the device as well as the specific application. The classification can also be based on the novelty of the technology or specific claims made on the device. Class I entails the lowest risk and class III the highest. Higher risk classes require more extensive testing to prove safety and efficacy. Tissue adhesives for topical wound closure are generally categorized as class II, whereas tissue adhesives for nontopical use, defined as the adhesion of internal tissues and vessels, are generally categorized as class III.[11]

As of this review, there are 17 FDA-approved surgical adhesives for internal use (see Table 8.2). This table does not include standalone hemostatic agents, topical adhesives, or nonliquid adhesives. The most common application of approved adhesives is for vascular sealing, which is often defined in the indication for use as an "adjunct to standard methods of hemostasis." The use of the word "adjunct" indicates that the company does not claim the device can reduce or eliminate the use of sutures or staples. Notably, these adhesives have applications that require relatively low adhesion strength. Thus, there is still an unmet need for high-strength surgical adhesives that can be used in applications other than sealants, fixation, and embolism.

Cyanoacrylate-based adhesives can achieve high-strength closure but have several negative aspects that preclude them from internal use. They have an exothermic curing reaction, which can cause thermal damage to tissue. They also can have mechanical mismatches with soft tissue and have toxic

---

[†]One notable exception is product code MGO, which is defined as "closure, wound, adhesive." This product is described as a "hydrophilic wound dress" whose main function is to absorb wound exudate and not to hold a wound closed in an adhesive manner. Therefore, despite the use of "adhesive" in the code name, the authors do not consider this product code to apply to surgical adhesive devices and have excluded the code from this review.

Table 8.2 FDA-approved surgical adhesives for internal use.

Device Name	Manufacturer	Product code[a]	Category	Composition	Indication for use[b]
Tisseel	Baxter Healthcare Corp	N/A	Sealant	Fibrin	For use as an **adjunct to hemostasis** in adult and pediatric patients (>1 month of age) undergoing surgery when control of bleeding by conventional surgical techniques (such as suture, ligature, and cautery) is ineffective or impractical. An adjunct to standard surgical techniques (such as suture and ligature) to **prevent leakage from colonic anastomoses** following the reversal of temporary colostomies.
Artiss	Baxter Healthcare Corp	N/A	Planar adhesive	Fibrin	To **adhere autologous skin grafts** to surgically prepared wound beds resulting from **burns** in adult and pediatric populations.
Evicel	OMRIX Biopharmaceuticals Ltd	N/A	Sealant	Fibrin	**Adjunct to hemostasis** in patients undergoing liver surgery, when control of bleeding by conventional surgical techniques is ineffective or impractical.
Fibrin Sealant (Human)	Instituto Grifols, S.A.	N/A	Sealant	Fibrin	**Adjunct to hemostasis** for mild to moderate bleeding in adults undergoing surgery when control of bleeding by standard surgical techniques (such as suture, ligature, and cautery) is ineffective or impractical. Effective in heparinized patients.
Raplixa	The Medicines Company	N/A	Sealant	Fibrin	**Adjunct to hemostasis** for mild to moderate bleeding in adults undergoing surgery when control of bleeding by standard surgical techniques (such as suture, ligature, and cautery) is ineffective or impractical.
BioGlue Surgical Adhesive	Cryolife, Inc.	MUQ	Sealant	Bovine serum albumin and glutaraldehyde	**Adjunct to standard methods of achieving hemostasis** (such as sutures and staples) in adult patients in open surgical **repair of large vessels** (such as the aorta, femoral, and carotid arteries).
CoSeal Surgical Sealant	Baxter Bioscience	NBE	Sealant	Polyethylene glycol	**Vascular reconstructions** to achieve **adjunctive hemostasis** by mechanically sealing areas of leakage.
Progel Pleural Air Leak Sealant	Neomend, Inc.	NBE	Sealant	Polyethylene glycol and human serum albumin	**Application to visceral pleura** during an open thoracotomy after standard visceral pleural closure with, for example, sutures or staples, of visible air leaks (≥2 mm) incurred during open resection of **lung parenchyma**.
	Baxter Healthcare Corporation	NBE	Sealant		**Vascular and cardiac reconstructions** (excluding application to arterial and venous grafts used in coronary artery bypass graft

Table 8.2 (*Continued*)

Device Name	Manufacturer	Product code[a]	Category	Composition	Indication for use[b]
PreveLeak Surgical Sealant				Bovine serum albumin and polyaldehyde	surgery) to achieve **adjunctive hemostasis** by mechanically sealing areas of potential leakage.
Tridyne Vascular Sealant	Neomend Inc.	NBE	Sealant	Human serum albumin and polyethylene glycol	**Aortic surgery** when **adjunctive measures to achieve hemostasis** are required by mechanically sealing areas of leakage.
Ethicon Omnex Surgical Sealant	Ethicon, Inc.	NBE	Sealant	2-Octyl cyanoacrylate and butyl lactoyl cyanoacrylate	**Vascular reconstructions** to achieve **adjunctive hemostasis** by mechanically sealing areas of leakage.
DuraSeal Dural Sealant System	Integra Lifesciences Corporation	NQR	Sealant	Polyethylene glycol ester and trilysine amine crosslinker	Adjunct to sutured **dural repair** during cranial surgery to provide **watertight closure**. Should only be used with autologous duraplasty material.
DuraSeal Exact Spine Sealant System	Integra Lifesciences Corporation	NQR	Sealant	Polyethylene glycol ester and trilysine amine crosslinker	Adjunct to sutured **dural repair** during spinal surgery to **provide watertight closure**.
Adherus AutoSpray Dural Sealant	Hyperbranch Medical Technology	NQR	Sealant	Polyethylene glycol ester and polyethyleneimine	For use in patients 13 years of age and older, as an adjunct to standard methods of **dural repair**, such as sutures, to provide **watertight closure** during cranial procedures.
ReSure Sealant	Ocular Therpeutix, Inc.	PFZ	Sealant	Polyethylene glycol	Intraoperative **management of clear corneal incisions** (up to 3.5 mm) **with a demonstrated wound leak** for which a temporary dry surface can be achieved, in order to prevent postoperative fluid egress from such incisions following cataract surgery with intraocular lens (IOL) placement in adults.
TissueGlu	Cohera Medical, Inc.	PJK	Planar adhesive	Polyurethane	**Approximation of tissue layers** where subcutaneous dead space exists between the tissue planes in **abdominoplasty**.
Trufill *n*-Butyl Cyanoacrylate Liquid Embolic System	Codman & Shurtleff, Inc.	KGG	Embolic agent	*n*-butyl cyanoacrylate	**Embolization of cerebral arteriovenous malformations** when presurgical devascularization is desired.

[a]Fibrin-based adhesives are regulated as a biologic product and thus are not assigned a product code.
[b]These are derived verbatim from FDA databases,[18,19] but the emphasis is by the authors.

breakdown products, including formaldehyde. Increasing the length of the monomer side chain and the molecular weight of the polymer leads to slower degradation and thus lower levels of toxic degradation products present at any given time.[16] Because of these drawbacks, the FDA has approved cyanoacrylate adhesives mostly for topical wound closure and has approved only two cyanoacrylate devices for long-term internal use. One of these cyanoacrylate adhesives is approved by the FDA for use in the embolization of cerebral arteriovenous malformations. However, the use of this cyanoacrylate adhesive has several adverse effects, including nontarget embolization, adhesive migration due to delayed polymerization, and premature polymerization causing catheter blockage and retention.[17]

8.1.4 FDA Guidance for Testing of Surgical Adhesives

The FDA often provides guidance documents to assist manufacturers in preparing device submissions and has published three guidance documents related to surgical adhesives. Two of the guidance documents pertain to the use of tissue adhesive for topical wound closure. This category includes the closure of surgical incisions, such as laparoscopic incisions and traumatic lacerations that have easily approximated edges, with (product code OMD)[20] or without (product code MPN)[21] an adjunct wound closure device that maintains temporary wound closure during adhesive application. For wounds that are not easily approximated, these adhesives can be used in conjunction with, but not in place of, deep dermal stitches. The FDA recommends extensive bench testing and validation, including animal testing, biocompatibility testing, sterility testing, and shelf-life testing (see Table 8.3), as appropriate for the clinical application and device design. The results of these tests may help designers mitigate the risks of unintentional bonding, wound dehiscence, adverse tissue reaction and chemical burns, infection, delayed polymerization, and weak bonding that leads to loss of approximation and applicator malfunction.[20,21] The FDA also recommends including information on the material properties of the adhesive, including viscosity, curing time, physical and mechanical testing, and component information. In addition, the FDA has published a guidance for efficacy studies of fibrin sealants, and the guidance is largely focused on hemostatic applications.[22] This guidance states that any clinical trials should use a variety of testing end-points, including time to hemostasis, blood loss, transfusion requirements, tissue sealing, and wound healing. There is not yet an FDA guidance document for tissue adhesives used internally for any application. In some cases, guidances for device design exist for specific clinical applications. It is the responsibility of the manufacturer to utilize these guidances where they exist and to otherwise develop themselves the requirements and testing needed to demonstrate device safety and effectiveness. The designers of internal adhesives may leverage the existing guidance documents as a starting point and include any additional testing that proves their adhesive is safe and effective.

Table 8.3 FDA guidance for testing of adhesives for topical wound closure.[a]

Performance property	Testing	Acceptance criteria
Strength (adhesive)	– Lap shear adhesion strength (ASTM F2255) – T-peel adhesion strength (ASTM F2256) – Tensile strength (ASTM F2258) – Wound closure strength (ASTM F2458)	– Sufficient strength for intended application as determined by *in vivo* and *ex vivo* testing.
Strength (adjunct)	– Peel adhesion strength (ASTM D330) – Shear adhesion (ASTM D3654) – Creep and tensile strength (ASTM D882)	– Sufficient to maintain wound approximation until adhesive is cured
Degradation rate and products	– Hydrolytic degradation (incubate in saline at 50 °C for 15 days). – Analyze degradation solution using gas and liquid chromatography for formulation additives, monomer impurities, and degradation products	– Final degradation products are of an appropriate size (<50–60 kDa) for renal clearance[23,b]
Curing and application/ usability	– Heat of polymerization – Cure time – Animal testing mimicking method of application – Histopathology of wound closure and healing	– Thermal output low enough to minimize patient discomfort – If fumes are generated, chemical burns are not caused – Reasonable cure time for surgical location and application – Minimal risk of delayed healing or dehiscence
Shelf life	– Performance – Cure time – Viscosity – Sterility – Purity – Water content – Appearance – Applicator functionality	– Minimal changes to functional properties after accelerated and real-time aging
Biocompatibility	– *In vivo* subchronic implantation that mimics intended use – Systemic toxicity – Macroscopic pathology – Histopathology – Endotoxin level – Other tests as necessary, dependent on device application, composition, and degradation[12]	– Minimal irritation or inflammation at the implantation site – No systemic toxicity – Endotoxin units (EU) less than 20 units per device for blood-contacting or implanted devices or less than 2.15 units per device for cerebrospinal fluid–contacting devices[24]
Sterility	– Sterilization validation testing – Sterility assurance level (SAL)	– SAL of at least 1×10^{-6} [24,c]

[a] The information in this table is derived from FDA guidance documents on tissue adhesives.[20,21]

[b] The renal system has reduced filtration of molecules larger than 50 kDa and an approximate molecular weight cut-off of 60 kDa. Degradation products that have a molecular weight larger than 50–60 kDa may accumulate in tissue instead of being cleared from the body.

[c] A SAL of 1×10^{-6} indicates that there is a probability of 1 in 1 000 000 of a viable organism on a product after sterilization has occurred.

The majority of current FDA-approved surgical adhesives function as adjuncts to mechanical closure (see Table 8.2). There is a clinical need for surgical adhesives that independently close wounds and mitigate potential clinical risks surrounding the current use of mechanical closure methods. These adhesives must meet the FDA criteria outlined in Table 8.3 that are relevant to the clinical application. Nature serves as an inspiration for new surgical adhesives because natural adhesives are generally low cost and inherently meet many FDA requirements.

8.2 Natural Biomimetic Adhesives

8.2.1 Biomimetic Adhesion

Biomimetics describes the use of synthetic products and devices to mimic biologic materials, processes, and mechanisms that occur in nature. To address the unmet clinical need for internal surgical adhesives, many natural adhesives are being studied for their attractive properties. Biological adhesives innately have high adhesive strength and compatibility, and some adhesives adhere in subaqueous environments. Organisms in nature utilize different macrostructures, microstructures, and crosslinking mechanisms to achieve biological adhesion for purposes that vary from defense to movement (Table 8.4). To achieve these high adhesive strengths, biological adhesives rely on numerous mechanisms of adhesion, including pressure-sensitive, mechanical, dry, and intermolecular interactions.

8.2.2 Pressure-sensitive Adhesion

Pressure-sensitive adhesives rely on polymer chain interactions and their physical states (*e.g.*, aqueous, hot melt, *etc.*) to achieve adhesion. These adhesives are often applied as a liquid to a transfer material or tape backing and remain viscous and tacky after solvent evaporation. These adhesives rely on intimate contact with smooth and dry surfaces and perform poorly when the surface is moist, irregular, or dirty. Adhesive wound dressings are a common example. Natural forms of pressure-sensitive adhesives are limited; however, adhesives derived from squid and Australian frogs (*Notaden bennettii*) exhibit pressure-sensitive properties.

Squid ring teeth (SRT) are attractive for adhesive applications because of their semireversible assembly, high extensibility, and energy-absorbing capacity. They line the interior ring of the suction cups in tentacles and aid in grip strengthening and anchoring[25] but do not naturally function as an adhesive. SRT proteins can be directly isolated, or recombinant SRT-based biopolymers can be expressed.[26–29] SRTs have been shown to be cytocompatible and demonstrate a wide elastic range.[30] The SRT protein complex has the semireversible ability to transition from a nonadhesive solid to an adhesive melt, which has an extensibility of $\leq$300% strain.[31] The protein complex consists of β-sheet regions, which are composed of alanine and histidine; amorphous regions, consisting of tyrosine and glycine; and proline to separate the regions.[31] Below the glass transition temperature ($T_g = 32$ °C),[31] SRT proteins form nonadhesive β-sheet networks, whereas, above the T_g, amorphous chains are mobile, β-sheet domains form physical crosslinks, and the SRT proteins are adhesive. Furthermore, higher adhesive strength is achieved when the SRT proteins are first processed above the T_g and then cooled below the T_g.[31,32] Recent applications include a recombinant SRT used as a wet adhesive[33] and an SRT-coated polypropylene (PP) mesh for the improvement of hernia repair procedures.[34] The SRT-coated mesh enhanced cell proliferation, decreased tissue contraction, and increased collagen deposition, which suggested better tissue integration.

Australian frogs secrete adhesives that function in both wet and dry environments, work on hydrophobic and hydrophilic substrates, and are thought to be pressure sensitive.[35,36] In particular, *N. bennettii* frogs secrete a substance, which transitions rapidly to a hydrogel, from their dorsal skin for use as a defense mechanism. The adhesives are 85–90% water by weight, and more than half the dry weight of the adhesive is protein, which is enriched in proline, glycine, and glutamine/glutamic acid.[35] The most abundant protein, Nb-1R, is large (350–500 kDa) and likely serves as a structural scaffold for the glue. The *N. bennettii* glue is largely unstructured with 64% random coil and 36% β-elements[35] and likely uses hydrophobic interactions for physical crosslinking.[35,36] Adding *N. bennettii* glue increased the strength of tendon repair and demonstrated its promise as a supplement to rotator cuff repair surgeries.[37]

8.2.3 Mechanical Adhesion

Mechanical adhesives create strong bonds by maximizing surface area and contact points for mechanical interlocking.[38] The physical properties of the adhesive (*e.g.*, viscosity, particle size) and surface (*e.g.*, porosity, texture) play important roles in facilitating strong bonding.[39,40]

English ivy uses mechanical adhesives to climb substrates, which enables the plant to compete for sunlight.[41] Ivy extends its roots in a parallel orientation to create contact points, secretes its adhesive, and then lignifies its roots to become rigid.[41] By doing so, ivy hooks itself into crevices and pores of the substrate. The ivy adhesive consists of arabinogalactan protein (AGP)-rich nanoparticles.[42] The spherical nanoparticles create low-viscosity solutions that increase surface area coverage and facilitate penetration of irregular surfaces (see Figure 8.1). Close packing of the nanoparticles is achieved by evaporation and facilitated by the structurally flexible protein backbone. Ionic crosslinking of calcium ions with the negatively charged carboxyl groups results in the cohesion of the adhesive. As the film hardens and cures, intimate interactions between the nanoparticles and the substrate are created.[42] Currently, the spheroidal nanoparticles of ivy are inspiring the creation of numerous nanomedicines, including nanocarriers for drug delivery and nanofillers for tissue scaffolds.[43]

Table 8.4 Adhesives in nature.

Origin	Adhesive strength	Macrostructure	Microstructure	Component type	Crosslinking	Function
Australian Frog *Notaden bennettii*	1.7 ± 0.3 MPa[35,a]	3D Mesh[35]	Spherical aggregates[32]	Water, Nb-1R[35]	Unknown	Defense
Squid Ring Teeth *Dosidicus gigas*	1.5 MPa,[32,b] 2.5 MPa[32,c]	Thermoplastic Film	Ring structure with aligned nanopores[32]	Proteins	Physical[31]	Native function not for adhesion[25]
English Ivy *Hedera helix*	3.81 ± 2.41 N[67,d]	Film[42]	Spherical nanoparticles[42]	Glycoproteins[42]	Ionic[42]	Vertical climbing[41]
Geckos *Gekkonidae* family	1 MPa[47,e]	Setae	Hooked fibrillar spatulae	Corneous β-proteins	None	Movement on surfaces
Caddisfly Larvae *Trichoptera* order	Unknown	Silk threads[56]	Silk fiber coating[56]	Glycoproteins[56,57]	Dityrosine and tyrosine to phenolic groups[57,58]	Building/ Protection[55]
Sandcastle Worms *Phragmatopoma californica*	Unknown	Foam[63]	Shell surrounding water-filled pores[63]	Polycationic and polyanionic proteins[63]	DOPA[63]	Building[63]
Mussels *Mytilus* genus	0.3 MPa,[68,f] 6 MPa[69,g]	Plaque	Hierarchical/ Coacervate	Mussel foot proteins	Iron & DOPA	Sessile

[a] Lap shear test with wooden substrates.
[b] Normal tensile adhesion strength of SRT, which has been processed above the glass transition temperature, on glass substrates tested underwater.
[c] Lap shear tests of SRT, which has been processed above the glass transition temperature, with glass substrates underwater.
[d] Maximum tensile force at failure with a bark substrate.
[e] Maximum pull-off force of an individual seta from a silicon cantilever.
[f] Average detachment force of an individual thread on an aluminium substrate.
[g] Estimated average plaque adhesive strength based on 2 mm diameter samples.

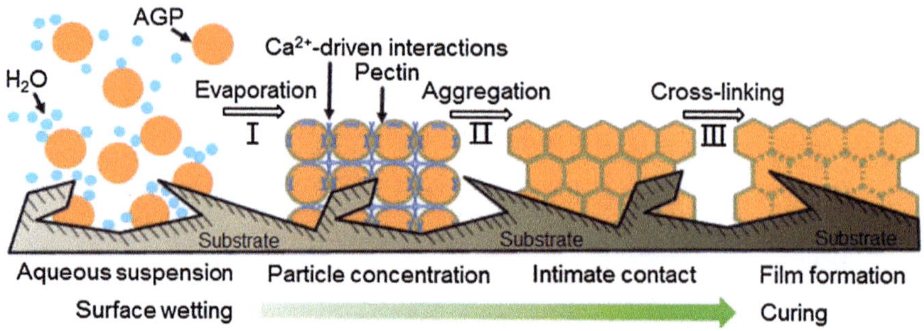

Figure 8.1 The proposed molecular basis for ivy adhesives. The wetting process commences as AGP-rich nanoparticles are secreted in the extracellular space. Next, the nanoparticles are concentrated during evaporation, and calcium ions drive ionic crosslinking to form a strong, cohesive film. Intimate contact with the substrate contributes to mechanical adhesion. Reproduced from ref. 42 with permission from the National Academy of Sciences.

8.2.4 Dry Adhesion

Dry adhesives rely on topographical structure rather than secretion for adhesion.[44] Dry adhesives create a multitude of nanoscale contact points with the substrate to form a network of van der Waals forces.[44] Gecko feet are well known as a natural dry adhesive. To climb vertical and inverted substrates, geckos rely on a hierarchical fibrillar structure[44] (see Figure 8.2). The feet are covered with blade-like lamellar structures, which have keratinous hairs named setae on the ends. The tips of the setae have nanoscale spatulae that create intimate contact with surfaces. The hairs are mainly composed of β-keratins, which were recently renamed corneous β-proteins.[45,46] Corneous β-proteins contribute to the hydrophobicity of the feet and have a β-sheet structure. The high density and orientation of the spatulae impact the adhesive ability of the gecko feet. The gecko is able to attach when its curved spatulae are engaged at an angle of $\sim 30°$ with respect to a surface and detach when the angle exceeds 30°.[47] To engage the spatulae, the gecko drags its foot along the surface.[44,48,49] Piezoresistive cantilever measurements yielded a maximum adhesive force due to one seta as 194 ± 25 μN[47] when pulled parallel to the substrate. One study translated the lessons from gecko feet to a biomedical application by combining the nanotopography of gecko feet with chemical crosslinking to create adhesives capable of bonding porcine intestine *in vitro* and abdominal subfascial tissue in a rat model.[50]

8.2.5 Intermolecular Interactions

Adhesives can also bind directly to tissue and device surfaces through a variety of chemical bonds and physical interactions. Covalent bonds can lead to strong bond strength and higher stiffness of the adhesive material. The natural entanglement of protein chains results in adhesives with low

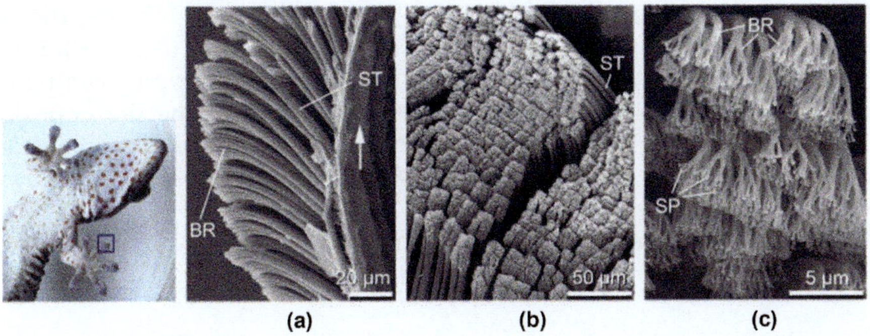

Figure 8.2 Structural hierarchy of gecko feet. Scanning electron micrographs of (a) and (b) rows of setae at different magnifications and (c) spatulae, the finest terminal branches of setae. ST: seta; SP: spatula; BR: branch. Reproduced from ref. 49 with permission from Elsevier, Copyright 2005.

stiffness and low adhesive strength. Hydrogen bonds can create weak interactions with many materials. Ionic bonds, such as those found in alginate-based adhesives with calcium ions, lead to gelation and weak adhesion. Physical crosslinks or weak ionic bonding allow the binding mechanism to break and reform passively and are often used in self-healing adhesives, which are attractive in applications where the material may be periodically fractured or disrupted. Natural organisms often use a combination of these intermolecular adhesive strategies. For example, it is hypothesized that barnacles use hydrophobic amyloid fibril assembly[51] and an enzyme-catalyzed curing mechanism[52] to form an adhesive cement underwater. Barnacle adhesion is an actively developing area of research.[53,54]

Caddisfly larvae produce silk fibers for use as permanent homes or to cover the abdominal area with sticks and rocks for protection.[55] Similar to the silk fibers used by silkworms and spiders, caddisfly silk fibers are secreted from silk storage glands and coated with an adhesive layer;[56] however, caddisfly silk must function underwater. The proposed caddisfly mechanism involves the adhesive coating, which contains glycoproteins,[56,57] initially adhering to a substrate through intermolecular interactions such as chain entanglement, electrostatic interactions, and hydrogen bonds.[57] Permanent interfacial adhesion is achieved through crosslinking of tyrosine residues in the coating to phenolic groups present on the substrate. Dityrosine crosslinking within the adhesive further contributes to cohesion and the stability of the silk fiber.[57,58] A recent study demonstrated that caddisfly larvae silk is cytocompatible, is thermally stable, and can be easily sterilized.[59] Thus, this material has potential for future applications such as a tissue scaffold for cell growth or as surgical sutures.

Sandcastle worms form tubular shell homes, which contain a natural chemical adhesive, in subaqueous environments. A mixture of at least six polycationic and polyanionic proteins make up the adhesive.[60–63] A few positively charged proteins contain L-3,4-dihydroxyphenylalanine

(DOPA) residues, which play a role in underwater adhesion.[64,65] Two discrete packets of oppositely charged proteins are secreted and mixed to form an insoluble foam.[63] Both packets contain catechol oxidase, which can rapidly oxidize DOPA to form a crosslinked network. Inspired by the oppositely charged package delivery system of the sandcastle worm, researchers developed an aqueous dispersion with low viscosity to create injectable tissue adhesives.[66]

Finally, chemical adhesives used by mussels are of great interest to many researchers and will be thoroughly discussed in Section 8.3.

8.3 Mussel-inspired Adhesion

8.3.1 Mussel Adhesive Chemistry

Marine mussel adhesive chemistry has been studied extensively over the last several decades and has supported advances in numerous adhesive applications. Mussels are found in turbulent and wet environments, yet they are able to strongly affix themselves to rock surfaces. A complex system of proteins with highly localized distributions confer adhesion in these conditions.[70–72] Mussels secrete a protein solution that solidifies to form a byssal thread and an adhesive plaque. The plaque is composed of several mussel foot proteins (mfps) (see Figure 8.3). Several other proteins, such as prepolymerized collagens and thread matrix proteins, play structural roles in thread development.

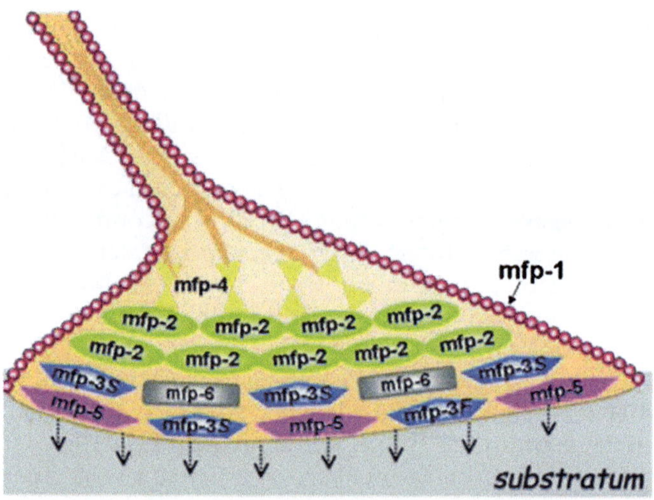

Figure 8.3 Structure of mussel adhesion. Mussel foot proteins (mfps) comprise the adhesive plaque. Prepolymerized collagens and thread matrix proteins contribute to the structure of the byssal thread. Reproduced from ref. 73 with permission from John Wiley & Sons, Copyright © 2014 WILEY-VCH Verlag GmbH & Co. KGaA, Weinheim.

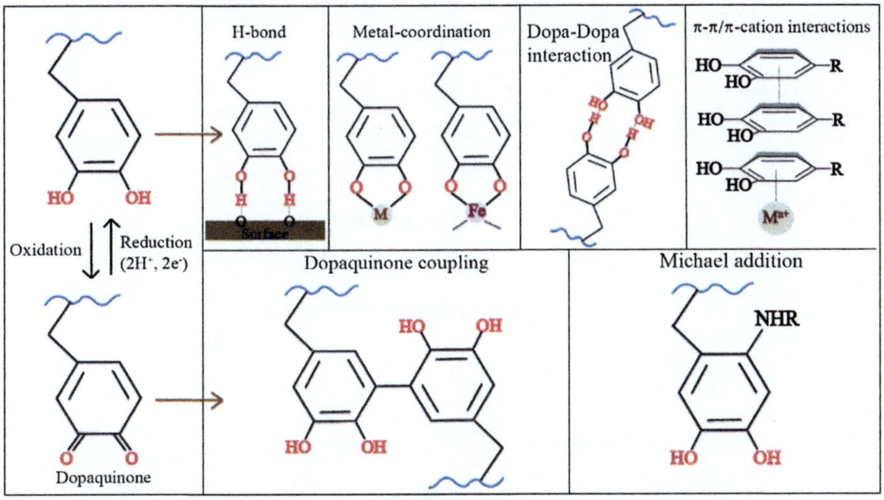

Figure 8.4 DOPA reactive chemistry. Reduced forms of DOPA participate in many different noncovalent interactions. DOPA can be easily oxidized to quinones and result in irreversible covalent crosslinking. Reproduced from ref. 71 with permission from the Royal Society of Chemistry.

The mfps contain various amounts of DOPA, a posttranslationally modified amino acid that is enzymatically converted from tyrosine. DOPA plays a key role in adhesion through multiple binding mechanisms. Reduced and oxidized forms of DOPA impart both strong adhesion to surfaces and strong cohesion within the plaque[74] (see Figure 8.4). Reduced DOPA can bind through metal chelation, hydrogen bonding, and interaction with other aromatic rings. DOPA is easily oxidized to quinones and semiquinones, which are highly reactive and participate in irreversible intermolecular covalent crosslinking. The DOPA oxidation state is spatiotemporally regulated by the microenvironment before secretion.[75] Therefore, control over the oxidation state of DOPA is crucial to utilizing both reduced DOPA and oxidized quinone adhesive chemistries.[70]

8.3.2 Mussel Foot Proteins (mfps)

Mfps vary widely in sequence and function.[69] Not all mfps contain significant levels of DOPA and instead provide structural support within the plaque or surround the plaque to form a protective layer around internal components.[76–78] Different mfps are localized to specific glands and are able to spontaneously self-assemble into complex architectures that mature over time.[79] Mfp-3, mfp-5, and mfp-6 are integral to plaque–substrate binding and play different roles in adhesion.[72] These proteins have varying amounts of DOPA and other posttranslationally modified amino acids, including phosphoserine and hydroxyarginine.[70] Whereas mfp-3 and mfp-5 have higher

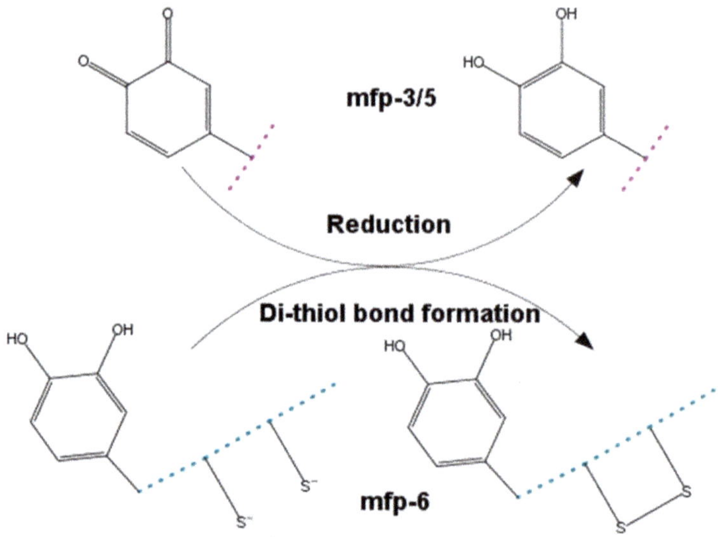

Figure 8.5 Proposed mechanisms for thiols in mfp-6 to reduce catechols in mfp-3 and mfp-5.

DOPA levels,[70] mfp-6 has higher levels of cysteine thiols, which are thought to control the reduction–oxidation chemistry of DOPA residues.[80] Mfp-6 thiols and potentially their catechols may act as sacrificial oxidation targets to reduce mfp-3 and mfp-5 catechols for direct interface adhesion (see Figure 8.5).

Mfp-5 is located at the mussel plaque–substrate interface, has the highest DOPA content at 30 mol%, and thus plays a key role in surface binding.[81] Mfp-3 is polymorphic with two main classes of variants, mfp-3f and mfp-3s.[72] Mfp-3f variants are DOPA rich with >20 mol% DOPA, whereas mfp-3s only has 5–10 mol% DOPA.[82,83] Mfp-3f has a high positive charge density, which aids in water and salt displacement from binding surfaces,[84] whereas mfp-3s has a lower charge density, which establishes a hydrophobic environment that shields the DOPA residues from oxidation by alkaline seawater.[85] In addition, a high-pH environment neutralizes the ionic charges of mfp-3s, and decreased electrostatic repulsion between the proteins results in coacervation.[86] Coacervation, or fluid–fluid phase separation of polymer-rich and polymer-poor solutions, enhances wetting properties by decreasing interfacial energy.[72] Coacervation was also shown to protect DOPA from oxidation in mfp-3 and mfp-6.[87] Furthermore, hydrophobic mfp-3s can also overcome hydration forces by repelling water from surfaces before adhesion[73] and may prime the surface for other DOPA-rich mfps. These surface-binding mfps, mfp-5 and mfp-3f, may coacervate when secreted as well.[88] In addition to the mechanisms we described, numerous other factors, such as catechol oxidation state and adhesive bridging, may contribute to mussel adhesion.[72] Overall, the complex and tightly controlled interplay of mfps results in multiple functions that range from structural support to plaque–substrate adhesion.

8.3.3 Commercial Production of mfps

Commercial production of mfps has been investigated. Initially, mfps were extracted directly from mussels to produce adhesives such as Cell-Tak, but this process was time consuming, costly, and required a large number of mussels. The Cha group has produced recombinant versions of mfp-3 and mfp-5, the mfps with the highest DOPA content, to evaluate their commercial adhesive potential.[89,90] In particular, mfp-5 and an mfp-3 variant based on sequences from *Mytilus galloprovincialis* were expressed in *Escherichia coli*. These recombinant variants include tyrosine residues, which are post-translationally converted to DOPA using tyrosinase. Atomic force microscopy (AFM) studies demonstrated that the adhesion force of the DOPA-modified mfp-5 was 550 nN and was significantly higher than that of tyrosinase-treated Cell-Tak (240 nN), whereas the DOPA-modified mfp-3 variant had adhesion forces (230 nN) similar to Cell-Tak. However, the yields of purified mfp-5 and the mfp-3 variant were only 7% and 18%, respectively, and were insufficient for macroscale adhesion studies. To address the low yield, a recombinant protein, fp-151, was constructed and fused repeats of mfp-1 and mfp-5.[91] When bulk adhesive testing of fp-151 was performed in a butt-joint configuration using polymethylmethacrylate substrates, fp-151 had a significantly higher tensile strength of 1.8 MPa compared to that of commercially available fibrin glue (0.3 MPa). These foundational studies provide motivation to develop other materials that utilize the adhesive capabilities of mussel adhesive protein.

8.4 Nature-inspired Surgical Adhesives for Soft Tissues

8.4.1 Materials for Surgical Adhesives

Systems discussed in Sections 8.2 and 8.3 have been utilized to meet the adhesive design requirements outlined in Section 8.1.2. Each design decision, from the choice of components to modifications, can be made to tune the material properties of the intended tissue adhesive to a specific application.

There are three general classes of potential adhesives: synthetic polymers, natural polymers harvested from plants or animals, and recombinant proteins. Of these, the first two classes are the lowest cost options. Although synthetic polymers offer tunable properties and design flexibility, they can have toxic degradation products and typically require organic solvents if used in a liquid form. Organic solvents can cause tissue inflammation and cell death and may result in difficulties in small animal testing, where the volume of organic solvent needed may easily exceed the lethal levels for the animal. Natural polymers, such as proteins or polysaccharides, harvested from animals or plants can be low cost but require additional testing[92] to receive FDA approval.[93,94] Recombinant proteins have increased cost, must be thoroughly purified to remove any potentially harmful byproducts, and can be difficult to produce at scale. However, if the intended application for the surgical adhesive has a niche market or low anticipated sale volume, a smaller production target may be tolerable.

8.4.2 Synthetic Polymer-based Adhesive Biomaterials

Synthetic polymers confer a large degree of design flexibility to tailor the material properties for specific tissues and applications. This section begins by highlighting a number of different polymer systems that utilize mussel-inspired catechol chemistry to achieve strong adhesion. By using different polymeric backbones, different mechanical properties and adhesion strengths are achieved. Furthermore, concerns with swelling, which can compress nearby tissues, are addressed through copolymerization. Finally, examples that combine catechol chemistry and sandcastle worm-inspired coacervation are presented.

The Wilker group incorporated catechol groups within a polystyrene backbone and reacted the material with various crosslinking agents, including metals for chelation as well as periodate and chromate to promote oxidation.[95] Each adhesive formulation was tested in a lap shear configuration by applying the material between two aluminum adherends and pulling the adherends apart at a constant speed until failure. When the adhesives were applied in air or underwater, the adhesion strengths exceeded that of commercial glues such as Super Glue.[96–98] Catechol chemistry was incorporated within other polymeric backbones, such as polylactic acid[99,100] and a copolymer of polymethylmethacrylate and oligo ethylene glycol,[101] and optimization of the adhesive formulations resulted in adhesion strengths that rivaled commercial glues.[100] These studies outlined a general approach to designing biomimetic adhesives by incorporating catechol chemistry into a polymeric backbone and optimizing adhesive strength by testing multiple catechol crosslinkers and different catechol incorporation percentages.

The Messersmith group has incorporated DOPA into a number of synthetic materials such as polyethylene oxide (PEO),[102,103] polyacrylate copolymers,[104–106] and a copolymer of hyaluronic acid and Pluronic F127.[107] One example is the development of a catechol-modified PEO-based adhesive[108–110] and a catechol-modified thermosensitive polypropylene oxide–PEO block copolymer[111] as tissue adhesives for fetal membrane repair.[112] Although both materials sealed fetal membrane defects and withstood pressures exceeding the baseline pressure experienced in the uterus, they did not tolerate repeated mechanical load well. For this application, the mechanical properties need to be resilient over repeated strain. By testing their material *ex vivo*, they identified a feature of their adhesives that needed to be adjusted before further testing for fetal membrane applications.

The Lee group has also created several synthetic adhesives that utilize catechol chemistry for surgical applications.[113–118] For example, a multiblock copolymer consisting of PEO and polycaprolactone (PCL), which is used to produce elastomeric polyurethanes, was functionalized with DOPA groups for use in hernia repair with surgical mesh.[119] The material was coated onto biological meshes and crosslinked by oxidizing DOPA moieties with sodium periodate. Measured lap shear and burst strengths of their adhesive were higher than those of Tisseel, a commercially available fibrin glue. However, it should be noted that the adhesives were weighted down for 2 hours before testing. Although high swelling ratios are a common concern with PEO, the addition of PCL in the polymer system led to lower equilibrium swelling. Adding PCL also increased the elastic modulus and toughness of the polymers. Although the cure conditions need to mimic a more physiological environment, these materials show promising adhesion and elastic character.

The Guo group has developed several bioinspired adhesives.[120–122] In one material, they combined catechol chemistry with coacervation, which mimics the way mussels and sandcastle worms secrete their adhesives without dissolution into aqueous environments.[123] They combined DOPA-grafted polyacrylic acid, or PAA, in a zinc solution for applications in tissue adhesion and orthopedic sealing. PAA complexed with zinc ions in a low-pH environment and coacervated (see Figure 8.6). Increasing the pH caused gelation by forming more catechol complexes. When used with aluminum substrates and cured underwater, the coacervate material achieved a higher lap shear adhesion strength of 1.6 MPa compared to <0.1 MPa with a cyanoacrylate adhesive. This study thus demonstrated a strong adhesive strength using a synthetic polymer that mimicked multiple aspects of natural mussel adhesion.

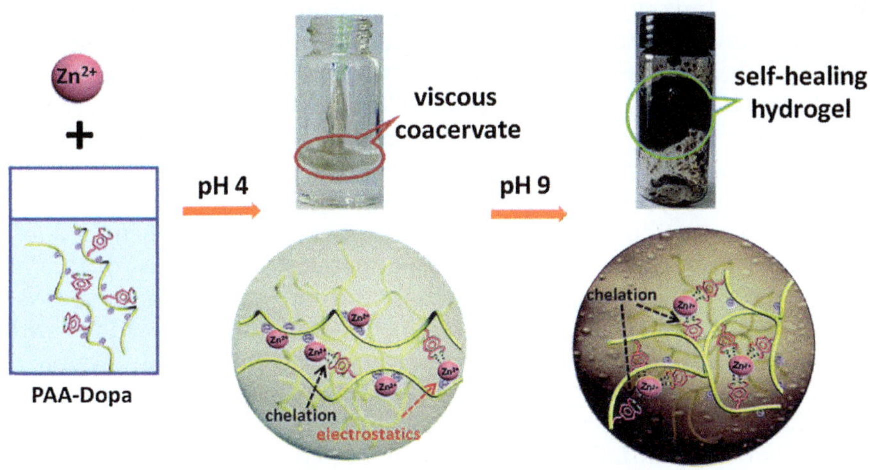

Figure 8.6 Adding DOPA-modified PAA to a Zn^{2+} solution resulted in a coacervate at low pH and a hydrogel at high pH. Reproduced from ref. 123 with permission from the Royal Society of Chemistry.

The Stewart group has developed many adhesive systems based on complex coacervates that mimic the sandcastle worm adhesives.[124–128] One interesting example also utilized mussel adhesive chemistry by combining synthetic polymers with natural proteins. Aminated gelatin, which served as a polycation, was combined with a polyanionic DOPA-containing polymer and divalent cations to form a coacervate adhesive.[129–131] Many of the coacervate formulations underwent a phase transition below the physiological temperature to form elastic solids.[130] Adding magnesium cations resulted in materials with a lower elastic modulus compared to calcium ions. Materials using either divalent cation demonstrated an increase in lap shear adhesion strength when brought above the coacervation temperature; however, materials with the magnesium cations resulted in the highest adhesion strengths. Overall, this study demonstrated that the combination of catechol chemistry and coacervate, which could further transition to a solid, enabled the design of materials with distinct material and adhesive characteristics.

8.4.3 Protein-based Adhesive Biomaterials

Natural and recombinant proteins have also been designed into different adhesives. This section begins with a discussion of fibrin, which is a biologic product used commonly in sealant applications. Several examples are presented that incorporate dityrosine crosslinking, which mimics the crosslinking chemistry utilized in different biological materials, such as silk fibers from caddisfly and some silkworms[58,132] and resilin, which is found in dragonfly wings and locust cuticles.[132,133] Native resilin, which naturally contains dityrosine crosslinking, has an elastic character with a Young's modulus of 600 kPa and elongation at break of 250–300%.[134] In addition, dityrosine crosslinking can confer mechanical stability to protein chains[135] This section ends with examples that utilize elastomeric proteins, such as gelatin and elastin, which are suitable for soft tissue adhesion.

Fibrinogen is enzymatically converted to fibrin *via* thrombin, and it plays a key role in the blood clotting process. Fibrin is commonly found in commercially available surgical sealants such as Tisseel and Evicel. The elastic modulus of fibrin is ~20–40 Pa[136] and is strongly influenced by fibrinogen concentration and clot structure and composition.[137–140] Elvin and coworkers created an elastic fibrinogen-based material that utilized dityrosine crosslinking based on ruthenium-based photochemistry.[141] Fibrinogen was chosen because it is tyrosine rich, and, because of its interactions with extracellular matrix proteins, it has the potential to function as a surgical tissue sealant. Crosslinking of the adhesive was complete after <30 seconds upon illumination with blue light. Controllable adhesive curing using visible light avoids the potentially damaging effects of UV illumination while still providing precise spatial and temporal control over curing. The material had a failure strain of 60% and achieved an adhesive strength more than fivefold stronger than Tisseel.

Recombinant mfps without DOPA modifications have also been developed as biomimetic adhesives. To overcome the challenges of low DOPA conversion and DOPA instability, the Cha group investigated an unmodified version of mfp-151, which was discussed in Section 8.3.[142] Because unmodified mfp-151 is tyrosine rich (~20 mol%), dityrosine photocrosslinking was utilized (see Figure 8.7). Increasing the degree of crosslinking (either by increasing protein or crosslinker concentration) decreased swelling and increased the elastic modulus of the material. The lap shear adhesion strength of the photocrosslinked mfp-151 appeared to be higher than that of Tisseel. Overall, the combination of mussel adhesive proteins with biologically inspired dityrosine crosslinking resulted in elastomeric materials with potential in surgical adhesive applications.

Gelatin is denatured collagen, which is the most abundant animal protein. Its wide availability and ease of modification result in its use in a variety of commercial applications, such as glues, gelling agents, and coatings. Gelatin crosslinked with formaldehyde was used in one of the earliest surgical adhesives.[143] These gelatin-based adhesives faced backlash for the known toxicity of formaldehyde. Gelatin has since been developed without formaldehyde to create low-cost, naturally derived surgical adhesives. Elvin and coworkers used ruthenium-based photochemistry to create an elastic gelatin-based adhesive for repairing soft tissues such as gastrointestinal or lung tissue.[144] Porcine gelatin with dityrosine crosslinks showed an adhesion strength that was ~5 times higher than that of Tisseel. Crosslinking gelatin that had been chemically altered to increase tyrosine content resulted in materials that were 5 times stiffer and swelled 24 times less than

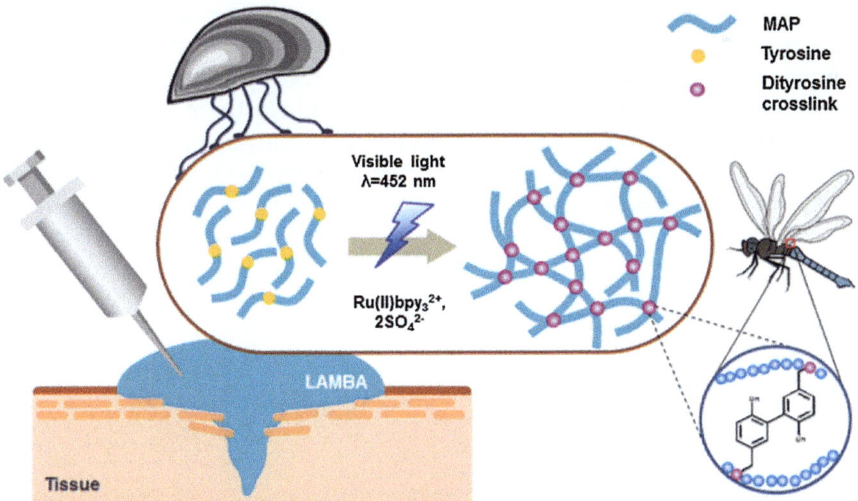

Figure 8.7 Recombinant mussel adhesive proteins (MAPs), such as mfp-151, form dityrosine crosslinks through ruthenium-based photochemistry. The product is named light-activated, mussel protein-based bioadhesive (LAMBA). Reproduced from ref. 142 with permission from Elsevier, Copyright 2015.

crosslinking unmodified gelatin. Low to moderate concentrations of the reactants and leachates of the uncrosslinked or photopolymerized gelatin were not cytotoxic. Finally, photopolymerized gelatin successfully sealed a lung incision *in vivo* in a sheep model.

The Annabi group created a photocrosslinkable gelatin-based surgical adhesive for corneal defect repair.[145] For optical repair, materials need to be highly transparent in addition to being adhesive. The gelatin was functionalized with methacryloyl groups, and visible light was used to activate crosslinking with Eosin Y (initiator), triethanolamine (coinitiator), and *N*-vinylcaprolactam (comonomer). The transparent adhesive was named GelCORE (gel for corneal regeneration). *In vitro* studies demonstrated that this adhesive not only was cytocompatible but also promoted cell migration and proliferation. GelCORE outperformed commercially available Evicel and CoSeal in terms of burst pressure strength and adhesive strength *in vitro*. *Ex vivo* burst pressure strength of GelCORE was higher than ReSure, a commercial ocular sealant. GelCORE demonstrated a tunable tensile and compressive modulus that is both soft and elastic. They also found that, in a corneal defect model in rabbits, GelCORE promoted the regeneration of stromal and epithelial layers that were similar in thickness to native cornea.

Elastin is a protein that gives tissues, such as skin and lungs, the ability to return to their original shape upon repetitive deformation. Elastin is composed of tropoelastin monomers that have coacervated and crosslinked together.[146] This structure results in an elastomeric material (Young's modulus of 300–1000 kPa) with high extensibility (100–150%).[147,148] The Khademhosseini group developed a methacryloyl-substituted recombinant tropoelastin (MeTro) surgical sealant that can be photocrosslinked.[149] They performed tensile, burst pressure, compression, and wound closure testing *via* a combination of *in vitro, ex vivo,* and *in vivo* testing to extensively characterize the material. At a medium methacryloyl substitution level, increasing the protein concentration increased the elastic modulus from 20 to 50 kPa and increased the lap shear adhesive strength, which varied from 30 to 80 kPa. Burst pressure testing indicated that MeTro with a high degree of methacryloyl substitution outperformed Progel, Evicel, and sutures *in vivo*.

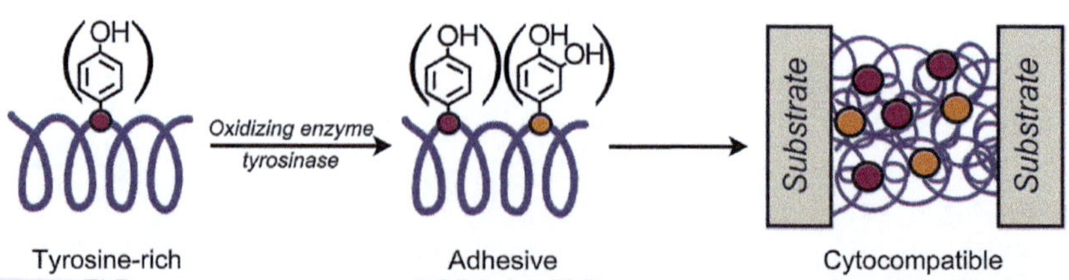

Figure 8.8 Schematic showing the design for obtaining a DOPA-modified elastin-like polypeptide (ELP). Reproduced from ref. 150 with permission from Elsevier, Copyright 2017.

Our group combined catechol chemistry from mussels and coacervates from sandcastles to design a recombinant elastin-like polypeptide (ELP) that was enzymatically modified to incorporate DOPA (see Figure 8.8).[150] Elastin naturally forms coacervates, and we designed the material so that it would coacervate in physiological environments and thus facilitate the underwater application of the glue. The unmodified ELP had a high yield (220 mg L^{-1} of culture with >98% purity). The DOPA-modified protein was cytocompatible and outperformed Tisseel in lap shear tests when applied to aluminum and cured in a humid environment or underwater. This material resulted in the strongest underwater bonds of any rationally designed protein. Further work identified that adhesion was directly affected by factors such as humidity, cure time, temperature, protein structure, and molecular weight.[151] Although the original work relied on DOPA autooxidation for crosslinking, we have investigated different crosslinking chemistries with this material to shorten cure times and increase adhesion strengths.[152,153] A recent study probed the contribution of DOPA and thiol chemistries, which are both found in mfps, to wet adhesion.[152] We found that DOPA-based adhesives had slower cure times and lower adhesion strengths but better recovery from disturbances to the curing process. Combining both chemistries within an adhesive and adding iron ions resulted in a superior adhesive with fast cure times, strong bonds, and good recovery from disturbances during curing. Overall, these elastin-based proteins not only show promise as a surgical adhesive but also serve as a model system to probe important contributions to adhesion.

8.4.4 Conclusions and Future Directions

Designing materials that strongly adhere to wet tissues and maintain natural tissue elasticity is a difficult challenge. Understanding the FDA criteria and the breadth of research utilizing different materials and adhesive chemistries is critical for future surgical adhesive development. Numerous guidelines and experimental tests for biomedical adhesives exist, and the use of standardized tests during development is important so that the performance of different materials can be easily compared. Many adhesives need a fuller understanding of their properties, including characterization of modulus, degradation kinetics, cure time kinetics, and wettability. These tests need to be performed in a physiological environment to effectively predict behavior *in vivo*. Many adhesive materials currently rely on some type of crosslinking chemistry, such as DOPA or dityrosine crosslinking, to achieve adhesion. However, there is a delicate balance to be achieved since more crosslinking generally increases adhesive strength but makes materials stiffer and can alter inherent elastomeric properties away from target tissue values. Although natural systems such as mussels and sandcastle worms provide inspiration for wet adhesion, their complex systems of regulation and precise control tend to be oversimplified in adhesive development. For example, most mussel-inspired adhesives focus only on the inclusion of DOPA without regard to mussel regulation of DOPA and the synergy between mfps. Thus, the focus has been on proteins, such as mfp-5 and mfp-3, that directly impart adhesion instead of on important regulatory proteins, like mfp-6, that maintain the adhesive environment. Using one mechanism for adhesion may not be enough. Our recent work demonstrates that incorporating multiple components of an adhesive system, such as thiol chemistry and coacervation, can better mimic natural adhesives and result in a better glue.[152] In addition, a better understanding of the full scope of mfp composition, location, and interaction will be imperative to fully utilize the unique abilities of marine mussels to adhere in wet, turbulent environments. Future adhesive designs need to creatively apply a full understanding of natural adhesive systems to develop materials that meet unmet clinical soft tissue adhesive needs.

Acknowledgements

This work was supported by a National Science Foundation graduate fellowship to J.E.T.

References

1. M. J. Hall, A. Schwartzman, J. Zhang, Z. Liu, 2010, Division of Health Care Statistics, U.S. Department of Health and Human Services, 2017.
2. J. Höer, U. Klinge, A. Schachtrupp, C. Töns and V. Schumpelick, *Langenbeck's Arch. Surg.*, 2001, **386**, 218.
3. A. J. Singer, J. E. Hollander, S. M. Valentine, H. C. Thode and M. C. Henry, *Acad. Emerg. Med.*, 1996, **3**, 378.
4. C. Cokluk and K. Aydin, *Minim. Invasive Neurosurg.*, 2007, **50**, 324.
5. A. Kaul, S. Hutfless, H. Le, S. A. Hamed, K. Tymitz, H. Nguyen and M. R. Marohn, *Surg. Endosc.*, 2012, **26**, 1269.
6. G. Biggs, J. Hafron, J. Feliciano and D. M. Hoenig, *Urology*, 2005, **66**, 882.e3.
7. C. W. Swenson, G. E. Kolenic, E. R. Trowbridge, M. B. Berger, C. Lewicky-Gaupp, R. U. Margulies, D. M. Morgan, D. E. Fenner and J. O. DeLancey, *Int. Urogynecol. J.*, 2017, **28**, 1377.
8. M. J. Albeck, S. E. Børgesen, F. Gjerris, J. F. Schmidt and P. S. Sørensen, *J. Neurosurg*, 1991, **74**, 597.
9. D. E. Ott, in *Practical Manual for Laparoscopic & Hysteroscopic Gynecological Surgery*, ed. T. Schollmeyer, L. Mettler, D. Rüther and I. Alkatout, Jaypee Brothers Medical Publishers, New Delhi, 2013, 2, 11, 143.
10. R. D. Shippert, *Am. J. Cosmet. Surg.*, 2005, **22**, 25.
11. 21CFR878.4010, Food and Drug Administration, 2019.
12. ISO 10993: 2018 Biological Evaluation of Medical Devices, International Standards Organization, 2018.
13. Food, Drug, and Cosmetic Act, Section 201(h), Food and Drug Administration.
14. Public Health Service, Section 351, Food and Drug Administration.

15. Medical Device Classification Product Codes – Guidance for Industry and Food and Drug Administration Staff, Food and Drug Administration, 2013.
16. G. J. Mattamal, *Expert Rev. Med. Devices*, 2008, **5**, 41.
17. H. Hill, J. F. B. Chick, A. Hage and R. N. Srinivasa, *Diagn. Interv. Radiol.*, 2018, **24**, 98.
18. Fibrin, https://www.fda.gov/vaccines-blood-biologics/approved-blood-products/fibrin, (accessed November 2019).
19. Premarket Approval (PMA), https://www.accessdata.fda.gov/scripts/cdrh/cfdocs/cfPMA/pma.cfm, (accessed November 2019).
20. Class II Special Controls Guidance Document: Tissue Adhesive with Adjunct Wound Closure Device Intended for the Topical Approximation of Skin - Guidance for Industry and FDA Staff, Center for Devices and Radiological Health, 2010.
21. Class II Special Controls Guidance Document: Tissue Adhesive for the Topical Approximation of Skin - Guidance for Industry and FDA Staff, Center for Devices and Radiological Health, 2008.
22. Efficacy Studies to Support Marketing of Fibrin Sealant Products Manufactured for Commercial Use, Center for Biologics Evaluation and Research, 1999.
23. G. Arturson and G. Wallenius, *Scand. J. Clin. Lab. Invest.*, 1964, **16**, 81.
24. Submission and Review of Sterility Information in Premarket Notification (510(k)) Submissions for Devices Labeled as Sterile: Guidance for Industry and Food and Drug Administration Staff, Food and Drug Administration, 2016.
25. A. Miserez, J. C. Weaver, P. B. Pedersen, T. Schneeberk, R. T. Hanlon, D. Kisailus and H. Birkedal, *Adv. Mater.*, 2009, **21**, 401.
26. A. Pena-Francesch and M. C. Demirel, *Front. Chem.*, 2019, **7**, 69.
27. P. A. Guerette, S. Hoon, Y. Seow, M. Raida, A. Masic, F. T. Wong, V. H. B. Ho, K. W. Kong, M. C. Demirel, A. Pena-Francesch, S. Amini, G. Z. Tay, D. Ding and A. Miserez, *Nat. Biotechnol.*, 2013, **31**, 908.
28. P. A. Guerette, S. Hoon, D. Ding, S. Amini, A. Masic, V. Ravi, B. Venkatesh, J. C. Weaver and A. Miserez, *ACS Nano*, 2014, **8**, 7170.
29. D. Ding, P. A. Guerette, S. Hoon, K. W. Kong, T. Cornvik, M. Nilsson, A. Kumar, J. Lescar and A. Miserez, *Biomacromolecules*, 2014, **15**, 3278.
30. D. Ding, P. A. Guerette, J. Fu, L. Zhang, S. A. Irvine and A. Miserez, *Adv. Mater.*, 2015, **27**, 3953.
31. A. Pena-Francesch, N. E. Domeradzka, H. Jung, B. Barbu, M. Vural, Y. Kikuchi, B. D. Allen and M. C. Demirel, *APL Mater.*, 2018, **6**, 010701.
32. A. Pena-Francesch, B. Akgun, A. Miserez, W. Zhu, H. Gao and M. C. Demirel, *Adv. Funct. Mater.*, 2014, **24**, 6227.
33. K. Deepankumar, C. Lim, I. Polte, B. Zappone, C. Labate, M. P. De Santo, H. Mohanram, A. Palaniappan, D. S. Hwang and A. Miserez, *Adv. Funct. Mater.*, 2020, **30**, 1907534.

34. A. N. Leberfinger, M. Hospodiuk, A. Pena-Francesch, B. Ayan, V. Ozbolat, S. V. Koduru, I. T. Ozbolat, M. C. Demirel and D. J. Ravnic, *Plast. Reconstr. Surg. Glob. Open.*, 2018, **6**, e1881.
35. L. D. Graham, V. Glattauer, M. G. Huson, J. M. Maxwell, R. B. Knott, J. W. White, P. R. Vaughan, Y. Peng, M. J. Tyler and J. A. Werkmeister, *Biomacromolecules*, 2005, **6**, 3300.
36. L. D. Graham, V. Glattauer, D. Li, M. J. Tyler and J. A. M. Ramshaw, *Comp. Biochem. Physiol. B.*, 2013, **165**, 250.
37. N. L. Millar, T. A. Bradley, N. A. Walsh, R. C. Appleyard, M. J. Tyler and G. A. C. Murrell, *J. Shoulder Elbow Surg.*, 2009, **18**, 639.
38. S. Ebnesajjad, in *Handbook of Adhesives and Surface Preparation*, ed. S. Ebnesajjad, Elsevier, Oxford, 2011, 1, 1.
39. J. G. Speight, *Asphalt Materials Science and Technology*, Elsevier, Oxford, 2016.
40. K. Modaresifar, S. Azizian and A. Hadjizadeh, *Polym. Rev.*, 2016, **56**, 329.
41. J. N. Burris, S. C. Lenaghan and C. N. Stewart, *Plant Cell Rep.*, 2018, **37**, 565.
42. Y. Huang, Y. Wang, L. Tan, L. Sun, J. Petrosino, M.-Z. Cui, F. Hao and M. Zhang, *Proc. Natl. Acad. Sci. U. S. A.*, 2016, **113**, E3193.
43. Y. Huang, Y.-J. Wang, Y. Wang, S. Yi, Z. Fan, L. Sun, D. Lin, N. Anreddy, H. Zhu, M. Schmidt, Z.-S. Chen and M. Zhang, *Acta Biomater.*, 2015, **25**, 268.
44. Y. Li, J. Krahn and C. Menon, *J. Bionic. Eng.*, 2016, **13**, 181.
45. M. Calvaresi, L. Eckhart and L. Alibardi, *J. Struct. Biol.*, 2016, **194**, 282.
46. N. W. Rizzo, K. H. Gardner, D. J. Walls, N. M. Keiper-Hrynko, T. S. Ganzke and D. L. Hallahan, *J. R. Soc., Interface*, 2006, **3**, 441.
47. K. Autumn, Y. A. Liang, S. T. Hsieh, W. Zesch, W. P. Chan, T. W. Kenny, R. Fearing and R. J. Full, *Nature*, 2000, **405**, 681.
48. K. Autumn, *Integr. Comp. Biol.*, 2002, **42**, 1081.
49. H. Gao, X. Wang, H. Yao, S. Gorb and E. Arzt, *Mech. Mater.*, 2005, **37**, 275.
50. A. Mahdavi, L. Ferreira, C. Sundback, J. W. Nichol, E. P. Chan, D. J. D. Carter, C. J. Bettinger, S. Patanavanich, L. Chignozha, E. Ben-Joseph, A. Galakatos, H. Pryor, I. Pomerantseva, P. T. Masiakos, W. Faquin, A. Zumbuehl, S. Hong, J. Borenstein, J. Vacanti, R. Langer and J. M. Karp, *Proc. Natl. Acad. Sci. U. S. A.*, 2008, **105**, 2307.
51. K. Kamino, *J. Adhes.*, 2010, **86**, 96.
52. C. R. So, J. M. Scancella, K. P. Fears, T. Essock-Burns, S. E. Haynes, D. H. Leary, Z. Diana, C. Wang, S. North, C. S. Oh, Z. Wang, B. Orihuela, D. Rittschof, C. M. Spillmann and K. J. Wahl, *ACS Appl. Mater. Interfaces*, 2017, **9**, 11493.
53. C. Liang, J. Strickland, Z. Ye, W. Wu, B. Hu and D. Rittschof, *Front. Mar. Sci.*, 2019, **6**, 565.
54. K. Kamino, *Biofouling*, 2013, **29**, 735.
55. G. B. Wiggins, *Caddisflies: The Underwater Architects*, University of Toronto Press, Toronto, 2004.

56. M. S. Engster, *Cell Tissue Res.*, 1976, **169**, 77.
57. C.-S Wang, H. Pan, G. M. Weerasekare and R. J. Stewart, *J. R. Soc., Interface*, 2015, **12**, 20150710.
58. C.-S. Wang, N. N. Ashton, R. B. Weiss and R. J. Stewart, Hysperophylax occidentalis, *Insect Biochem. Molec.*, 2014, **54**, 69.
59. M. Tszydel, A. Zabłotni, D. Wojciechowska, M. Michalak, I. Krucińska, K. Szustakiewicz, M. Maj, A. Jaruszewska and J. Strzelecki, *J. Mech. Behav. Biomed.*, 2015, **45**, 142.
60. R. J. Stewart, *J. Exp. Biol.*, 2004, **207**, 4727.
61. B. J. Endrizzi and R. J. Stewart, *J. Adhes.*, 2009, **85**, 546.
62. H. Zhao, C. Sun, R. J. Stewart and J. H. Waite, *J. Biol. Chem.*, 2005, **280**, 42938.
63. C. S. Wang and R. J. Stewart, *Biomacromolecules*, 2013, **14**, 1607.
64. R. A. Jensen and D. E. Morse, *J. Comp. Physiol., B*, 1988, **158**, 317.
65. J. H. Waite, R. A. Jensen and D. E. Morse, *Biochemistry*, 1992, **31**, 5733.
66. Y. Lee, C. Xu, M. Sebastin, A. Lee, N. Holwell, C. Xu, D. Miranda Nieves, L. Mu, R. S. Langer, C. Lin and J. M. Karp, *Adv. Healthcare Mater.*, 2015, **4**, 2587.
67. T. Steinbrecher, E. Danninger, D. Harder, T. Speck, O. Kraft and R. Schwaiger, *Acta Biomater.*, 2010, **6**, 1497.
68. J. R. Burkett, J. L. Wojtas, J. L. Cloud and J. J. Wilker, *J. Adhes.*, 2009, **85**, 601.
69. B. P. Lee, P. B. Messersmith, J. N. Israelachvili and J. H. Waite, *Annu. Rev. Mater. Res.*, 2011, **41**, 99.
70. J. H. Waite, *J. Exp. Biol.*, 2017, **220**, 517.
71. Q. Guo, J. Chen, J. Wang, H. Zeng and J. Yu, *Nanoscale*, 2020, **12**, 1307.
72. P. Kord Forooshani and B. P. Lee, *J. Polym. Sci. Part A: Polym. Chem.*, 2017, **55**, 9.
73. Y. Akdogan, W. Wei, K.-Y. Huang, Y. Kageyama, E. W. Danner, D. R. Miller, N. Martinez Rodriguez, J. H. Waite and S. Han, *Angew. Chem. Int. Ed.*, 2014, **53**, 11253.
74. S. C. T. Nicklisch and J. H. Waite, *Biofouling*, 2012, **28**, 865.
75. T. Priemel, R. Palia, M. Babych, C. J. Thibodeaux, S. Bourgault and M. J. Harrington, *Proc. Natl. Acad. Sci. U. S. A.*, 2020, **117**, 7613.
76. C. V. Benedict and J. H. Waite, *J. Morphol.*, 1986, **189**, 171.
77. L. M. Rzepecki, K. M. Hansen and J. H. Waite, *Biol. Bull.*, 1992, **183**, 123.
78. H. Zhao and J. H. Waite, *Biochemistry*, 2006, **45**, 14223.
79. T. Priemel, E. Degtyar, M. N. Dean and M. J. Harrington, *Nat. Commun.*, 2017, **8**, 14539.
80. J. Yu, W. Wei, E. Danner, R. K. Ashley, J. N. Israelachvili and J. H. Waite, *Nat. Chem. Biol.*, 2011, **7**, 588.
81. H. Zhao and J. H. Waite, *J. Biol. Chem.*, 2006, **281**, 26150.
82. V. Papov, T. Diamond, K. Biemann and J. H. Waite, *J. Biol.*, 1995, **270**, 20183.

83. H. Zhao, N. B. Robertson, S. A. Jewhurst and J. H. Waite, *J. Biol. Chem.*, 2006, **281**, 11090.
84. G. P. Maier, M. V. Rapp, J. H. Waite, J. N. Israelachvili and A. Butler, *Science*, 2015, **349**, 628.
85. W. Wei, J. Yu, C. Broomell, J. N. Israelachvili and J. H. Waite, *J. Am. Chem. Soc.*, 2013, **135**, 377.
86. W. Wei, Y. Tan, N. R. Martinez Rodriguez, J. Yu, J. N. Israelachvili and J. H. Waite, *Acta Biomater.*, 2014, **10**, 1663.
87. E. Valois, R. Mirshafian and J. H. Waite, *Sci. Adv.*, 2020, **6**, eaaz6486.
88. B. Yang, S. Jin, Y. Park, Y. M. Jung and H. J. Cha, *Small*, 2018, **14**, 1803377.
89. D. S. Hwang, H. J. Yoo, J. H. Jun, W. K. Moon and H. J. Cha, *Appl. Environ. Microbiol.*, 2004, **70**, 3352.
90. D. S. Hwang, Y. Gim and H. J. Cha, *Biotechnol. Prog.*, 2005, **21**, 965.
91. D. S. Hwang, Y. Gim, H. J. Yoo and H. J. Cha, *Biomaterials*, 2007, **28**, 3560.
92. ISO 22442:2015 Medical Devices Utilizing Animal Tissues and Their Derivatives, International Standards Organization, 2015.
93. Regulatory Considerations for Human Cells, Tissues, and Cellular and Tissue-Based Products: Minimal Manipulation and Homologous Use, Food and Drug Administration, 2017.
94. Current Good Tissue Practice (CGTP) and Additional Requirements for Manufacturers of Human Cells, Tissues, and Cellular and Tissue-Based Products (HCT/Ps), Food and Drug Administration, 2011.
95. G. Westwood, T. N. Horton and J. J. Wilker, *Macromolecules*, 2007, **40**, 3960.
96. C. R. Matos-Pérez, J. D. White and J. J. Wilker, *J. Am. Chem. Soc.*, 2012, **134**, 9498.
97. M. A. North, C. A. Del Grosso and J. J. Wilker, *ACS Appl. Mater. Interfaces*, 2017, **9**, 7866.
98. H. J. Meredith, C. L. Jenkins and J. J. Wilker, *Adv. Funct. Mater.*, 2014, **24**, 3259.
99. C. L. Jenkins, H. M. Siebert and J. J. Wilker, *Macromolecules*, 2017, **50**, 561.
100. H. M. Siebert and J. J. Wilker, *ACS Sustainable Chem. Eng.*, 2019, **7**, 13315.
101. H. J. Meredith and J. J. Wilker, *Adv. Funct. Mater.*, 2015, **25**, 5057.
102. C. E. Brubaker and P. B. Messersmith, *Biomacromolecules*, 2011, **12**, 4326.
103. P. B. Messersmith, J. L. Dalsin, B. P. Lee and S. A. Burke, *U. S. Pat.*, 8 673 286, 2008
104. B. D. B. Tiu, P. Delparastan, M. R. Ney, M. Gerst and P. B. Messersmith, *ACS Appl. Mater. Interfaces*, 2019, **11**, 28296.
105. B. D. B. Tiu, P. Delparastan, M. R. Ney, M. Gerst and P. Messersmith, *Angew. Chem. Int. Ed.*, 2020, **59**, 16616.
106. H. Lee, B. P. Lee and P. B. Messersmith, *Nature*, 2007, **448**, 338.

107. Y. Lee, H. J. Chung, S. Yeo, C.-H. Ahn, H. Lee, P. B. Messersmith and T. G. Park, *Soft Matter*, 2010, **6**, 977.
108. A. Kivelio, P. Dekoninck, M. Perrini, C. E. Brubaker, P. B. Messersmith, E. Mazza, J. Deprest, R. Zimmermann, M. Ehrbar and N. Ochsenbein-Koelble, *Eur. J. Obstet. Gyn. R. B.*, 2013, **171**, 240.
109. G. Bilic, C. Brubaker, P. B. Messersmith, A. S. Mallik, T. M. Quinn, C. Haller, E. Done, L. Gucciardo, S. M. Zeisberger, R. Zimmermann, J. Deprest and A. H. Zisch, *Am. J. Obstet. Gynecol.*, 2010, **202**(85), e1.
110. C. E. Brubaker, H. Kissler, L.-J. Wang, D. B. Kaufman and P. B. Messersmith, *Biomaterials*, 2010, **31**, 420.
111. D. G. Barrett, G. G. Bushnell and P. B. Messersmith, *Adv. Healthc. Mater.*, 2013, **2**, 745.
112. M. Perrini, D. Barrett, N. Ochsenbein-Koelble, R. Zimmermann, P. Messersmith and M. Ehrbar, *J. Mech. Behav. Biomed.*, 2016, **58**, 57.
113. A. R. Narkar, B. Barker, M. Clisch, J. Jiang and B. P. Lee, *Chem. Mater.*, 2016, **28**, 5432.
114. R. Pinnaratip, H. Meng, R. M. Rajachar and B. P. Lee, *Biomed. Mater.*, 2018, **13**, 025003.
115. M. Cencer, M. Murley, Y. Liu and B. P. Lee, *Biomacromolecules*, 2015, **16**, 404.
116. A. R. Narkar and B. P. Lee, *Langmuir*, 2018, **34**, 9410.
117. Y. Li, H. Meng, Y. Liu, A. Narkar and B. P. Lee, *ACS Appl. Mater. Interfaces*, 2016, **8**, 11980.
118. R. Pinnaratip, P. Kord Forooshani, M. Li, Y. H. Hu, R. Rajachar and B. P. Lee, *ACS Biomater. Sci. Eng.*, 2020, **6**, 4502.
119. J. L. Murphy, L. Vollenweider, F. Xu and B. P. Lee, *Biomacromolecules*, 2010, **11**, 2976.
120. Z. Yuan, J. Wang, X. Niu, J. Ma, X. Qin, L. Li, L. Shi, Y. Wu and X. Guo, *Ind. Eng. Chem. Res.*, 2019, **58**, 9390.
121. A. Li, Z. Xu, N. Sun, Z. Si, Y. Xu and X. Guo, *J. Appl. Polym. Sci.*, 2019, **136**, 47126.
122. A. Li, Y. Jia, S. Sun, Y. Xu, B. B. Minsky, M. A. C. Stuart, H. Cölfen, R. von Klitzing and X. Guo, *ACS Appl. Mater. Interfaces*, 2018, **10**, 10471.
123. W. Wang, Y. Xu, A. Li, T. Li, M. Liu, R. von Klitzing, C. K. Ober, A. B. Kayitmazer, L. Li and X. Guo, *RSC Adv.*, 2015, **5**, 66871.
124. H. Shao, K. N. Bachus and R. J. Stewart, *Macromol. Biosci.*, 2009, **9**, 464.
125. H. Shao, G. M. Weerasekare and R. J. Stewart, *J. Biomed. Mater. Res.*, 2011, **97A**, 46.
126. R. J. Stewart, H. Shao, *U. S. Pat.*, 8 283 384, 2012.
127. I. T. Song and R. J. Stewart, *Soft Matter*, 2018, **14**, 379.
128. L. K. Mann, R. Papanna, K. J. Moise, R. H. Byrd, E. J. Popek, S. Kaur, S. C. G. Tseng and R. J. Stewart, *Acta Biomater.*, 2012, **8**, 2160.
129. S. Kaur, G. M. Weerasekare and R. J. Stewart, *ACS Appl. Mater. Interfaces*, 2011, **3**, 941.
130. H. Shao and R. J. Stewart, *Adv. Mater.*, 2010, **22**, 729.
131. B. D. Winslow, H. Shao and R. J. Stewart, *Biomaterials*, 2010, **31**, 9373.

132. B. P. Partlow, M. B. Applegate, F. G. Omenetto and D. L. Kaplan, *ACS Biomater. Sci. Eng.*, 2016, **2**, 2108.
133. R. S.-C. Su, Y. Kim and J. C. Liu, *Acta Biomater.*, 2014, **10**, 1601.
134. T. Weis-Fogh, *J. Exp. Biol.*, 1960, **37**, 889.
135. K. Stewart, P. L. Spedding, M. S. Otterburn and D. M. Lewis, *J. Appl. Polym. Sci.*, 1995, **66**, 2365.
136. L. Urech, A. G. Bittermann, J. A. Hubbell and H. Hall, *Biomaterials*, 2005, **26**, 1369.
137. J. D. Ferry and P. R. Morrison, *J. Am. Chem. Soc.*, 1947, **69**, 388.
138. M. W. Mosesson, J. P. DiOrio, M. F. Muller, J. R. Shainoff, K. R. Siebenlist, D. L. Amrani, G. A. Homandberg, J. Soria, C. Soria and M. Samama, *Blood*, 1987, **69**, 1073.
139. B. Blomback, *Thromb. Res.*, 1996, **83**, 1.
140. M. W. Mosesson, K. R. Siebenlist and D. A. Meh, *Ann. N. Y. Acad. Sci.*, 2006, **936**, 11.
141. C. M. Elvin, A. G. Brownlee, M. G. Huson, T. A. Tebb, M. Kim, R. E. Lyons, T. Vuocolo, N. E. Liyou, T. C. Hughes, J. A. M. Ramshaw and J. A. Werkmeister, *Biomaterials*, 2009, **30**, 2059.
142. E. Y. Jeon, B. H. Hwang, Y. J. Yang, B. J. Kim, B.-H. Choi, G. Y. Jung and H. J. Cha, *Biomaterials*, 2015, **67**, 11.
143. J. M. Albes, C. Krettek, B. Hausen, R. Rohde, A. Haverich and H.-G. Borst, *Ann. Thorac. Surg.*, 1993, **56**, 910.
144. C. M. Elvin, T. Vuocolo, A. G. Brownlee, L. Sando, M. G. Huson, N. E. Liyou, P. R. Stockwell, R. E. Lyons, M. Kim, G. A. Edwards, G. Johnson, G. A. McFarland, J. A. M. Ramshaw and J. A. Werkmeister, *Biomaterials*, 2010, **31**, 8323.
145. E. Shirzaei Sani, A. Kheirkhah, D. Rana, Z. Sun, W. Foulsham, A. Sheikhi, A. Khademhosseini, R. Dana and N. Annabi, *Sci. Adv.*, 2019, **5**, eaav1281.
146. S. Mithieux and A. Weiss, *Adv. Protein Chem.*, 2005, **70**, 437.
147. J. Gosline, M. Lillie, E. Carrington, P. Guerette, C. Ortlepp and K. Savage, *Phil. Trans. R. Soc. Lond. B.*, 2002, **357**, 121.
148. Y.-C. Fung, *Biomechanics: Mechanical Properties of Living Tissues*, Springer, New York, 1993, 2.
149. N. Annabi, Y.-N. Zhang, A. Assmann, E. S. Sani, G. Cheng, A. D. Lassaletta, A. Vegh, B. Dehghani, G. U. Ruiz-Esparza, X. Wang, S. Gangadharan, A. S. Weiss and A. Khademhosseini, *Sci. Transl. Med.*, 2017, **9**, eaai7466.
150. M. J. Brennan, B. F. Kilbride, J. J. Wilker and J. C. Liu, *Biomaterials*, 2017, **124**, 116.
151. M. J. Brennan, S. E. Hollingshead, J. J. Wilker and J. C. Liu, *R. Soc. Open Sci.*, 2018, **5**, 171225.
152. C.-Y. Lin and J. C. Liu, *ACS Appl. Bio. Mater.*, 2020, **3**, 3894.
153. S. E. Hollingshead, J. E. Torres, J. J. Wilker and J. C. Liu, *ACS Appl. Bio. Mater.*, 2022, **5**, 630.

CHAPTER 9

Silk-elastin-like Protein-based Polymers for Controlled Delivery Applications

M. W. TALBOT,[a,b] D. B. STEINHAUFF,[a,b] K. J. ISAACSON,[a,b] J. CAPPELLO[c] AND H. GHANDEHARI*[a,b,c]

[a] Utah Center for Nanomedicine, Nano Institute of Utah, University of Utah, Salt Lake City, UT, USA; [b] Department of Biomedical Engineering, University of Utah, Salt Lake City, UT, USA; [c] Department of Pharmaceutics and Pharmaceutical Chemistry, University of Utah, Salt Lake City, UT, USA
*Email: hamid.ghandehari@utah.edu

9.1 Introduction

Prior to the development of recombinant protein-based polymers, in-depth studies were conducted elucidating the independent physicochemical properties of constituent structural motifs of silk and elastin. Characterization of the 84 000 g mol^{-1} isolated crystalline fraction of *Bombyx mori* (silkworm) silk fibroin pointed to repeats of SG(AG)$_n$ as a minimum structural repeat sequence.[1] Synthetic elastin-like polypeptides (ELPs) recapitulated the thermoresponsivity of tropoelastin, transitioning from conformationally flexible, soluble polymers to a dense, viscoelastic phase upon heating.[2–4]

Capitalizing upon such work, Cappello, Ferrari, and coworkers were the first to design recombinantly produced structural protein-based polymers, namely silk-like protein polymers (SLPs) and silk-elastin-like protein

Biomaterials Science Series No. 10
Biomimetic Protein Based Elastomers: Emerging Materials for the Future
Edited by Namita Roy Choudhury, Julie C. Liu and Naba K. Dutta

Published by the Royal Society of Chemistry, www.rsc.org

polymers (SELPs).[5–7] SELPs integrate the structural strength of the beta-sheet forming motif found in *Bombyx mori* silk fibroin (GAGAGS) and thermo-responsive viscoelasticity of human tropoelastin and ELPs (GVGVP) (see Figure 9.1). SELP genes were produced from monomeric repeats containing both silk- and elastin-like motifs later linked together repeatedly in a process called concatemerization. Researchers have illustrated the dependence of physicochemical properties on primary amino acid sequence, length of silk and elastin motifs, and the silk- to elastin-like motif ratio. Variation of these primary sequences has allowed us to produce and study SELPs with varying structures and resulting functions.[5,8–11]

9.1.1 Structure–Function Relationships

A central theme of this chapter will be the correlation between the primary amino acid sequences, resulting supramolecular structures, and functionality of SELPs for given applications. For most SELPs, structural properties are governed by their sequence (see Figure 9.1) and environment (*e.g.*, temperature and cure time), which dictate crosslinking, network density, and tortuosity.[9]

Silk-like motifs form highly crystalline structures in nature *via* the establishment of physical crosslinks in extensively aligned beta-sheets.[5,6] SELPs of approximately 65 000–85 000 g mol^{-1} with silk motifs of more than two repeats per monomer allow time-, temperature- and concentration-dependent self-assembly, facilitating the adoption of a 3D hydrogel network.[5] Thermo-responsive hydrogel-forming SELPs present an injectable, localized matrix from which a bioactive agent, including protein or macromolecule,[6] drug with a low molecular weight (MW),[14–18] cell,[19,20] plasmid,[21–23] or adenovirus[13,24,25] may be released. The rate of release may be controlled by diffusion through pores and the partitioning behavior of bioactive agents to polymer strands.[14,15]

SELP MW and sequence influence sensitivity to environmental stimuli. Temperature, pH,[26] presence of enzymes,[13,23] and 3D network properties, such as crosslinking density, influence the amount of absorbed water—swelling ratio, soluble fraction,[9] and mechanical stiffness.[11,12] Therapeutic agents may interact with constituent hydrophobic or charged residues of the polymer backbone, observed in partitioning behavior.[18]

This chapter will focus on recent developments in the use of SELP hydrogels for the controlled release of adenoviral gene delivery systems, chemotherapeutics, imaging agents, synthetic glycosaminoglycans (GAGs), and developments in self-assembling SELP nanoparticles and the use of SELPs as liquid embolics. We will build upon previous work with SELPs in controlled release, reviewed elsewhere.[27–31]

9.1.2 Physical Characterization of Hydrogel-forming Silk-elastin-like Protein Polymers (SELPs)

In seminal studies conducted by Cappello *et al.*, gelation and beta-sheet formation were characterized by X-ray diffraction, Fourier-transform

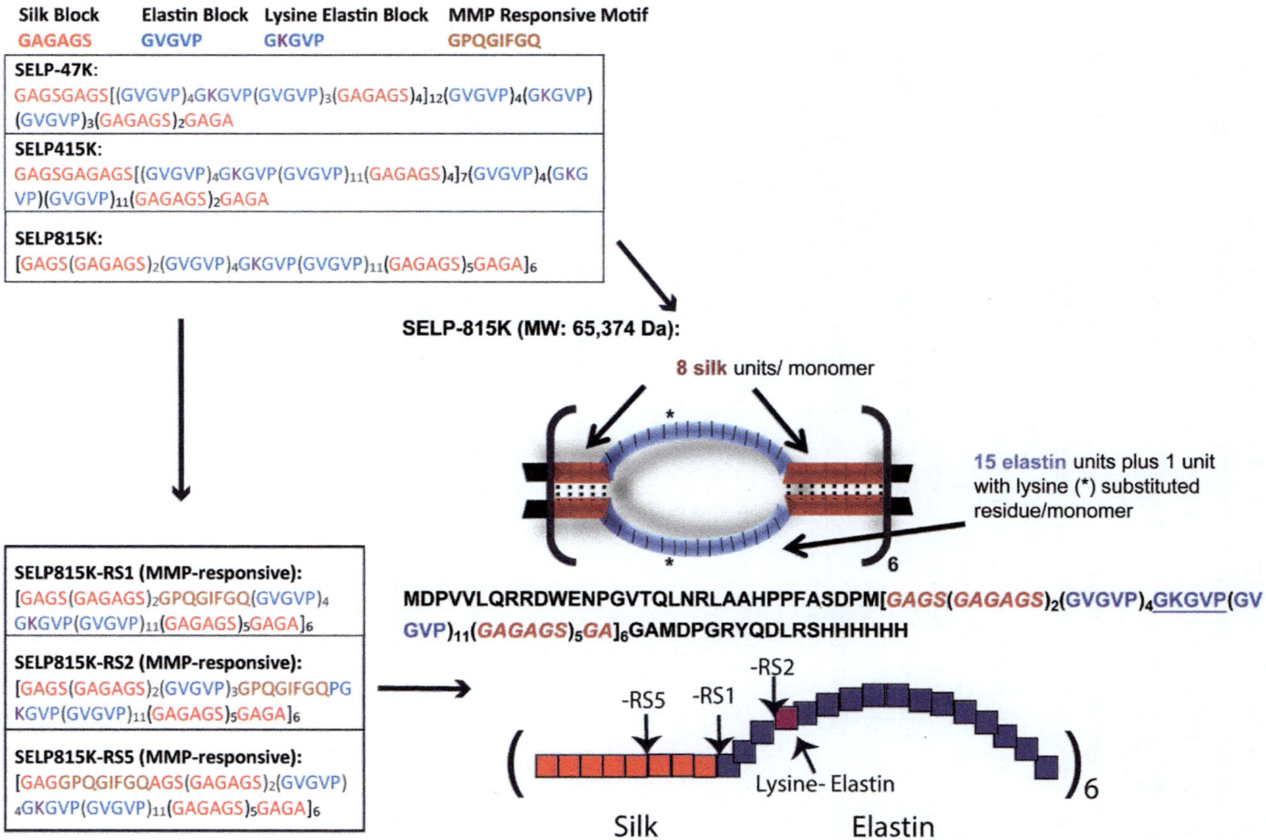

Figure 9.1 Schematics and amino acid sequences of some of the hydrogel-forming SELPs used in controlled release applications. Single-letter amino acid abbreviations are used. MMP: Matrix metalloproteinase, RS: Responsive sequence. Arrows for -RS5, -RS1, and -RS2 indicate the location of matrix metalloproteinase responsive sequences in SELP815K-RS5, SELP815K-RS1, and SELP815K-RS2, respectively. Generally, red indicates silk blocks (GAGAGS) and blue indicates elastin blocks (GVGVP). Adapted from ref. 12 and 13 with permission from Elsevier, Copyright 2015.

infrared spectroscopy (FTIR), differential scanning calorimetry (DSC), and viscometry for SELPs of varying silk and elastin motif lengths.[5,6] Together, results pointed to the degree of hydrogen bonding of silk motifs as the basis for structural strength.[5,6] Further physicochemical characterizations, including variations in the polymer structure, concentration, and cure time upon bound water;[32] degree of swelling;[33] and soluble fraction[11] for SELP hydrogels, have been discussed in relation to effects upon controlled release.[28,33]

Crosslinking is initiated through nucleation-like events and is followed by nanofiber self-assembly from beta-sheets.[34,35] Changes to SELP network structure have been characterized by techniques including small-angle neutron scattering (SANS)[9] and atomic force microscopy (AFM).[34,35] SANS, which measures elastically scattered neutrons from SELP, was utilized to evaluate matrix structure during the sol–gel transition and postgelation.[9] When fit to a combined Debye–Bueche/power law model, data indicate that additional noncovalent crosslinks form over time, reducing the fibril correlation length, which indicates matrix density, from ~68 nm to ~34 nm in the sol–gel transition.[9] SELPs form nanofibers at concentrations below the minimum hydrogel-forming concentration as evidenced by AFM.[34,35] We hypothesize these nanofibers are precursors to full hydrogel formation at appropriate polymer concentration, facilitated by thermally driven conformational sampling. Even at nongelling concentrations, we hypothesize that crosslinks form between neighboring nanofibers, the strength and length of which show dependence upon the silk-like to elastin-like sequence and ratio; increased nanofiber density was visualized for SELP-47K over SELP-415K by tapping mode AFM.[34]

9.1.3 Recombinant Synthesis, Expression, and Purification

For a more extensive review of recombinant synthesis and scale-up of SELPs, readers are encouraged to refer to additional articles.[30,31,36] Briefly, monomer genes are chemically synthesized or oligonucleotide inserts ligated into previously designed monomer genes.[5,9,10] The monomer gene is then multimerized by random concatemerization to produce the polymer gene, which is ligated into an expression vector and transformed into microbes. Polymer genes are screened and selected by colony growth on antibiotic-selective plates and the subsequent polymerase chain reaction of isolated colonies. The isolated plasmid is then retransformed into bacteria, expression induced, and protein polymer product purified and screened by sodium dodecyl sulfate–polyacrylamide gel electrophoresis (SDS–PAGE).[5,9,10]

For recent SELPs incorporating matrix metalloproteinase (MMP) bioresponsive motifs, monomer gene pSY1378 was digested at three unique restriction sites (see Figure 9.1, arrows), corresponding to -RS1 (RS, responsive sequence), -RS2, and -RS5 constructs, and random concatemerization utilized to produce a 6 mer polymer plasmid.[10,13] These polymers have an enzymatically degradable sequence located between silk-like and elastin-like motifs in -RS1, between elastin-like motifs in -RS2, and between silk-like motifs in -RS5. Accessibility and degradation by MMPs, in addition to physicochemical properties, are unique to each construct, as demonstrated by *in vitro* and *in vivo* degradation and release from MMP-RS SELPs in the presence of MMP-2 and MMP-9.[11,13]

For all SELPs generated within our laboratory, pPT317 expression vectors were transformed into EcR3 cells and expression induced by a temperature-sensitive promoter, CI857.[7] Following several passages and inoculation of larger cultures, induction occurs at 42 °C, followed by fermentation at 40 °C. Typical yields range from 1 to 1.5 kg of biomass from a 6 L fermentation. Following bacterial lysis, purification steps include cationic polyethylenimine precipitation, multiple ammonium sulfate precipitations, tangential flow filtration, and both cationic and anionic chromatography. The sequential process separates SELP based upon charge, isoelectric point, and size. The removal of endotoxins from SELP solutions is a critical aspect of purification typically accomplished in the final steps and is confirmed by Limulus amebocyte lysate (LAL) pyrogen testing. The final product is then freeze dried and stored at −80 °C pending postpurification processing.[30]

9.1.3.1 Postpurification Processing

Shear processing, we hypothesize, linearizes lyophilized SELPs.[11] As a final step in the biomanufacturing process, shearing may normalize batch to batch properties, reduce the swelling ratio, and lower the polymer-soluble fraction (see Figure 9.2). Shearing at 17 000 psi increases gelation rate and polymer stiffness (storage modulus, G') and reduces the minimum gel-forming concentration (MGFC) over lyophilized SELPs.[11,12] Reduction of the MGFC, particularly for SELP815K-RS5, is believed to be a result of loss of secondary and tertiary structures in polymer strands.[11] The placement of the MMP-RS between silk motifs of -RS5 disrupts the formation of beta-sheets, which is ameliorated by shearing. Shearing restored some crystallization potential lost due to long-range structural disruption by the MMP-RS location between silk motifs. This has benefits in controlled, sustained release to prevent a burst response.[11,13]

The influence of shearing over the macromolecular assembly and mechanical properties is further illustrated by accelerated gelation kinetics of both sheared SELP-47K and SELP-815K, reducing the time for hydrogel to reach their final modulus value (G') compared to lyophilized formulations.[12] Shearing significantly shortens gelation time in SELP-47K, which forms shorter beta-sheets compared to SELP-815K, as SELP-815K has twice as many silk-like repeats with hydrogen bond formation potential.[12] The pore size and swelling ratio of SELP-815K were both reduced upon shearing facilitated by increased accessibility and intermolecular interactions of SELP-815K's long silk-like motifs.[11]

Other postpurification processes utilized to generate unique mechanical and functional properties of SELPs include wet spinning;[37] electrospinning;[38]

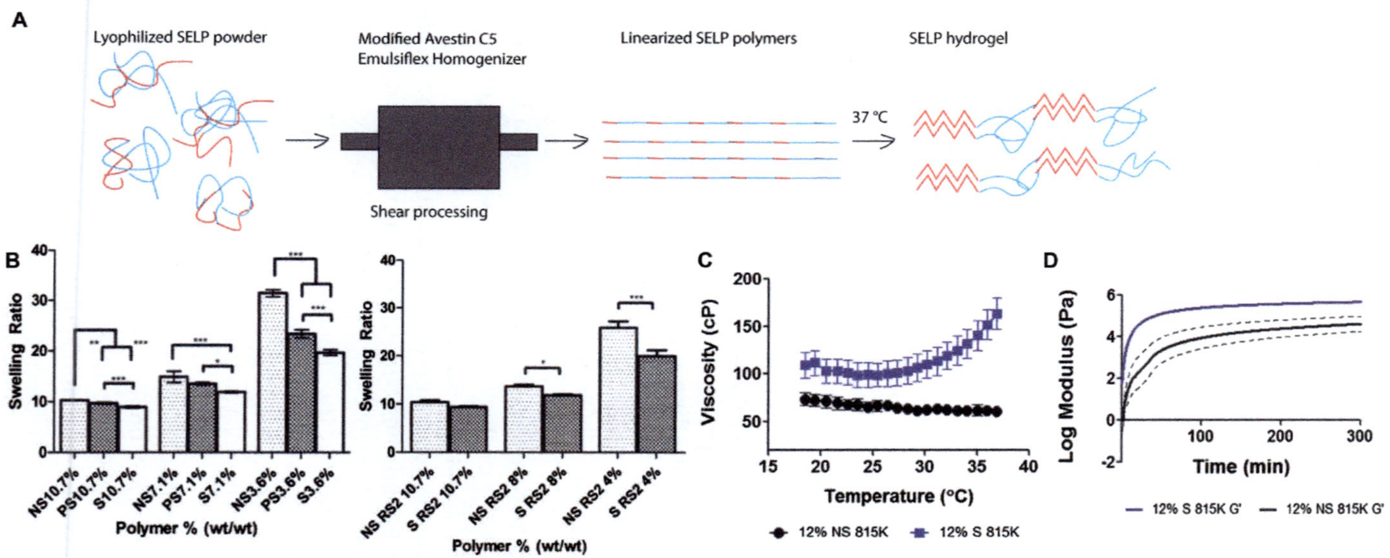

Figure 9.2 Effects of shear processing upon SELP swelling ratio. (A) Lyophilized powdered SELP is resuspended in phosphate-buffered saline (PBS) and linearized by shear processing at 17 000 psi (~112 000 kPa), increasing resultant polymer–polymer interactions, (B) Swelling ratio is reduced by shear processing for both SELP-815K (left) and SELP815K-RS2 (right); nonsheared (NS), partially-sheared (PS, 4000 psi), and fully-shear processed (S, 17 000 psi). (C) Viscosity and (D) storage modulus of sheared SELP-815K show increased values in comparison to nonsheared; dashed lines represent standard error of the mean for $n = 3$. Adapted from ref. 11 and 12 with permission from Elsevier, Copyright 2014 and 2015.

and methanol, ethanol, and glutaraldehyde chemical treatment.[39,40] Wet spinning produced SELP fibers with diameters ranging from less than 10 μm to over 60 μm with failure strains of up to 700% and elastic moduli of 20 MPa following methanol/water treatment.[37] Electrospinning was employed for the increased attention this technique has received to generate flexible substrates with micron and nanoscale topologies. Formic acid was used to dissolve SELP, and a field of 2.2–2.4 $kV\,cm^{-1}$ was applied with a feed rate of 0.25 $mL\,h^{-1}$. This technique produced fibers of approximately 100–300 nm, dependent upon SELP sequence and concentration.[38] Chemical treatments were applied to fibers and have also been used to cure SELP thin films. Thin films treated by methanol, ethanol, and glutaraldehyde have shown tunability of light transmittance and beta-sheet assembly, providing hydrogels with desired optical properties, stiffness, swelling, and release characteristics.[39,40]

9.2 Controlled Release From SELPs

Sustained and controlled release rates from hydrogels can prolong the diffusion of drug or bioactive agents, be tuned to achieve therapeutically relevant concentrations *in vivo*, localize release, and reduce toxicity-based events.[30] Release is affected by pore size and tortuosity and drug–polymer, drug–solvent, drug–drug, and polymer–solvent chemical interactions.[14–16] Examples of mechanisms by which SELPs control release include ionic or hydrophobic interactions of agents with polymers, diffusion through SELP pores, and encapsulation within polymers in pockets.[16] Early work in drug delivery from SELPs focused on analyzing the dependence of release upon the tunable hydrogel parameters of concentration, cure time, and polymer sequence, as well as the effects of soluble fraction, swelling ratio, and physicochemical properties of the drug.[6,14,15] Recent advances include the release of a GAG, extensive characterization of the effects of drug formulation upon mechanical properties, and release for embolics incorporating chemotherapeutics or fluorescent dyes.[12,16,17]

The release of low-MW drugs, proteins, macromolecules, and plasmid DNA from SELP matrices has been the subject of multiple review articles.[27,29–31] Dinerman *et al.* determined that SELP-47K hydrogels loaded with the low-MW, hydrophilic drug theophylline (180 $g\,mol^{-1}$) display release kinetics dominated by first-order Fickian diffusion and nearly 100% release in less than 4 hours.[14] In contrast to higher-MW compounds, cure time did not show significant effects upon solute partitioning, but longer cure times showed an apparent increase in tortuosity of the diffusional path.[14] Further results demonstrated that surface functionalization can alter drug uptake and diffusion within the SELP structure, as the extent of hydrophobic drug functionalization showed a dominant effect on the release of labeled poly(amidoamine) dendrimers.[15] Cappello *et al.* conducted early-release studies using 20–40 wt% SELP-47K hydrogels, observing that fluorescently labeled proteins also follow characteristics of first-order Fickian diffusion.[6] The observed release rate was not influenced by the MW of the drug during these assessments, possibly due to the pore size of tested SELPs or confounding properties of model compounds.[6]

SELPs have also been investigated for localized delivery of plasmid DNA.[21–23] These studies articulate that the ionic interactions of DNA with positively charged amino acid residues in the SELP backbone and SELP crosslinking density attenuate release. The length of DNA between 2.60 kbp and 11.00 kbp and DNA conformation were shown to influence release and calculated effective diffusivity.[22] *In vivo* results in an ectopic murine tumor model indicated localized expression of pRL-CMV-*luc Renilla* luciferase plasmid release from intratumorally injected 100 μL SELP-47K, 4 and 8 wt% hydrogels, was greatly enhanced during a 21 day period in comparison to naked DNA.[22]

9.2.1 Adenoviral Delivery

Most lung and head and neck cancer patients are resistant to chemotherapy, brachytherapy, and external radiation treatments.[41] Increasing the dose of chemotherapeutic or radiation results in unwanted off-target toxicity, incurring damage to healthy tissue. Furthermore, these treatments lack specificity for tumor cells. Gene therapy is an alternative approach with great potential in the treatment of these aggressive cancers, which targets mutations causing the malignancy or genes responsible for cancer progression and metastases. Adenoviral gene delivery systems take advantage of viral mechanisms for improved gene transfection efficiency and expression. Clinical trials of adenoviral gene therapy restoring the function of the *p*53 tumor suppressor pathway demonstrated gene transfection, expression, and subsequent mediation of apoptosis and clinical responses.[41] Unfortunately, adenoviral gene therapy is hindered by patient immunogenicity, off-target toxicity, and frequency of administration. The delivery of adenovirus in a SELP matrix serves to both enhance efficacy and bioactivity, localizing and stabilizing adenovirus, and potentially shielding detection by immune cells.[28,30]

AFM of virus–SELP interactions suggests that viral loading may alter the assembly of SELPs, and the polymers, in turn, may increase viral stability and bioactivity (see Figure 9.3A).[42] In nongelling concentrations of SELP-47K and -415K, charged ~100 nm adenoviral particles acted as nucleation centers with a greater localized crosslinking density of SELP nanofibers. Negatively charged amino acid residues, including glutamate and aspartate of the adenoviral capsid protein hexon, were posited to interact with positively charged SELP residues, attracting polymer strands that accumulated at the capsid interface, fostering interpolymer crosslinks. SELPs encapsulate viral particles, which may conceal their detection by immune cells and reduce the acute immune response.[42] SELP also may stabilize viral particles, as greater preservation of bioactivity was seen of adenovirus delivery by SELP hydrogels compared to PBS formulations.[43] Further investigation is needed to understand the extent and mechanism of polymer–adenovirus interactions improving stability.

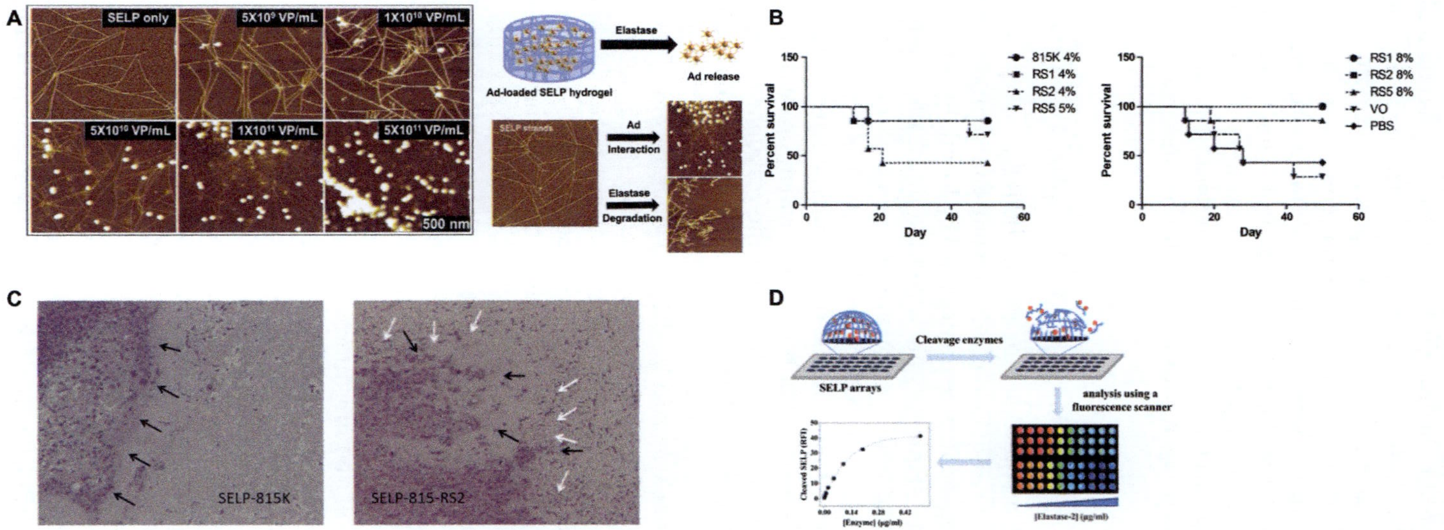

Figure 9.3 (A) Atomic force microscopy of adenovirus interacting with SELP and schematic of the matrix-mediated release of adenovirus from SELPs, VP: viral particles, Ad: Adenovirus. (B) Survival of mice receiving adenoviral gene therapy delivered by virus only (VO), phosphate-buffered saline (PBS), SELP-815K and SELP-815K with MMP degradable sequences (see Figure 9.1) -RS1, -RS2, at 4 and 8 wt% and -RS5 at 5 and 8 wt%. Survival was defined by lack of mortality, pain or distress, weight loss greater than 10%, tumor burden greater than 2000 mm^3, or ulceration of tumor site; $p \leq 0.05$ by Mantel–Cox test. (C) Histology (hematoxylin and eosin staining) 50 days postinjection shows reduced fibrotic encapsulation in MMP-responsive SELPs, indicative of hydrogel degradation. Black arrows point to the tumor border, while white arrows indicate vascular infiltration. (D) A Cy5-conjugated C-peptide array designed to assess SELP-enzymatic biodegradation as a function of polymer sequence and enzyme. Reproduced from ref. 42 and 44 with permission from American Chemical Society, Copyright 2015 and 2016. Adapted from ref. 13 with permission from Elsevier, Copyright 2015.

Adenoviral release from SELP hydrogels is dictated by parameters affecting the rate of diffusion, though some physical entrapment of adenoviruses was apparent in studies.[24,25] This entrapment resulted in the incomplete release from matrices of higher polymer content. Lesions with granulomatous cellular infiltration were seen in SELP–adenovirus hydrogels localized to the injection site for the treatment of head and neck cancer xenografts. Following week 12, all signs of the injection were absent in poloxamer and free virus groups. In contrast, SELP formulations persisted and were encapsulated by a fibrotic layer, indicating that the biodegradability of SELPs could be improved.[25]

9.2.1.1 Matrix Metalloproteinase-responsive (MMP-RS) SELPs

Regarding the use of SELPs as matrices for *in vivo* delivery, a key development from our lab was the genetic design and recombinant production of SELPs with enzymatically degradable sequences at key locations within the polymer backbone.[10,13] These sequences are degraded by MMPs, a group of serine proteinases upregulated within solid tumor microenvironments and sites of inflammation.[45] We showed that MMP-RS SELP-815K constructs are afforded *in vivo* degradation by MMPs in the tumor microenvironment.[13] In addition to allowing natural clearance of the matrix, which prevents fibrotic encapsulation, the degradable sequences facilitate the responsive release of entrapped adenoviruses not afforded by SELP-815K without MMP-RS. MMP degradation of the matrix increases the swelling ratio and soluble fraction, exposing adenoviruses previously restrained by polymer strands.[10,11,13] By varying the location of the MMP responsive sequence, these SELPs offer unique degradation rates and physicochemical properties. Due to this responsive mechanism, nearly all viral vectors can be released from higher concentration hydrogels, which allows for a greater ability to tune the release rate of adenovectors without sacrificing mechanical properties.[13]

In vitro degradation profiled by SDS–PAGE of SELP-815K and MMP-responsive SELPs digested with 40 nM MMP-2 demonstrated the inability to digest SELP-815K, whereas near-100% digestion of MMP-responsive constructs occurred by 1 hour.[10,13] *In vitro* degradation, demonstrated by protein loss and the release of 100 nm fluorescent beads in the presence of MMP-2 and MMP-9, is best demonstrated by SELP815K-RS1 and -RS2 sequences; SELP815K-RS2 degradation in the presence of MMP-9 is thus highlighted. Following 14-day incubation in the presence of 40 nM MMP-9, SELP-RS2 hydrogel swelling increased 40%, and in comparison to no-MMP control, cumulative release of beads increased 95% and polymer lost as soluble fraction was increased to 35% from 24%.[10] Protein loss from polymers in the presence of MMPs indicated that MMPs had the lowest accessibility to MMP-RS when MMP-RS motifs were inserted in silk-like motifs (-RS5).[10,13] For SELP815K-RS5, the presence of the degradable motif came at the cost of mechanical properties, but poor mechanical properties were largely mitigated by shear processing.[11] It was posited that the insertion of a mostly hydrophilic MMP-degradable motif between hydrophobic silk-like motifs disrupted SELP beta-sheet assembly.[11]

Biodegradation effects upon release and survival outcomes were compared between SELP-815K constructs with and without MMP-RS, PBS control, and free virus.[13] Of all groups, only SELP815K-RS1 and SELP815K-RS2 at 8 wt% displayed 100% survival over the 50-day study and the greatest reductions to tumor size (Figure 9.3B). By comparison, the survival of mice in the control and free virus groups were between 29 and 43%. The robustness of MMP-responsive SELP hydrogels was compared with SELP-815K by histology following *in vivo* studies. Price *et al.* found that the border between hydrogel and tumor was more tortuous for MMP-responsive SELP groups, indicating less robust gels. Nevertheless, MMP-responsive SELPs demonstrated greater cellular infiltration compared to SELP-815K parent polymer (Figure 9.3C) while also providing a localized depot such that off-target effects were reduced.[13] Increasing polymer concentration from 4 to 8 wt% and the introduction of shear processing, as discussed previously, drastically improved mechanical properties of MMP-responsive SELPs.[11] SELP-815K 8 wt% was not evaluated, as our previous studies indicated polymer concentrations above 4 wt% give rise to viral entrapment and subsequent incomplete release.[24] Overall, MMP-responsive SELPs demonstrated exquisite control over the release of entrapped adenovirus, improving transfection and sustained expression of tumor suppressor. These studies clearly demonstrate that control over polymer structure using recombinant techniques, in this case over the location of the MMP-responsive sequence in the monomer repeat, has a profound impact on bioactivity and preclinical outcome, *i.e.*, the treatment of solid tumors in animal models.[13]

To rapidly assess the degradation of SELPs as a function of polymer sequence, enzyme, and enzyme concentration, a Cy5-conjugated C-peptide fluorescent assay was developed.[44] By the incorporation of Cy5-conjugated C-peptide in SELP-815K, -RS1, -RS2, and -RS5, the degradation rates of eight protease enzymes on SELP-815K and analog constructs were quantified from released Cy5. Fluorescence intensity of released Cy5 was quantified on a microchip array to rapidly evaluate relative matrix degradation and release (see Figure 9.3D).[44]

9.2.2 Controlled Release of Synthetic Glycosaminoglycans

The ability of SELP hydrogels to localize and control the release of therapeutics has led to the development of a thermoresponsive enema system.[46] In this regard, we have used SELP hydrogels to prolong the residence times of antiinflammatory synthetic GAG ethers (SAGEs), GM-0111 (5400 $g\,mol^{-1}$, see Scheme 9.1) in the rectum for the treatment of radiation-induced proctitis (RIP). RIP is a complication arising from radiation treatment of the lower abdominal area, in which irradiation of healthy intestinal tissue damages the mucosal layer, producing an inflammatory response.[46] For the more than 200 000 individuals receiving lower abdominal radiotherapy each

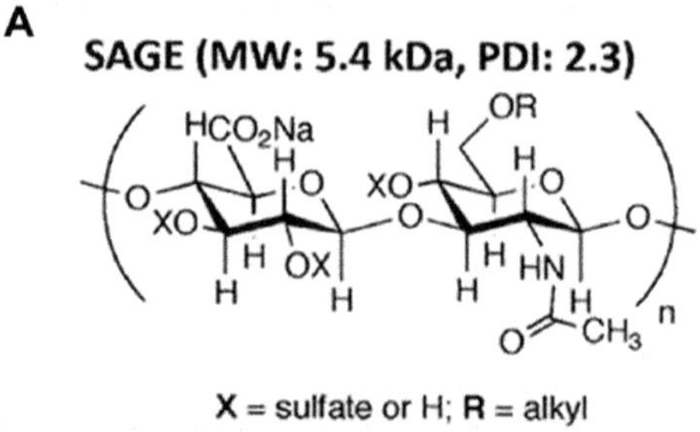

Scheme 9.1 (A) Chemical structure of the semisynthetic glycosaminoglycan ether, GM-0111, 5400 g mol^{-1} average MW, PDI 2.3. Reproduced from ref. 51 with permission from Elsevier, Copyright 2019.

year, >75% develop acute RIP, and in 5–10% of cases chronic RIP.[47] Chronic RIP is characterized by mucoid discharge, bleeding, and pain and can result in life-threatening sepsis, fistula formation, perforation, and internal bleeding.[48] No current prophylactic options exist; rather, clinical options seek to alleviate discomfort. SAGE is a negatively charged molecule based on sulfated hyaluronic acid that previous studies determined are significantly better at preventing mucosal inflammation than clinically available antiinflammatory agents.[49,50]

In scanning electron microscopy (SEM) images of flash-frozen, lyophilized hydrogels, Jensen *et al.* showed SAGE alone naturally self-assembled into hydrophilic films.[46] When mixed with SELP hydrogels at a concentration of 100 mg mL^{-1}, SAGE films interpenetrated crosslinked SELP hydrogels without any apparent phase separation. SELP viscosity and gelation kinetics were altered by the addition of SAGE in sheared SELPs, indicating ionic interactions between SELP and SAGE. The release rate of polyanionic SAGE was slowed by interactions of its sulfate groups with positively charged lysine residues in the SELP backbone. Unlike adenovirus, size-exclusion partitioning of SAGE was not apparent, as average pore sizes far exceeded that of SAGE.[14] SELP–SAGE ionic interactions were reported due to the apparent displacement of SAGE by ions present in the high osmolality simulated intestinal fluid.[46] The *in vitro* release of 2.5 kDa (2500 g mol^{-1}) SAGE GM-0111 from SELP was seen out to 24-hours, with a 15-minute burst release of 39 ± 11% of GM-0111 from 4 wt% SELP-815K and 10 ± 2% release from 11 wt% gels, though later time points were not statistically significant. At 11 wt% SELP, more lysine residues were present for a stronger interaction with polyanionic SAGE and slower release. The release profile illustrated that the effects of SELP hydrogel concentration on SAGE release are mostly seen in the burst response.[46]

Bioaccumulation results demonstrated improved accumulation of SAGE in SELP–SAGE groups in comparison to PBS formulations at both 3 and 12 hours, as the fluorescent micrographs showed a greater intensity of fluorescently labeled GM-0111 in the rectal walls of mice that received

SELP-815K 4 wt% and 11 wt% loaded with SAGE.[46] The efficacy of this system was evaluated using a murine model for proctitis within clinical guidelines.[52] Briefly, mice were instilled with therapeutic formulations and then exposed to 35 Gy radiation targeted at the lower abdominal area. Pain responses were evaluated at days 3, 7, and 21 after administration with successive abdominal stimulation by von Frey filaments (10 repeats of ~1 second of stimulation with 3 second pauses) over a 21 day period following irradiation.[46] The von Frey filaments allow the application of quite a steady force dependent upon filament diameter.[53] Tactile sensitivity was evaluated based upon response rate indicative of pain to stimulatory forces between 0.04 g to 4 g force; 0.4 g stimulation was selected as this group demonstrated the greatest differential in pain response between therapeutic formulations. Over the duration of this study, response rates were substantially lowered in SELP-SAGE groups over PBS–SAGE formulations. SELP groups allowed a 50% release within 1 hour and accumulation in the lumen, potentially coating rectal tissue for long-term sustained release. SELP was hypothesized to form a barrier in the rectal lining, inhibiting mucosal penetration of reactive oxygen species while sustaining SAGE release for the recovery of its antiinflammatory and cytokine modulation of cellular signaling. Gelation prolongs SELP residence times within the intestine, slowing clearance by peristalsis.[46] The prophylactic action of SAGE and its incorporation within SELP hydrogels holds promise, and this work is further being explored for variation in concentration and sequence of SELP polymer.

Seeing the success of SELP-SAGE in the treatment of RIP, we then evaluated SELP-SAGE hydrogels of varying polymer sequences and concentrations in a model of interstitial cystitis (IC),[54,55] also known as painful bladder syndrome. IC affects between 3.3 and 7.9 million people in the United States annually and, like RIP, has an unmet clinical need for prophylactic treatment.[56] Recent results show IC may be characterized by mast cell infiltration and the subsequent inflammation of the mucosal barrier of the bladder.[55] LL-37, a peptide of the innate immune system, induces apoptosis and leukocyte infiltration characteristic to IC and was thus used to induce the IC model in mice.[49] GAGs, predominantly pentosan sulfate, have seen extensive use in IC treatments but fail to show immediate effects or consistency in improvements to patient quality of life.[49] This is perhaps due to poor residence time and rapid clearance disallowing therapeutic effects to take hold. SELP formulations were compared with traditional polymeric delivery systems, poly(lactic-*co*-glycolic acid) (PLGA)–poly(ethylene glycol) (PEG) copolymers (PLGA–PEG–PLGA), and Poloxamer 407 for treatment of IC.[51]

For all polymers, *in situ* formulations displayed successful intravesical accumulation of fluorescently labeled SAGE at 10 mg mL^{-1} in mice bladders to ameliorate insult caused by LL-37.[51] PEG–PLGA–PEG formulations, which form thin films apparently stabilized by hydrophobic/hydrophilic interactions with SAGE, underwent macroscopic phase separation, likely causing the near-total release of SAGE upon administration. Poloxamer 407 triblock copolymer rapidly assembled into a highly viscous gel, also stabilized by hydrophobic/hydrophilic interactions with SAGE. Poloxamer 407 at 10 and 20 wt% (low and high concentration, respectively) released >80% SAGE within 1 minute, indicative of networks showing little restriction of SAGE diffusion. In 12 wt% SELP-415K and SELP-815K groups, >50% SAGE was released at 24 hours. This prolonged release (compared to 100 $mg\,mL^{-1}$ formulations used in RIP applications and other polymer groups) imparts a greater dependence on ionic interactions with the lysine residues.[46,51]

For SELP-SAGE, the SEM of lyophilized samples indicated the incorporation of SAGE into SELP matrices likely lowered matrix density, but beta-sheet formation was not prohibited by the presence of SAGE.[51] Rheological characterization showed accelerated gelation of SELP-815K but a lower maximum elastic modulus value. SELP-815K-SAGE formed larger pores than SELP-415K-SAGE and exhibited faster gelation and higher elastic modulus. Consequently, SELP-815K showed the greatest risk of obstructing the bladder outlet and urethra of mice, which was not evident in non-SELP groups. SELP-415K-SAGE hydrogels are softer than SELP-815K and appeared to have molded into a thin coating on the bladder wall upon necropsy. In addition to mechanical tests, bladder obstruction was evaluated by a void spot assay. Briefly, filter paper was placed under wire mesh cages to collect urine and imaged under UV light (365 nm) on an Alpha Innotech FluorChem SP Digital Imaging System, and the number of voiding events, mean voiding volume, and total urine volume were calculated from stained areas using ImageJ (Figure 9.4A–C). Bladder voiding of polymer–SAGE formulations was compared to healthy mice.[51]

Polymer formulations were tested for the amelioration of pain responses following an established model.[54] In SELP-415K 12 wt% and SELP-815K groups, the release of SAGE and ensuing analgesic effects and reduced inflammation of the urothelium were demonstrated out to 24 hours. Increased analgesic effect (Figure 9.4E) and decreased hyperalgesia, a condition marked by heightened sensitivity to pain, were observed in SELP groups. SELP-415K also demonstrated decreased allodynia or pain responses resulting from typically nonpainful stimulation. Although SELP-415K and PEG–PLGA–PEG showed limited accumulation of SAGE in the urothelium at 3 hours, in comparison to more rapidly gelling poloxamer and SELP-815K, all demonstrated improved accumulation over PBS control at the 12- and 24 hour timepoints and reduced edema compared to controls (Figure 9.4D). SELP-815K hydrogels proved most prohibitive to SAGE release and to hydrogel clearance from the bladder; hydrogel-mediated occlusion of the urethra was observed in some animals of the SELP-815K group. The favorable effects on pain responses, prolonged retention, and slow release within the bladder afforded by the soft, clearable SAGE-loaded SELP-415K hydrogels resulted in their selection as the best formulation for the amelioration of discomfort.[51] The understanding of polymer, therapeutic, and environmental parameters upon hydrogel matrix properties and release may be used to predict the relative rates of release from SELP hydrogels.

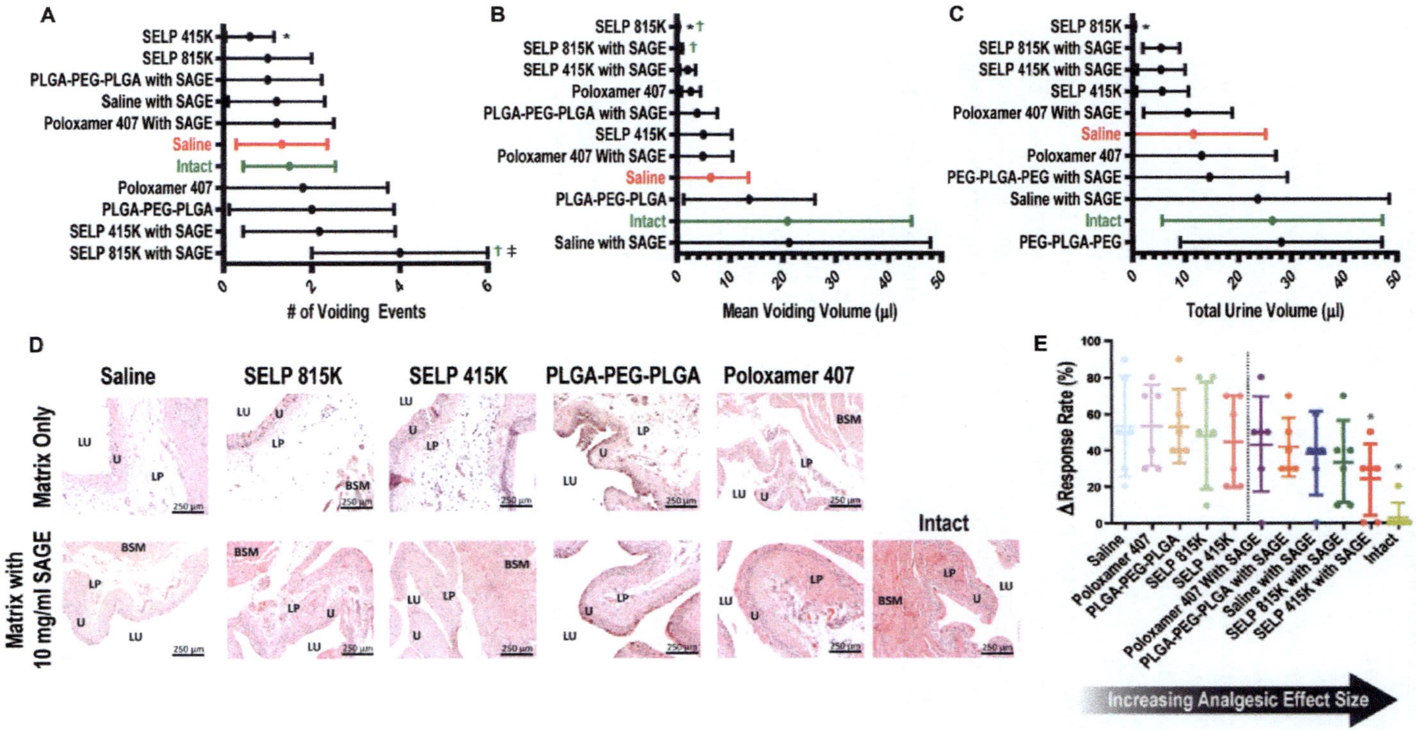

Figure 9.4 *In vivo* function and pain outcomes for polymer-SAGE formulations in comparison to saline sham control and intact (healthy mice receiving no therapy; each group consisted of six animals): (A) Functional evaluation of polymers based upon the number of mouse bladder voiding events with greater than 0.5 µL calculated volume, (B) mean voiding volume, and (C) total volume of urination in a 4 hour period; *, †, and ‡ indicate a statistical difference with a p-value of less than 0.05 to the saline control group, intact mice, and the polymer without SAGE, respectively. One-way ANOVA with Bonferroni Post-Test was completed for each sample. (D) Histology of bladders shows SAGE reduced edema and vessel dilation in polymer groups, with anatomical features labeled as bladder lumen (LU), urothelium (U), lamina propria (LP), and bladder smooth muscle (BSM). 250 µm scale bars. (E) Analgesic effect of polymer-SAGE as determined by von Frey filament stimulation; * and *** indicate p-values of less than 0.05 or 0.001 when compared to saline sham control by one-way ANOVA with Bonferroni Post-Test. Reproduced from ref. 51 with permission from Elsevier, Copyright 2019.

9.3 SELPs as Embolics, Chemoembolics, and Embolization Visualization Systems

Embolic agents are commonly utilized in interventional radiology, administered to selectively occlude blood vessels feeding solid, often hypervascular tumors in patients excluded from surgical resection procedures.[57] Chemoembolics incorporate chemotherapeutic, commonly doxorubicin in drug-eluting beads, which have shown 30-day release and minimal arterial blood inflow following a single administration.[12,58] However, nondegradable poly(vinyl alcohol) drug-eluting beads form a fibrotic layer embedded in the tissue intima and also occlude proximal to tumor capillaries.[59] By contrast, thermoresponsive SELP-815K and SELP-47K liquid embolic formulations exhibited potential to penetrate greater depths for locoregional drug release at the tumor capillary level.[12] Moreover, sheared SELPs have enhanced gelation and increased stiffness to lyophilized SELPs, affording their potential as embolic materials.[12]

The SELP embolic system was initially evaluated for the treatment of hepatocellular carcinoma (HCC) by transarterial chemoembolization (TACE).[12] HCC is among the leading causes of cancer worldwide and the foremost primary tumor malignancy of the liver.[57,60] In a microfluidic device model of tumor vasculature, SELPs were compared for their injectability and ability to occlude the model vasculature.[12] At 20 °C, the viscosity of sheared 12 wt% SELP-815 K is ~150 mPa s, and thus readily injectable through a commercially available 2.8 F microcatheter. As temperature increases to 37 °C, viscosity rises roughly 50 mPa s. Shear storage modulus (stiffness, G') increases above 1×10^3 Pa by 5 minutes, reaching the target of $>1\times10^5$ Pa by 5 hours necessary to embolize the hepatic artery. While both 12 wt% sheared SELP-47K and SELP-815K showed acceptable gelation rates and storage moduli, the extended length of silk-like blocks in SELP-815K afforded more rapid gelation. Sheared 12 wt% SELP-815K's 5-minute G' value was 10 times that of 12 wt% sheared SELP-47K, prompting the selection of SELP-815K as the formulation for further development. SELP-815K embolic was injected proximally to the branch point of hepatic arteries in New Zealand white rabbits (see Figure 9.5A–C). *In vivo* angiography showed a stable embolus within 10 minutes, occluding with less than half the volume of material of clinically used microspheres and importantly, off-target embolization was not observed.[12]

9.3.1 Chemoembolic SELPs

The incorporation of sorafenib tosylate, sorafenib base, doxorubicin hydrochloride (Dox-HCl), and doxorubicin base was tested as powdered formulations (at 9 and 12 wt% polymer) and in dimethyl sulfoxide (DMSO) (9 wt% polymer) formulations, for both single-agent and combination release (see Figure 9.5D,E).[16] Physicochemical properties were evaluated to gain an

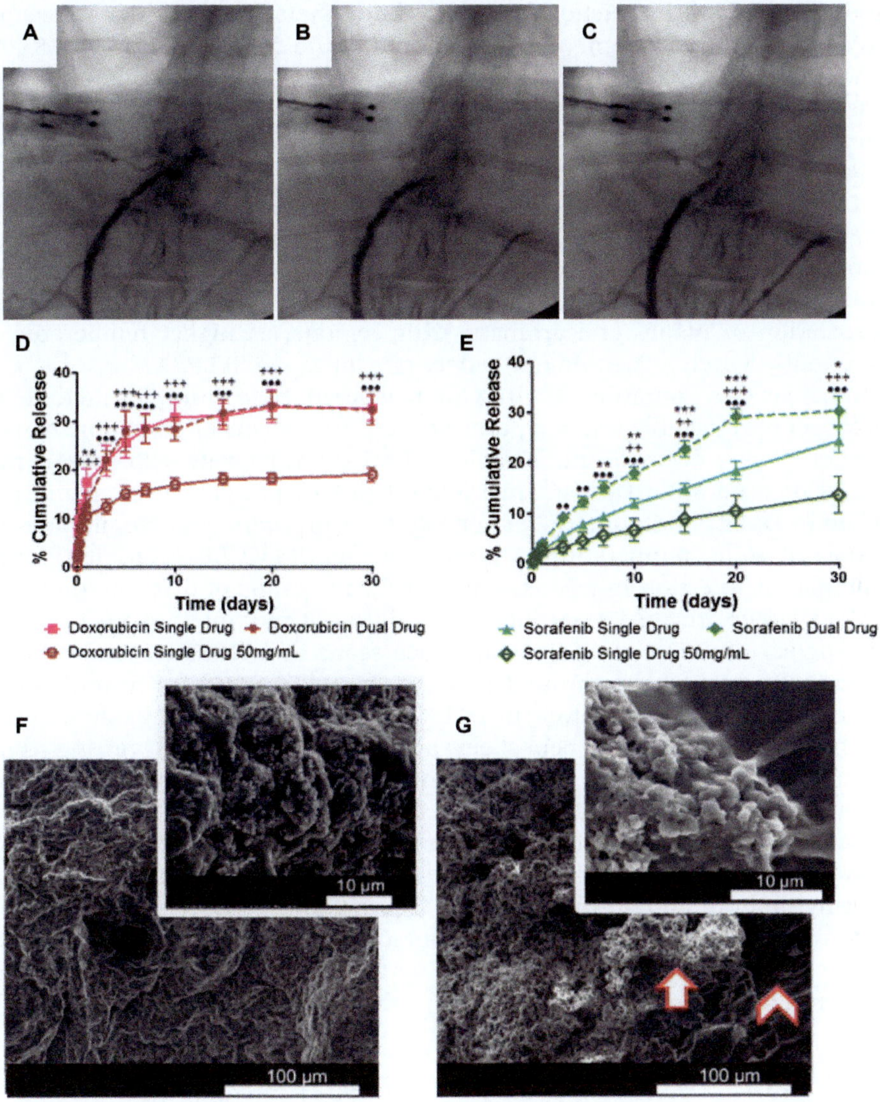

Figure 9.5 (A) Angiography of New Zealand white rabbit hepatic artery embolized with 12 wt% SELP-815K pre-SELP injection following the addition of contrast agent, (B) 5 minutes post-SELP injection showing occlusion, and (C) 10 minutes postembolization confirming continued stasis. (D) *In vitro* release of doxorubicin powder and (E) sorafenib powder single and dual formulations from SELP-815K chemoembolics. (F) Representative SEM of SELP incorporating doxorubicin and sorafenib in dimethyl sulfoxide (DMSO) shows hydrogel pores lined with star polyhedron doxorubicin and cubic sorafenib crystals. (G) SEM of SELP with powdered base drug formulation. Polymer (arrowhead) encapsulates drug clusters (arrow) during gelation. Reproduced from ref. 12 with permission from Elsevier, Copyright 2015 and from ref. 16 with permission from American Chemical Society, Copyright 2016.

understanding of interactions between chemotherapeutic and hydrogels to predict the efficacy of embolus formation and the release of drug. At 18 °C, 9 wt% SELP-815K incorporating Dox-HCl has approximately double the viscosity of drug-free SELP hydrogels; Dox-HCl in DMSO formulation and Dox-HCl powder result in initial viscosities of 349 ± 149 mPa s and 250 ± 114 mPa s, respectively. Similar trends of increased viscosity are seen for SELP-sorafenib tosylate and sorafenib base in DMSO, but the effect is slightly less pronounced. Sorafenib tosylate and Dox-HCl both exhibit temperature dependence of viscosity, believed to arise from the reorganization of drug within the matrix upon heating. This reorganization may be a self-association of planar and aromatic drug regions, as higher temperatures potentially lessen a "salting out" effect of drug upon SELP.[16]

Whereas incorporation of drug in powdered base form resulted in a hydrogel of injectable viscosity, other tested drug forms of doxorubicin and sorafenib were deemed uninjectable.[16] Viscosity values are >200 mPa s for SELPs incorporating Dox-HCl in DMSO, Dox-HCl powder, or sorafenib tosylate in DMSO and ~175 mPa s for SELPs incorporating sorafenib base in DMSO at room temperature. The viscosity of DMSO formulations is not suitable for injection in all cases. In salt formulations of drug dissolved in DMSO, either drug or DMSO could affect interpolymer interactions to cause the increase in viscosity. Secondary trace scans of attenuated total reflectance Fourier-transform infrared spectroscopy (ATR-FTIR) for formulations with sorafenib base dissolved in DMSO, single and dual drug, showed an upshift in the primary beta-sheet absorbance peak,[61] suggesting that dissolved sorafenib forms hydrogen bonds with polymer networks. The influence upon viscosity was not seen in either doxorubicin base or sorafenib base powdered forms, supporting their use as chemoembolic formulations. Powdered base formulations of both drugs clustered in organized crystals, coating inner SELP pores in SEM (see Figure 9.5F,G). The *in vitro* release of Dox from 12 wt% hydrogels showed a burst release for all drug formulations, achieving approximately 20% release by 24 hours. Sustained release was observed over 15 days, reaching an approximate 30% cumulative release.[16]

9.3.2 SELP Embolics for Fluorescence Image-guided Surgery

Fluorescence-guided visualization during robotic and endoscopic surgical resection aids surgeons in identifying tumor margins and surrounding healthy tissue.[62,63] Core to oncological surgical resection is that tumor-free, negative surgical margins reduce rates of cancer recurrence and improve both surgical and survival outcomes. However, intraoperative bleeding presents challenges during surgical resection of hypervascular tumors or tumors not suitable for haptic identification of margins, such as juvenile nasopharyngeal angiofibroma (JNA) of the sinonasal cavity.[64,65] JNA tumors are particularly difficult to resect, owing to substantial obscuration of surgeons' field of view from intraoperative bleeding, minimal accessibility, and proximity to the carotid arteries, trigeminal nerve, and optic nerve.[65]

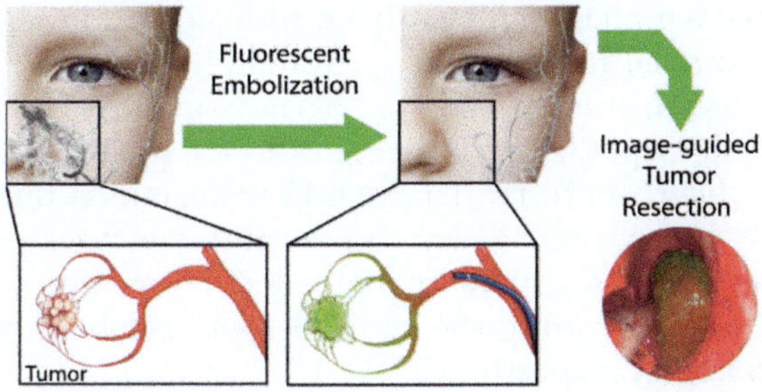

Figure 9.6 Fluorescence image-guided tumor resection of a JNA tumor by dual-functional SELP embolic. Adapted from ref. 17, https://doi.org/10.7150/thno.39700, under the terms of the CC BY 4.0 license https://creativecommons.org/licenses/by/4.0/.

As SELP-815K demonstrated rapid gelation, lack of organic solvent, and lasting occlusion of the vasculature, it was selected for its potential as an occlusive depot from which imaging agent indocyanine green (ICG, 775 g mol^{-1}) may diffuse to demarcate tumor margins (see Figure 9.6).[17] ICG is a near-infrared dye that strongly binds to circulating albumin, and as albumin extravasates across blood vessels, ICG-bound albumin may accumulate within tumoral tissue due to poor lymphatic drainage.[66,67] Although near-infrared dyes have seen extensive use in image-guided surgery to delineate tumors, rapid dilution and clearance prohibit substantial accumulation of free ICG within tumor tissue to guide surgeons.[66]

SELP-ICG mechanical properties were assessed by rheology and microfluidic device occlusion, and release was evaluated from an agar tissue phantom.[17] ICG affects hydrogel assembly, seemingly promoting polymer–polymer interactions, larger pore size, and reduced swelling ratio and soluble fraction; at 10 mg mL^{-1} ICG, SELP-815K swelling ratio was reduced by 15.2%. Interestingly, SELP-ICG displays heterogeneity in multiscale pore sizes, as at 0, 0.5, 2.5, and 10 mg mL^{-1} loaded ICG, SELPs exhibit pore sizes of 21 ± 38, 15 ± 23, 4 ± 24, and 25 ± 75 μm^2, respectively. SELP-ICG is injectable through 2.3 F microcatheters, successfully penetrating the model microvasculature, which was strongly correlated with fluorescence detected by a clinically utilized endoscope.[17]

9.4 Ophthalmic Application

The utility of SELP films for ophthalmic drug release was recognized and demonstrated by SELP-47K 5 wt% hydrogel thin films cast from molds within contact lens dimensions.[39,40] These thin films display approximately 96% transmittance in the visible region following methanol treatment, employed as a noncovalent crosslinking agent to improve film transparency and reduce the soluble fraction. Comparison of surface roughness by AFM and hydrophilicity by contact angle measurement with contact lens materials such as Asmoficon and Enfilcon indicate SELPs show similar hydrophilicity but surface roughness approximately one order of magnitude greater than a polished silicon wafer.[68,69] To assess the use of the material as an ophthalmic drug depot, ciprofloxacin antibiotic (MW = 331 g mol^{-1}) was mixed with SELP hydrogel films. To extend the sustained release of drug, a three-layer SELP composite of dry SELPs and hydrogel was generated. Ethanol and methanol treatment of the outer layers prove beneficial for slowing release rate, extending the time to 50% release from 15 hours to 48 hours and 46 hours, respectively.[39]

9.5 Micellar-like SELP Particles

Micelles increase drug solubility; hydrophobic drugs may interact with the hydrophobic inward-assembling polymer block, while the outer hydrophilic polymer region can afford improved bioavailability.[70] Seeing the potential for the development of a micellar-like SELP delivery system, Xia *et al.* designed new SELPs for the uptake of hydrophobic drugs.[71,72] The naming convention for these SELPs deviates from that previously described. S2E8Y, for example, contains two silk motifs, eight elastin motifs, and one tyrosine-substituted elastin. SE8Y contains only one silk, while S4E8Y contains four (refer to Figure 9.7 for sequences).

At 0.5 mg mL^{-1}, SE8Y and S2E8Y demonstrate reversible temperature-dependent assembly with two transition phases by UV–vis absorbance.[71] DSC curves exhibit an initial exothermic transition for these two SELPs, believed to be due to the crystallization of silk motifs within micellar-like nanoparticles. Coacervates may form at the second transition temperature above 60 °C, as a rise in optical density was seen. Circular dichroism (CD) spectra reveal type II beta-turn secondary structure reminiscent of ELPs for SE8Y and S2E8Y, whereas S4E8Y exhibits spectra characteristic of beta-sheets. At higher temperatures, gel-like states and aligned nanofibers were reported as present in SEM of S4E8Y, deemed a result of silk motif crosslinking.[71]

Fluorescence spectra taken of variable concentrations of these SELPs incubated with the solvatochromic fluorophore 8-anilinonaphthalene-1-sulfonic acid (ANS) revealed an apparent critical micellar concentration (CMC) following attenuation of fluorescence spectra with uptake.[74] No transition temperature was observed in S4E8Y micellar-like assembly between 20 °C and 40 °C, in contrast to SE8Y and S2E8Y, for which 241 ± 13 and 212 ± 16 nm nanospheres form at 40 °C in SEM.[71] Spectra for both SE8Y and S2E8Y illustrated micellar-like self-assembly triggered by hydrophobic solute or increased temperature, allowing controlled uptake of fluorescent probe and model hydrophobic drug, doxorubicin.[72] Average hydrodynamic radii of doxorubicin-loaded SELP nanoparticles at 25 °C were 50 ± 10, 72 ± 11, and 142 ± 10 nm for SE8Y, S2E8Y, and S4E8Y, respectively. The loading efficiencies of micellar SELP particles were 6.5% for SE8Y, 6% for S2E8Y, and only 4% for S4E8Y. Poorest loading was seen in S4E8Y due to spontaneous, temperature- and drug-independent micellar self-assembly.[72]

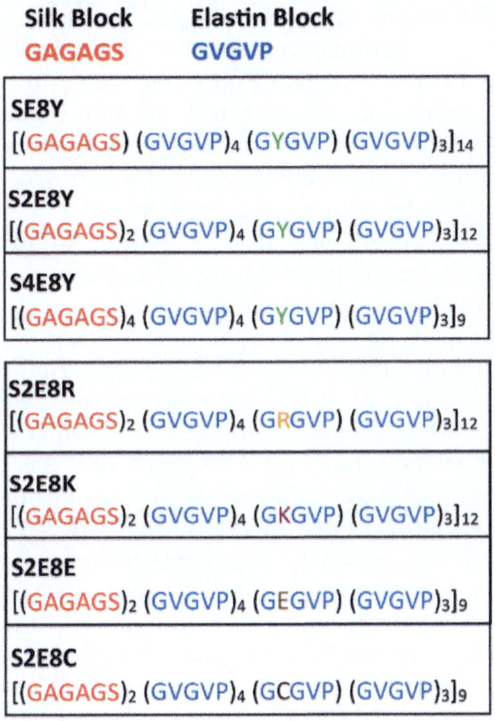

Figure 9.7 Schematics and amino acid sequences of top: SELPs reported to form micellar-like nanoparticles; and bottom: those evaluated for mucoadhesive characteristics. Single-letter abbreviations are used. Procedures for generating these SELPs may be found in ref. 71–73.

Computer models match experimental results for SE8Y and S4E8Y, indicating that the reduced length of silk-like blocks in SE8Y affords flexibility and conformational sampling even as the temperature is increased.[75] Above the described second transition T_t for SE8Y, the hydrogen bonding network is broken as an apparent coacervate forms. Meanwhile, S4E8Y SELPs display longer silk-like units that self-assemble below physiological temperatures.[75]

Confocal microscopy and fluorescence-activated cell sorting (FACS) indicated that drug-loaded SE8Y particles were endocytosed by HeLa cells, facilitating the controlled release of Dox within the cytoplasm.[72] Dox bioactivity was not inhibited by release from thermally responsive micelles, successfully translocating to the nucleus and resulting in observed cytotoxicity.[72]

9.5.1 Mucoadhesive SELP Nanoparticles

Oral delivery is a common and noninvasive mode of drug administration but can suffer from rapid mucosal turnover and subsequent clearance, effected by physicochemical properties of the drug formulation, which govern interactions with the body's mucosal linings.[73] To improve the efficacy of oral delivery, some formulations employ electrostatic interactions between polymers and mucins for mucoadhesion and retention, improving absorption in the gastrointestinal tract.[76] Similarly, nanocarriers incorporating thiol groups have been shown to form disulfide bridges with mucins also containing thiol groups.[76] Thus, additional S2E8X constructs with variable substitutions at the "X" position of elastin (GXGVP) in arginine (R), lysine (K), cysteine (C), and glutamic acid (E) (*i.e.*, S2E8R, arginine substitution) were produced to evaluate the potential for mucoadhesive nanoparticle delivery with variation in physicochemical properties.[73]

Retention of SELP particle constructs within a biosimilar mucus formulation was observed to characterize particle–mucosa interactions *in vitro*.[73] SELP particles were seeded on thin biosimilar mucosal layers, which were coincubated for 24 hours at 37 °C. Samples were washed three times with PBS and centrifuged at 5000 rpm, and the supernatant was removed prior to observations. S2E8C particles exhibited retention within the biosimilar mucin, believed to result from disulfide bonding between S2E8C's cysteine thiol group and mucins. However, when retention was evaluated in the mucus-secreting HT29-MTX cell line, S2E8C particles were not retained. Parker *et al.* hypothesized this to be a result of mucolytic properties, as S2E8C samples lacked mucus following coincubation, though additional tests will need to be completed. S2E8K, S2E8R, and S2E8E SELP constructs demonstrated retention within biosimilar mucin, Caco-2, and HT29-MTX cell lines. Of these three constructs, S2E8E were retained least, while S2E8K demonstrated the greatest retention in the HT29-MTX cell line and an approximate 32-fold relative increase over S2E8K retention in the Caco-2 cell line, which is not mucus secreting. These findings may be explained by the negative charge of glutamic acid residues, as the positively charged arginine and lysine residues display stronger ionic interaction with negatively charged mucins.[73,76]

9.6 Three-dimensional Nanoparticle Networks

Directing chemotherapy to tumor tissue is a significant challenge due to the dynamic nature of the *in vivo* environment.[67] Nanoparticle formulations may improve the delivery of hydrophobic drugs by means including increased surface area, solubility, and improved tissue and cellular uptake.[67,77] Others have investigated mild hyperthermia of tissue for improving localized therapeutic uptake from the bloodstream by increased fenestration size[78] and blood vessel diameter[79] and promoting small molecule transport across endothelial cells.[80] Given the combined promise of these delivery strategies, Isaacson *et al.* proposed an externally triggered SELP nanogel delivery system, through which nanogels might undergo a size reduction in response to localized hyperthermia.[81] We propose future responsive nanomedicines of this nature may allow increased extravasation, or leakage across the endothelial wall, and deeper tissue penetration prior to drug release.[81]

SELPs self-assemble into nanostructures when present in aqueous environments above a critical concentration, typically between 1 and 2 mg mL^{-1}, without destabilizing in excess water following particle formation.[81] Longer silk motifs (SELP-47K, SELP-415K, and SELP-815K) promote self-assembly into 3D nanoparticle networks with characteristics distinct from micellar-like structures, which are assembled from SELPs with shorter silk blocks.[71,72] Unlike micelles, which disassemble below a critical formation concentration, SELP-based nanoassemblies maintain structural integrity despite dilution, indicating an architecture that is not held together solely through hydrophobic interactions. Instead, it was hypothesized that interpolymeric crosslinking between silk-like blocks provide the stabilizing forces for these structures.[81]

These stabilizing hydrogen bond crosslinks greatly affect overall SELP behavior.[81] When heated above their precipitation temperature, T_p, SELPs exhibit only minor thermal reversibility due to the crosslinks inhibiting water penetration into the precipitated structures. However, certain procedures can be taken to address this lack of reversibility. Postpurification shearing of SELPs has been shown to increase thermal reversibility following cooling. Shearing was essential for SELP self-assembly and was hypothesized to allow greater participation in interpolymeric crosslinking. Interestingly, nonsheared SELP samples resulted in the formation of nanofibers rather than the globular particles described.[81]

A potential advantage of the SELP nanoparticle networks includes their response to external stimuli (see Figure 9.8).[81] Like their single polymer counterparts, SELP nanoparticle networks retain their thermoresponsive properties, yet the response is manifested differently. Due to the 3D nature of the nanoparticle network, they undergo a "size collapse" in response to temperature increase, which can result in up to eightfold changes in volume.[18] Isaacson *et al.* hypothesized that this size-changing property could be combined with mild tumor hyperthermia to improve drug extravasation.[81] In short, peripherally induced hyperthermia can deliver localized heating to the desired tumor vasculature, which could facilitate regional particle size collapse and extravasation. In tissues outside of the hyperthermic region, unstimulated SELP nanostructures would remain in an expanded state and be too large to readily cross the vascular wall.

Additionally, SELP-815K analogs with MMP-responsive sequences were investigated for nanoparticle self-assembly due to their potential to increase the rate of drug release upon reaching the tumor microenvironment.[18,45] The degradation of these structures with the SELP815K-RS1, -RS2, and -RS5 (see Figure 9.1) sequences showed no effect on release rates of doxorubicin. Rather, it was likely the inherent differences in backbone hydrophilicity of these polymers allowed variance in water molecule penetration into the nanoconstructs that resulted in differences in drug release rates.[18]

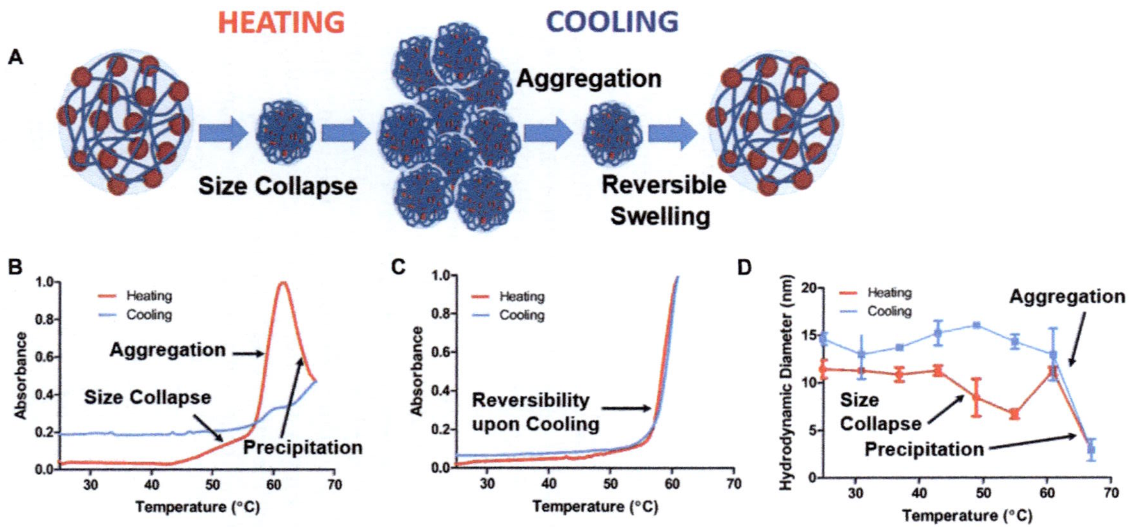

Figure 9.8 Thermal response behavior of SELP-815K 3D nanoparticle networks. (A) Schematic of self-assembly and reversible swelling. (B) Absorbance turbidimetry profile from 25 to 67 °C. (C) Absorbance turbidimetry from 25 to 59 °C. (D) Thermal-ramp dynamic light scattering of nanoassemblies at 25–67 °C. Adapted from ref. 81 with permission from John Wiley & Sons, Copyright © 2017 WILEY-VCH Verlag GmbH & Co. KGaA, Weinheim.

9.7 Cellular Delivery

SELP scaffolds hold great promise for reparative tissue engineering implants and models due to micron-scale pore sizes and robust hydrogel assembly.[19,20] SELP fiber mats[37,38] and the incorporation of bioresponsive cell-attachment motifs such as RGD[19] incorporate traditional tissue engineering strategies to control cell fate but are afforded molecular level definition to control assembly and motif density. Mesenchymal stem cell response and viability in SELPs has been the topic of multiple papers, as SELP elastic modulus and gelation may be favorable for cartilage repair.[19,20] Tissue regeneration was tested for the SELP incorporating RGD motifs mixed with chondrocytes in an *ex vivo* model of osteochondral defect, a 4 mm pig knee biopsy punch.[19] This work demonstrated that SELP hydrogels may act as scaffolds to allow chondrocyte attachment and elicit hyaline cartilage production, with 10 times more DNA and GAG present than PEG or alginate alternatives. However, the optimization of mechanical properties and injectability may be improved in future SELP scaffolds incorporating motifs for attachment.[19]

9.8 Summary and Future Directions

SELP hydrogels have a promising future as robust embolic materials, tissue scaffolds, and matrices for delivery of therapeutic agents, largely due to the controllable structure–function relationships. SELP structural modifications have garnered advantageous material properties, including variable gelation kinetics, material stiffness, and degradation rates. Limitations to the expanding use of SELPs lie in the ability to produce genes expressing new polymers to meet design criteria and the nature of SELP gelation, including irreversibility and dependence upon concentration and beta-sheet formation. Notable achievements include the design of MMP-responsive sequences within the polymer backbone, resulting in SELP degradation and the release of entrapped adenovirus in MMP-upregulated environments such as solid tumors. These sequences have been shown to form mechanically robust hydrogels following shear processing, irreversibly gelling at physiological temperature. Additionally, SELP embolic systems from which chemotherapeutic drug or fluoroscopy imaging agent was released show promise for rapid embolus formation and biocompatible alternatives to current technologies.

Moving forward, we see the potential for optimizing SELP sequences and formulations for tissue engineering and cell delivery, embolics, and as drug and gene delivery systems. Future investigations of SELP material properties and kinetics of gelation may include SELPs of varied MWs and new lengths of silk and elastin motifs. Specific to tissue engineering and cell delivery applications, motifs of the extracellular matrix may be inserted into the SELP backbone for the evaluation of cell fate determination and cell attachment.

Abbreviations

Ad	Adenovirus
AFM	Atomic force microscopy
ANS	8-Anilinonaphthalene-1-sulfonic acid
ATR-FTIR	Attenuated total reflectance Fourier-transform infrared
CD	Circular dichroism
CMC	Critical micelle concentration
DLS	Dynamic light scattering
DMSO	Dimethyl sulfoxide
Dox	Doxorubicin
DSC	Differential scanning calorimetry
ELP	Elastin-like protein polymer
FTIR	Fourier-transform infrared spectroscopy
GAG	Glycosaminoglycan
GAGAGS	Single-letter amino acid abbreviation of silk motif
GVGVP	Single-letter amino acid abbreviation of elastin motif
HCC	Hepatocellular carcinoma
HCl	Hydrochloride
IC	Interstitial cystitis
ICG	Indocyanine green
kbp	Kilobase-pair
LAL	Limulus amebocyte lysate
LPS	Lipopolysaccharide
MMP	Matrix metalloproteinase
MGFC	Minimum gel-forming concentration
MW	Molecular weight
PAGE	Polyacrylamide gel electrophoresis
PBS	Phosphate-buffered saline
PEG	Poly(ethylene glycol)
PDI	Polydispersity index
PLGA	Poly(lactic-*co*-glycolic acid)
Poloxamer	Poloxamer 407, a nonionic triblock copolymer of poly(oxypropylene) and poly(oxyethylene)
RIP	Radiation-induced proctitis
RS	Responsive sequence (incorporate motifs degraded by matrix metalloproteinases)
SAGE	Semisynthetic glycosaminoglycan ether
SANS	Small-angle neutron scattering
SELP	Silk-elastin-like protein polymer
SEM	Scanning electron microscopy
TACE	Transarterial chemoembolization
SDS	Sodium dodecyl sulfate
UV–vis	Ultraviolet-visible spectrophotometry
VP	Viral particles
VO	Virus-only

Acknowledgements

This work was supported in part by The University of Utah Nanotechnology Training program and NIH grants R01CA227225, R42 CA168123, and R41NS100184.

References

1. F. Lucas, J. Shaw and S. Smith, *Biochem. J.*, 1957, **66**, 468.
2. D. Urry, *J. Protein Chem.*, 1984, **3**, 403.
3. B. Cox, B. Starcher and D. Urry, *J. Biol. Chem.*, 1974, **249**, 997.
4. D. Urry, C. Luan, T. Parker, D. Gowda, K. Prasad and M. Reid, *et al.*, *J. Am. Chem. Soc.*, 1991, **113**, 4346.
5. J. Cappello, J. Crissman, M. Dorman, M. Mikolajczak, G. Textor and M. Marquet, *et al.*, *Biotechnol. Prog.*, 1990, **6**, 198.
6. J. Cappello, J. Crissman, M. Crissman, F. Ferrari, G. Textor and O. Wallis, *et al.*, *J. Controlled Release*, 1998, **53**, 105.
7. Protein Polymer Technologies Inc, *U.S. Pat.*, 5514581A, 1996.
8. M. Haider, V. Leung, F. Ferrari, J. Crissman, J. Powell and J. Cappello, *et al.*, *Mol. Pharm.*, 2005, **2**, 139.
9. R. Dandu, A. Cresce, R. Briber, P. Dowell, J. Cappello and H. Ghandehari, *Polymer*, 2009, **50**, 366.
10. J. Gustafson, R. Price, J. Frandsen, C. Henak, J. Cappello and H. Ghandehari, *Biomacromolecules*, 2013, **14**, 618.
11. R. Price, A. Poursaid, J. Cappello and H. Ghandehari, *J. Controlled Release*, 2014, **195**, 92.
12. A. Poursaid, R. Price, A. Tiede, E. Olsen, E. Huo and L. McGill, *et al.*, *Biomaterials*, 2015, **57**, 142.
13. R. Price, A. Poursaid, J. Cappello and H. Ghandehari, *J. Controlled Release*, 2015, **213**, 96.
14. A. Dinerman, J. Cappello, H. Ghandehari and S. Hoag, *J. Controlled Release*, 2002, **82**, 277.
15. A. Dinerman, J. Cappello, M. El-Sayed, S. Hoag and H. Ghandehari, *Macromol. Biosci.*, 2010, **10**, 1235.
16. A. Poursaid, M. Jensen, I. Nourbakhsh, M. Weisenberger, J. Hellgeth and S. Sampath, *et al.*, *Mol. Pharm.*, 2016, **13**, 2736.
17. M. Jensen, Z. Barber, N. Khurana, K. Isaacson, D. Steinhauff and B. Green, *et al.*, *Theranostics*, 2020, **10**, 4530.
18. K. Isaacson, M. Jensen, D. Steinhauff, J. Kirklow, R. Mohammadpour and J. Grunberger, *et al.*, *J. Drug Target.*, 2020, **28**, 766.
19. F. Cipriani, M. Kruger, I. de Torre, L. Sierra, M. Rodrigo, L. Kock and J. Rodriguez-Cabello, *Biomacromolecules*, 2018, **19**, 4333.
20. M. Haider, J. Cappello, H. Ghandehari and K. Leong, *Pharm. Res.*, 2008, **25**, 692.
21. Z. Megeed, J. Cappello and H. Ghandehari, *Pharm. Res.*, 2002, **19**, 954.

22. Z. Megeed, M. Haider, D. Li, B. O'Malley, J. Cappello and H. Ghandehari, *J. Controlled Release*, 2004, **94**, 433.
23. D. Hwang, V. Moolchandani, R. Dandu, M. Haider, J. Cappello and H. Ghandehari, *Int. J. Pharm.*, 2009, **368**, 215.
24. J. Gustafson, K. Greish, J. Frandsen, J. Cappello and H. Ghandehari, *J. Controlled Release*, 2009, **140**, 256.
25. R. Price, J. Gustafson, K. Greish, J. Cappello, L. McGill and H. Ghandehari, *Int. J. Pharm.*, 2012, **427**, 97.
26. A. Nagarsekar, J. Crissman, M. Crissman, F. Ferrari, J. Cappello and H. Ghandehari, *J. Biomed. Mater. Res.*, 2002, **62**, 195.
27. Z. Megeed, J. Cappello and H. Ghandehari, Genetically engineered silk-elastinlike protein polymers for controlled drug delivery, *Adv. Drug Delivery Rev.*, 2002, **54**, 1075.
28. J. Gustafson and H. Ghandehari, *Adv. Drug Delivery Rev.*, 2010, **62**, 1509.
29. J. Frandsen and H. Ghandehari, *Chem. Soc. Rev.*, 2012, **41**, 2696.
30. R. Price, A. Poursaid and H. Ghandehari, *J. Controlled Release*, 2014, **190**, 304.
31. W. Huang, A. Rollett and D. Kaplan, *Expert Opin. Drug Delivery*, 2014, **12**, 779.
32. Z. Megeed, J. Cappello and H. Ghandehari, *Biomacromolecules*, 2004, **5**, 793.
33. A. Dinerman, J. Cappello, H. Ghandehari and S. Hoag, *Biomaterials*, 2002, **23**, 4203.
34. J. Chang, X. Peng, K. Hijji, J. Cappello, H. Ghandehari and S. Solares, *et al.*, *J. Am. Chem. Soc.*, 2011, **133**, 1745.
35. S. Johnson, Y. Ko, N. Varongchayakul, S. Lee, J. Cappello and H. Ghandehari, *et al.*, *Chem. Commun.*, 2012, **48**, 10654.36.
36. T. Collins, M. Barroca, F. Branca, J. Padrão, R. Machado and M. Casal, *Biomacromolecules*, 2014, **15**, 2701.
37. W. Qiu, W. Teng, J. Cappello and X. Wu, *Biomacromolecules*, 2009, **10**, 602.
38. R. Machado, A. da Costa, V. Sencadas, C. Garcia-Arevalo, C. Costa and J. Padrão, *et al.*, *Biomed. Mater.*, 2013, **8**, 065009.
39. W. Teng, J. Cappello and X. Wu, *J. Controlled Release*, 2011, **156**, 186.
40. W. Teng, Y. Huang, J. Cappello and X. Wu, *J. Phys. Chem.*, 2011, **115**, 1608.
41. C. Moon, Y. Oh and J. A. Roth, *Clin. Cancer Res.*, 2003, **9**, 5055.
42. S. Jung, J. Choi, C. Yun, S. Kim, I. Kwon and H. Ghandehari, *Mol Pharm.*, 2015, **12**, 1673.
43. A. Hatefi, J. Cappello and H. Ghandehari, *Pharm. Res.*, 2007, **24**, 773.
44. H. Jeon, S. Jung, Y. Jung, Y. Kim, H. Ghandehari and K. Ha, *Anal. Chem.*, 2016, **88**, 5398.
45. K. Isaacson, M. Jensen, N. Subrahmanyam and H. Ghandehari, *J. Controlled Release*, 2017, 62.
46. M. Jensen, W. Jia, K. Isaacson, A. Schults, J. Cappello and G. Prestwich, *et al.*, *J. Controlled Release*, 2017, **263**, 46.

47. M. Hauer-Jensen, J. Wang, M. Boerma, Q. Fu and J. Denham, *Curr. Opin. Support Palliat. Care*, 2007, **1**(1), 23.
48. A. Shadad, F. Sullivan, J. Martin and L. Egan, *World J. Gastroenterol.*, 2013, **19**, 185.
49. W. Lee, J. Savage, J. Zhang, W. Jia, S. Oottamasathien and G. Prestwich, *PLoS One*, 2013, **8**, 1.
50. A. Pulsipher, X. Qin, A. Thomas, G. Prestwich, S. Oottamasathien and J. Alt, *Int. Forum Allergy Rhinol.*, 2017, **7**, 177.
51. M. Jensen, W. Jia, A. Schults, K. Isaacson, D. Steinhauff and B. Green, *et al.*, *Biomaterials*, 2019, **217**, 19293.
52. R. Stacey and J. Green, *Ther. Adv. Chronic Dis.*, 2014, **5**, 15.
53. J. Laird, L. Martinez-Caro, E. Garcia-Nicas and F. Cervero, *Pain*, 2001, **92**, 335.
54. W. Jia, A. Schults, M. Jensen, X. Ye, J. Alt and G. Prestwich, *et al.*, *Am. J. Clin. Exp. Urol.*, 2017, **5**, 10.
55. M. Jensen, W. Jia, A. Schults, X. Ye, G. Prestwich and S. Oottamasathien, *Cytokine*, 2018, **110**, 420.
56. A. Elbadawi, *Urology*, 1997, **49**, 14.
57. J. Bruix, M. Reig and M. Sherman, *Gastroenterology*, 2016, **150**, 835.
58. M. Gonzalez, Y. Tang, G. Phillips, A. Lloyd, B. Hall and P. Stratford, *et al.*, *J. Mater. Sci.: Mater. Med.*, 2008, **19**, 767.
59. A. Laurent, *Tech. Vasc. Interv. Radiol.*, 2007, **10**, 248.
60. J. Bruix and M. Sherman, *Hepatology*, 2011, **53**, 1020.
61. X. Hu, D. Kaplan and P. Cebe, *Macromolecules*, 2006, **39**, 6161.
62. P. Low, S. Singhal and M. Srinivasarao, *Curr. Opin. Chem. Biol.*, 2018, **45**, 64.
63. N. Zhu, S. Mondal, S. Gao, S. Achilefua, V. Gruev and R. Liang, *Opt. Lett.*, 2014, **39**, 3830.
64. E. Duffis, C. Gandhi, C. Prestigiacomo, R. Abruzzo, F. Albuquerque and K. Bulsara, *et al.*, *J. Neurointerv. Surg.*, 2012, **4**, 251.
65. Z. Liu, D. Wang, X. Sun, J. Wang, L. Hu and H. Li, *et al.*, *Int. J. Pediatr. Otorhinolaryngol.*, 2011, **75**, 1088.
66. J. Chan, R. Tsang, S. Wong and W. Wei, *Head Neck*, 2015, 169.
67. H. Maeda, H. Nakamura and J. Fang, *Adv. Drug Delivery Rev.*, 2013, **65**, 71.
68. M. Read, P. Morgan and C. Maldonado-Codina, *J. Biomed. Mater. Res. - Part B Appl. Biomater.*, 2009, **91**, 662.
69. C. Teichert, J. MacKay, D. Savage, M. Lagally, M. Brohl and P. Wagner, *Appl. Phys. Lett.*, 1995, **66**, 2346.
70. R. Duncan, *Nat. Rev. Drug Discovery*, 2003, **2**, 347.
71. X. Xia, Q. Xu, X. Hu, G. Qin and D. Kaplan, *Biomacromolecules*, 2011, **12**, 3844.
72. X. Xia, M. Wang, Y. Lin, Q. Xu and D. Kaplan, *Biomacromolecules*, 2014, **15**, 908.
73. W. Kim, J. Thévenot, E. Ibarboure, S. Lecommandoux and E. Chaikof, *Angew Chem. Int Ed.*, 2010, 49, 4257.

74. J. Yeo, W. Huang, A. Tarakanova, Y. Zhang, D. Kaplan and M. Buehler, *J. Mater. Chem. B*, 2018, **6**, 3727.
75. J. Leal, H. Smyth and D. Ghosh, *Int. J. Pharm.*, 2017, **532**, 555.
76. R. Parker, W. Wu, T. Mckay, Q. Xu and D. Kaplan, *J. Funct. Biomater.*, 2019, **10**, 18.
77. S. Moktan, C. Ryppa, F. Kratz and D. Raucher, *Invest. New Drugs*, 2012, **30**, 236.
78. D. Kirui, E. Koay, X. Guo, V. Cristini and H. Shen, *et al.*, *Nanomed.: Nanotechnol. Biol. Med.*, 2014, **10**, 1487.80.
79. B. Wong and C. Hollowed, *Temperature*, 2017, **4**, 41.
80. G. Kong, R. Braun and M. Dewhirst, *Cancer Res.*, 2001, **61**, 3027.
81. K. Isaacson, M. Jensen, A. Watanabe, B. Green, M. Correa and J. Cappello, *et al.*, *Macromol. Biosci.*, 2018, **18**, 1700192.

CHAPTER 10

Processing and Properties of Regenerated Bombyx mori *Silk and Recombinant Spider Silk Fibers*

G. LANG* AND D. SONNLEITNER

Biopolymer Processing Group, Faculty of Engineering, University of Bayreuth, Ludwig-Thoma-Straße 36A, 95447 Bayreuth, Germany
*Email: gregor.lang@uni-bayreuth.de

10.1 Introduction

Silk materials and their utilization in nature are some of the most astonishing masterpieces of evolution that originated more than 400 million years ago. Thousands of arthropod species produce such high-performance materials, which are highly specialized for an outstanding diversity of purposes such as hunting, gluing, construction, protection, and reproduction.[1,2] Consequently, it is not surprising that mankind had started to exploit the potential of silk materials thousands of years ago. Aboriginal Australian made use of the outstanding mechanical properties of spider dragline silk, applying it as fishing lines, whereas New Guinean natives used complete spider webs as fishing nets.[3] First medical applications were realized by the ancient Greeks using cobwebs as wound coverage.[4] Based on an almost unparalleled success story of textile material development, starting with the domestication of *Bombyx mori* about 5000 years ago,[5] the term silk is

Biomaterials Science Series No. 10
Biomimetic Protein Based Elastomers: Emerging Materials for the Future
Edited by Namita Roy Choudhury, Julie C. Liu and Naba K. Dutta

Published by the Royal Society of Chemistry, www.rsc.org

commonly associated with textile fibers obtained by unraveling cocoons of silkworms (*e.g.*, *B. mori*), which are nowadays produced in quantities of kilotons per year.[6] Strikingly, the early findings of the Aborigines and Greeks (mechanics and biocompatibility) are still the motivation of ongoing research. Advanced biochemical understanding based on the progressive development and improvement of analytical tools in combination with interdisciplinary approaches extending into, *e.g.*, engineering or physics, have boosted the scientific outcome on silk materials, especially over the last two decades (Figure 10.1). Furthermore, in 1993, the US Food and Drug Administration (FDA) recognized silk fibroin (SF) as a biomaterial, opening the door for commercial use in the medical sector. Today, there are various companies such as Sofregen Inc. (USA), Vaxess Inc. (USA), Silk Technology Ltd. (USA), Oxford Biomaterials Ltd. (UK), Orthox Ltd. (UK), and Daewoong Bio (South Korea) working on the development of silkworm silk-based products. Besides its wide use as a suture material (*e.g.*, for vessel ligation or mucosal wound closure)[7–9] represented by various different products (*e.g.*, Sofsilk™, Surusil®), different medical devices (*e.g.*, surgical meshes Seriscaffold®) and injectables (soft tissue bulk agent: Silk Voice®)[10] were approved by the FDA.[11]

Interestingly, although spider dragline silk can be stronger than *B. mori* silk (toughness: *Araneus diadematus* dragline ~160 MJ m^{-3}, *B. mori* cocoon silk ~70 MJ m^{-3})[12] and is also known to provide excellent biocompatibility, the quantity of publications referring to SF is significantly higher than those on spider silk (Figure 10.1). This reflects the problem of availability that spider silk entails. Unlike silkworms, which can effectively be farmed and the silk of whom can be easily harvested from their cocoons, spider silk must be actively pulled out of the spiders and farming is inefficient due to their cannibalistic nature. To produce a single golden cape from spider silk,

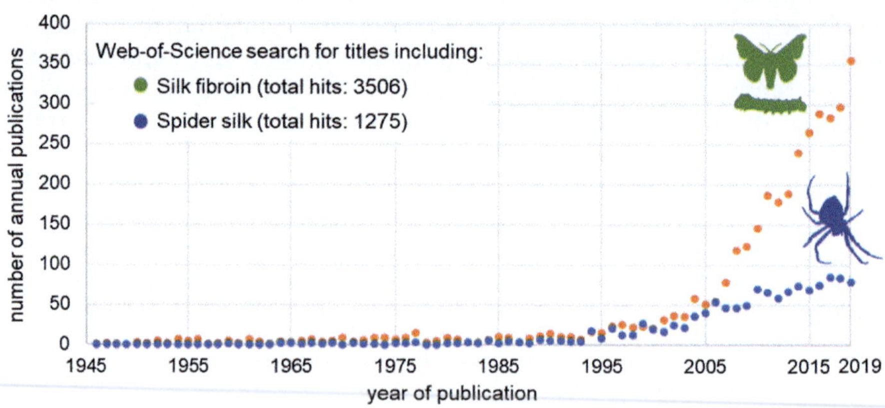

Figure 10.1 Number of annual publications including "Silk fibroin" or "Spider silk" in their title as listed in the Web of Science Core Collection database. (date of search: 05.08.2020, year range: 1945–2019).

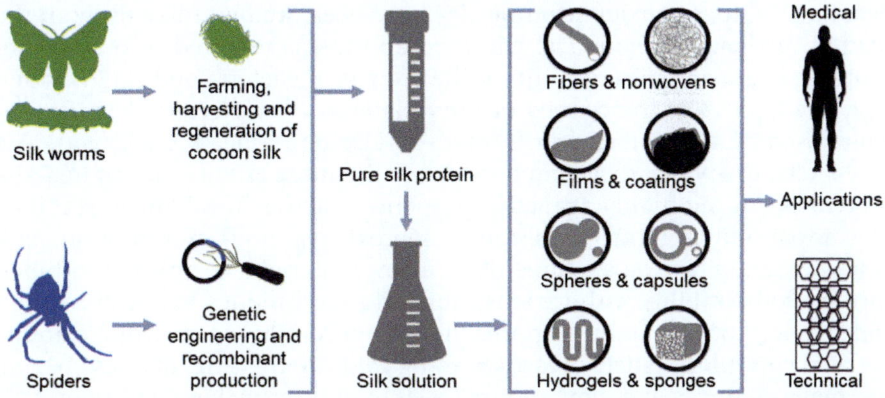

Figure 10.2 Schematic overview of the different steps required to render silk materials employable for various technical and medical applications: harvesting/production, preparation/conditioning, and processing into the relevant morphologies.

it took 8 years and 1.2 million golden orb spiders, demonstrating the time and cost-consuming aspects of harvesting natural spider silk fibers.[13]

Nevertheless, modern biotechnology has opened the door to produce spider silk proteins *via* genetic engineering and recombinant production in host systems such as bacteria, yeast, insects, mammalian cell lines, plants, and transgenic animals.[13] Like *B. mori* fibroin, these spider silk proteins can be dissolved in different solvent systems, making it possible to process them into various morphologies such as fibers and nonwovens, films and coatings, spheres and capsules, or hydrogels and sponges.[14] Combining specific morphologies with the outstanding properties of silk materials, they become highly attractive candidates for various applications in the medical and technical sectors (Figure 10.2). Today, there are several biotech companies such as AMSilk (Germany), Spidey Tek (USA), Kraig Biocraft Laboratories (USA), Bolt Threads (USA), Spiber Technologies (Sweden), and Spiber Inc. (Japan) working on the production and commercialization of spider silk-inspired materials and products all over the world.

10.2 Production/Preparation and Preconditioning of Silk Materials for Technical Processing

10.2.1 Regeneration and Preconditioning of *B. mori* Fibroin

Bombyx mori silk fibers are built up from double filaments composed of fibroin forming the mechanically stable cores and a protective outer shell made of sericin (about 20–30 wt% of the total cocoon mass) connecting the fibroin filaments.[15] Sericins (molecular weight [MW]: 20–400 kDa)[16] are glue-like glycoproteins that are subjected to induce inflammation in the human body but were also shown to potentially have a positive impact on wound healing.[16–19] Its antioxidant properties have been studied intensively in the field of medicines, cosmetics, and food and it was reported to provide antifouling characteristics, inhibiting the growth of bacteria and fungi, which is, *e.g.*, attractive in the context of filter materials.[20–23] Nevertheless, fibroin (composed of a light chain with MW ~26 kDa and a heavy chain with MW ~390 kDa, linked *via* disulfide bonds)[24] is the most extensively studied silk protein providing highly attractive properties such as mechanical stability, biocompatibility, biodegradability, nontoxicity, and nonimmunogenicity.[25–28] Sericulture as a source of excellent material availability combined with its compatibility with various fabrication techniques has resulted in a wide variety of processing approaches. Importantly, apart from extreme thermal conditions (such as fast scanning calorimetry with 2000 $K\,s^{-1}$),[29] SF thermally degrades at a temperature lower than its crystal melting point.[30,31] However, a recent study by Guo *et al.* successfully applied high pressure (632 MPa) to first densify amorphous silk nanomaterials with ultralow β-sheet content ($T < T_G$) and then induce structural changes by increasing the temperature to above glass transition ($T > T_G$).[32] The resulting structures displayed a mechanically stable, homogeneous fused state, which was assigned to thermal fusion *via* molecular rearrangement and self-assembly assisted by bound water. Nevertheless, continuous processing such as fiber spinning of SF is usually performed from solution, and purification and dissolution of sufficient concentrations are a prerequisite to applying established processing techniques. To prepare the so-called regenerated silk fibroin (RSF) in a pure and soluble form, different steps of chemical treatment must be performed. The predominantly applied standard procedure is schematically shown in Figure 10.3.

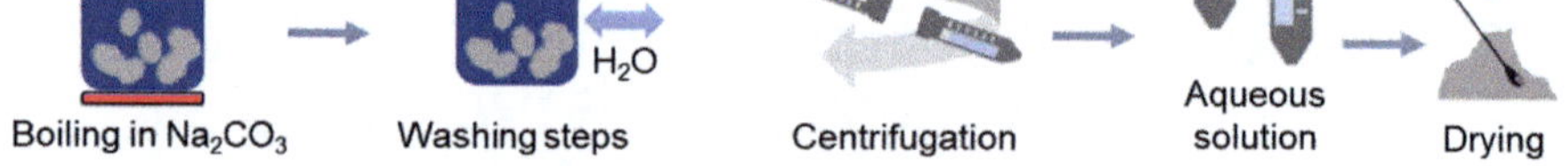

Figure 10.3 Schematic procedure of silk fibroin preparation – in the first step, stifled and cleaned cocoons are degummed (removal of sericin), *e.g.*, by boiling in Na_2CO_3. After several washing steps, silk fibers are dissolved in chaotropic salt solutions (mostly LiBr) and subsequently dialyzed against water to remove salts. Undissolved fragments are removed *via* centrifugation resulting in purely aqueous solutions that can be directly processed or freeze-dried for storage and later dissolution in organic solvents. Finally, solutions are ready to be processed applying various techniques.

After stifling treatment (kill the pupae for storage of cocoons), which can have a significant impact on the fibroin's secondary structure and thus later performance,[33,34] the cocoons (without pupae) are usually boiled in aqueous sodium carbonate (Na_2CO_3) or sodium bicarbonate ($NaHCO_3$) solution. This so-called degumming process dissolves sericin while the fibroin fibers remain stable. Importantly, it must be considered that dissolution of sericin leads to the exposal of fibroin to sodium carbonate, which will start hydrolyzing fibroin.[35,36] The resulting degradation of fibroin reduces its MW depending on boiling times and salt concentration. Wray *et al.* performed degumming in 0.2 M Na_2CO_3, showing that 30 minutes of boiling time resulted in a mean MW of ~150 kDa, while after 60 minutes, the MW was significantly reduced to ~50 kDa.[36] Consequently, this impact on fibroin properties not only weakens, *e.g.*, the mechanical properties of the degummed fibroin fibers[37] but further affects all subsequent processing steps. This includes the rheological properties of aqueous solutions that are highly affected by degumming conditions determining the processing conditions and further propagates to, *e.g.*, mechanical characteristics of the product. Dou and Zuo demonstrated the strong impact of Na_2CO_3 concentration (0.05%–5%) during degumming.[38] With increasing Na_2CO_3 concentration, the shear viscosity of fibroin dissolved in 98% formic acid (FA) significantly decreased along with the loss of MW, leading to reduced electrospinnability. Alladyce *et al.* showed that films, which were cast after dissolving fibroin in aqueous solutions, displayed highly varying mechanical properties depending on the degumming conditions (film strength after degumming with mild conditions: 8.9 MPa and aggressive conditions: 4.3 MPa).[39] Furthermore, degumming also affects the biodegradation of SF. Wang *et al.* performed a comparative study with Na_2CO_3 and $NaHCO_3$, showing that Na_2CO_3 provides higher degumming efficiency, resulting in a lower MW and thus mechanical strength than $NaHCO_3$-degummed samples.[40] This also affected the enzymatic degradation rate (protease XIV solution), which was higher for Na_2CO_3-treated silk samples. Consequently, the degumming procedure in the regeneration of SF must be considered essential for all subsequent steps, including dissolution, processing, and application, all of which will inherently affect the resulting properties. Alternatively, substituting sodium carbonate by degumming with 8.0 M urea was found to maintain the fibroin's molecular structure.[41]

To obtain pure fibroin solutions, the degummed fibers are washed several times to remove any residual sericin and then dissolved in lithium bromide (LiBr, commonly 9.3 M),[42–44] a chaotropic salt that partially denatures the protein's secondary structure resulting in a transition from β-sheet-rich native structures into amorphous dissolved structures.[45] Importantly, the concentration of silk fibers in LiBr solution can also affect the MW of fibroin. Greving *et al.* applied gel electrophoresis to demonstrate a reduction of the fibroin's MW, when the ratio of fiber to LiBr was lower than 2 g per 10 mL 9 M LiBr.[46] Thus, besides the degumming conditions, also the dissolution in chaotropic agents must be considered as an irreversible impact on the material properties. Consequently, research on cost-efficient alternatives avoiding metals by, *e.g.*, applying Ajisawa's reagent, is ongoing.[47] Generally, dialysis against water and subsequent removal of undissolved protein fractions *via* centrifugation has proven to be a sufficient method to remove chaotropic agents resulting in purely aqueous protein solutions.[48] Importantly, these solutions were shown to assemble essentially different than the natural spinning dope of silkworms. Hossain *et al.* found that fibroin solutions extracted from the middle division of the *B. mori* silk gland form large complexes, building clusters that undergo dissociation and assemble into larger structures connected *via* divalent metal ions.[49] Such complexes could not be seen in purely aqueous regenerated silk solutions and also not induced by dialysis against metal ion (Ca^{2+}, Mg^{2+}, K^{+})-containing buffers.[50] Nevertheless, various approaches were published applying such aqueous silk solutions as the basis for different processing techniques. Furthermore, freeze-drying of aqueous solutions is broadly performed for storage and subsequent dissolution in, *e.g.*, organic solvents that enable fast and easy processing due to the stability of the solutions and the high volatility of solvents as compared to purely aqueous solutions.

10.2.2 Recombinant Production of Spider Silk Proteins

In the last two decades, a huge variety of different approaches to producing artificial spider silk proteins has been established based on the genetic analysis of various spider species and the identification of the respective silk genes. Although the diversity of artificially produced silk types emerges with the ongoing investigation of silk encoding genes, the respective literature is predominated by advanced experimental data of recombinantly produced *Nephila clavipes* Major ampullate Spidroin (MaSp), MaSp1 and MaSp2, and *A. diadematus* fibroin (ADF), ADF3 and ADF4. For the purpose of clarity and comparability, this chapter will mainly focus on these silk types.

In 1990, Xu and Lewis[51] characterized a 2.4 kb segment of cDNA partially encoding *N. clavipes* MaSp1 (the full-length transcript is ~12 kb).[52] Furthermore, Hinman and Lewis[53] published a 2 kb segment (full-length transcript ~11.5 kb)[52] of the MaSp2 protein from *N. clavipes* dragline silk. The prediction of the amino acid sequence revealed highly conserved repetitive motifs (approximately 34 and 45 amino acid residues long) combining short crystal-forming polyalanine blocks (8–10 amino acid residues) and amorphous glycine-rich domains.[12] Analyzing the predicted amino acid sequences of *A. diadematus* fibroin genes, Guerette *et al.*[54] identified comparable modular characteristics of alternating crystalline and amorphous regions. It was concluded that although a very diverse set of genes encodes spider silk proteins, almost all of them can be attributed to a single gene family.[54,55] Considering their semicrystalline secondary structure incorporated in an elastomeric block copolymer-like setup, it is not surprising that most silk fibers that fulfill mechanical purposes display comparable mechanical characteristics in terms of a combination of high tensile strength combined with extensibility resulting in high toughness.[56,57] Furthermore, this is also reflected in the presence of the carboxy-[58,59] and amino-terminal[60] domains (C- and N-terminal domains), flanking the repetitive core sequence. These relatively short spidroin regions (<150 amino acids in length) play a crucial role in protein storage and assembly upon fiber spinning.[60–62] They are known to trigger mesophases (in the storage) and protein assembly for liquid–solid transition during fiber spinning in a pH-dependent manner.[63,64] The availability of cDNA libraries and the knowledge of a highly repetitive protein architecture with highly conserved non-repetitive termini are the basis for ongoing biotechnological spider silk design, production, purification, and processing (Figure 10.4).

Importantly, these artificially generated spider silk proteins usually are produced not by the expression of cloned natural silk genes but by analog genes that result in mimicking amino acid sequences. The reason for the challenging aspect of recombinant spider silk (rSS) production is based on the highly repetitive character of the underlying genes (*e.g.*, in bacterial host: homolog recombination),[65,66] their high GC-content, the high MW of the resulting constructs, and the specific codon usage of spiders, meaning the prevalence of specific codons for each amino acid, which is different to that of the host organism.[64,67] Nevertheless, the recombinant production of spider silk-inspired proteins was established for a huge variety of different host systems, including *Escherichia coli*, *Pichia pastoris*, transgenic plants, animals, and different cell lines (Table 10.1).

Furthermore, genetic engineering of spider silk genes opens the door to not only aim for the reproduction of the natural blueprint but incorporate functional modifications for specific purposes. Thereby, silk proteins can be modified by changing their amino acid sequence, the addition of functional peptides, or the design of chimeric proteins.[93] Biofunctionalization for medical applications has intensively been targeted by applying genetically adapted silk proteins. To improve the adhesion of mammalian cells on

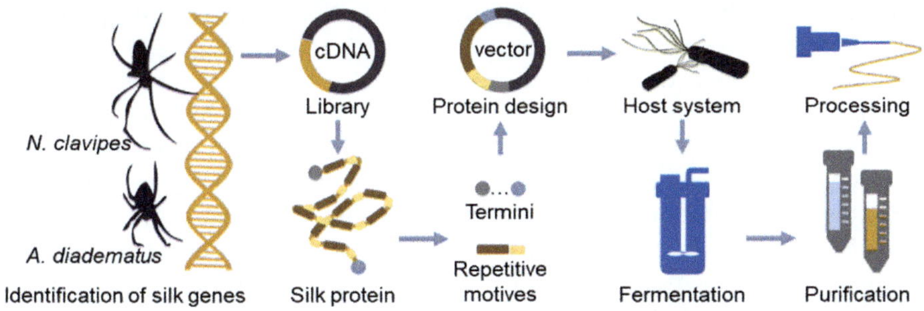

Figure 10.4 Schematic overview of the production of recombinant spider silk, starting with gene identification and functional analysis of the pure proteins. With this information, proteins can be designed and produced (*e.g.*, *via* the fermentation of a bacterial host system). After purification, various processing technologies can be applied to generate spider silk-based materials for, *e.g.*, biomedical applications.

Table 10.1 Overview of various recombinantly produced spider silk-inspired proteins.

Protein origin	MW [kDa]	Host system	References
N. clavipes			
MaSp2	31–112	*E. coli*	Lewis *et al.* (1996)[68]
MaSp2	8–32	*E. coli*	Lewis *et al.* (1996)[68]
MaSp1	65	*P. pastoris*	Fahnestock & Bedzyk (1997)[69]
MaSp1, MaSp2	65–163	*E. coli*	Fahnestock & Irwin (1997)[70]
MaSp1	43	*E. coli*	Arcidiacono *et al.* (1998)[66]
MaSp1	13–100	Tobacco and Potato	Scheller *et al.* (2001)[71]
MaSp1, MaSp2	59–140	Mammalian cell	Lazaris *et al.* (2002)[72]
MaSp1	60	Mammalian cell	Service (2002)[73]
MaSp1, MaSp2	60	Tobacco	Menassa *et al.* (2004)[74]
MaSp1	64	Thale cress	Yang *et al.* (2005)[75]
FLAG	37	*E. coli*	Miao *et al.* (2006)[76]
FLAG	37	*B. mori*	Miao *et al.* (2006)[77]
MaSp1	31–66	Mice	Xu *et al.* (2007)[78]
MaSp1	285	*E. coli*	Xia *et al.* (2010)[79]
MaSp1	83	*B. mori*	Wen *et al.* (2010)[80]
FLAG with MaSp2	106–130	*B. mori*	Teufle *et al.* (2012)[81]
MaSp1	44	*P. pastoris*	Gaines & Marcotte (2013)[82]
FLAG	47–250	Nicotiana plant	Hauptmann *et al.* (2013)[83]
MaSp1	250	Tobacco	Weichert *et al.* (2014)[84]
MaSp1, MaSp2	65	Goat	Copeland *et al.* (2015)[85]
MaSp1, MaSp2	73–108	Tobacco	Peng *et al.* (2016)[86]
MaSp1 (192-mer)	556	*E. coli*	Bowen *et al.* (2018)[87]
A. diadematus			
ADF3	60	Mammalian cell	Lazaris *et al.* (2002)[72]
ADF3, ADF4	60	Insect cells (Sf9)	Hümmerich *et al.* (2004)[88]
ADF3, ADF4	35–60	*E. coli*	Hümmerich *et al.* (2004)[89]
ADF4	50–60	Insect cells (Sf9)	Ittah *et al.* (2006)[90]
ADF1, ADF2, ADF3	31–56	Salmonella	Widmaier *et al.* (2009)[91]
ADF3 (Dimer)	286	*E. coli*	Heidebrecht *et al.* (2015)[92]

recombinant spider silk surfaces, engineered *A. diadematus* (eADF) proteins were functionalized with cell-binding RGD motifs (eADF4(C16)-RGD).[94] This spider silk construct was shown to be applicable in different morphologies with a high potential in the field of tissue engineering and biofabrication.[95–97] Furthermore, a recently published study by Kumari *et al.* demonstrated that eADF4(C16)-RGD displays bacterial and fungal repellent properties simultaneously promoting mammalian cell adhesion.[98] This intrinsic bioselective behavior is a unique feature further boosting the attractiveness of such materials for medical applications. Other studies were performed modifying recombinant MaSp1 (*N. clavipes*) with RGD.[99] Moreover, the spider silk–based protein 4RepCT (obtained from *Euprosthenops australis*) was functionalized with different cell-binding peptides (RGD, IKVAV, and YIGSR), increasing the adhesion of fibroblast, keratinocytes, endothelial, and Schwann cells.[100] To produce scaffolds for bone regeneration, an RGD-containing MaSp1-protein (*N. clavipes*) was genetically equipped with an R5 sequence, successfully inducing biomineralization in the presence of silicic acid solution.[101] Other modifications were performed for, *e.g.*, gene and drug delivery,[102–107] multilayer coatings of medical devices,[108] improved wound healing,[109] and achieving antimicrobial activity.[110]

10.2.3 Preparation and Conditioning of Silk Solutions for Technical Processing

Processing of silk materials is generally simple, as most solution processing techniques can be directly applied. However, it gets very complex if the high-performance characteristics of natural silk are meant to be achieved, as these are based on highly specific hierarchical structural features generated throughout the whole process of natural silk fiber spinning. Consequently, it is indispensable to consider the natural blueprints of silkworms and spiders in the preparation and conditioning of the respective silk solutions for technical processing. As dragline spider silk is the most investigated spider silk type, the following part will focus on the dragline-producing major ampullate gland. Interestingly, although the silk glands of silkworms have evolved from salivary glands and fibers are pulled from the animal's head region,[111] whereas spider silk glands presumably originate from epidermal invaginations of the abdomen,[112] both gland types display a high degree of similarities in terms of structure and function (Figure 10.5). They have a tail, in which the main components (*B. mori*: heavy and light chain fibroin, spiders: MaSp1 and MaSp2)[113,114] are expressed and transported to the ampulla for storage at high concentrations of ~30% in *B. mori* and ~50% in the major ampullate gland of orb-weaving spiders.[115–117] Although it is not fully described, the predominating theory hypothesizes that highly concentrated silk proteins assemble into micellar structures mainly driven by hydrophobic interactions

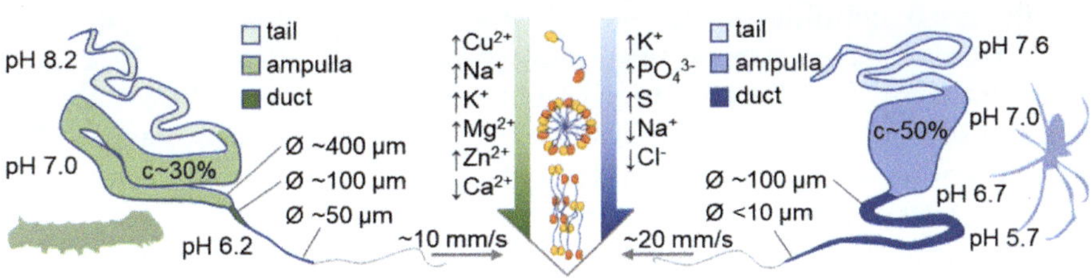

Figure 10.5 Comparison of *B. mori* spinning gland (left) and major ampullate gland of orb-weaving spiders (right). Both glands can be subdivided into tail (silk protein-producing cells), ampulla (storage of highly concentrated micellar silk complexes), and duct (fiber assembly). Fiber formation is induced *via* a multistep assembly process, applying pH-gradients,[123,124] ion exchange,[125,126] shear forces, dehydration,[128] and stretching. Silk proteins thereby undergo assembly of single molecules into micellar complexes that elongate into anisotropic physically (due to crystal formation) interconnected structures when exposed to the shear forces within the narrowing duct,[111,128] and finally solidify upon pulling, stretching, and drying.

and their pH-responsive nonrepetitive terminal domains (especially N-terminal) containing hydrophilic blocks that form the micellar shell.[118–122] When entering the duct, silk proteins are exposed to a further pH drop and ion exchange occurs while shear forces arise due to the narrowing duct geometry.[111,123–129] These factors induce transition in the secondary structure of the silk proteins from helical and random coil structures to β-sheet-rich, crystalline structures,[129–131] resulting in a final crystallinity of the solidified fibers of ∼28–37% (for dragline of *Nephila* spiders)[132–134] and ∼50–55% (for *B. mori* fibers).[134,135] Interestingly, prevalent assumptions that this transition predominantly occurs within the animal could be disproved for silkworms by applying *in situ* Fourier-transform infrared spectroscopy (FTIR) studies on as-spun fibers. It was demonstrated that fibers are initially spun in a highly hydrated state with less than 15% of the maximum potential β-sheet content and structural maturation occurs upon *ex vivo* drying, an effect that could be inhibited by applying low humidity.[136] Furthermore, it was found that particularly ion exchange might play a crucial role in the conditioning of the silk dope to match the rheological requirements of the spinning process. A recent study by Laity *et al.* gave reason to suspect that competitional utilization of divalent (mainly Ca^{2+}) and monovalent (mainly K^{+}) ions forming different strengths of salt bridges between acidic amino acids (Asp, Glu) could be the silkworm's mechanism to trigger the viscosity of its spinning dope achieving ideal flow conditions within the duct.[137] In this context, it might be important to regard the common natural spinning speed that these systems are optimized for (*B. mori* silk ∼10 mm s^{-1}, *Nephila* dragline ∼20 mm s^{-1}).[138–140] Considering such highly specific aspects in the dynamics of silk material processing might be the key to achieving biomimetic high-performance materials in a technical process.

In general, aqueous and organic solutions are two basic strategies to dissolve silk proteins for subsequent processing, both of which have their advantages depending on the aimed product. Organic solutions are often made from 1,1,1,3,3,3-hexafluoro-2-propanol (HFIP), FA or trifluoroacetic acid (TFA).[14,141–143] These strong solvents destroy β-sheets while stabilizing helical structures and thus result in a high solubility of silk proteins.[144] FA is thereby known to allow for confirmation of β-sheets during drying rather than HFIP, although the degree of crystallinity seems to be dependent on the drying time and relative humidity.[145–147] Furthermore, it was demonstrated that short storage times (a few days) of SF in FA did not result in significant molecular decomposition.[147,148] In another study, degumming conditions were adapted to either receive nanoparticles (resulting from degumming in 5% Na_2CO_3) or nanofibrils (resulting from degumming in 0.005 wt% Na_2CO_3), which remained stable after dissolving them in 98% FA.[38] Rheological examinations were applied to demonstrate the presence of fibrillar structures within the organic solutions, resulting in comparable shear-dependent viscosities as in the natural spinning dope. Considering that preassembly of micellar or fibrillar structures might be an important prerequisite to enable efficient processing of, *e.g.*, mechanically stable fibers in a biomimetic manner, other approaches were established, such as employing different drying strategies of aqueous fibroin solutions to allow for the assembly of fibrillar structures and redissolving them in FA. Slowly drying aqueous silk solutions over 4 days[149] or sonication of fibroin hydrogels resulted in the assembly of nanofilaments,[150] which could serve as precursor structures in the processing of 98% FA or HFIP solutions. Comparable results were achieved by the dissolution of degummed fibroin in mixtures of calcium chloride anhydrous ($CaCal_2$, 1–10%) with FA and subsequent drying of these solutions.[151] After washing away $CaCal_2$ in water, the films were redissolved in FA, resulting in nanofibrillar structures within the solution. Alternatively, fibers were directly spun from FA-$CaCl_2$ solutions preserving nanofibrillar features.[152] Although such strategies resulted in desired rheological impact and improved spinnability, they do not fully cover the biological mechanisms of fiber assembly. It is hypothesized that structural integrity can only be achieved by externally triggered multistep self-assembly balancing nanostructure and conformational composition (as in the silk glands of silkworms and spiders).[153] To achieve biomimetic assembly, nanostructure assembly and crystallization were studied considering molecular mobility, charge, hydrophilic interactions, and concentration as the key factors determining silk protein interactions.[153] It was shown that increasing concentrations of aqueous silk solutions (by simply drying) first resulted in the formation of micellar structures that subsequently assembled into fibrillar structures. Micellar structures are thermodynamically described as a result of the fibroin's hydrophilic–hydrophobic–hydrophilic arrangement. The formation of nanofibrils instead of, *e.g.*, bigger globules can be attributed to the isoelectric point (4.2) of silk, so its negative surface charges cause repulsion of the micelles. Consequently, the fibril formation is the preferred assembly path because an approaching micelle only experiences repulsive forces from one other micelle instead of multiple ones, as in the case of globule formation. To form stable structures, fibril assembly must be accompanied by the structural transition of silk I (ordered, noncrystalline) to silk II (β-sheet containing). Using aqueous regenerated fibroin solutions, it could be demonstrated that these correlative aspects result from thermodynamic as well as kinetic mechanisms.[153] In another study, to mimic the impact of hydrophilic sericin in the gland, polyethylene oxide (PEO) was added to aqueous fibroin solutions absorbing water and increasing the local fibroin concentration, which positively resulted in the formation of silk I structures.[154] Besides the impact of concentration, pH was also demonstrated to be suitable to control micelle and fibril formation in aqueous fibroin solutions. At pH 6.8, micellar structures assembled, whereas further decrease to pH 4.8 resulted in the formation of nanofibrils comparable to the natural spinning process.[155] Related considerations were also performed applying spider silk–based materials. It could be shown that the N-terminal domain of recombinant MaSp1 mini-spidroins was responsible for immediate charge-dependent protein assembly at pH 6.3, whereas aggregation at pH 7 was significantly delayed.[156] The findings are consistent with the natural model of pH drop within a spider's major ampullate gland, where protein aggregation must be inhibited within the ampulla (pH7). Furthermore, phosphate-induced liquid–liquid phase separation was achieved using recombinant eADF4 proteins comprising the C-terminal domain. It could be demonstrated that even low concentrations of silk solutions resulted in the formation of a self-concentrated high-density micellar phase composed of noncovalently linked oligomers.[157,158] Comprehensively, these data make clear that the recombinant strategy must not only reproduce the long repetitive protein core domains but also include the nonrepetitive terminal domains (important for silk storage, assembly, dimerization),[61,159,160] explicitly, when aiming for biomimetic material properties of spider silk fibers.

10.3 Artificial Silk Fibers and Nonwovens

Though there are a lot of different methods to produce fibers from polymers, most of the large-scale commercial ones require the polymer to be in a molten state. This is problematic when natural polymers like proteins should be processed since their degradation temperature is usually below their (theoretical) melting temperature. Consequently, fiber spinning of natural biopolymers is commonly performed from solution. The following section discusses the most prominent methods suitable for the processing of silk proteins, considering their general benefits and drawbacks. Figure 10.6 gives an overview of typical fiber diameter ranges resulting from different processing methods.

While electrospinning is the technique that can produce fibers with the smallest diameters providing a comparably high production rate (10–100 $m\,s^{-1}$), it is also the most uncontrollable process in terms of

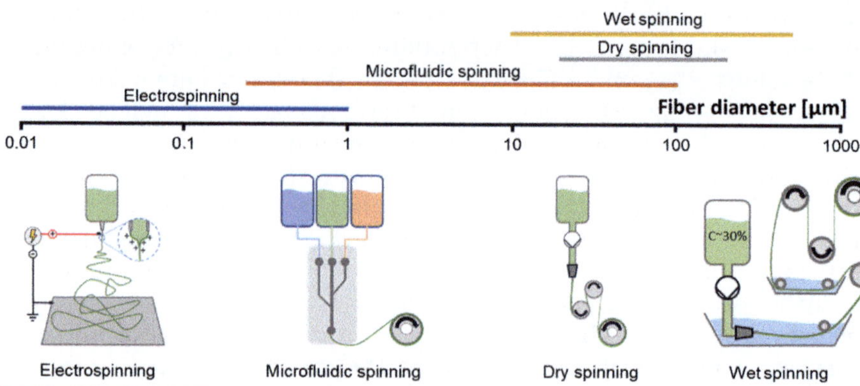

Figure 10.6 Schematic overview of different fiber-spinning techniques and their typical range of possible fiber diameters based on solution processing, including electrospinning, microfluidic spinning, dry spinning, and wet spinning.[161–164]

targeted fiber deposition as it generally produces nonwovens or aligned fiber mats (if appropriate collector systems are applied).[165–167] Therefore, defined continuous postprocessing of single fibers such as in-line drawing is better adjustable for microfluidic, dry, and wet spinning. Considering dry spinning is the simplest of the presented techniques, it enables fast and efficient fiber production by solvent evaporation but lacks control over important biomimetic parameters such as ion exchange or pH drop and is therefore mostly dependent on the preparation of the spinning dope.[168] Since wet spinning is a well-established method for fiber production, it provides easy scale-up and good knowledge of the process, though the mechanism of fiber formation in a coagulation bath is very different from the natural process and has only a limited potential of continuously producing fibers in a real biomimetic manner.[165] The most sophisticated approach of mimicking the natural process is microfluidic spinning, where biomimetic parameters (funnel geometry, shear forces, gradients in pH, ion composition) can be controlled by designing a complex microfluidic chip with several sheath and coagulation flows.[169,170] However, this complexity makes the process less robust and so far results in relatively low production rates (1–10 mm s^{-1}).[165] The presented processes are only clearly distinguished in theory. In practical approaches, various combinations were reported to combine benefits and compensate for the drawbacks of each method. There are examples of electrospinning into coagulation baths, using microfluidic chips as spinneret for dry or electrospinning or introducing an air gap to wet spinning for additional polymer chain orientation before coagulation in the bath.[166,171,172]

10.3.1 Wet Spinning

Wet spinning is one of the oldest known techniques to produce fibers on an industrial scale. The dissolved polymer is extruded into a coagulation bath containing an antisolvent. Due to the diffusion-driven removal of the solvent, the polymer starts to precipitate in the bath and can be drawn into a fiber by collecting drums. This was first documented for SF in a patent by Esselen in 1933,[173] using an inorganic salt system as a solvent for the spinning dope and sodium bisulfate as an antisolvent. Yazawa *et al.* reported the first mechanical measurements on wet-spun RSF fibers produced from aqueous salt solutions, but the resulting fibers were significantly inferior as compared with their natural blueprint (tenacity of 2.5 g per den and extensibility of 20–25%). This might be assigned to the large fiber diameters of more than 100 μm, indicating insufficient shear stress during processing.[174] More promising was the development of using organic solvents[175] such as hexafluoroacetone hydrate (HFA),[176,177] HFIP,[142,177] and trifluoroacetic acid (TFA),[178] leading to fibers with some mechanical properties similar or even slightly superior to that of natural silkworm silk (strength up to 408 ± 80 MPa and extensibility up to $21 \pm 3\%$).[177] By analogy, there was a tendency toward using polar organic antisolvents such as methanol, ethanol, acetone, propanol, isopropanol, or tetrahydrofuran for the coagulation bath.[175,179] In contrast, a recently published work addressing 3D printing of RSF into a coagulation bath displayed the suitability of a biomimetic antisolvent system.[180] Applying a mixture of inorganic salts (0.5 m dipotassium phosphate and 4 m sodium chloride) combined with a slightly acidic pH (~pH6), the authors demonstrated the transferability of natural protein assembly mechanisms toward the technical application of 3D printing, which resembles wet spinning with spatial fiber deposition.

The next generation of spinning dopes was based on biomimetic aqueous solutions, making use of reverse dialysis and forced airflow technique to increase RSF concentrations up to 30%, which resulted in reported values of 450 ± 20 MPa maximum strength and $28 \pm 4\%$ extensibility.[181–184] Recent trends tend to move away from the idea of completely dissolving the SF within the spinning dope applying solvents that preserve the intrinsic microfibrillar structure instead. This was demonstrated by Ki *et al.*[185] with an FA–phosphoric acid mixture and later adapted by Zhang *et al.*[186] using a $CaCl_2$–FA mixture as a solvent system. They could achieve the highest strength (470.4 ± 53.5 MPa) and extensibility ($38.6 \pm 6.3\%$) reported from such wet spinning approaches so far. The good mechanical properties were attributed to well-developed crystalline structures and a higher degree of orientation than in natural fibers.[186] A variation from this process is air-gap spinning, where the spinning dope is not directly spun into the coagulation bath but must pass a small air gap (~1 mm) before entering the coagulate.[179,187] This technique was most prominently used to develop Kevlar fibers and provides the benefit of structural preorientation due to the gravitational forces and a reduced risk of needle blocking and polymer aggregation, making the process more robust.[188] In the case of silk spinning, this technique can produce fibers with 2.5 fold higher extensibility, resulting in 1.5 fold tougher silk fibers as compared to native *B. mori* silk fibers.[179]

For recombinantly produced spider silk, the general conception is that the naturally occurring protein structures with high MW must be mimicked to unlock the full potential of the spider silk's mechanical properties. Adrianos *et al.* demonstrated this by compositional variation of the motifs (GPGGX, GGX, and a noniterated spacer) in a wet-spinning process of recombinant *N. clavipes* flag-like silk proteins.[189] Strikingly, the presence of the noniterated spacer improved the fiber toughness by a factor of ~5. Mechanical properties close to natural spider dragline silk were reported by Bowen *et al.*, increasing the MW of their construct to 556 kDa.[87] Other groups demonstrated the importance of the terminal domains for the preassembly in the spinning dope, having an even greater influence on the toughness of the final fibers than solely increasing the MW.[92,190,191] Apart from the solvent/coagulate system, several studies confirmed that poststretched fibers display superior strength, extensibility, or toughness as compared to as-spun fibers with a significant increase of up to a factor of five for the resulting toughness.[92,178,181,190,192] In this context, Sun *et al.* emphasized that the draw ratio is more effective in improving the mechanical properties than the draw rate.[193]

Additional steps in posttreatment include washing baths,[174] storage in coagulation baths of different compositions[194,195] with and without stress applied to the fibers,[178,185] temperature treatment in air or submerged,[181,184] and steam annealing in water vapor.[184,196] Although all of these strategies were demonstrated to improve the mechanical properties of silk fibers, the achieved effects were significantly smaller than that of the draw ratio displaying the need for holistic concepts also in the field of posttreatment strategies.

10.3.2 Dry Spinning

Dry spinning resembles the natural system in a very simplified manner. As the silk dope is extruded through a spinneret/needle into air, the solvent evaporates, creating a solid fiber, which is taken up by a collector system. Variations in the uptake speed and the number of guiding rolls create different degrees of poststretching. Dry spinning of RSF started much later than wet spinning, with the first study about solvent removal in the elongational flow being published in 2004 by Kojic *et al.*[197] Since dry spinning works without the need of a coagulation bath but by simple diffusion of the volatile solvent to the vapor phase, it is more closely related to the natural spinning process, energy saving, and environmentally friendly than wet spinning in general.[198] The field showed rapid improvements of the mechanical properties of RSF elastomer fibers toward increased strength (357.3 MPa) and more than twice the breaking elongation (34.1%) as compared to degummed silkworm silk fibers (314.2 MPa and 14.2%).[198,199] Subsequent publications stated that the trend in improved mechanical properties is based on increased silk dope concentrations (from 20% to more than 50%), generated by forced air flow, with a simultaneous decrease in diameter down to 2 μm.[193,198–200]

Recently, new approaches emerged preserving the native nanofibrillar structure. Therefore, degummed SF was not exposed to LiBr but directly dissolved in $CaCl_2$–FA[201] or HFIP[202] to generate fibers with a higher degree in a hierarchical organization comparable to the natural blueprint. This resulted in moderate mechanical improvements (strength of 340 MPa and breaking elongation of 36% for $CaCl_2$–FA).[201] The fibers derived from direct dissolution in HFIP showed an increased modulus (11 GPa) but reduced strength (133 MPa) and extensibility (14%).[202] Nevertheless, these approaches provide the benefit of a simplified spinning dope preparation and further development might result in superior mechanical properties due to their high potential of hierarchical structuring.

To reinforce artificial silk fibers, different filler materials were introduced to RSF solutions. The addition of nanoanatase (TiO_2)[203] and graphene oxide (GO)[204] led to a similar degree of crystallinity (natural silk: 55%, RSF with TiO_2: 40.4%, RSF with GO: 43.7%), smaller crystallite size, and pronounced mesophase formation (natural silk: 10%, RSF with TiO_2: 16.6%, RSF with GO: 18.6%). The resulting mechanical properties, though, showed great differences. While the addition of TiO_2 led to a significant increase in breaking elongation (up to 88.3%) and reduced strength (218.5 MPa),[203] graphene oxide showed only moderate increased breaking elongation (35.9%) but a significant increase in strength (435.5 MPa)[204] as compared to the natural counterpart.

As for the other processing methods, posttreatment is of great importance for the final fiber properties. Sun *et al.* studied the postdrawing process in detail and found a decrease in the crystallite size with increased draw rate and ratio (up to 4).[193] This was attributed to the faster rearrangement of the molecular chains and therefore the reduction of crystallite defects, with the draw ratio having a greater effect than the draw rate. Furthermore, chemical posttreatment was applied for most of the fibers using water or ethanol baths, with and without external stress for several hours and additional drying steps to further trigger conformational transition toward crystal formation.[193,199,200,203,204]

10.3.3 Microfluidic Spinning

Microfluidic spinning allows for the greatest control of the technical silk-spinning process. Depending on the complexity of the chip design, varying numbers of spinning dopes, coagulates, and other solutions can be applied to create sheath flows, core shell fibers, or gradients within the microfluidic channels.

Since the spinning apparatus of silkworms and spiders take advantage of several microfluidic effects,[205] the technique of microfluidic fiber formation provides a powerful tool to mimic the natural spinning in a technical setup.[206] The ease of microfluidic chip design allows a high level of control over geometrical parameters such as precise and dynamic adjustment of the composition of the fluids involved.[207] The first microfluidic approach to producing fibers from recombinant spider silk was introduced by Rammensee *et al.* in 2008.[208] Studying the self-assembly of eADF4 and eADF3 proteins, they found that high shear rates ($800\ s^{-1}$) and colloidal aggregates are required to allow for shear-induced fiber assembly. The most straightforward chip designs simply mimic the geometry of the hyperbolically shaped spinning funnel at the end of the silk worm's spinning gland.[205] Luo *et al.* used a similar single-channel design combined with an aqueous RSF solution to create dry-spun fibers resulting in smaller crystallite size, higher β-sheet content, and a higher degree of molecular orientation such as in natural silk.[207] These fibers showed excellent mechanical properties with a breaking stress of 614 MPa, a breaking strain of 27%, and a breaking energy of $101\ kJ\ kg^{-1}$, whereas degummed cocoon silk displayed 354 MPa breaking stress, 16% breaking strain, and $34\ kJ\ kg^{-1}$ breaking energy. A comparable approach was reported by Peng *et al.* spinning rSS dragline protein (47 kDa) into an ethanol coagulation bath with postspin drawing to produce elastomeric fibers with breaking strength and strain of up to 505 MPa and 44%, respectively.[170] More sophisticated multichannel designs allowed for the production of complex morphologies and tunable compositions of the fibers.[206] Designs of three inlets and one outlet were applied to create a hydrodynamic focusing of the spinning dope,[208] which could also be used to create multiple nanometer- and micrometer-scale fibers at the same time by taking advantage of Kelvin–Helmholtz instabilities at the dehydrating interface between core and sheath fluids.[209] Interestingly, fluid simulations recently demonstrated to provide a powerful tool for predicting the diameters of microfluidically spun RSF and regenerated spider silk fibers, as shown by experimental verification.[169,210] Increasing the number of channels enables the application of even more complex conditions by gradually changing multiple parameters, such as the pH and the ion composition along the spinning axis. Together with controlled hydrodynamic focusing, this is predicted to lead to a further increase in mechanical performance and a higher degree of molecular anisotropy in the generated fibers by creating a system that closely resembles the natural spinning process.[169,210] Another interesting development is the use of a sacrificial sericin-like polymer phase to induce phase separation and entrap the liquid SF at mild conditions to more closely mimic the natural composition of silk fibers. This was shown by Cheng *et al.* by applying alginate as a sericin substitute.[211] Alginate was subsequently removed in sodium citrate and meter-long fibers were spun, marking an important step toward a continuous spinning system.

Reinforcement with a filler material was performed by Lu *et al.* introducing cellulose nanofibers in the microfluidic dry-spinning process of RSF. They created fibers with higher crystallinity, higher mesophase content, and smaller crystallite size, leading to improved mechanical performance.[212] The largest average tensile strength they reported was 487 MPa, together with a maximum breaking stress of 686 MPa, which was 2.5 times that of degummed silk fibers (265 MPa) and 2.2 times that of pure RSF fibers (307 MPa). Moreover, the breaking elongation was extended to 16%.[212]

As for the other approaches, posttreatment can be applied to further adjust the mechanical properties but so far has not been extensively studied for microfluidic spinning. Nevertheless, postdrawing was shown to induce a conformation transition from random coil/α-helix to a β-sheet content higher than that of degummed cocoon silk.[207]

10.3.4 Electrospinning

Electrospinning enables the production of fibers with diameters ranging from micrometers down to a few nanometers by exposing the spinning dope to an electrical field applying high voltages (0–100 kV).[213] The most common setup comprises a syringe pump, forcing the spinning dope from the syringe through an electrically charged needle until a droplet is formed.[167] Due to the strong electrostatic field at the surface of the spinning dope, Coulomb forces induce the formation of a Taylor cone, from which a thin jet of polymer solution is directed toward the collector electrode as the electrostatic forces overcome the surface tension. In order to produce fibers instead of spray, the molecular cohesion and chain entanglement (mostly dependent on MW and concentration) must be sufficiently high.[166] The solidifying fiber experiences whipping instabilities due to the high concentration of surface charges and is further stretched, leading to the final fiber diameter as it is deposited as a nonwoven mesh on the collector.[214]

Though at first glance, it is the least biomimetic spinning method (application of an electric field, very high production rate of several $m\,s^{-1}$, predominant use of organic solvent), electrospinning is the simplest and most effective method for the fabrication of continuous fibers.[213] The diameter and morphology of the fibers can be adjusted by altering the spinning dope properties (polymer/protein concentration, MW, viscosity, conductivity, volatility, and surface tension of the solvent), the process parameters (voltage, distance between needle and collector, flow rate, and manipulation of the electrical field) or the ambient conditions (temperature and relative humidity).[215,216] It is therefore only reasonable to internally compare results from one protein/solvent system and a defined set of spinning parameters.

There have been several reports about the optimization of different parameters of the electrospinning process. While Cao *et al.* found that the distance has the biggest influence on the diameter distribution,[217] Park *et al.* investigated the impact of the electric field strength on the electrospinning rate, concluding that for a 13% RSF solution in FA (98%), the spinning rate could be significantly increased (from 0.25 $mL\,h^{-1}$ to 3.0 $mL\,h^{-1}$) when the electric field strength was raised from 0.5 $kV\,cm^{-1}$ to 2.5 $kV\,cm^{-1}$. Other groups concluded that the concentration has the greatest influence on the final fiber diameter, with a minimal concentration for fiber formation around 5% and optimal conditions between 10 wt% and 30 wt% depending on the solvent system.[217–220] Common solvent systems for RSF and spider silk for electrospinning are mostly based on organic solvents such as FA,[219,221–223] HFIP,[224–227] and HFA,[228] with the trend also going toward aqueous systems[172,217,229–231] to more closely resemble the natural model and avoid the use of toxic agents. Since obtaining single fibers with this method is a challenge, electrospinning is often used to produce nonwovens that provide a high surface area, high porosity, and, dependent on the material, good biocompatibility. Consequently, these structures are often studied in the context of their specific morphological features, which, *e.g.*, can provide good cell adhesion in tissue engineering.[232]

Several groups have investigated the mechanics of RSF nonwovens, finding that not only the spinning parameters but also the posttreatment has a big influence on the final properties. Due to the rapid evaporation of the solvent, the proteins solidify in a state of less hierarchical structuring and orientation than the natural fiber. Therefore, additional posttreatment steps must be applied to induce crystallization, render the structures water insoluble, and increase mechanical properties.[233] Cao *et al.* reported that posttreatment in ethanol led to water-insoluble fiber mats with increased tensile strength and elongation at break from 5.6 ± 0.7 MPa and $5.0 \pm 0.5\%$

to 11.1 ± 0.7 MPa and 10.2 ± 1.6%, respectively.[217] Also, introducing the step of postdrawing the mats with different draw ratios resulted in increased tensile strength but reduced elasticity, from 1.8 MPa and 8.37% (as spun) to 8.6 MPa and 4.6%, respectively, after drawing at a ratio of 1.4×.[234] Apart from ethanol, often used procedures for posttreatment of silk nonwovens are annealing treatment with methanol,[227] glutaraldehyde,[235] or water vapor in a desiccator[236] performed at room temperature or at elevated temperatures between 60 and 100 °C.[227] These additional processing steps induce transition from random coil structures to a higher degree of β-sheets, which is usually detected by X-ray diffraction (XRD) analysis or FTIR measurements, where a shift in the absorption peaks for random coils at 1655 cm^{-1} (amide I), 1540 cm^{-1} (amide II), and 1241 cm^{-1} to lower wavenumbers can be observed.[172,237] Furthermore, the surface properties of the mats can be altered by plasma treatment, effectively improving the wettability as displayed by a decrease of the water contact angle from, *e.g.*, 78° to 26.34° after being exposed to oxygen plasma at 70 mTorr for 120 seconds.[222]

Depending on the targeted application, many different natural and synthetic polymers have been blended with silk proteins to produce elastomeric fibers and mats with superior tailored properties. The addition of silk to relatively brittle chitosan showed increased tensile strength from 1.0 MPa to 10.3 MPa with increasing SF content,[235] while the addition of collagen and gelatin improved cell adhesion and proliferation on nonwovens used as scaffolds for tissue engineering.[238,239] Other approaches with natural polymers such as hyaluronic acid and keratin could show an improvement in mesenchymal stem cell proliferation and improved spinning performance with the promotion of network structures, respectively.[240–242] Besides blending silk with other materials, modifying the electrospinning setup can produce more complex morphologies, like core shell systems or side-by-side setups, providing unique characteristics.[233] Peng *et al.* used a side-by-side arrangement of two syringes to create fibers consisting of a distinct SF and poly(l-lactide) (PLLA) phase. They showed good mechanical properties (tensile strength 16.5 ± 1.4 MPa, modulus 205 ± 20.6 MPa, and elongation at break 53 ± 8%), displaying the potential to synergize the surface chemistry of the two sides with different functionality.[227] A different kind of morphology, apart from nonwovens, was presented by Aghaei *et al.*, who reported the production of a highly twisted yarn consisting of a silk/tropoelastin mixture supplied from two syringes.[243] An electric field was applied to the needles of both syringes and fibers were spun on a rotating funnel from which the yarn was drawn continuously to a collector bar. The generated twisted yarn provided varying mechanical properties, depending on the state of hydration. Dry samples were rather rigid (tensile strength 11.9 ± 3.9 MPa, modulus 354 ± 180 MPa, and elongation at break 9.0 ± 6.4%) as compared to hydrated yarns (tensile strength 4.0 ± 0.1 MPa, modulus 8.6 ± 0.9 MPa, and elongation at break 73 ± 15%).

Electrospun silk fibers are mainly used for applications in the field of tissue engineering as scaffolds for cell growth, for example, as nerve guide conduits[244] or vascular grafts,[245] and are already applied in *in vitro* and *in vivo* experiments. Additionally, their potential is examined in a wide range of technical applications, from a multifunctional interlayer for lithium–sulfur batteries[246] to a bioresorbable, high-performance air filter system.[226]

10.4 Conclusion and Future Prospects

The processing of silk proteins to fibers with mechanical properties similar to or even better than the natural system is still challenging and thus not established in a broad industrial range. Although for RSF fibers, it was shown that the mechanical properties could be reproduced *via* technical processing, so far, no continuously processed artificial silk fiber fully reflects the mechanical gold standard of natural spider dragline silk fibers. It is known that their outstanding combination of high extensibility and strength, leading to superior toughness, can be assigned to the nanoconfinement of β-sheet crystals that make the most efficient use of their hydrogen bonds. At first glance, it might be surprising that these physical networks can form structures that are mechanically superior to covalently crosslinked elastomers such as rubber, but simulations demonstrated the underlying uniform shear deformations and the emergence of dissipative molecular stick-slip deformation creating such outstanding toughness.[247,248] This behavior is highly dependent on crystal size and distribution, which on the one hand explains why spider silk (smaller crystals highly distributed) exceed the mechanical performance of SF fibers and on the other hand emphasizes the theory that the reproduction of such properties can only be achieved by controlling the level of secondary and tertiary structure in an anisotropic manner. Based on the here presented findings, we conclude with five biomimetically inspired hypotheses that simultaneously must be synergized to make use of the extraordinary potential of elastomeric artificial silk fibers:

Genetic engineering can unlock the mechanical potential: Recombinantly produced spider silk proteins are mostly built of crystal-forming polyalanine blocks and amorphous glycine-rich domains based on simplified repetitive core constructs. Several groups demonstrated that increasing the MW of these constructs can significantly increase the mechanical performance of the resulting fibers by strengthening the intermolecular interaction.[87,92] Additionally, modifying artificial spidroins with nonrepetitive terminal domains as well as applying composite approaches with different proteins can facilitate protein storage and induce biomimetic structural assembly mechanisms.

Aqueous solvent systems allow for nanofibrilization: Though organic solvents are often used to dissolve silk proteins, these solvents were demonstrated to potentially degrade the MW over time and effectively inhibit the formation of nanofibrils in the spinning dope. This initial nanofibrilization, though, is a key element in creating the complex hierarchical structure of small crystal size that leads to a distinct mesophase formation needed to provide the mechanical stability on a macroscale. Aqueous solvent systems provide the benefits of long-time storage and the capability of inducing mesophase formation by allowing high concentrations of the protein, leading to the conclusion that water management during the spinning process is a crucial factor for producing high-performance fibers.

Ion and pH gradients must be applied: Most processes are based either on the evaporation of the volatile organic solvent or coagulation in an anti-solvent. This neglects the effects of gradual ion exchange and pH drop playing an important role in the assembly mechanisms of the natural spinning apparatus. Using microfluidics, advanced control can be applied and pH and chaotropic salt composition can be precisely controlled in a gradual manner, leading to better control over the coagulation process and a higher degree of hierarchical, anisotropic structuring.

Shear forces induce orientation: The spinning geometry is usually neglected and standard cylindrical or, less frequently, conical spinnerets are used to extrude the spinning dope into a coagulation medium. To precisely mimic the shear forces occurring in natural spinning glands, the narrowing funnel diameter should be incorporated in the process design to gradually increase shear forces and induce stretching of the preassembled micellar structures to achieve mechanically favorable molecular orientation.

Poststretching increases the toughness: The easiest way to significantly increase the mechanical performance of silk fibers was demonstrated to be poststretching and the respective decrease in fiber diameter.[168] Strikingly, comparing various results (Figure 10.7), it becomes obvious that high toughness is only achievable with fiber diameters below 50 μm and draw

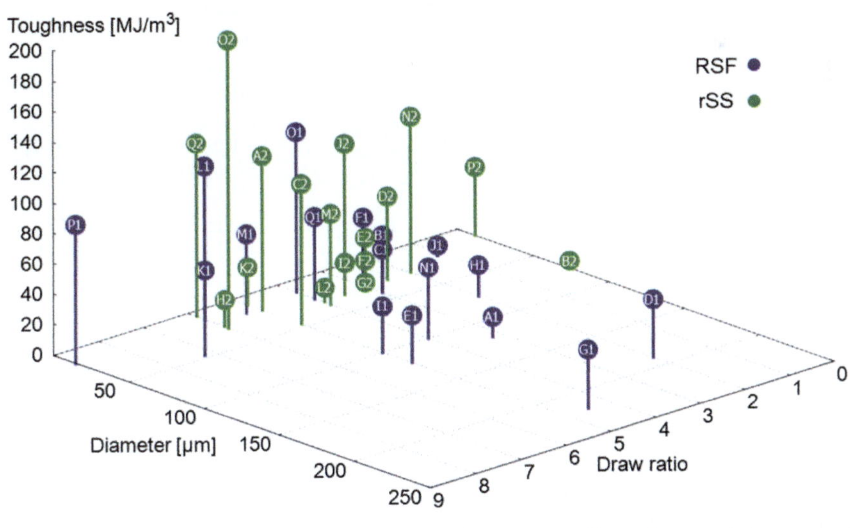

Figure 10.7 3D plot of a selected data set of wet-spun fibers from Table 10.2, with diameter and draw ratio plotted against the toughness. The highest concentration of fibers with high toughness values can be found for diameters smaller than 50 μm and draw ratios between 2 and 6.

Table 10.2 Selection of reported mechanical data of fibers generated from wet-spun regenerated silk fibroin and recombinant spider silk.

Label	Draw ratio	Ø [μm]	Toughness [MJ m^{-3}]	References
Wet spinning of regenerated silk fibroin				
A1	3.2	118.5	12.9	Matsumoto *et al.* (1996)[195]
B1	3	40	37.6	Yao *et al.* (2002)[176]
C1	3	40	28.2	Zhao *et al.* (2003)[196]
D1	2	189	38	Um *et al.* (2004)[249]
E1	5	119	30.6	Um *et al.* (2004)[250]
F1	2.7	18.5	38.9	Marsano *et al.* (2005)[251]
G1	4.5	220	38.7	Lee *et al.* (2007)[252]
H1	2	73	20.3	Corsini *et al.* (2007)[253]
I1	5	100	30.4	Ki *et al.* (2007)[185]
J1	1.1	20	6.8	Sohn & Gido (2009)[254]
K1	7.2	47	55.5	Plaza *et al.* (2009)[255]
L1	6	10.8	100.6	Yan *et al.* (2010)[181]
M1	5.3	18.4	51.9	Plaza *et al.* (2012)[256]
N1	4	100	46.4	Ling *et al.* (2012)[182]
O1	4	12.8	105.3	Zhang *et al.* (2015)[186]
P1	9	15	91	Fang *et al.* (2016)[183]
Q1	4	25	53.5	Chen *et al.* (2016)[257]
Wet spinning of recombinant spider silk				
A2	5	20	101.4	Lazaris *et al.* (2002)[72]
B2	0	74.1	4.7	Brooks *et al.* (2008)[258]
C2	5	46	91.7	Elices *et al.* (2011)[259]
D2	2.5	28.3	54.6	Teule *et al.* (2012)[81]
E2	2.5	14	22.6	Teule *et al.* (2012)[81]
F2	3	29.1	17.4	An *et al.* (2012)[260]
G2	3	29.1	2.5	An *et al.* (2012)[260]
H2	6	24.5	17.4	Gnesa *et al.* (2012)[261]
I2	3.5	30.5	20.7	Gnesa *et al.* (2012)[261]
J2	3	15.1	89.1	Adrianos *et al.* (2013)[189]
K2	5	10	24.4	Lin *et al.* (2013)[262]
L2	4	31.5	9.3	Albertson *et al.* (2014)[263]
M2	4	36	59.3	Alberston *et al.* (2014)[263]
N2	2	29	102.5	Copeland *et al.* (2015)[85]
O2	6	27	189	Heidebrecht *et al.* (2015)[92]
P2	0	12	45	Andersson *et al.* (2017)[264]
Q2	6	5.7	114	Bowen *et al.* (2018)[87]

ratios between 2 and 6, underlining the hypothesis that the natural fiber diameter should be matched for high-performance fibers. The application of postdrawing enhances the degree of molecular orientation and can further increase the β-sheet content in the processed fibers to values close to that of natural silk.

Although the toughness cannot be taken as the sole factor to quantify the mechanical properties of a fiber, it provides an indication of the mechanical potential since it is calculated as the area under the stress–strain curve and therefore takes both values (strength and extensibility) into account.

Table 10.3 Comparison of key mechanical values of natural dragline silk and synthetic constructs.

Source		Modulus [GPa]	Strength [MPa]	Extensibility [%]	Toughness [MJ m^{-3}]
Dragline silk[92]	*A. diadematus*	8 ± 2	1183 ± 334	24 ± 8	167 ± 65
Dragline silk[265]	*N. edulis*	7.9 ± 1.8	1150 ± 20	39 ± 8	214.5 ± 39
Dragline silk[266]	*N. clavipes*	13.8 ± 3.6	1215 ± 233	17 ± 3.5	111.2 ± 30.5
N1L(AQ)12NR3[92]	*A. diadematus*	4 ± 1	370 ± 59	110 ± 25	189 ± 33
Masp1 192-mer[87]	*N. clavipes*	13.7 ± 3	1031 ± 111	18 ± 6	114 ± 51

Comparing two high-performing synthetic spider silk fibers with their natural counterparts (Table 10.3) indicates the diversity of mechanical compositions that can be found in silk fiber systems resulting in the same toughness range.

While Heidebrecht *et al.* achieved the highest published toughness of 189 MJ m^{-3} *via* an extremely high extensibility (110%) using an aqueous solvent system combined with an isopropanol-containing coagulation bath, they observed comparably low values of strength (370 MPa) and modulus (4 GPa).[92] In contrast, Bowen *et al.* reported excellent values of strength and modulus of 1031 MPa and 13.7 GPa, respectively, for fibers spun from HFIP, but these fibers displayed a lower extensibility of 18%, resulting in a toughness of 114 MJ m^{-3}.[87]

Nevertheless, although these promising results demonstrate that application of such high-performance fibers in a broad industrial field is within reach (also indicated by the presence of biotech companies working on their commercialization), there are still some remaining challenges such as process continuity and scalability or the storage of spinning dopes without loss in quality (such as MW). It is highly probable that this can be reached by a systematical application of biomimetic principles. Though simultaneously fulfilling these hypotheses might be challenging, it will not only affect the mechanical properties of the resulting fibers but will also lead to desirable low-pressure and low-temperature processes, avoiding the use of organic solvents. Importantly, regarding potential applications, it must be considered that the mechanical properties of such silk fibers are highly dependent on humidity with water acting as a softener.[192,267] In the natural system, the mechanically stable inner core is protected by an outer shell consisting of varying mixtures of spidroins, glycoproteins, and lipids, which is suspected to also balance the internal state of hydration.[268] Hydration is known to play a crucial role in the energy dissipation function of many bioelastomers/elastomeric proteins (*e.g.*, collagen) and silk fibers have evolved to function even under varying environmental conditions.[269] To ensure a reliable performance of artificial silk fibers and make use of their elastomeric properties, further examinations will be required to address this aspect.

Their multifunctional, highly advanced properties undoubtedly render silk materials exceptionally desirable for numerous future applications.

Considering the tireless scientific efforts invested into their exploration and exploitation, it becomes clear that this trend will proceed, and silk might play a pioneering role in the field of the biologization of material science. This means, making use of bottom-up strategies that apply low energy self-assembly mechanisms to result in highly functional and programmable mesostructured materials not only might revolutionize the field of biopolymer processing but also offers opportunities to redefine the established paradigms of various manufacturing principles. Driven by ancient fascination and promoted by the economic potential in various sectors (medicine, textiles, photonics, electronic devices, *etc.*), unraveling the remaining secrets of silk and translating them to human technologies is still ongoing but will be evolved iteratively lab by lab.

References

1. T. D. Sutherland, J. H. Young, S. Weisman, C. Y. Hayashi and D. J. Merritt, *Annu. Rev. Entomol.*, 2010, **55**, 171–188.
2. T. A. Blackledge, N. Scharff, J. A. Coddington, T. Szuts, J. W. Wenzel, C. Y. Hayashi and I. Agnarsson, *Proc. Natl. Acad. Sci. U. S. A.*, 2009, **106**, 5229–5234.
3. C. Holland, K. Numata, J. Rnjak-Kovacina and F. P. Seib, *Adv. Healthcare Mater.*, 2019, **8**(1), 1800465.
4. R. Lewis, *Bioscience*, 1996, **46**, 636–638.
5. J. Nagaraju and M. R. Goldsmith, *Curr. Sci.*, 2002, **83**, 415–425.
6. B. D. Liu, C. Feng, L. Fang and B. Nie, *Textile Bioengineering and Informatics Symposium Proceedings*, 2010, vol. 1–3, pp. 1096–1100.
7. G. Banche, J. Roana, N. Mandras, M. Amasio, C. Gallesio, V. Allizond, A. Angeretti, V. Tullio and A. M. Cuffini, *J. Oral Maxillofac. Surg.*, 2007, **65**, 1503–1507.
8. D. Gogoi, A. J. Choudhury, J. Chutia, A. R. Pal, M. Khan, M. Choudhury, P. Pathak, G. Das and D. S. Patil, *Biopolymers*, 2014, **101**, 355–365.
9. Y. Y. Jo, H. Kweon, D. W. Kim, M. K. Kim, S. G. Kim, J. Y. Kim, W. S. Chae, S. P. Hong, Y. H. Park, S. Lee and J. Y. Choi, *Sci. Rep.*, 2017, **7**, 42441.
10. C. P. Gulka, J. E. Brown, J. E. M. Giordano, J. E. Hickey, M. P. Montero, A. Hoang and T. L. Carroll, *Laryngoscope*, 2019, **129**, 1856–1862.
11. T. Yucel, M. L. Lovett and D. L. Keplan, *J. Controlled Release*, 2014, **190**, 381–397.
12. J. M. Gosline, P. A. Guerette, C. S. Ortlepp and K. N. Savage, *J. Exp. Biol.*, 1999, **202**, 3295–3303.
13. O. Tokareva, V. A. Michalczechen-Lacerda, E. L. Rech and D. L. Kaplan, *Microb. Biotechnol.*, 2013, **6**, 651–663.
14. G. Lang, H. Herold and T. Scheibel, *Subcell. Biochem.*, 2017, **82**, 527–573.
15. V. R. Silva, M. Ribani, M. L. Gimenes and A. P. Scheer, *Chisa*, 2012, **42**, 833–841.

16. R. I. Kunz, R. M. C. Brancalhao, L. D. C. Ribeiro and M. R. M. Natali, *BioMed Res. Int.*, 2016, 8175701.
17. T. V. Chirila, S. Suzuki and N. C. McKirdy, *Prog. Biomater.*, 2016, **5**, 135–145.
18. P. Aramwit, S. Kanokpanont, W. De-Eknamkul and T. Srichana, *J. Biosci. Bioeng.*, 2009, **107**, 556–561.
19. P. Aramwit, P. Towiwat and T. Srichana, *Nat. Prod. Commun.*, 2013, **8**, 501–504.
20. S. Sarovart, B. Sudatis, P. Meesilpa, B. P. Grady and R. Magaraphan, *Rev. Adv. Mater. Sci.*, 2003, **5**, 193–198.
21. V. K. Verma and S. Subbiah, *J. Chem. Technol. Biotechnol.*, 2019, **94**, 3637–3649.
22. R. Pedregal-Cortes, G. Toriz, E. Delgado and G. H. Pollack, *Biofouling*, 2019, **35**, 732–741.
23. C. M. Zhou, Y. L. Shi, C. S. Sun, S. C. Yu, M. H. Liu and C. J. Gao, *J. Membr. Sci.*, 2014, **471**, 381–391.
24. Y. Qi, H. Wang, K. Wei, Y. Yang, R. Y. Zheng, I. S. Kim and K. Q. Zhang, *Int. J. Mol. Sci.*, 2017, **18**(3), 237.
25. B. Kundu, R. Rajkhowa, S. C. Kundu and X. Wang, *Adv. Drug Delivery Rev.*, 2013, **65**, 457–470.
26. A. E. Thurber, F. G. Omenetto and D. L. Kaplan, *Biomaterials*, 2015, **71**, 145–157.
27. Y. Cao and B. C. Wang, *Int. J. Mol. Sci.*, 2009, **10**, 1514–1524.
28. W. W. Huang, S. J. Ling, C. M. Li, F. G. Omenetto and D. L. Kaplan, *Chem. Soc. Rev.*, 2018, **47**, 6486–6504.
29. P. Cebe, X. Hu, D. L. Kaplan, E. Zhuravlev, A. Wurm, D. Arbeiter and C. Schick, *Sci. Rep.*, 2013, **3**(1), 1–7.
30. X. Hu, D. Kaplan and P. Cebe, *Macromolecules*, 2006, **39**, 6161–6170.
31. X. Hu, D. Kaplan and P. Cebe, *Thermochim. Acta*, 2007, **461**, 137–144.
32. C. C. Guo, C. M. Li, H. V. Vu, P. Hanna, A. Lechtig, Y. M. Qiu, X. Mu, S. J. Ling, A. Nazarian, S. J. Lin and D. L. Kaplan, *Nat. Mater.*, 2020, **19**, 102.
33. S. D. Aznar-Cervantes, A. Pagan, B. Monteagudo Santesteban and J. L. Cenis, *Sci. Rep.*, 2019, **9**, 6703.
34. E. K. Nguku, S. K. Raina, K. G. Mburugu and O. M. Mugenda, *ITC&DC: 4th International Textile Clothing & Design Conference, Book of Proceedings*, 2008, p. 122.
35. G. Freddi, R. Mossotti and R. Innocenti, *J. Biotechnol.*, 2003, **106**, 101–112.
36. L. S. Wray, X. Hu, J. Gallego, I. Georgakoudi, F. G. Omenetto, D. Schmidt and D. L. Kaplan, *J. Biomed. Mater. Res. B*, 2011, **99b**, 89–101.
37. K. Nultsch, L. K. Bast, M. Naf, S. El Yakhlifi, N. Bruns and O. Germershaus, *Macromol. Mater. Eng.*, 2018, **303**(12), 1800408.
38. H. Dou and B. Q. Zuo, *J. Nanomater.*, 2014, 646021.
39. B. J. Allardyce, R. Rajkhowa, R. J. Dilley, M. D. Atlas, J. Kaur and X. G. Wang, *Text. Res. J.*, 2016, **86**, 275–287.

40. L. P. Wang, Z. W. Luo, Q. Zhang, Y. P. Guan, J. Y. Cai, R. C. You and X. F. Li, *Fiber Polym.*, 2019, **20**, 45–50.
41. H. Y. Wang and Y. Q. Zhang, *Soft Matter*, 2013, **9**, 138–145.
42. D. N. Rockwood, R. C. Preda, T. Yucel, X. Wang, M. L. Lovett and D. L. Kaplan, *Nat. Protoc.*, 2011, **6**, 1612–1631.
43. M. Sah and K. Pramanik, *Proceedings of 2010 International Conference on Biotechnology and Food Science (Icbfs 2010)*, 2010, pp. 206–212.
44. M. Rodel, K. Baumann, J. Groll and U. Gbureck, *J. Tissue Eng.*, 2018, **9**, 2041731418788509.
45. H. L. Xu, D. L. ZhuGe, P. P. Chen, M. Q. Tong, M. T. Lin, X. Jiang, Y. W. Zheng, B. Chen, X. K. Li and Y. Z. Zhao, *Drug Delivery*, 2018, **25**, 364–375.
46. I. Greving, C. Dicko, A. Terry, P. Callow and F. Vollrath, *Soft Matter*, 2010, **6**, 4389–4395.
47. Z. Zheng, S. Guo, Y. Liu, J. Wu, G. Li, M. Liu, X. Wang and D. Kaplan, *J. Biomater. Appl.*, 2016, **31**, 450–463.
48. D. Pawcenis, M. A. Koperska, J. M. Milczarek, T. Lojewski and J. Lojewska, *Appl. Phys. A Mater.*, 2014, **114**, 301–308.
49. K. S. Hossain, A. Ochi, E. Ooyama, J. Magoshi and N. Nemoto, *Biomacromolecules*, 2003, **4**, 350–359.
50. K. S. Hossain, E. Ohyama, A. Ochi, J. Magoshi and N. Nemoto, *J. Phys. Chem. B*, 2003, **107**, 8066–8073.
51. M. Xu and R. V. Lewis, *Proc. Natl. Acad. Sci. U. S. A.*, 1990, **87**, 7120–7124.
52. C. Y. Hayashi, N. H. Shipley and R. V. Lewis, *Int. J. Biol. Macromol.*, 1999, **24**, 271–275.
53. M. B. Hinman and R. V. Lewis, *J. Biol. Chem.*, 1992, **267**, 19320–19324.
54. P. A. Guerette, D. G. Ginzinger, B. H. Weber and J. M. Gosline, *Science*, 1996, **272**, 112–115.
55. J. E. Garb, T. DiMauro, V. Vo and C. Y. Hayashi, *Science*, 2006, **312**, 1762.
56. M. McGill, G. P. Holland and D. L. Kaplan, *Macromol. Rapid Commun.*, 2019, **40**(1), e1800390.
57. T. A. Blackledge, J. Perez-Rigueiro, G. R. Plaza, B. Perea, A. Navarro, G. V. Guinea and M. Elices, *Sci. Rep.*, 2012, **2**, 782.
58. R. Beckwitt and S. Arcidiacono, *J. Biol. Chem.*, 1994, **269**, 6661–6663.
59. A. Sponner, E. Unger, F. Grosse and K. Weisshart, *Biomacromolecules*, 2004, **5**, 840–845.
60. A. Rising, G. Hjalm, W. Engstrom and J. Johansson, *Biomacromolecules*, 2006, **7**, 3120–3124.
61. J. E. Garb, N. A. Ayoub and C. Y. Hayashi, *BMC Evol. Biol.*, 2010, **10**, 243.
62. M. A. Collin, T. H. Clarke, 3rd, N. A. Ayoub and C. Y. Hayashi, *Int. J. Biol. Macromol.*, 2018, **113**, 829–840.
63. A. A. Walker, C. Holland and T. D. Sutherland, *Proc. R. Soc. B*, 2015, **282**, 20150259.
64. T. Scheibel, *Microb. Cell Fact.*, 2004, **3**, 14.

65. C. Vendrely and T. Scheibel, *Macromol. Biosci.*, 2007, **7**, 401–409.
66. S. Arcidiacono, C. Mello, D. Kaplan, S. Cheley and H. Bayley, *Appl. Microbiol. Biotechnol.*, 1998, **49**, 31–38.
67. M. Humenik, A. M. Smith and T. Scheibel, *Polymers*, 2011, **3**, 640–661.
68. R. V. Lewis, M. Hinman, S. Kothakota and M. J. Fournier, *Protein Expression Purif.*, 1996, **7**, 400–406.
69. S. R. Fahnestock and L. A. Bedzyk, *Appl. Microbiol. Biotechnol.*, 1997, **47**, 33–39.
70. S. R. Fahnestock and S. L. Irwin, *Appl. Microbiol. Biotechnol.*, 1997, **47**, 23–32.
71. J. Scheller, K. H. Guhrs, F. Grosse and U. Conrad, *Nat. Biotechnol.*, 2001, **19**, 573–577.
72. A. Lazaris, S. Arcidiacono, Y. Huang, J. F. Zhou, F. Duguay, N. Chretien, E. A. Welsh, J. W. Soares and C. N. Karatzas, *Science*, 2002, **295**, 472–476.
73. R. F. Service, *Science*, 2002, **295**, 419–+.
74. R. Menassa, Z. Hong, C. N. Karatzas, A. Lazaris, A. Richman and J. Brandle, *Plant Biotechnol. J.*, 2004, **2**, 431–438.
75. J. J. Yang, L. A. Barr, S. R. Fahnestock and Z. B. Liu, *Transgenic Res.*, 2005, **14**, 313–324.
76. Y. G. Miao, Y. S. Zhang, K. Nakagaki, T. F. Zhao, A. C. Zhao, Y. Meng, M. Nakagaki, E. Y. Park and K. Maenaka, *Appl. Microbiol. Biotechnol.*, 2006, **71**, 192–199.
77. Y. G. Miao, A. C. Zhao, Y. S. Zhang, K. Nakagaki, Y. Meng, T. F. Zhao and M. Nakagaki, *J. Appl. Entomol.*, 2006, **130**, 297–301.
78. H. T. Xu, B. L. Fan, S. Y. Yu, Y. H. Huang, Z. H. Zhao, Z. X. Lian, Y. P. Dai, L. L. Wang, Z. L. Liu, J. Fei and N. Li, *Anim. Biotechnol.*, 2007, **18**, 1–12.
79. X. X. Xia, Z. G. Qian, C. S. Ki, Y. H. Park, D. L. Kaplan and S. Y. Lee, *Proc. Natl. Acad. Sci. U. S. A.*, 2010, **107**, 14059–14063.
80. H. Wen, X. Lan, Y. Zhang, T. Zhao, Y. Wang, Z. Kajiura and M. Nakagaki, *Mol. Biol. Rep.*, 2010, **37**, 1815–1821.
81. F. Teule, Y. G. Miao, B. H. Sohn, Y. S. Kim, J. J. Hull, M. J. Fraser, R. V. Lewis and D. L. Jarvis, *Proc. Natl. Acad. Sci. U. S. A.*, 2012, **109**, 923–928.
82. W. A. Gaines and W. R. Marcotte, *AATCC Rev.*, 2011, **11**, 75–79.
83. V. Hauptmann, N. Weichert, M. Menzel, D. Knoch, N. Paege, J. Scheller, U. Spohn, U. Conrad and M. Gils, *Transgenic Res.*, 2013, **22**, 369–377.
84. N. Weichert, V. Hauptmann, M. Menzel, K. Schallau, P. Gunkel, T. C. Hertel, M. Pietzsch, U. Spohn and U. Conrad, *Plant Biotechnol. J.*, 2014, **12**, 265–275.
85. C. G. Copeland, B. E. Bell, C. D. Christensen and R. V. Lewis, *ACS Biomater. Sci. Eng.*, 2015, **1**, 577–584.
86. C. A. Peng, J. Russo, C. Gravgaard, H. McCartney, W. Gaines and W. R. Marcotte, *Transgenic Res.*, 2016, **25**, 517–526.
87. C. H. Bowen, B. Dai, C. J. Sargent, W. Bai, P. Ladiwala, H. Feng, W. Huang, D. L. Kaplan, J. M. Galazka and F. Zhang, *Biomacromolecules*, 2018, **19**, 3853–3860.

88. D. Huemmerich, T. Scheibel, F. Vollrath, S. Cohen, U. Gat and S. Ittah, *Curr. Biol.*, 2004, **14**, 2070–2074.
89. D. Huemmerich, C. W. Helsen, S. Quedzuweit, J. Oschmann, R. Rudolph and T. Scheibel, *Biochemistry*, 2004, **43**, 13604–13612.
90. S. Ittah, S. Cohen, S. Garty, D. Cohn and U. Gat, *Biomacromolecules*, 2006, **7**, 1790–1795.
91. D. M. Widmaier, D. Tullman-Ercek, E. A. Mirsky, R. Hill, S. Govindarajan, J. Minshull and C. A. Voigt, *Mol. Syst. Biol.*, 2009, **5**, 309.
92. A. Heidebrecht, L. Eisoldt, J. Diehl, A. Schmidt, M. Geffers, G. Lang and T. Scheibel, *Adv. Mater.*, 2015, **27**, 2189–2194.
93. T. Deptuch and H. Dams-Kozlowska, *Materials*, 2017, **10**(12), 1417.
94. S. Wohlrab, S. Muller, A. Schmidt, S. Neubauer, H. Kessler, A. Leal-Egana and T. Scheibel, *Biomaterials*, 2012, **33**, 6650–6659.
95. K. Schacht, T. Jungst, M. Schweinlin, A. Ewald, J. Groll and T. Scheibel, *Angew. Chem., Int. Ed.*, 2015, **54**, 2816–2820.
96. F. Bauer, S. Wohlrab and T. Scheibel, *Biomater. Sci.*, 2013, **1**, 1244–1249.
97. E. DeSimone, K. Schacht, A. Pellert and T. Scheibel, *Biofabrication*, 2017, **9**(4), 044104.
98. S. Kumari, G. Lang, E. DeSimone, C. Spengler, V. T. Trossmann, S. Lücker, M. Hudel, K. Jacobs, N. Krämer and T. Scheibel, *Mater. Today*, 2020, **41**, 21–33.
99. E. Bini, C. W. Foo, J. Huang, V. Karageorgiou, B. Kitchel and D. L. Kaplan, *Biomacromolecules*, 2006, **7**, 3139–3145.
100. M. Widhe, U. Johansson, C. O. Hillerdahl and M. Hedhammar, *Biomaterials*, 2013, **34**, 8223–8234.
101. C. W. P. Foo, S. V. Patwardhan, D. J. Belton, B. Kitchel, D. Anastasiades, J. Huang, R. R. Naik, C. C. Perry and D. L. Kaplan, *Proc. Natl. Acad. Sci. U. S. A.*, 2006, **103**, 9428–9433.
102. K. Numata, B. Subramanian, H. A. Currie and D. L. Kaplan, *Biomaterials*, 2009, **30**, 5775–5784.
103. M. Zorko and U. Langel, *Adv. Drug Delivery Rev.*, 2005, **57**, 529–545.
104. K. Numata, A. J. Mieszawska-Czajkowska, L. A. Kvenvold and D. L. Kaplan, *Macromol. Biosci.*, 2012, **12**, 75–82.
105. M. B. Schierling, E. Doblhofer and T. Scheibel, *Biomater. Sci.*, 2016, **4**, 1515–1523.
106. M. Lucke, I. Mottas, T. Herbst, C. Hotz, L. Romer, M. Schierling, H. M. Herold, U. Slotta, T. Spinetti, T. Scheibel, G. Winter, C. Bourquin and J. Engert, *Biomaterials*, 2018, **172**, 105–115.
107. E. Doblhofer and T. Scheibel, *J. Pharm. Sci.*, 2015, **104**, 988–994.
108. C. B. Borkner, S. Wohlrab, E. Moller, G. Lang and T. Scheibel, *ACS Biomater. Sci. Eng.*, 2017, **3**, 767–775.
109. D. Chouhan, P. Das, N. Thatikonda, S. K. Nandi, M. Hedhammar and B. B. Mandal, *ACS Biomater. Sci. Eng.*, 2019, **5**, 3537–3548.
110. H. A. Currie, O. Deschaume, R. R. Naik, C. C. Perry and D. L. Kaplan, *Adv. Funct. Mater.*, 2011, **21**, 2889–2895.

111. T. Asakura, K. Umemura, Y. Nakazawa, H. Hirose, J. Higham and D. Knight, *Biomacromolecules*, 2007, **8**, 175–181.
112. C. L. Craig, *Annu. Rev. Entomol.*, 1997, **42**, 231–267.
113. Z. M. Dong, P. Zhao, Y. Zhang, Q. R. Song, X. L. Zhang, P. C. Guo, D. D. Wang and Q. Y. Xia, *Sci. Rep.*, 2016, **6**, 21158.
114. J. E. Garb, R. A. Haney, E. E. Schwager, M. Gregoric, M. Kuntner, I. Agnarsson and T. A. Blackledge, *Commun. Biol.*, 2019, **2**, 275.
115. D. H. Hijirida, K. G. Do, C. Michal, S. Wong, D. Zax and L. W. Jelinski, *Biophys. J.*, 1996, **71**, 3442–3447.
116. H. C. Wu, S. R. Wu, T. C. K. Yang and J. C. Yang, *Materials*, 2018, **11**(10), 1951.
117. T. Asakura, K. Okushita and M. P. Williamson, *Macromolecules*, 2015, **48**, 2345–2357.
118. H. J. Jin and D. L. Kaplan, *Nature*, 2003, **424**, 1057–1061.
119. Y. X. He, N. N. Zhang, W. F. Li, N. Jia, B. Y. Chen, K. Zhou, J. Zhang, Y. Chen and C. Z. Zhou, *J. Mol. Biol.*, 2012, **418**, 197–207.
120. L. Romer and T. Scheibel, *Prion*, 2008, **2**, 154–161.
121. L. Eisoldt, C. Thamm and T. Scheibel, *Biopolymers*, 2012, **97**, 355–361.
122. Z. Lin, W. D. Huang, J. F. Zhang, J. S. Fan and D. W. Yang, *Proc. Natl. Acad. Sci. U. S. A.*, 2009, **106**, 8906–8911.
123. M. Andersson, G. Chen, M. Otikovs, M. Landreh, K. Nordling, N. Kronqvist, P. Westermark, H. Jornvall, S. Knight, Y. Ridderstrale, L. Holm, Q. Meng, K. Jaudzems, M. Chesler, J. Johansson and A. Rising, *PLoS Biol.*, 2014, **12**, e1001921.
124. L. J. Domigan, M. Andersson, K. A. Alberti, M. Chesler, Q. Xu, J. Johansson, A. Rising and D. L. Kaplan, *Insect Biochem. Mol. Biol.*, 2015, **65**, 100–106.
125. L. Zhou, X. Chen, Z. Z. Shao, Y. F. Huang and D. P. Knight, *J. Phys. Chem. B*, 2005, **109**, 16937–16945.
126. D. P. Knight and F. Vollrath, *Naturwissenschaften*, 2001, **88**, 179–182.
127. J. Sparkes and C. Holland, *Nat. Commun.*, 2017, **8**, 594.
128. D. P. Knight and F. Vollrath, *Proc. R. Soc. B*, 1999, **266**, 519–523.
129. C. Dicko, F. Vollrath and J. M. Kenney, *Biomacromolecules*, 2004, **5**, 704–710.
130. X. Chen, D. P. Knight, Z. Z. Shao and F. Vollrath, *Biochemistry*, 2002, **41**, 14944–14950.
131. D. H. Hijirida, K. G. Do, C. Michal, S. Wong, D. Zax and L. W. Jelinski, *Biophys. J.*, 1996, **71**, 3442–3447.
132. S. Sampath, T. Isdebski, J. E. Jenkins, J. V. Ayon, R. W. Henning, J. P. R. O. Orgel, O. Antipoa and J. L. Yarger, *Soft Matter*, 2012, **8**, 6713–6722.
133. G. M. Gray, A. van der Vaart, C. Guo, J. Jones, D. Onofrei, B. R. Cherry, R. V. Lewis, J. L. Yarger and G. P. Holland, *Int. J. Mol. Sci.*, 2016, **17**, 2023.
134. T. Lefevre, M. E. Rousseau and M. Pezolet, *Biophys. J.*, 2007, **92**, 2885–2895.

135. T. Asakura and J. M. Yao, *Protein Sci.*, 2002, **11**, 2706–2713.
136. M. Boulet-Audet, C. Holland, T. Gheysens and F. Vollrath, *Biomacromolecules*, 2016, **17**, 3198–3204.
137. P. R. Laity, E. Baldwin and C. Holland, *Macromol. Biosci.*, 2019, **19**(3), e1800188.
138. M. Kiyosawa, E. Ito, K. Shirai, R. Kanekatsu, M. Miura and K. Kiguchi, *Zool. Sci.*, 1999, **16**, 215–223.
139. X. Chen, Z. Z. Shao and F. Vollrath, *Soft Matter*, 2006, **2**, 448–451.
140. Y. Liu, Z. Shao and F. Vollrath, *Chem. Commun.*, 2005, 2489–2491.
141. S. W. Ha, A. E. Tonelli and S. M. Hudson, *Biomacromolecules*, 2005, **6**, 1722–1731.
142. Z. H. Zhu, K. Ohgo and T. Asakura, *eXPRESS Polym. Lett.*, 2008, **2**, 885–889.
143. A. Koeppel and C. Holland, *ACS Biomater. Sci. Eng.*, 2017, **3**, 226–237.
144. C. H. Zhao, J. M. Yao, H. Masuda, R. Kishore and T. Asakura, *Biopolymers*, 2003, **69**, 253–259.
145. L. Jeong, K. Y. Lee and W. H. Park, *Adv. Biomater. Vii*, 2007, **342–343**, 813.
146. K. Spiess, R. Ene, C. D. Keenan, J. Senker, F. Kremer and T. Scheibel, *J. Mater. Chem.*, 2011, **21**, 13594–13604.
147. I. C. Um, H. Y. Kweon, K. G. Lee and Y. H. Park, *Int. J. Biol. Macromol.*, 2003, **33**, 203–213.
148. I. C. Um, H. Y. Kweon, Y. H. Park and S. Hudson, *Int. J. Biol. Macromol.*, 2001, **29**, 91–97.
149. F. Zhang, B. Q. Zuo, Z. H. Fan, Z. G. Xie, Q. Lu, X. G. Zhang and D. L. Kaplan, *Biomacromolecules*, 2012, **13**, 798–804.
150. H. J. Zhao, X. Ren, Y. Zhang and L. Huang, *Bio-Med. Mater. Eng.*, 2015, **26**, 89–94.
151. Z. Liu, F. Zhang, J. F. Ming, S. Y. Bie, J. J. Li and B. Q. Zuo, *J. Appl. Polym. Sci.*, 2015, **132**, 41236.
152. X. X. Yue, F. Zhang, H. Y. Wu, J. F. Ming, Z. H. Fan and B. Q. Zuo, *Mater. Lett.*, 2014, **128**, 175–178.
153. Q. Lu, H. S. Zhu, C. C. Zhang, F. Zhang, B. Zhang and D. L. Kaplan, *Biomacromolecules*, 2012, **13**, 826–832.
154. H. J. Jin and D. L. Kaplan, *Nature*, 2003, **424**, 1057–1061.
155. P. Chen, H. S. Kim, C. Y. Park, H. S. Kim, I. J. Chin and H. J. Jin, *Macromol. Res.*, 2008, **16**, 539–543.
156. G. Askarieh, M. Hedhammar, K. Nordling, A. Saenz, C. Casals, A. Rising, J. Johansson and S. D. Knight, *Nature*, 2010, **465**(7295), 236–238.
157. A. Heidebrecht, L. Eisoldt, J. Diehl, A. Schmidt, M. Geffers, G. Lang and T. Scheibel, *Adv. Mater.*, 2015, **27**(13), 2189–2194.
158. L. Eisoldt, J. G. Hardy, M. Heim and T. R. Scheibel, *J. Struct. Biol.*, 2010, **170**, 413–419.
159. K. Zheng and S. Ling, *Biotechnol. J.*, 2019, **14**, e1700753.
160. N. Kronqvist, M. Otikovs, V. Chmyrov, G. Chen, M. Andersson, K. Nordling, M. Landreh, M. Sarr, H. Jornvall, S. Wennmalm,

J. Widengren, Q. Meng, A. Rising, D. Otzen, S. D. Knight, K. Jaudzems and J. Johansson, *Nat. Commun.*, 2014, **5**, 3254.
161. A. Kumar and S. Sinha-Ray, *Fibers*, 2018, **6**(3), 45.
162. R. Vasireddi, J. Kruse, M. Vakili, S. Kulkarni, T. F. Keller, D. C. F. Monteiro and M. Trebbin, *Sci. Rep.*, 2019, **9**, 14297.
163. Y. Jun, E. Kang, S. Chae and S. H. Lee, *Lab Chip*, 2014, **14**, 2145–2160.
164. C. M. Clarkson and J. P. Youngblood, *Green Mater.*, 2018, **6**, 47.
165. L. Shang, Y. Yu, Y. Liu, Z. Chen, T. Kong and Y. Zhao, *ACS Nano*, 2019, **13**, 2749–2772.
166. A. Greiner and J. H. Wendorff, *Angew. Chem., Int. Ed. Engl.*, 2007, **46**, 5670–5703.
167. J.-G. Zhang and X.-M. Mo, *Front. Mater. Sci.*, 2013, **7**, 129–142.
168. A. Koeppel and C. Holland, *ACS Biomater. Sci. Eng.*, 2017, **3**, 226–237.
169. D. Li, M. M. Jacobsen, N. Gyune Rim, D. Backman, D. L. Kaplan and J. Y. Wong, *Biofabrication*, 2017, **9**, 025025.
170. Q. Peng, Y. Zhang, L. Lu, H. Shao, K. Qin, X. Hu and X. Xia, *Sci. Rep.*, 2016, **6**, 3295.
171. Q. Peng, H. Shao, X. Hu and Y. Zhang, *J. Visualized Exp.*, 2017, **127**, 56271.
172. B. N. Singh, N. N. Panda and K. Pramanik, *Int. J. Biol. Macromol.*, 2016, **87**, 201–207.
173. G. J. Esselen, *U. S. Pat.*, US1934413 A, 1933.
174. S. Yazawa, *J. Chem. Soc. Jpn*, 1960, 1428–1430.
175. X. Chen, H. Cai, S. Ling, Z. Shao and Y. Huang, *Appl. Spectrosc.*, 2012, **66**, 696–699.
176. J. Yao, H. Masuda, C. Zhao and T. Asakura, *Macromolecules*, 2002, **35**, 6–9.
177. Z. Zhu, Y. Kikuchi, K. Kojima, T. Tamura, N. Kuwabara, T. Nakamura and T. Asakura, *J. Biomater. Sci., Polym. Ed.*, 2010, **21**, 395–411.
178. N. J. Fredin, J. T. Zhang and D. M. Lynn, *Langmuir*, 2005, **21**, 5803–5811.
179. K. Yazawa, A. D. Malay, N. Ifuku, T. Ishii, H. Masunaga, T. Hikima and K. Numata, *Biomacromolecules*, 2018, **19**, 2227–2237.
180. X. Mu, Y. Wang, C. C. Guo, Y. M. Li, S. J. Ling, W. W. Huang, P. Cebe, H. H. Hsu, F. De Ferrari, X. C. Jiang, Q. B. Xu, A. Balduini, F. G. Omenetto and D. L. Kaplan, *Macromol. Biosci.*, 2020, **20**(1), e1900191.
181. J. Yan, G. Zhou, D. P. Knight, Z. Shao and X. Chen, *Biomacromolecules*, 2010, **11**, 1–5.
182. S. Ling, L. Zhou, W. Zhou, Z. Shao and X. Chen, *Mater. Lett.*, 2012, **81**, 13–15.
183. G. Fang, Y. Huang, Y. Tang, Z. Qi, J. Yao, Z. Shao and X. Chen, *ACS Biomater. Sci. Eng.*, 2016, **2**, 1992–2000.
184. G. Zhou, Z. Shao, D. P. Knight, J. Yan and X. Chen, *Adv. Mater.*, 2009, **21**, 366–370.
185. C. S. Ki, K. H. Lee, D. H. Baek, M. Hattori, I. C. Um, D. W. Ihm and Y. H. Park, *J. Appl. Polym. Sci.*, 2007, **105**, 1605–1610.

186. F. Zhang, Q. Lu, X. Yue, B. Zuo, M. Qin, F. Li, D. L. Kaplan and X. Zhang, *Acta Biomater.*, 2015, **12**, 139–145.
187. X. Zhang and Z. Pan, *Polymers*, 2019, **11**(10), 1622.
188. M. Jassal, *Indian J. Fibre Text. Res.*, 2002, 290–306.
189. S. L. Adrianos, F. Teulé, M. B. Hinman, J. A. Jones, W. S. Weber, J. L. Yarger and R. V. Lewis, *Biomacromolecules*, 2013, **14**, 1751–1760.
190. B. An, M. B. Hinman, G. P. Holland, J. L. Yarger and R. V. Lewis, *Biomacromolecules*, 2011, **12**, 2375–2381.
191. X.-X. Xia, Z.-G. Qian, C. S. Ki, Y. H. Park, D. L. Kaplan and S. Y. Lee, *Proc. Natl. Acad. Sci. U. S. A.*, 2010, **107**, 14059–14063.
192. G. Lang, H. Herold and T. Scheibel, *Sub-cellular Biochem.*, 2017, **82**, 527–573.
193. M. Sun, Y. Zhang, Y. Zhao, H. Shao and X. Hu, *J. Mater. Chem.*, 2012, **22**, 18372.
194. H. Ishizaka, *J. Srric. Sci. Jpn.*, 1989, 87–95.
195. K. Matsumoto, H. Uejima, T. Iwasaki, Y. Sano and H. Sumino, *J. Appl. Polym. Sci.*, 1996, **60**, 503–511.
196. C. Zhao, J. Yao, H. Masuda, R. Kishore and T. Asakura, *Biopolymers*, 2003, **69**, 253–259.
197. N. Kojic, M. Kojic, S. Gudlavalleti and G. McKinley, *Biomacromolecules*, 2004, **5**, 1698–1707.
198. W. Wei, Y. Zhang, Y. Zhao, J. Luo, H. Shao and X. Hu, *Mater. Sci. Eng.: C*, 2011, **31**, 1602–1608.
199. W. Wei, Y. Zhang, H. Shao and X. Hu, *J. Mater. Res.*, 2011, **26**, 1100–1106.
200. Y. Jin, Y. Zhang, Y. Hang, H. Shao and X. Hu, *J. Mater. Res.*, 2013, **28**, 2897–2902.
201. X. Yue, F. Zhang, H. Wu, J. Ming, Z. Fan and B. Zuo, *Mater. Lett.*, 2014, **128**, 175–178.
202. S. Ling, Z. Qin, C. Li, W. Huang, D. L. Kaplan and M. J. Buehler, *Nat. Commun.*, 2017, **8**, 1387.
203. H. Pan, Y. Zhang, H. Shao, X. Hu, X. Li, F. Tian and J. Wang, *J. Mater. Chem. B*, 2014, **2**, 1408.
204. C. Zhang, Y. Zhang, H. Shao and X. Hu, *ACS Appl. Mater. Interfaces*, 2016, **8**, 3349–3358.
205. D. N. Breslauer, L. P. Lee and S. J. Muller, *Biomacromolecules*, 2009, **10**, 49–57.
206. E. Kang, G. S. Jeong, Y. Y. Choi, K. H. Lee, A. Khademhosseini and S.-H. Lee, *Nat. Mater.*, 2011, **10**, 877–883.
207. J. Luo, L. Zhang, Q. Peng, M. Sun, Y. Zhang, H. Shao and X. Hu, *Int. J. Biol. Macromol.*, 2014, **66**, 319–324.
208. S. Rammensee, U. Slotta, T. Scheibel and A. R. Bausch, *Proc. Natl. Acad. Sci. U. S. A.*, 2008, **105**, 6590–6595.
209. S.-K. Chae, E. Kang, A. Khademhosseini and S.-H. Lee, *Adv. Mater.*, 2013, **25**, 3071–3078.

210. B. Hoffmann, C. Gruat-Henry, P. Mulinti, L. Jiang, B. D. Brooks and A. E. Brooks, *PLoS One*, 2018, **13**, e0195522.
211. J. Cheng, D. Park, Y. Jun, J. Lee, J. Hyun and S.-H. Lee, *Lab Chip*, 2016, **16**, 2654–2661.
212. L. Lu, S. Fan, Q. Niu, Q. Peng, L. Geng, G. Yang, H. Shao, B. S. Hsiao and Y. Zhang, *ACS Sustainable Chem. Eng.*, 2019, **7**, 14765–14774.
213. X. Zhang, M. R. Reagan and D. L. Kaplan, *Adv. Drug Delivery Rev.*, 2009, **61**, 988–1006.
214. K. Garg and G. L. Bowlin, *Biomicrofluidics*, 2011, **5**, 13403.
215. B. K. Park and I. C. Um, *Int. J. Biol. Macromol.*, 2017, **95**, 8–13.
216. B. K. Park and I. C. Um, *Int. J. Biol. Macromol.*, 2018, **106**, 1166–1172.
217. H. Cao, X. Chen, L. Huang and Z. Shao, *Mater. Sci. Eng., C*, 2009, **29**, 2270–2274.
218. J. Ayutsede, M. Gandhi, S. Sukigara, M. Micklus, H.-E. Chen and F. Ko, *Polymer*, 2005, **46**, 1625–1634.
219. S. Sukigara, M. Gandhi, J. Ayutsede, M. Micklus and F. Ko, *Polymer*, 2003, **44**, 5721–5727.
220. S. Sukigara, M. Gandhi, J. Ayutsede, M. Micklus and F. Ko, *Polymer*, 2004, **45**, 3701–3708.
221. B.-M. Min, G. Lee, S. H. Kim, Y. S. Nam, T. S. Lee and W. H. Park, *Biomaterials*, 2004, **25**, 1289–1297.
222. S. Arasteh, S. Kazemnejad, S. Khanjani, H. Heidari-Vala, M. M. Akhondi and S. Mobini, *Methods*, 2016, **99**, 3–12.
223. M. Zhu, J. Gu, L. He, F. K. Mahar, I. Kim and K. Wei, *Fibers Polym.*, 2019, **20**, 1850–1856.
224. S. Zarkoob, R. K. Eby, D. H. Reneker, S. D. Hudson, D. Ertley and W. W. Adams, *Polymer*, 2004, **45**, 3973–3977.
225. A. Leal-Egaña, G. Lang, C. Mauerer, J. Wickinghoff, M. Weber, S. Geimer and T. Scheibel, *Adv. Eng. Mater.*, 2012, **14**, B67–B75.
226. G. Lang, S. Jokisch and T. Scheibel, *J. Visualized Exp.*, 2013, e50492.
227. L. Peng, S. Jiang, M. Seuß, A. Fery, G. Lang, T. Scheibel and S. Agarwal, *Macromol. Mater. Eng.*, 2016, **301**, 48–55.
228. K. Ohgo, C. Zhao, M. Kobayashi and T. Asakura, *Polymer*, 2003, **44**, 841–846.
229. H. Jin, *Biomaterials*, 2004, **25**, 1039–1047.
230. E. Yu, H.-Y. Mi, J. Zhang, J. A. Thomson and L.-S. Turng, *J. Biomed. Mater. Res., Part A*, 2018, **106**, 985–996.
231. E. DeSimone, T. B. Aigner, M. Humenik, G. Lang and T. Scheibel, *Mater. Sci. Eng. C Mater. Biol. Appl.*, 2020, **106**, 110145.
232. R. Konwarh, P. Gupta and B. B. Mandal, *Biotechnol. Adv.*, 2016, **34**, 845–858.
233. M. Humenik, G. Lang and T. Scheibel, *Wiley Interdiscip. Rev.: Nanomed. Nanobiotechnol.*, 2018, **10**, e1509.
234. S. Fan, Y. Zhang, H. Shao and X. Hu, *Int. J. Biol. Macromol.*, 2013, **56**, 83–88.
235. Z.-X. Cai, X.-M. Mo, K.-H. Zhang, L.-P. Fan, A.-L. Yin, C.-L. He and H.-S. Wang, *Int. J. Mol. Sci.*, 2010, **11**, 3529–3539.

236. W. Qiu, Y. Huang, W. Teng, C. M. Cohn, J. Cappello and X. Wu, *Biomacromolecules*, 2010, **11**, 3219–3227.
237. G. Lang, B. R. Neugirg, D. Kluge, A. Fery and T. Scheibel, *ACS Appl. Mater. Interfaces*, 2017, **9**, 892–900.
238. J. Zhou, C. Cao, X. Ma and J. Lin, *Int. J. Biol. Macromol.*, 2010, **47**, 514–519.
239. M. Okhawilai, R. Rangkupan, S. Kanokpanont and S. Damrongsakkul, *Int. J. Biol. Macromol.*, 2010, **46**, 544–550.
240. M. Zoccola, A. Aluigi, C. Vineis, C. Tonin, F. Ferrero and M. G. Piacentino, *Biomacromolecules*, 2008, **9**, 2819–2825.
241. M. Garcia-Fuentes, A. J. Meinel, M. Hilbe, L. Meinel and H. P. Merkle, *Biomaterials*, 2009, **30**, 5068–5076.
242. K. Zhang, L. Fan, Z. Yan, Q. Yu and X. Mo, *J. Biomater. Sci., Polym. Ed.*, 2012, **23**, 1185–1198.
243. B. Aghaei-Ghareh-Bolagh, S. M. Mithieux, M. A. Hiob, Y. Wang, A. Chong and A. S. Weiss, *Acta Biomater.*, 2019, **91**, 112–122.
244. S. Shrestha, B. K. Shrestha, J. Lee, O. K. Joong, B.-S. Kim, C. H. Park and C. S. Kim, *Mater. Sci. Eng. C Mater. Biol. Appl.*, 2019, **102**, 511–523.
245. H.-Y. Mi, X. Jing, Z.-T. Li, Y.-J. Lin, J. A. Thomson and L.-S. Turng, *J. Biomed. Mater. Res., Part B*, 2019, **107**, 2397–2408.
246. K. Wu, Y. Hu, Z. Cheng, P. Pan, L. Jiang, J. Mao, C. Ni, X. Gu and Z. Wang, *J. Membr. Sci.*, 2019, **592**, 117349.
247. D. Lopez Barreiro, J. Yeo, A. Tarakanova, F. J. Martin-Martinez and M. J. Buehler, *Macromol. Biosci.*, 2019, **19**, e1800253.
248. S. Keten, Z. Xu, B. Ihle and M. J. Buehler, *Nat. Mater.*, 2010, **9**, 359–367.
249. I. C. Um, H. Kweon, K. G. Lee, D. W. Ihm, J. H. Lee and Y. H. Park, *Int. J. Biol. Macromol.*, 2004, **34**, 89–105.
250. I. C. Um, C. S. Ki, H. Kweon, K. G. Lee, D. W. Ihm and Y. H. Park, *Int. J. Biol. Macromol.*, 2004, **34**, 107–119.
251. E. Marsano, P. Corsini, C. Arosio, A. Boschi, M. Mormino and G. Freddi, *Int. J. Biol. Macromol.*, 2005, **37**, 179–188.
252. K. H. Lee, D. H. Baek, C. S. Ki and Y. H. Park, *Int. J. Biol. Macromol.*, 2007, **41**, 168–172.
253. P. Corsini, J. Perez-Rigueiro, G. V. Guinea, G. R. Plaza, M. Elices, E. Marsano, M. M. Carnasciali and G. Freddi, *J. Polym. Sci., Part B: Polym. Phys.*, 2007, **45**, 2568–2579.
254. S. Sohn and S. P. Gido, *Biomacromolecules*, 2009, **10**, 2086–2091.
255. G. R. Plaza, P. Corsini, E. Marsano, J. Pérez-Rigueiro, L. Biancotto, M. Elices, C. Riekel, F. Agulló-Rueda, E. Gallardo, J. M. Calleja and G. V. Guinea, *Macromolecules*, 2009, **42**, 8977–8982.
256. G. R. Plaza, P. Corsini, E. Marsano, J. Pérez-Rigueiro, M. Elices, C. Riekel, C. Vendrely and G. V. Guinea, *J. Polym. Sci., Part B: Polym. Phys.*, 2012, **50**, 455–465.
257. Z. Chen, H. Zhang, Z. Lin, Y. Lin, J. H. van Esch and X. Y. Liu, *Adv. Funct. Mater.*, 2016, **26**, 8978–8990.

258. A. E. Brooks, S. M. Stricker, S. B. Joshi, T. J. Kamerzell, C. R. Middaugh and R. V. Lewis, *Biomacromolecules*, 2008, **9**, 1506–1510.
259. M. Elices, G. V. Guinea, G. R. Plaza, C. Karatzas, C. Riekel, F. Agulló-Rueda, R. Daza and J. Pérez-Rigueiro, *Macromolecules*, 2011, **44**, 1166–1176.
260. B. An, J. E. Jenkins, S. Sampath, G. P. Holland, M. Hinman, J. L. Yarger and R. Lewis, *Biomacromolecules*, 2012, **13**, 3938–3948.
261. E. Gnesa, Y. Hsia, J. L. Yarger, W. Weber, J. Lin-Cereghino, G. Lin-Cereghino, S. Tang, K. Agari and C. Vierra, *Biomacromolecules*, 2012, **13**, 304–312.
262. Z. Lin, Q. Deng, X. Y. Liu and D. Yang, *Adv. Mater.*, 2013, **25**, 1216–1220.
263. A. E. Albertson, F. Teule, W. Weber, J. L. Yarger and R. V. Lewis, *J. Mech. Behav. Biomed. Mater.*, 2014, **29**, 225–234.
264. M. Andersson, Q. Jia, A. Abella, X. Y. Lee, M. Landreh, P. Purhonen, H. Hebert, M. Tenje, C. V. Robinson, Q. Meng, G. R. Plaza, J. Johansson and A. Rising, *Nat. Chem. Biol.*, 2017, **13**, 262–264.
265. F. Vollrath and D. P. Knight, *Nature*, 2001, **410**, 541–548.
266. B. O. Swanson, T. A. Blackledge, A. P. Summers and C. Y. Hayashi, *Evolution*, 2006, **60**, 2539–2551.
267. B. Aghaei-Ghareh-Bolagh, S. M. Mithieux, M. A. Hiob, Y. W. Wang, A. Chong and A. S. Weiss, *Acta Biomater.*, 2019, **91**, 112–122.
268. A. Sponner, W. Vater, S. Monajembashi, E. Unger, F. Grosse and K. Weisshart, *PLoS One*, 2007, **2**(10), e998.
269. F. Vollrath and D. Porter, *Soft Matter*, 2006, **2**, 377–385.

CHAPTER 11

Protein, Biomimetic Protein, and Designer Peptide-directed Synthesis of Metal Nanoparticles, Metal Nanoclusters and Nanobioconjugates, and Their Potential Applications

ANKIT K. DUTTA[a] AND NABA K. DUTTA*[b]

[a] Department of Medical Oncology, Dana-Farber Cancer Institute, Harvard Medical School, Boston, MA 02115, USA; [b] Chemical and Environmental Engineering, School of Engineering, RMIT University, Melbourne, VIC 3000, Australia
*Email: naba.dutta@rmit.edu.au

11.1 Introduction

Natural organisms produce a diverse range of nano-, micro- and macroscale structures as well as exquisitely complex hierarchically organized functional materials in an energy-efficient and highly reproducible manner. The properties of these materials have been controlled and optimized for their functions through evolution over millions of years. Inspired by these

Biomaterials Science Series No. 10
Biomimetic Protein Based Elastomers: Emerging Materials for the Future
Edited by Namita Roy Choudhury, Julie C. Liu and Naba K. Dutta

Published by the Royal Society of Chemistry, www.rsc.org

extraordinary natural phenomena, an evolving concept is to create organized functional nanomaterials and nanobioconjugates that can mimic such natural objects using engineered molecules, including proteins, peptides, DNAs, sequence-defined synthetic molecules (*e.g.*, peptoids), and other polymers and biomimetic polymers.[1,2] In this endeavor, the synthesis of nanoparticles (NPs), nanoclusters (NCs) and nanobioconjugates of controlled composition, size, shape, architecture, morphology, organization, and functions have been of major focus.

NPs and NCs may be classified broadly into three categories: (i) organic, (ii) inorganic, and (iii) carbon-based.[3] The chapter will be mainly focused on inorganic NPs and NCs. Almost all the metals can be synthesized into their NPs/NCs either by destructive, *i.e.*, 'top-down' or constructive, *i.e.*, 'bottom-up' approaches. The most commonly used metals for NP and NC synthesis include gold (Au), platinum (Pt), palladium (Pd), silver (Ag), aluminum (Al), cobalt (Co), copper (Cu), iron (Fe), zinc (Zn), lead (Pb), manganese (Mn), and cadmium (Cd). The specific focus of this chapter is on the noble metal NPs and NCs, their alloys, conjugates, and bioconjugates due to the extensive amount of research performed on these classes of particles and their potential applications in a wide range of fields, including nanotechnology and nanobiotechnology, medicine and drug delivery, catalysis and sensors, and optical and electronics.

The use of sustainable approaches for the synthesis and manufacturing of functional nanostructures, including metal nanoparticles (m-NPs) and metal nanoclusters (m-NCs) of predictable characteristics and functions for a variety of applications, represents a grand challenge. The use of energy-neutral and environmentally benign processes instead of traditional highly energy-consumptive processes for synthesis has the potential not only to reduce costs associated with the chemical and materials generation but also to reduce environmental impacts. For a specific m-NP of a given composition, morphology, and structure, the influence of the size, shape, and dielectric environment is the most critical parameters that dictate their characteristics and function. The protein/peptide-directed regulation and synthesis of m-NPs/m-NCs are essentially a bioinspired strategy, which is akin to the natural biomineralization process. The mineral formation approach in biology is unique in that it occurs under mild physiological conditions however, produces a wide range of products with a variety of compositions, morphologies, and properties. They are not only capable of forming NPs/NCs of different sizes, shapes, and compositions but also generate functional composites, hybrids, and incredible hierarchical structures, including elegant graded structures. In this chapter, protein-directed synthesis of inorganic NPs and NCs, their nanobioconjugates, synthesis, characteristics, functions, and applications will be the main focus. This chapter will focus on the synthesis and functionalization strategies for m-NPs and m-NCs; m-NP/m-NC-alloys; and hybrid m-NPs/m-NCs involving specifically protein/peptide as either the host or the directing agent, the reducing agent, or the stabilizing agent, or having it perform multiple of the above functions concurrently. The role of the biohost molecule, the amino acid sequence, and surface functional groups in the m-NP/m-NC formation, stability, and physical properties will be presented.

11.2 Nanoparticles (NPs) and Nanoclusters (NCs): Effect of the Size and the Shape

The characteristic properties of the bulk materials change suddenly when their size approaches the nanometer scale due to the dramatic increase in the fraction of atoms at and near the surface with the decrease in the size. The properties of m-NPs and m-NCs, including electronic, optical, electrochemical, magnetic, and catalytic activities, are vastly different from those of the bulk metal or metallic particles (size > mm) (Figure 11.1). The characteristics and properties exhibited by m-NPs are also significantly dependent on the geometric factors (such as the size and the shape), environment (such as dielectric, solvent, temperature, pH), and the capping layer used to stabilize them (Figure 11.1b–i). Nanoscale metals may be roughly classified into three size domains corresponding to three different major characteristic length scales: (i) large NPs (size $\gg$ 1 μm), (ii) NPs, m-NP ($\leq$100 nm), and (iii) nanoclusters, m-NCs ($<\sim$2 nm). The distinction between m-NPs and m-NCs is essential as they do not exhibit the same electronic and optical properties. Individual entities with at least one dimension on the length scale of 1–100 nm are referred to as NPs. NCs, on the other hand, are ultrasmall particles made up of a few to a hundred metal atoms. The size of NCs may range from a subnanometer to ~2 nm. Depending on the size, the same material can display a completely different set of properties and applications and thus different significance to its application. One of the ideal examples is gold (Au); for millennia, bulk gold has fascinated humanity, and it enjoyed extraordinary value among all the civilizations – as disparate as ancient Egypt to Ancient India – due to its scarcity, natural beauty, indestructible yellow luster, and its association with the golden divine brilliance (Figure 11.1a).[4] On the other hand, Au-NPs/Au-NCs, the most celebrated and researched of all the m-NPs/m-NCs of modern time, show a variety of optical properties depending on the size, shape, aspect ratio (AR), and dielectric environment (Figure 11.1b–d). The intense color of Au-NPs is caused by the collective excitation of their conduction electrons, or surface resonance modes, resulting in photon absorption at wavelengths that vary depending on the AR (Figure 11.1d),[5] shell thickness, and/or galvanic displacement.[6,7] In anisotropic Au-NPs, the plasmon resonance band varies with AR, which is reflected in the change in color and the UV–Vis absorbance spectra of Au-nanorods with varied ARs (Figure 11.1c,d). Panels e–i of Figure 11.1 show the corresponding transmission electron microscopy (TEM) images of the gold nanorods of different ARs presented in Figure 11.1c and d.[8] Furthermore, the color and optical features of Au-NPs

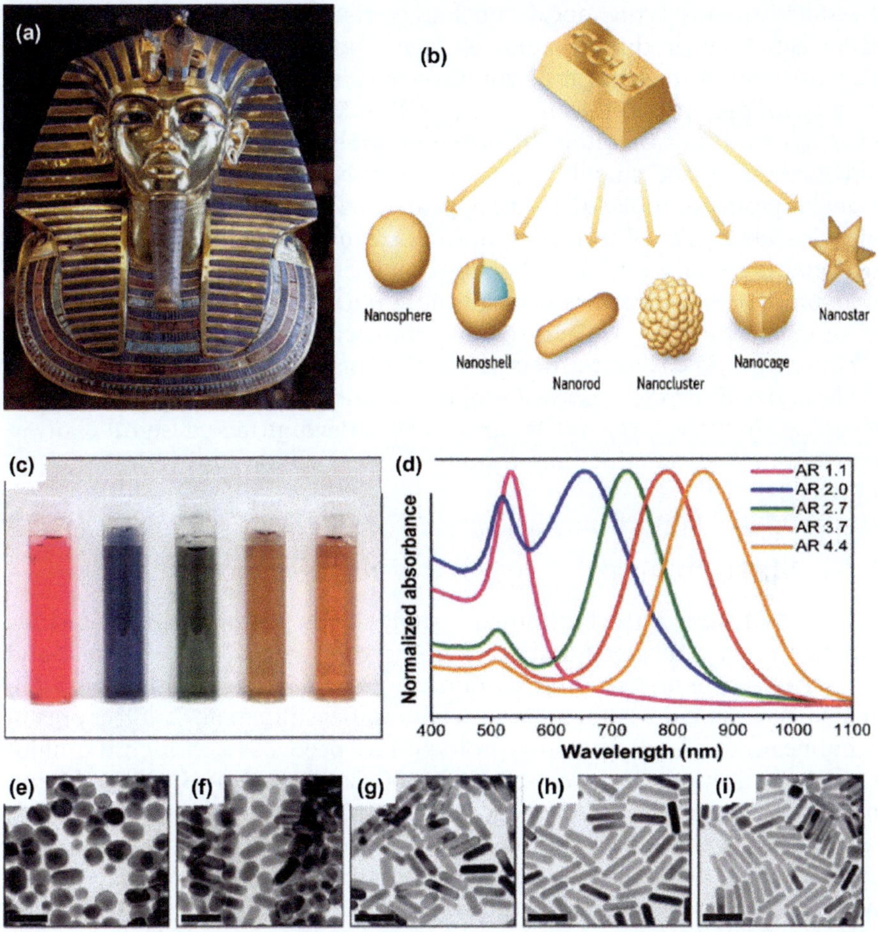

Figure 11.1 (a) Ancient Egyptian gold: Golden burial mask of Egyptian Pharaoh Tutankhamun (King Tut) of the 18th Dynasty (*ca.* 1323 BC). (b) The most common gold nanoparticle assemblies and morphologies. (c) Photographic image of gold nanorod solutions with increasing aspect ratio (AR) left to right. (d) UV–Vis absorbance spectra of gold nanorods with varied ARs. Transmission electron microscopy images of gold nanorods of different ARs: (e) AR 1.1, (f) AR 2.0, (g) AR 2.7, (h) AR 3.7, and (i) AR 4.4. Scale bars = 50 nm. Reproduced with permission: (a) from ref. 4, under the terms of the CC BY-SA 3.0 license https://creativecommons.org/licenses/by-sa/3.0/; (b) from ref. 6, https://doi.org/10.3390/nano8110939, under the terms of the CC BY 4.0 license https://creativecommons.org/licenses/by/4.0/; (c, d): from ref. 7 with permission from Springer Nature, Copyright 2007; (e–i) from ref. 8 with permission from American Chemical Society, Copyright 2014.

are highly reliant on aggregation; for example, Au-NP appears red when in a homogeneous suspension and turns blue/gray when in the aggregated state.

In contrast, Au-NCs do not exhibit characteristic red colors seen of Au-NPs; neither can they produce a color change when aggregated. Au-NCs with diameters similar to electron Fermi wavelengths (<2 nm) act like molecules and display discrete electronic features like fluorescence.[9] The emission wavelength and thus the fluorescent color correlates to the size or number of atoms present in the cluster (Figure 11.2). m-NCs appear to be fine aggregations of atoms or molecules that appear to be a transitional state between molecules and NPs. Fluorescent m-NCs are an emerging class of multifunctional materials engineered at the single-atom level that may compete with typical semiconductor quantum dots (QDs) in terms of functionality like tunable emission, ease of conjugation, extended photostability, and high quantum yield. Due to their distinctive qualities, including the display of unique size-dependent optical properties, and prospective applications in interdisciplinary sectors, m-NPs and m-NCs have attracted significant attention from the scientific community, medical sector, and industry in recent decades.[10–16]

11.3 Statistics and Network Visualization

Figure 11.3 illustrates the bibliometric analysis (Scopus database) of articles published every year since 2003 containing the search terms "metal" and "nanoparticle"; demonstrating a dramatic growth in the number of publications from 2005 to 2015. Figure 11.4 summarizes the major subject area of the publications. A color-coding protocol has been used to identify major areas of publications in groups. It appears that biomedical, environment, sensors, electronics, and analytical chemistry are the major areas of focus. The network visualization tool (Figure 11.5) was used to uncover the dynamics and cluster countries behind the publications. The countries are represented by circles in the diagram, and the size of the circle defines each element's weight. The cluster to which an item belongs determines the color of the object. Links are represented by lines between items. By default, only the strongest 1000 linkages between objects are displayed, which is limited to 1000 lines. There are contributions from 61 countries, which may be identified into 7 network clusters. Of the seven networks, the three major clusters were as follows: Cluster 1, the largest cluster of 19 countries with mainly the member states of the European Union (EU) (Argentina, Belarus, Belgium, Bulgaria, Croatia, Denmark, Finland, France, Germany, Italy, Luxemburg, Netherlands, Norway, Poland, Portugal, Russia, Spain, Sweden); second-largest Cluster 2, led by China and the United States (Australia, Canada, England, Israel, Japan, Jordan, Mexico, New Zealand, Pakistan, People's Republic of China, Scotland, Singapore, South Korea, Taiwan, Thailand, Turkey, United States, Wales), which contributed the most articles; Cluster 3, which consists of 10 countries from Asia, Middle East, and Africa (Egypt, Ethiopia, India, Iraq, Iceland, Malaysia, Saudi Arabia, Tunisia, United Arab Emirates). Out of these, ~50% of the articles have been published by the United States (21%) and China (27%) (Figure 11.5). In general,

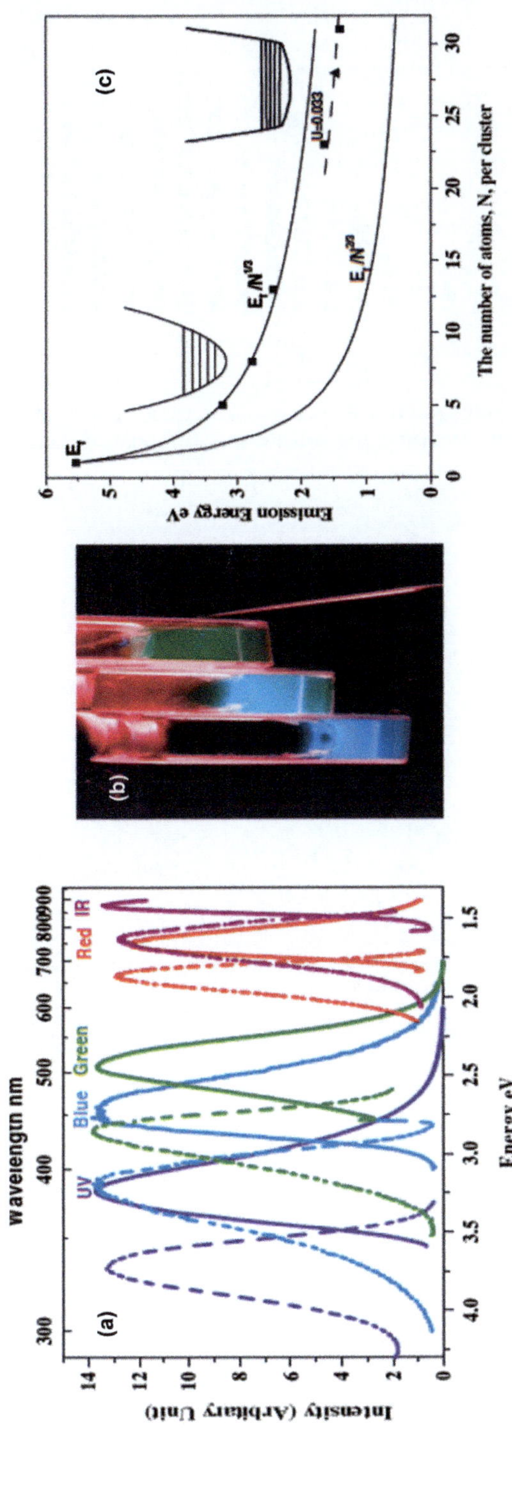

Figure 11.2 (a) Excitation (dashed) and emission (solid) spectra of different AuNCs. Excitation and emission maxima shift to longer wavelengths with increasing size leads to lower energy emission; (b) Emission from the three shortest wavelengths emitting AuNC solutions (from left to right) under long-wavelength UV lamp irradiation (366 nm); (c) Correlation of the number of atoms, N, per cluster with emission energy. Emission energy decreases with increasing number of atoms. The correlation of emission energy with N is quantitatively fit with $E_{\text{fermi}}/N^{1/3}$. Reproduced from ref. 9 with permission from American Physical Society, Copyright 2004.

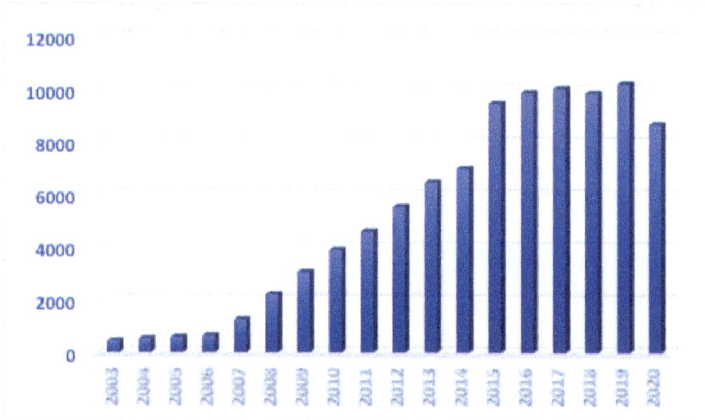

Figure 11.3 Number of Scholar references per year since 2003 containing the search terms metal and nanoparticles. Note that publication data in (a) is not cumulative.

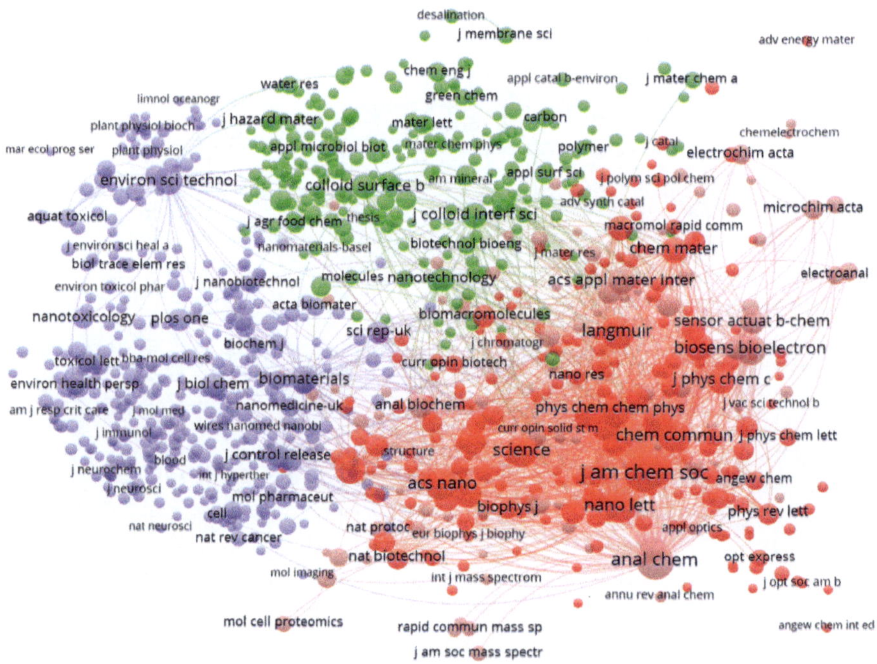

Figure 11.4 Summary of the major subject area of the publications. Color coding used: Blue – biomedical and environment, green – surface, interface, and nanotechnology, Red – sensor, analytical, and physical chemistry.

the closer two countries are to each other, the more strongly they are linked. Lines also show the strongest ties between the countries. Combined, the data shows that the United States and China are the largest contributors to the field.

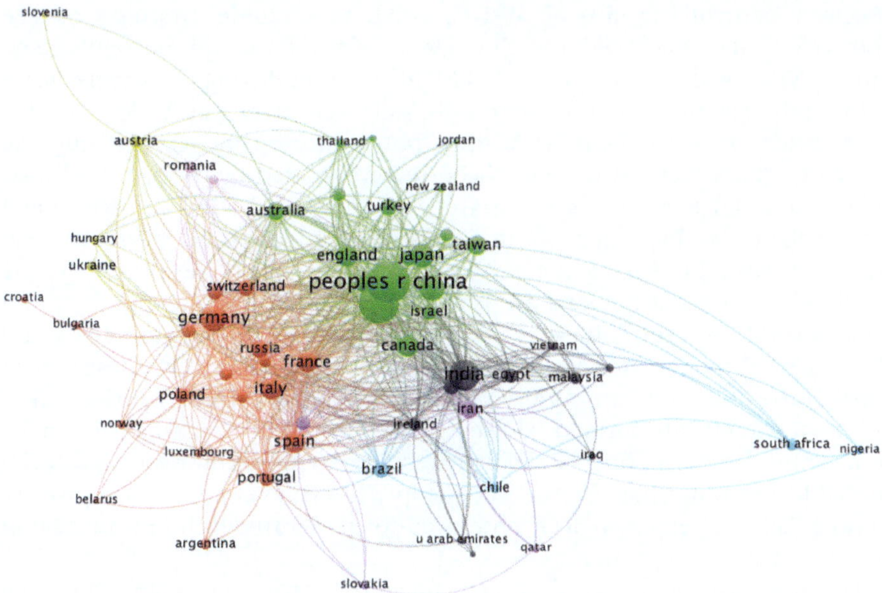

Figure 11.5 Network cluster of the world behind the research activities in the areas of functional metal nanoparticles and their applications.

11.4 Nanoparticles and Nanoclusters—A Historical Perspective and the Fundamental Principles

11.4.1 Metal Nanoparticles (m-NPs) and Localized Surface Plasmon Resonance

The creation and manipulation of colors through the use of m-NPs have been of significant interest and used over centuries. For example, the brilliant red color displayed in the medieval stained-glass windows is essentially created by the use of Au-NPs, whereas silver NPs are responsible for yellow color formation. Michael Faraday[17] initiated the first systematic scientific investigation on the synthesis of colloidal NPs and reported the formation of "beautiful ruby fluid" from the reduction of gold chloride precursor to Au-NPs using phosphorus. In 1908, Mie[18] revealed that the dazzling colors from Faraday's gold solutions were primarily from the size-dependent absorption and scattering of light from Au-NPs in the solution. The first detailed systematic structural investigation of Au-NPs, formed using varying synthetic conditions, was reported in the early 1950s.[19] They used imaging of NPs using electron microscopy for detailed characterization and visualization of the particle size, shape, and morphology. In the 1970s, the citrate-mediated growth method was developed successfully to synthesize monodisperse spherical Au-NPs over a wide range of sizes from 16 to 150 nm.[20] The ratio of citrate to the precursor $AuCl_4$ ions utilized was confirmed to be able to precisely control the size of Au-NPs. Such predictable synthesis of size-controlled Au-NPs allowed the elucidation of fundamental mechanistic details of NPs. From the studies, it was identified that during the synthesis, the color of the reaction mixture progressively changes from pale yellow ($AuCl_4^-$) to colorless (Au-atoms) to very dark blue, purple, and finally ruby red (Au-NP).[21] This method has now been recommended by the National Institute of Standards for the synthesis of size-controlled NPs. The NP growth mechanism was hypothesized to follow spontaneous nucleation and isotropic growth, known as the LaMer growth process.[22] According to this mechanism, the $AuCl^{-1}$ ions are first reduced to atomic Au^0 rapidly to the supersaturation level; following the collisions between Au^0 atoms is a sudden burst of nuclei formation (discussed in detail later). On these nuclei, most of the other gold atoms are added progressively (growth) until the final colloidal Au-NPs are formed. However, the proposed mechanism is unable to account for the nature of the transient dark coloration observed during Au-NP formation. Apart from spectroscopy, the UV–Vis absorption property of Au-NPs has been extensively employed to extract in-depth information on the size and shape of Au-NPs.[23]

The presence of localized surface plasmon resonance (LSPR) has been linked to the exhibition of distinct absorption features by Au-NPs. When electromagnetic waves (*e.g.*, light) are irradiated onto Au-NPs, the imposed oscillating electric field of light causes the conduction electrons of the NPs to oscillate coherently (Figure 11.6).[24] When Au-NPs' conduction electrons are restricted to dimensions smaller than the electron mean free pathlength (20 nm), this occurrence results in a size- and shape-dependent plasmon absorption spectrum (Figure 11.1d).[25,26] The frequency of the oscillation of

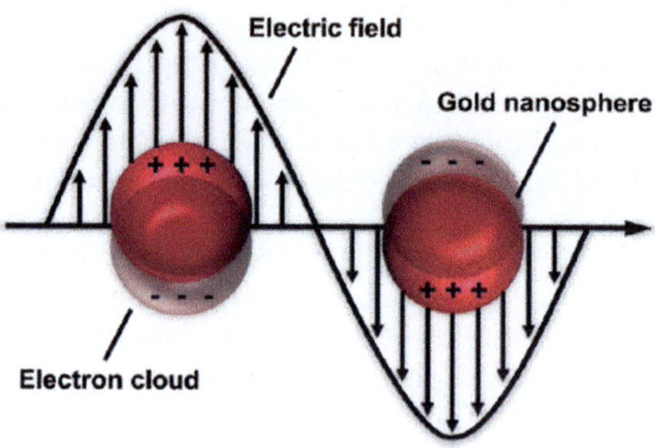

Figure 11.6 Schematic representation of the localized surface plasmon resonance on AuNPs showing the displacement of the conduction electron charge cloud relative to the nuclei. Upon illumination at resonant wavelengths, conduction band electrons on the AuNP surface are delocalized and undergo collective oscillation. Reproduced from ref. 24 with permission from American Chemical Society, Copyright 2016.

the LSPR of the m-NP is influenced by four main factors: (i) electron density, (ii) effective electron mass, (iii) charge distribution shape, and (iv) charge distribution size. Only the free electrons on the Au-NP surface, not the electrons in the core are affected by the electric field of photons. As a result, when AuNPs and bulk gold (particle size > mm) are exposed to electromagnetic waves, their valence bond electrons encounter different energies and their distributions. Hence, bulk gold does not display LSPR; nonetheless, when the particle size is reduced to < ~2 nm, the intensity of the LSPR diminishes substantially, rendering such colloids less brightly colored or colorless. While spherical Au-NPs have a single SPR band, anisotropic metal NPs like Au-nanorods have two LSPR bands, the transverse and longitudinal modes, the latter of which redshifts as the AR increases (Figure 11.1c,d).[27]

11.4.2 Metal Nanoclusters (m-NCs) and Photoluminescence

m-NCs, which are made up of a few to a hundred atoms, have lately gained a lot of attention as a new form of photoluminescent (PL) nanomaterial.[14–16,28–34] Though photoluminescence from bulk gold was first identified as early as 1969,[35] the quantum yield was very low (10^{-10}) and the emission was produced under severe conditions. Intense photoluminescence from ultrasmall m-NCs was reported in 1998.[36] Similar to semiconductor quantum dots (QDs), the m-NCs have the potential to operate as multielectron artificial atoms with discrete, size-tunable electronic transitions over the visible to near-infrared (NIR) spectrum. Many variables influence the PL of NCs, including cluster size, surface ligand architecture, cluster assembly structure, and environment (pH, temperature, dielectric environment). Figure 11.7[37] shows a schematic illustration of different routes for fluorescent Au-NCs synthesis and clearly depicts the influence of the type of ligands on determining the size of NCs and their resulting fluorescence characteristics. The figure illustrates that proteins and peptides are the most versatile material for the synthesis and tailoring of m-NCs. They can act not only as the directing agents but may perform multiple functions in the process, such as acting concurrently as the templating agent, reducing agent, and stabilizer once m-NCs are formed. The full understanding of the underlying mechanism for the PL of m-NCs at the fundamental atomic level is still lacking.[38] Three major hypotheses that have been generally accepted are the following: (i) metal quantum confinement effect (QCE);[39–41] (ii) ligand-to-metal charge transfer (LMCT);[42] and (iii) aggregation-induced enhancement (AIE) of fluorescence in NCs.[43]

According to QCE, the continuous band of the energy level of m-NPs breaks up into discrete energy levels when the size of m-NCs approaches the Fermi wavelength of metals (*i.e.*, de Broglie's wavelength of an electron at the Fermi energy, 1 nm). In comparison to NPs and bulk metal, this energy level breakup results in the formation of molecule-like features with significant PL features. Only a few to tens of metal atoms make up fluorescent m-NCs

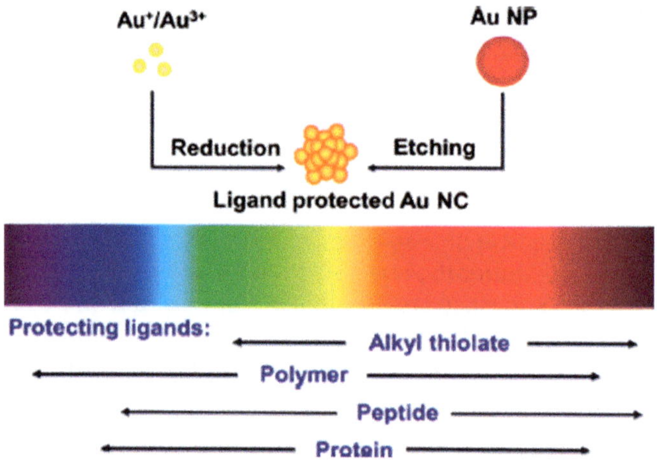

Figure 11.7 Schematic illustration of synthetic strategies for fluorescent Au-NCs and the effect of ligands on their fluorescence. Reproduced from ref. 37 with permission from American Chemical Society, Copyright 2015.

with discrete and size-tunable electronic transitions.[44,45] The number of atoms per cluster, N, is correlated to the emission energy, which decreases as the number of atoms rises. As shown in Figure 11.2c, the correlation of emission energy with N is quantitatively fit with $E_{\mathrm{fermi}}/N^{1/3}$. Figure 11.8[46] depicts general principles and fluorescence basics for molecular luminescence, as initially envisaged by Alexander Jablonski in the 1930s. A three-level model involving singlet ground states and excited states, as well as a triplet excited state (TS), describes the electronic structure of m-NCs. Borrowing the concept from molecular luminescence, TS should play a crucial role in determining the photophysical features of m-NCs, in particular, in terms of molecular brightness, on/off blinking, and photon stability. Strong PL, along with good photostability, substantial Stokes shift, and high emission rates, is a distinguishing property of these highly polarizable, few-atom m-NCs. By combining solid-state principles as well as a molecular model, it was proposed that PL in NCs originates from intraband (sp–sp) and interband (sp–d) transitions.[47] It was also revealed that the PL wavelength could be varied from the UV to NIR region by controlling the sizes of metal NCs; and a simple scaling correlation relation, $E_{\mathrm{Fermi}}/N^{1/3}$, between the size of the Au-NP and PL emission energies was established (Figure 11.2C).[40] The finding is consistent with the Jellium model, which predicts that as cluster size grows, the gap between the discrete 5d valence band and the 6sp conduction band shrinks (Figure 11.8). However, the display of size-independent PL characteristics of a series of glutathione-protected Au-NCs[48] and the ability to fine-tune the PL properties with the same number of core atoms through tuning different surface ligand coverage contradicts[49–51] the view of the classical QCE mechanism. On the contrary, the concept of LMCT

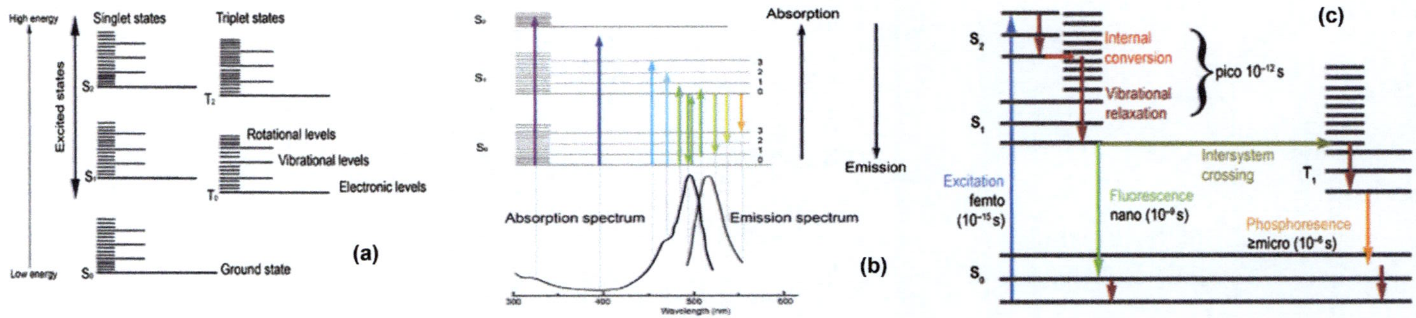

Figure 11.8 Fluorescence fundamentals. (a) Jablonski diagram displaying the energy states of a molecule. Molecules not absorbing energy are mainly confined to the lowest vibrational states of the ground state S0. (b) The spectral characteristics related to the absorption and emission of energy by a molecule can be related to the size of the energy steps needed to bring a molecule from one energy level to another. (c) A timeframe that the various steps in fluorescence excitation and emission and phosphorescence take. Reproduced from ref. 46 with permission from Springer Nature, Copyright 2005.

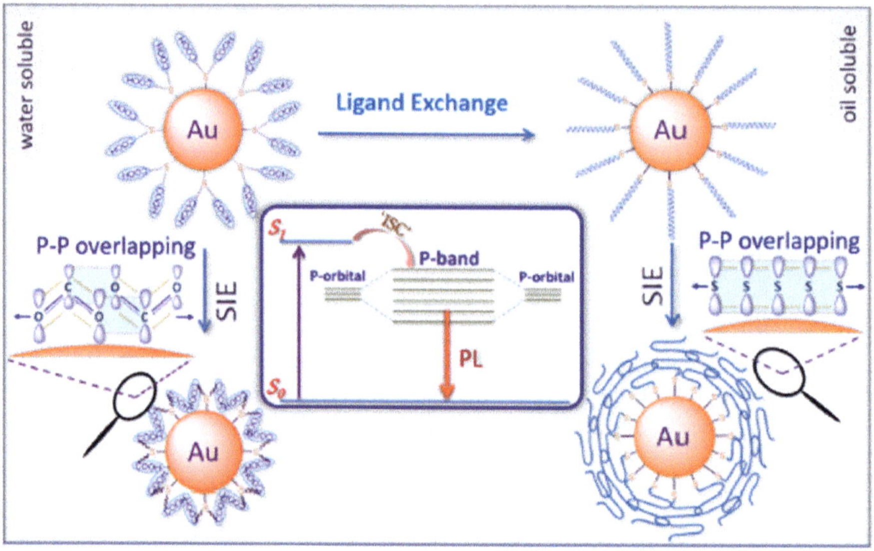

Figure 11.9 Schematic illustration of the ligand exchange process and solvent-induced emission (SIE) properties of Au NCs (inset: the energy-level structure of Au NCs in water and ethanol mixed solution). The p-band formed by the overlapping p-orbitals of electron-rich heteroatoms of well-organized surface ligands is used as an intermediate state or dark state to tune the optoelectronic properties. Reproduced from ref. 52 with permission from Springer Nature, Copyright 2019.

hypothesizes that the charge transfer on the shell of m-NCs occurs due to the interaction between the functional ligands and the m-NC core. The hypothesized LMCT mechanism is based on the notion of transition organometallic complexes emitting PL and is restricted in its ability to explain the PL features of m-NC. Recently, a hypothesis based on a p-band intermediate state (PBIS) has been hypothesized to play a major role in customizing the PL at the nanoscale interface (Figure 11.9),[52] arising from the substantial overlapping of p orbitals of the heteroatoms on the ligands through-space interactions. The main cause for the AIE effects of fluorescence in metal NCs is attributed to the restriction of intramolecular motion (RIM) and rotation of the ligand's shell on the NCs' surface after aggregation, which enable radiative energy transfer *via* restraining.[43]

11.5 Synthesis of m-NPs and m-NCs

11.5.1 Biomineralization and Biomimetic Synthesis of NPs and NCs

Biological systems provide numerous examples of highly controlled biomineralization processes that result in stabilized NPs with unique characteristics and functionalities. Nature has evolved over millennia to master the skill of creating finely ordered functional inorganic nanostructures and hierarchically organized complex structures from hard inorganic materials for a range of functions ranging from structural support to bioremediation.[53–55] In nature, biomineralization is ubiquitous and occurs in a variety of organisms, both vertebrates and invertebrates. This may be best exemplified by the selective mineralization of ferrihydrite ($FeOOH{\cdot}3H_2O$) as a labile source of iron within the iron storage protein ferritin. Ferritin is a universal intracellular protein that stores and releases iron in a controlled fashion. Biomineralizations are observed from gold biomineralization in unicellular bacteria[56] to silicon-based mineral deposition in diatoms[57] to organized calcium carbonate phases in shales[58] to hierarchical complex structures in vertebrate bones and teeth containing the calcium phosphate, hydroxyapatite (HA).[59] The unique opportunity to create advanced nanoscale materials and functional structure through facile environmentally benign protocol using mild conditions has led to a growing interest in the area of biomineralization. Organic matrix proteins and/or other biological macromolecules regulate the nucleation and development of the inorganic structure during the biomineralization process, which is critical for mineral deposition regulation.[60]

Many proteins and protein mimics found in living creatures have been used as enzymes or templates for *in vitro* material production. Both classical and combinatorial techniques have been used to identify the inorganic binding proteins. The combinatorial approaches can identify peptides that selectively recognize inorganic surfaces and have the potential to serve as templates for inorganic growth and nucleation.[61,62] A thorough structural and functional analysis of organic molecules regulating biomineralization has recently been reviewed by Suzuki.[63] Fundamental molecular research on silica, iron oxide, and calcium carbonate biomineralization is summarized by Arakaki *et al.*[64] Intriguingly, a significant study has revealed that proteins, notably intrinsically disordered proteins (IDPs), play a regulating role in the biomineralization process of both vertebrates and invertebrates. Indeed, the Swiss Protein Database classifies all proteins related to mineralization as being IDPs or having major intrinsically disordered regions (IDRs).[65] The principles of biomineralization, experimental techniques, and current engineering breakthroughs were recently reviewed by Avramenko *et al.*, who focused on the synthesis of semiconductor, magnetic, and metal(loid) oxide nanocrystals utilizing a biomimetic methodology.[66]

11.5.2 Artificial Synthesis of m-NPs and m-NCs

For the design and synthesis of nanomaterials, several approaches have been developed. However, the cost-effective, environmentally friendly synthesis of nanomaterials with regulated compositions, sizes, and shapes is the "Holy Grail" for materials scientists. Chemical synthesis and bioinspired engineering are two of the most used methods for fabricating nanoscale materials. Using these methods, m-NPs and m-NCs have been synthesized artificially using a variety of processes, which can be broadly categorized as (i) bottom-up and (ii) top-down. A simplified representation of the steps involved in the process is presented in Figure 11.10.

The reduction of a bulk substance into nanoscale particles is known as a top-down or destructive approach. Mechanical milling, nanolithography, laser ablation, sputtering, and thermal breakdown are some of the most extensively utilized NP creation processes following a top-down strategy. The approach of the build-up of material from single atoms/molecules through clusters to NPs is known as the bottom-up or constructive process. The most prevalent bottom-up processes for m-NP and m-NC generation include sol–gel, spinning disk reactor (SDR), chemical vapor deposition (CVD), pyrolysis, and biosynthesis. Bottom-up procedures to synthesizing m-NP/m-NCs utilizing wet chemicals are simple and frequently utilized, and the preparation fundamentally depends on the chemical reduction of metal salts using electrochemical routes or controlled breakdown of metastable organometallic compounds. A wide range of chemicals, including donor ligands, surfactants, polymers, ionomers, protein, peptides, and other biomolecules, are utilized in the process to regulate the nucleation and development of the originally generated m-NPs and m-NCs, as well as to stabilize them and to prevent agglomeration. Biosynthesis or biomimetic synthesis of m-NPs and m-NCs analogous to biomineralization is a green, ecologically friendly, sustainable, and energy-efficient methodology. Instead of using standard chemicals for bioreduction and capping, biosynthesis employs bacteria, plant extracts, fungus, proteins, DNA, and other precursors to generate and/or stabilize NPs/NCs. Many biosynthesized NPs/NCs have unique and improved features that find their way into a variety of applications, and they represent novel methods for nanomaterial manufacturing and activation. Bottom-up synthesis using the colloidal precursor method is the widely used solution-based chemical reduction approach to synthesize size, shape, structure, and atomic composition-controlled m-NPs and m-NCs.

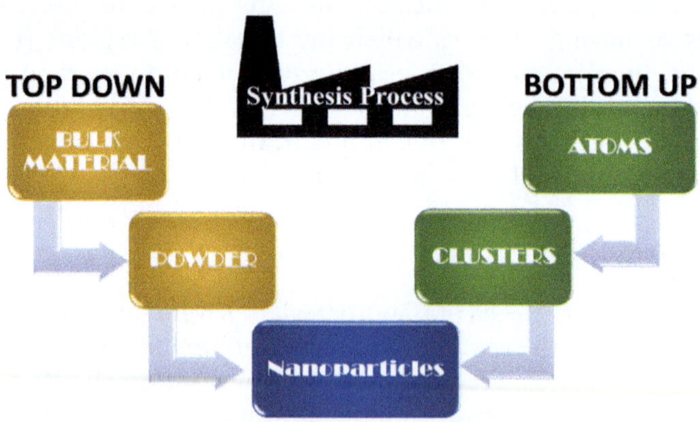

Figure 11.10 Nanoparticles and Nanocluster synthesis by various methods that are categorised into bottom-up or top-down method.

Turkevich *et al.*[19] produced the first repeatable standard methods for the synthesis of gold colloids (*e.g.*, for 20 nm gold by the reduction of gold salt as a precursor with sodium citrate). The technique began with stoichiometric mixing of a metallic precursor and surfactant, followed by the reduction of the metallic salt using a reducing agent. Bönnemann *et al.*[67] suggested a simple universal approach for the manufacturing of metal colloids of Groups VI–XI elements in the organic phase in 1991. In this method, typically, the metal salts are suspended in solvents such as tetrahydrofuran (THF) and treated with tetraalkylammonium hydrotriorganoborates. It has been hypothesized that the ammonium salt that is being formed during the process of reduction functions as a protective atmosphere for the metal particles. This process eliminates the need for the addition of external stabilizers. Bimetallic m-NPs have also been proposed through this method by co-reduction of different metallic salts.[68]

A wide range of directing/capping molecules, including thiolates, dendrimers, oligonucleotide, proteins, polyelectrolyte, and synthetic polymers, has been used to synthesize m-NCs utilizing wet chemical techniques.[69–76] In summary, this method for synthesizing diverse m-NPs and m-NCs has become one of the most potent and widely utilized in the field. The particle size generated by this method is strongly dependent on the directing/capping material employed and can be fine-tuned substantially using complex ligand[77,78] and reductants employed.[79–86] Recently, new approaches are being developed for continuous and scalable production of m-NPs and m-NCs by switching from batch to continuous synthesis protocols.[87–93]

11.5.2.1 Bottom-up Synthesis: Reduction, Nucleation and Growth Process, and Thermodynamics and Kinetics

Understanding nucleation and growth processes is crucial for controlling the size, shape, structure, morphology, and composition of m-NPs/m-NCs synthesized using a bottom-up approach. Although the NP/NC formation appears to be a complex and multistep process with various hypotheses, the chemical reduction of a salt precursor by a reductant is essentially an electrochemical process that involves electron transfer between two chemical species and may typically be represented as follows:[94]

$$\text{Oxidation half-reaction: } R_{\text{red}} \rightarrow R_{\text{oxi}} + ne^{-} \tag{11.1}$$

$$\text{Reduction half-reaction: } M^{n+} + ne^{-} \rightarrow M^{0} \tag{11.2}$$

$$\text{Global redox reaction: } R_{\text{red}} + M^{n+} \rightarrow R_{\text{oxi}} + M^{0} \tag{11.3}$$

From the thermodynamic point of view, any hypothetical redox reaction is spontaneous only when the "change in free energy" (ΔG):

$$\Delta G = -nF\Delta E \tag{11.4}$$

is negative. In eqn (11.4) 'n' is the number of electrons transferred in the balanced equation and F is the Faraday's constant (*ca.* 96 485 C mol^{-1}).

The hypothetical redox reaction is only favorable in terms of thermodynamics when E is positive, that is, when the reductant's oxidation potential is greater than that of the metal ion.

The equilibrium constant (K) of a redox reaction is also dependent on the magnitude of ΔE, as follows:[95]

$$\ln(K) = \frac{nF\Delta E}{RT}, \tag{11.5}$$

where R is the ideal gas constant and T is the absolute temperature. In general, the difference in ΔE should be >0.3 −0.4 V for a process to be of any practical importance. Strongly electropositive metals such as Au, Pt, Pd, Ag, and Rh react with moderate reducing agents under normal settings; however, more electronegative metals require extremely powerful reducing agents and, in many cases, severe circumstances. An increase in the value of E, as shown in eqn (11.5), potentially results in a more spontaneous reaction and the formation of a higher number of nuclei, resulting in smaller metal particles. It has been noted that a metal combined with reducing agents of varying strengths can produce particles of similar size but with various interior structures and degrees of crystallinity. Therefore, to obtain uniform nanocrystals, it is necessary to use the right reducing agent and operating circumstances. The two additional factors, namely complexation and solution pH, may change the reactivity of the metal through ΔE modifications. Other aspects of the system, such as ionic strength, particle charge, and the presence of surfactants that potentially alter aggregation, might all play a role in defining the final characteristics of the NPs.

11.5.2.2 Nucleation and Formation of Uniform Colloidal Particles in Solution

The role of nucleation in guiding the development of colloidal NP/NC is crucial, which may take place in either a homogeneous or a heterogeneous manner. Many models have been developed to describe the kinetics of nucleation; however, there is a current debate over the presence, role, and involvement of foreign and spectator elements in the general kinetics of nucleation.[96–100] Because the metal atoms produced by the reduction of metal precursors in homogenous solutions are essentially insoluble in liquid, they progressively congregate into clusters, forming embryos. As the system continues to generate additional metal atoms, embryos develop to a critical size, separating from the solution as solid particles and producing nuclei, which serve as templates for crystal development. At a given $\Delta\mu$ ($\Delta\mu$ = chemical potential difference between the actual state and the equilibrium state) in the nucleation process, an embryo must overcome a free energy barrier known as the nucleation barrier ΔG^* before it can grow into a stable crystal in the system. Finally, the nuclei so formed are evenly grown by

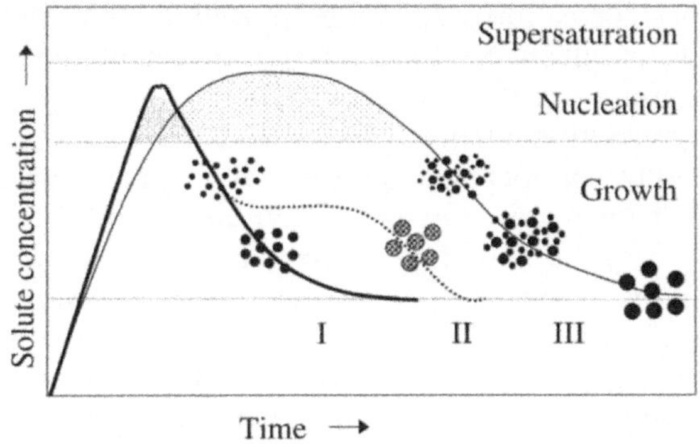

Figure 11.11 Mechanism of the formation of uniform particles in solution: curve I – single nucleation and uniform growth by diffusion (classical model of LaMer and Dinegar); curve II – nucleation, growth, and aggregation of smaller subunits; curve III – multiple nucleation events and Ostwald ripening growth. Reproduced from ref. 100 with permission from IOP Publishing, Copyright 2003.

diffusion of solutes from the solution to their surfaces until they reach their full size. A prerequisite for achieving monodispersity in NPs/NCs is that nucleation does not occur throughout the growth stage. Curve I in Figure 11.11 represents the classic model provided by LaMer and Dinegar for such a growth process.[101] In practice, the solute content, the redox potential of the reduction reaction, temperature, the kind and concentration of the surfactant/stabilizing agent applied, solvent viscosity, and surface tension are all factors that influence the quality and the quantity of the nuclei.

Heterogeneous nucleation is thought to be prevalent at low supersaturations, whereas homogeneous nucleation occurs at high supersaturations. Generally, nucleation stops when the concentration falls below the critical level, and the clusters continue to develop by molecular addition until the precipitated species' equilibrium concentration is attained. In addition to molecular addition, clusters can grow through aggregation with other particles, known as secondary growth (Figure 11.11 curve II)[102] or Ostwald ripening (Figure 11.11 curve III).[103]

11.5.2.3 Size and Shape Control during the Formation of NPs

During the nucleation and growth process, multiple factors influence the size and shape of the NPs in a solution. The shape and forms of the NPs are determined by both thermodynamic and kinetic parameters, which are influenced by the NPs' intrinsic structural features as well as the reaction environment that includes the solvent, capping agent, and reducing agent. Other reaction conditions such as concentration, time, and temperature are also critical. During the growth process, m-NPs, if conditions are favorable, can form facets to minimize surface energy and total excess free energy.[104]

The adsorptive organic capping agents employed to stabilize the NPs might play a key role in m-NPs/m-NCs size and shape regulation.[105] Due to the adsorption of capping agents, the total excess free energy of m-NPs is reduced, thereby preventing further growth and Ostwald ripening. Furthermore, the shape of the nanocrystals may also be regulated when the capping agents adsorb selectively onto certain m-NP surfaces.[106] Because distinct facets of m-NPs may have distinct electronic structures and atomic arrangements, different capping agents may absorb into these surfaces differently. As a result of the favored adsorption onto one set of surfaces over the others, distinct growth rates and crystallographic orientations should arise. Understanding the proper interaction between guest molecules and different metal crystal facets is thus a precondition for selecting effective shape control-capping agents. For example, long hydrocarbon chains in organic capping agents are hydrophobic and have a steric hindrance effect, which prevents direct contact with m-NPs' relatively high-energy surfaces, therefore stabilizing them.

The three most critical factors that regulate the nucleation process are (i) surface free energy, (ii) reaction temperature, and (iii) degree of supersaturation, as stated by Gibbs free energy.[98,107]

The excess free energy (ΔG_r) can be described by:

$$\Delta G_{\mathrm{r}} = 4\pi r^2 \gamma - \left[\frac{4}{3}\pi r^2 \cdot \frac{RT\ln(S)}{V_{\mathrm{m}}}\right] \tag{11.6}$$

where 'γ' is the surface free energy per unit area, 'r' is the radius of the clusters, 'R' is the ideal gas constant, 'T' is the temperature, 'S' is the ratio between solute concentrations at saturation and equilibrium conditions, and 'V_m' is the molar volume of the bulk crystal. As a result, when the solute is undersaturated and $S \leq 1$, ΔG_r stays positive and nucleation is inhibited. ΔG_r can only decrease with the rise of cluster radius when the solute is supersaturated, *i.e.*, $S > 1$, and clusters become stable with growth. Figure 11.12 illustrates the relationship between ΔG_r or N_r and r under the supersaturated condition. The expressions for critical radius (r^*) and the number of clusters that reaches the critical size (N_{r^*}) is given by

$$r^* = \frac{2\gamma V_{\mathrm{m}}}{RT\ln(S)} \tag{11.7}$$

$$N_{\mathrm{r}^*} = N_{\mathrm{A}}[A]_{\mathrm{eq}} S \cdot \exp\left\{\frac{-16\pi\gamma^3 V_m^2}{3(RT)^3[\ln(S)]^2}\right\} \tag{11.8}$$

where 'N_A' is the Avogadro's constant. When the solution is undersaturated, *i.e.*, $S \leq 1$, ΔG_E remains positive and increases with the growth of clusters.[98] Nucleation is not favored in such systems, whereas, when the solution is supersaturated, *i.e.*, $S > 1$, ΔG_E decreases with the increase of cluster radius,

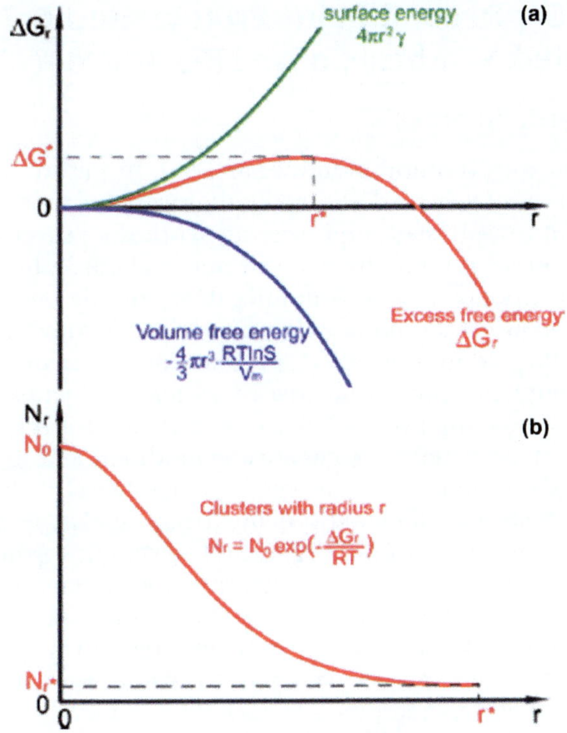

Figure 11.12 Illustration of (a) overall excess free energy ΔG_r and (b) average number of clusters N_r as a function of cluster size, r. The number of clusters decreases exponentially with increase in ΔG_r. When the radius of the cluster is smaller than the critical value, the system lowers its free energy by dissolving clusters, and at the same time new ones form due to spontaneous collisions. When cluster size is greater than the critical radius, the system lowers its free energy through continuous growth of clusters. Reproduced from ref. 98 with permission from Elsevier, Copyright 2009.

and clusters become stable with the growth. The rate of nucleation ($\mathrm{d}N_{r^*}/\mathrm{d}t$) from eqn (11.7) can be written as

$$\frac{\mathrm{d}N_{r^*}}{\mathrm{d}t} = f_0 N_A [A]_{eq} S \cdot \exp\left\{\frac{-16\pi\gamma^3 V_m^2}{3(RT)^3[\ln(S)]^2}\right\} \tag{11.9}$$

where 'f_0' is the ratio of the number of clusters having the critical size.[98] From the preceding equations, it can be shown that thermodynamic and kinetic variables, which are determined by the inherent structural characteristics of the metals and reaction parameters, can govern the size and structure of the crystals during nucleation. In the context of known biopolymer-mediated inorganic crystallizations, Feldheimet and Eaton[108] proposed a simple kinetic model. Different mechanisms of nucleation and growth of NPs in solution have been reviewed by Nguyen *et al.*[99]

11.6 Protein, Recombinant Protein, and Polypeptide-directed Synthesis of m-NPs and m-NCs

11.6.1 Biogenic Synthesis

Intracellular and intratumoral biomineralization in nature has been perfected over millions of years of the evolution process for specific functions. The procedure is complicated, and existing synthesis procedures make it impossible to gain such tight control over material characteristics. *Delftia acidovorans*, for example, is a Gram-negative, motile, nonsporulating, rod-shaped bacterium that generates a secondary metabolite capable of converting soluble gold to solid gold form. The bacteria demonstrate that a secreted metabolite can protect against toxic gold and cause gold biomineralization.[109] Exploiting marine microbial diversity for the biosynthesis of m-NPs has been reported.[110] A genetically modified M13 bacteriophage template was used to biomineralize ZnO.[111] Pulsipher *et al.* have reported the potential of thermophilic ferritin using *Archaeoglobus fulgidus* (AfFtn), which can provide templates for inorganic NP and their further growth.[112] Magnetotactic bacteria have evolved intricate subcellular machinery that creates linear chains of magnetite nanocrystals, which allow the host cell to sense direction.[113] As a result of this understanding, the biomineralization protein Mms6 has been used to stimulate the creation of homogenous magnetic nanocrystals *in vitro*.[114] Zhou *et al.*[115] showed site-selective biomineralization of Au-NPs in the inner channel of tobacco mosaic virus (TMV) by mutating threonine 103 to cysteine (T103C-TMV) to establish a strong coordination connection between arrayed sulfhydryl (–SH) ligands and gold species. Hosomomi *et al.*[116] described a novel method for producing Au-NPs inside *Escherichia coli* cells, in which an artificial oxidoreductase was overexpressed in the cytoplasm. Glasgow discovered that a genetically engineered *Salmonella enterica* self-assembling filament protein, PrgI, can bind Au-NPs and chemically reduce Au-conjugated filaments *in vitro* to create wires and networks with lengths of several micrometers.[117] When metal precursors (*e.g.*, chloroauric acid) are delivered to various cancer cells' cellular microenvironment, they can also generate NPs.[118,119] Cancer cells exposed to gold ions (grown in serum/serum-free conditions) release shock and stress-related proteins, including a unique gold-binding/reducing polypeptide (PP). The intracellular production mechanism has also been used to synthesize m-NPs inside live human cells.[120–122] Drescher *et al.*[122] studied the properties of *in situ* produced Au-NPs in the cellular setting. It has also been established that intracellular biomineralization of Au-NP with ion-binding peptides with cell-penetrating capacity may occur.[123] Advances in the synthesis of metal NPs using green technology have been summarized in a recent review,[124] which explains the synthetic method, the role of the enzyme in the production of the NPs, and its influence on the final characteristics of these NPs. Although significant progress has been made in biological nanomaterial production, the molecular process of biomineralization in bacteria remains mostly unknown.

Das *et al.*[125] studied the biosynthesis mechanism of Au-NP in *Rhizopus oryzae* and observed that $AuCl^{4-}$ [Au(III)] is reduced to AuNPs in both the cell wall and the cytoplasm. The adsorption and initial reduction of Au(III) to Au(I), followed by complexation and ultimate reduction to Au[0], has been reported to occur during biomineralization. Two proteins of 45 and 42 kDa have been found to engage in gold reduction, whereas an 80 kDa protein acts as a capping agent in Au-NP production. Nam *et al.*[126] discovered that yeast surface-displayed glutamic acid (E6) and aspartic acid (D6) peptides that can spontaneously reduce silver ions into Ag-NPs. They used the approach further to the M13 virus to create crystalline Ag-nanowires. He *et al.* reviewed the structural features of Ferritin family proteins and their capacity to sequester metal within their cavities, the prospect of using them as templates for metal NPs, and semiconductor manufacturing.[127] These new findings on the spontaneous reduction of metal ions on biological scaffolds emphasize the role of enzymes, proteins, and amino acid residues in biogenic nanomaterial creation and should aid in the development of new processes.

11.6.2 Role of Amino Acids in the Process of the Formation and Structuring of Nanocrystals

Extensive research into protein, recombinant protein, and polypeptide-based biomimetic mineralization has been focused that have demonstrated the compositional, structural, and organizational versatility of peptides, as well as their ability to provide a simple and environmentally friendly strategy and promising route for the synthesis of controlled inorganic/metal nanomaterials.[128] Because of the wide range of amino acid sequences accessible, peptides offer a lot of promise as morphology-specific ligands. Proteins/peptides are biopolymers made up of a linear polymer chain made up of up to 20 distinct L-amino acid residues in various combinations and sequences. The essential amino acid residues are distinguished from one another by their side chain (Figure 11.13), which specifies the sorts of interactions that can occur in a specific peptide.[129] The primary structure of a specific protein/peptide is formed by a unique set of amino acid residues assembled in a particular order and peptide bonds are generated by carboxyl acid (–COOH) functional groups and amino (–NH_2) groups of neighboring amino acid residues that bind individual amino acid residues together. Even a single amino acid replacement in a protein's sequence can alter its overall structure and function. Depending on the nature of intra- and intermolecular interactions, proteins/peptides can exhibit four levels of structure, namely primary, secondary, tertiary, and quaternary. Local folded structures that occur inside a polypeptide generate secondary structure due to interactions between backbone atoms, specifically between the carbonyl –O– of

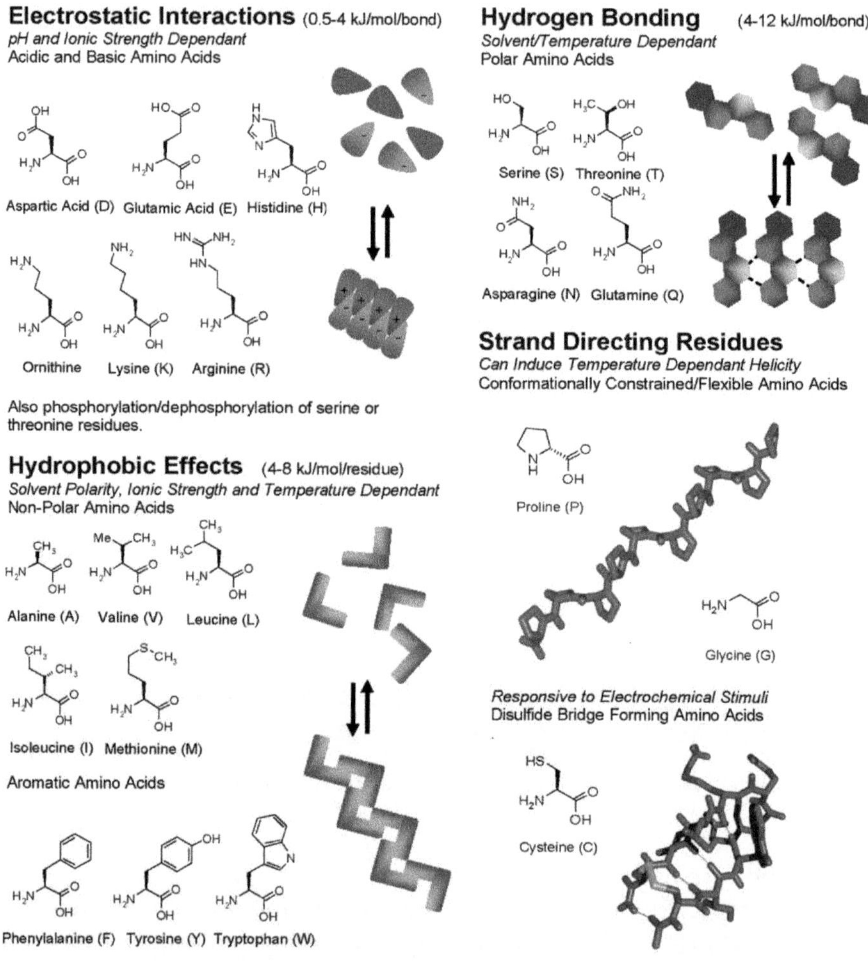

Figure 11.13 Schematic representation of different amino acids and the types of interactions in which they are involved in peptides. Reproduced from ref. 129 with permission from the Royal Society of Chemistry.

one amino acid and the amino –H of another. The α-helix and the β-pleated sheet are the two most prevalent secondary structures. The overall 3D structure of a PP is referred to as the tertiary structure. The tertiary structure is largely determined by interactions between the sidechain, R groups of the amino acids that make up the protein. The primary bonds and forces that hold the tertiary structure of protein/peptides are hydrogen bonding (between amino acid "R" groups nearby), hydrophobic interactions (hydrophobic "R" groups that seek to avoid water and position themselves toward the center of the protein), ionic bonding (bond formed between the positively and negatively charged "R" groups that come in close contact with one another), disulfide bridges (bond formed between the "R" groups of cysteine or methionine amino acids), and van der Waals forces (attractive and repulsive forces that occur between molecules that become polarized).[129]

Figure 11.13 presents a schematic representation of different amino acid residues and the types of interactions in which they are involved in protein/peptides. The 3D structure of a protein defines not only its size and shape but also its function. Interestingly, each type of interaction mentioned between amino acid residues of peptides/protein molecules has specific requirements. For example, hydrogen bonding requires precise positioning of the donor and acceptor of the residues approximately 2.8 Å apart. In contrast, hydrophobic interactions are less geometrically constrained. The π–π stacking interactions require the overlap of two π systems ~3.4 Å apart. Electrostatic interactions strongly depend on the ionic strength of the solution and are generally nondirectional. Therefore, they tend to be more flexible in terms of the distance between the participating charges. These requirements are considered important in the design and synthesis of synthetic proteins/PPs of a desired structure and function. In nature, proteins have a combination of these individually weak interactions, which work cooperatively to organize their structures and functions.[129,130] Most proteins fold into distinct 3D structures, and the shape into which they fold naturally is referred to as their native conformation. Proteins are frequently produced directly from genes in living organisms by translating messenger ribonucleic acid (mRNA). The amino acid sequence of a native protein is defined by the gene sequence contained in the live organism's genetic code. Recent advances in biosynthesis and recombinant DNA technology have made it possible to produce polypeptides with unprecedented control over composition, amino acid sequence, size, and structure, and hence potential control over characteristics and functions.[130] Modular protein engineering, which makes the use of a large library of peptide domains and templates, has lately emerged as a possible path for developing novel and innovative PPs.[131] Several notable examples of recombinant biomimetic polypeptides include variants of the green fluorescent protein,[132] spider silk,[133] resilin,[134] and other membrane and globular proteins.[135] Because of their substrate recognition qualities and sequence-specific self-assembly features, polypeptides are especially suitable molecules for the production of a variety of novel functional materials. The production of shape-controlled metal nanostructures in an aqueous medium at ambient temperature under moderate conditions has been described using carefully designed short peptides of a regulated sequence containing a facet-selective binding region.

Proteins play vital roles in regulating the construction of key soft matter and managing the biomineralization of inorganic nanostructures in natural systems. Proteins/peptides are currently being used by scientists from a variety of fields to control the creation of novel inorganic nanostructures and the construction of soft biomaterials. Among the proteins, stimuli-responsive protein/peptides are most important for the synthesis of controlled NPs of predetermined composition, shape, size, and morphology. Peptides with a preference for specific crystal facets have been frequently used to modify the morphologies of colloidal nanocrystals, but the processes behind facet selectivity remain a mystery. The progress has been slow due to a lack of peptide design principles, a possibility of a large number of potential amino acid sequences for a given composition, and the intricacy of interactions between amino acid residues. Due to the plethora of random and inactive amino acid sequences in peptides extracted from biological sources, it is also challenging to establish sequence reactivity connections.

11.6.3 Native Protein-directed Synthesis of m-NPs

Among many different native proteins, serum albumins (SAs) have been the most widely used protein for protein-mediated synthesis of m-NPs and m-NCs. SA is a globular protein that is the most abundant in plasma. It has many important functions, including regulating blood pH and osmotic pressure. It is nontoxic, water soluble, and stable in organic solvents at pH 4–9 and has several binding sites. It also acts as a fatty acid solubilizer and detoxifies blood plasma. SA can be derived from a variety of sources, but the most generally utilized SAs are egg (ovalbumin, OVA), bovine serum albumin (BSA), and human serum (I). They have been employed to enhance the biological characteristics of NPs as a control release carrier and capping agent.[136–139] Strategies for leveraging SA for drug delivery can be classified broadly into exogenous and *in situ* binding formulations, which use covalent attachment, noncovalent association, or encapsulation in albumin-based NPs. These approaches have had outstanding preclinical and clinical results. Singh *et al.*[140] advanced a facile method for the synthesis of Au, Ag, and their alloy Au–Ag-NPs in a BSA-based foam matrix, where the metal precursors are reduced *in situ* to yield NPs. Xie *et al.*[141] reported controlling the size and shape of Au-NPs in aqueous solutions through biomineralization using SA through altering the pH and temperature of the solution. Proteins are thought to be preferentially adsorbed on the {111} facets in acidic circumstances, resulting in anisotropic nanoplates. On the other hand, at higher pH, BSA unfolds and grows in all directions, resulting in smaller spherical-shaped NPs. Because of its zwitterionic character, BSA has been shown to attach to both cationic silver (Ag^+) and anionic gold ($AuCl_4^-$) ions in the foam. BSA-directed synthesis of ceria-based nanostructures, including ceria nanoclusters (C-NCs), nanoparticles (C-NPs), and nanochains (C-NHs), has been demonstrated by Yang *et al.*[142] using a simple, one-pot green synthetic approach. Such ceria-based NPs can be synthesized in a shape and size controllable manner by tuning the reaction time, temperature, and molar ratio. Zhang *et al.*[143] recently revealed that when a cancer patient is treated with the drug cisplatin, Pt-NPs with SA-coronas are autologously generated (biosynthesized in the blood) *in vivo* in human blood. They showed that this protein-coated Pt-NP can serve as an anticancer drug by inhibiting the development of chemotherapy-resistant tumors. Hartings[144] reported a method for making new material with gold NPs trapped inside an unfolded protein fiber using $HAuCl_4$ as the precursor and BSA as the protein. BSA was found to be an excellent capping and stabilizing agent for the production as well as size control of bioconjugated AuNPs.[145] Urchin-like shapes of Au-NPs have been synthesized using gelatine protein as a shape control agent.[146]

Shape-controlled Ag-NPs may be customized, according to Chakraborty *et al.*,[147] by selecting the proteins utilized as the ligand during a two-step manufacturing process. Depending on the proteins used, the wavelength of the surface plasmon resonance peak shown by the NP may be regulated from 400 to 700 nm (Figure 11.14). The use of lysozyme created spherically shaped NPs, while proteins such as BSA and catalase created anisotropic nanoplates. Aminated proteins produced better anisotropy than succinylated proteins, indicating that lysine residues play a role in anisotropy determination. Protein structure also plays an important function in the growth phase. Protein assemblies, for example, cages and crystalline protein structures, can provide confined chemical environments, which allow the immobilization of metal complexes.[148] Without the use of any reducing or stabilizing chemicals, AuNP–collagen core/shell nanoconjugates were synthesized in an aqueous solution at ambient temperature by Xing *et al.*[149] The formation is thought to be caused by electrostatic interaction between Au ions and collagen proteins, as well as *in situ* reduction by hydroxyproline residues.[149] The capacity to produce both anatase and rutile TiO_2 NPs smaller than 10 nm in diameter has been achieved by Bedwell *et al.*[150] using a biotemplated synthesis of TiO_2 NPs inside a protein cage-based system.

In the trypsin molecule, the presence of specific amino acid residues such as cysteine, methionine, and tyrosine, along with the unique spatial structures, allow it to bind and reduce the $AuCl_4^-$ ions while simultaneously forming and stabilizing Au-NP/Au-NC composite nanostructures using trypsin as linking and reducing agents in one-pot synthesis.[151] The unique structure and the optical property of the Au-NP-Trys-Au-NC nanocomposite enable it to be used in heavy metal ions sensing and elimination. Recently, Chakraborty *et al.* presented a progress report on the protein-induced shape control of noble metal NP synthesis.[152]

11.6.4 Recombinant Protein and Polypeptide-directed Synthesis of m-NPs

The use of a carefully chosen peptide or a well-designed recombinant protein to manufacture well-dispersed NPs in an aqueous solution at room temperature with excellent size and size distribution control has been recognized.[153] Despite the fact that the procedure is simple and it is recognized that the chosen peptide molecules may perform a single function (*e.g.*, as a stabilizing agent) or work as a multifunctional agent, the procedure is mostly empirical and achieved mostly by trial and error. Le *et al.*[154] used a phage display (PD) technology to specifically identify the facet-specific Pt-binding peptides to have a better understanding. The peptide sequence *Thr-Leu-His-Val-Ser-Ser-Tyr* (*TLHVSSY*, P7A) was

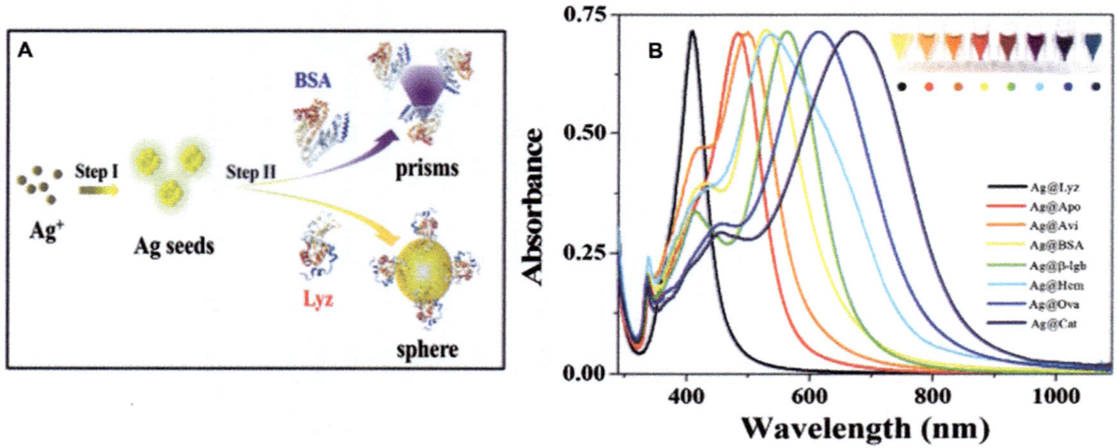

Figure 11.14 (A) A two-step process to synthesise shape-controlled Ag-NP. (B) UV–Vis absorption spectra of protein-coated Ag nanoparticles as synthesized at neutral pH using a selection of proteins. The inset shows the corresponding photographs of the nanoparticle solutions. Lysozyme: Lyz; Apotransferrin: Apo; Avidin: Avi; Bovine serum albumin:BSA; β-lactoglobulin: lgB; Hemoblobin: Hem; Ovalbumin: Ova; Catalase: Cat. Reproduced from ref. 147 with permission from American Chemical Society, Copyright 2018.

identified as the specific binder as a result of this research. Such understanding created an opportunity to produce free P7A using a solid-phase peptide synthesis approach for controlled m-NP generation. Though the exact origin of P7A's high Pt binding capacity is unknown, interactions between hydroxyl (from *threonine*, *tyrosine*, and *serine*) and/or polar (*i.e.*, *histidine*) amino acid residues and the Pt surface are considered to be involved. Some molecular geometrical factors may also play a role. Li *et al.*[154] clearly demonstrated specific peptide P7A-regulated rational synthesis of monodisperse Pt-NPs. During such synthesis, the rapid reduction of the Pt-precursor produces extremely tiny Pt-clusters; however, the growth rate of Pt-NC is considerably slowed in the presence of P7A, compared to the blank reaction, resulting in significantly smaller NC sizes. Furthermore, when P7As bind to the Pt-NC surface, the growth kinetics change, resulting in the observed morphologies that differ from blank reactions.[155–157] Tan *et al.*[157] investigated the scientific basis for the rational design of peptides for the synthesis of size- and shape-controlled m-NPs in depth. They offered a set of rudimentary rules based on the understandings. By strategically positioning specific amino acid residues in the peptide sequence, many multifunctional peptides with varied morphologies (nanospheres and nanoplates) and size tunability were generated. Lauren *et al.*[158] showed that a single peptide sequence may be used to make monodisperse and monomorphic Pt-NPs of various sizes and morphologies under ambient circumstances. The capacity of rationally designed facet-specific peptide sequences to function as regulating agents for the predictable synthesis of shape-controlled Pt-nanocrystals with selectively exposed crystal surfaces was demonstrated by Chiu *et al.*[159] They demonstrated the production of Pt-nanocubes and Pt-nanotetrahedrons using Pt-{100} and Pt-{111} facet-binding peptides. Pt-{100} binding peptide *Ac-Thr-Leu-Thr-Leu-Thr-Leu-Thr-Asn-*$CONH_2$ (T7) created cubes with six {100} facets, whereas the Pt-{111} binding peptide *Ac-Ser-Ser-Phe-Pro-Gln-Pro-Asn-*$CONH_2$ (S7) yielded tetrahedrons enclosed by four {111} facets (Figure 11.15). Figure 11.15c–f presents TEM images of Pt-NC obtained from different peptide-directed syntheses.

Under continuous $NaBH_4$ injection, Li *et al.*[160] verified the synthesis of Pt-multipods utilizing peptide, *Ac-Thr-Leu-His-Val-Ser-Ser-Tyr-*$CONH_2$ (BP7A) as a capping agent. Ruan *et al.*[161] demonstrated that Pt-binding peptide BP7A may directly regulate nucleation and facilitate and stabilize single-twinned Pt-seed production during the nucleation stage. Subsequently, the use of single-twinned planes as seeds in the growth process may allow the synthesis of Pt-NCs in a wider variety of forms. Based on this concept, Ruan *et al.*[161] demonstrated a two-step technique for the synthesis of shape-controlled m-NP. In the process, BP7A peptide was first used to create single-twinned seeds, which were subsequently rationally tuned to various forms using a second peptide to regulate m-NP development. Knecht and colleagues[162–165] proposed that crystallographic recognition controls peptide binding for biobased nanomaterials, and they demonstrated that materials directing peptides can be used to tune the size,

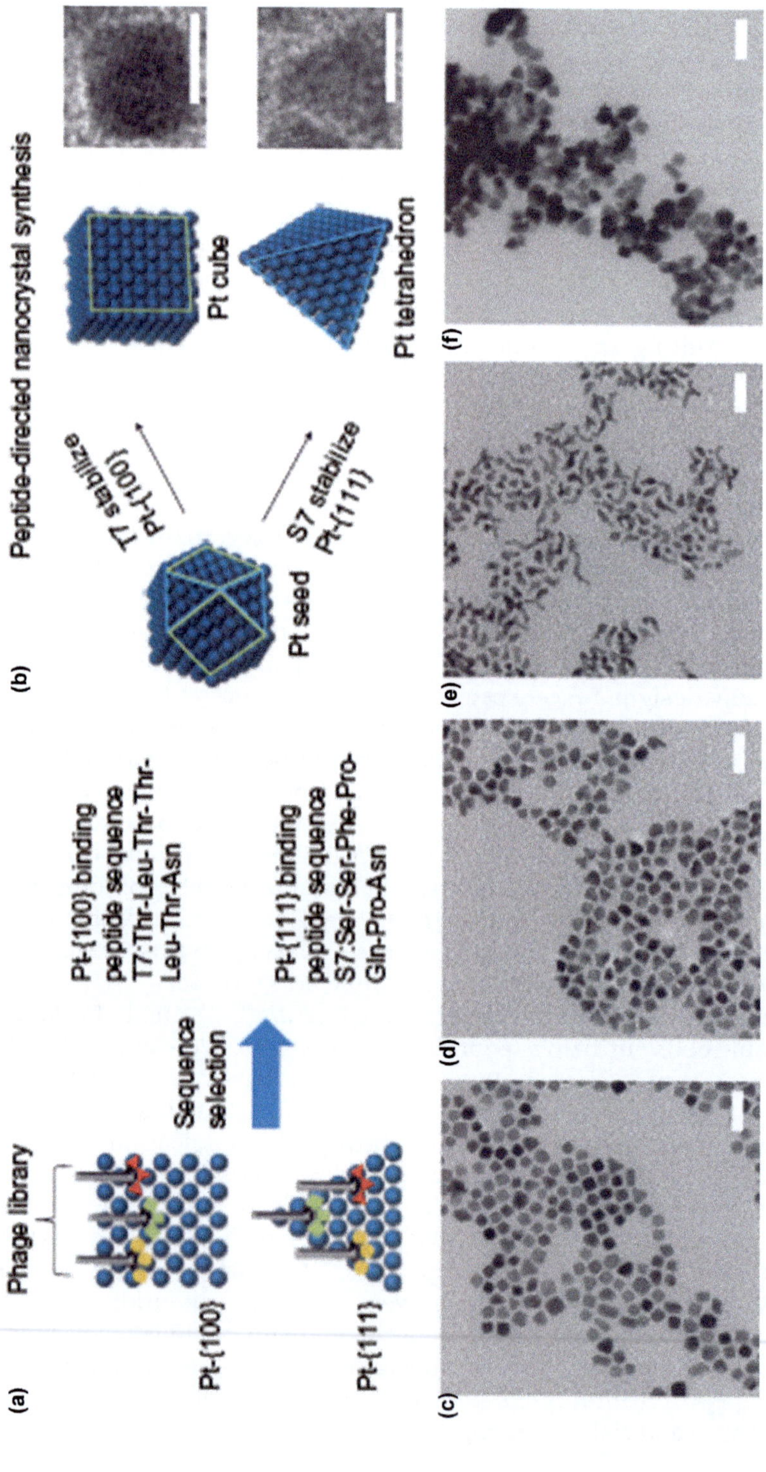
(a)
Phage library
Pt-{100}
Pt-{111}
Sequence selection
Pt-{100} binding peptide sequence T7:Thr-Leu-Thr-Thr-Leu-Thr-Asn
Pt-{111} binding peptide sequence S7:Ser-Ser-Phe-Pro-Gln-Pro-Asn
(b)
Peptide-directed nanocrystal synthesis
Pt seed
T7 stabilize Pt-{100}
S7 stabilize Pt-{111}
Pt cube
Pt tetrahedron
(c)
(d)
(e)
(f)

surface structure, and functionality of Pd-NPs by selectively changing the peptide sequence to change the binding motif. They investigated the production procedures of peptide-templated Ag-NP[166] and Au-NPs[167,168] in addition to Pd catalysts.

Functionalized and responsive Au-NPs were prepared using thiol-capped short elastin-like peptides (ELPs) with the thermally responsive ELP-sequence *VPGVG*, which displayed thermally triggered reversible aggregation at 40 °C.[169] Such ELP–AuNP conjugates were shown to be capable of reversible, thermally driven aggregation at near-physiological temperatures (transition temperature of 40 °C at pH = 7.4) and paving the way for innovative photothermal cancer treatment and diagnostic.[169] Alvarez-Rodríguez *et al.* reported a one-pot synthesis of photoresponsive Au-NP by reducing auric acid precursor in the presence of elastin-like polymer, AzoGlu15, with the following amino acid sequence: $C\text{-}[(VPGVG)_2(VPGEG)(VPGVG)_2]_{15}$.[170] The dynamic self-aggregation/disaggregation of silk-elastin modular protein–gold nanobioconjugates, Au-SELP-NPs [SELP-containing amino acid repeats of $(GVGVP)_4(GGGVP)(GVGVP)_3(GAGAGS)_4$], with regulated size and pattern was also reported.[171] In this study, the temperature-reversible conformational switching between the hydrophilic random coils and the hydrophobic β-turns in the elastin blocks was programmed between 50 and 60 °C. The ability to construct and synthesize responsive plasmonic nanostructures by individually modifying the refractive index and size of the SELPs through genetic engineering was demonstrated in this study.

Dutta and coworkers recently reported a simple but elegant environmentally friendly water-based method to synthesize Pt-NPs[172] and optically coupled Au-NP[16] using multistimuli-responsive resilin-mimetic peptides (RMPs) with a consensus sequence of *GGRPSDSYGAPGGGN*; that acts concurrently as the directing and stabilizing agent. They were able to control particle size from sub-5 nm to subnano (0–1 nm). The key question surrounding how the RMPs bring control and order to the metal particle remains unclear. However, it was identified that redox reactions of tyrosine

Figure 11.15 Biomimetic approach to the predictable synthesis of shaped nanocrystals. (a, b) Schematic illustration of facet-specific peptide sequence selection and nanocrystal synthesis: Pt-{100} binding peptide sequence T7 and Pt-{111} binding peptide sequence S7 are selected against {100}-faceted and {111}-faceted substrates (a); facet-specific peptides are then used to direct the synthesis of platinum nanocrystal cubes and tetrahedrons, respectively (b). TEM images of platinum nanocubes obtained from the T7-regulated reaction and platinum nanotetrahedrons obtained from the S7-regulated reaction; scale bars, 5 nm. (c–f) TEM images of platinum nanocrystals obtained from different peptide-regulated reactions: T7 (c), S7 (d), BP7A (e), and blank reaction (f); scale bars, 20 nm. All samples were conducted under the same conditions and collected after reacting for 10 minutes. Reproduced from ref. 159 with permission from Springer Nature, Copyright 2011.

amino acid residue (Y–OH/YO• redox couple) in RMP play key roles in the process. It has been hypothesized that engineered peptide sequence based on *GGRPSDSYGAPGGGN* consensus sequence with the inclusion of specific metal-recognition residue/sequence (*e.g.*, amino acid residue/sequences that recognize specific Pt-{100} and Pt-{111} planes) can be used as a powerful tool for directing crystal nucleation, phase, morphology, and growth dynamics and ultimately predictable control of nanocrystal shape through biomineralization process. Overall, there was significant success in the design of peptides of the selected structure and composition that can effectively control the structure and morphology of monometallic NPs at the atomic level.[173,174] However, limited attempts have been made to synthesize bimetallic NPs as they are far more complex than monometallic ones.[175–178] Aside from metallic NPs, biomimetic approaches have been used to produce oxide materials such as silica, Germania, zinc oxide, iron oxide, titanium dioxide,[179] functional materials,[180,181] and nanocrystals.[182]

11.6.5 Peptoid-directed Synthesis on m-NPs

Peptoids, a novel class of synthetic peptide and protein mimics that combine the advantages of both synthetic polymers and biomolecules (*e.g.*, excellent chemical stability and easy production), have recently received much interest; and the link between peptoid sequences and m-NP shape and growth kinetics has been investigated.[183] Due to the lack of chirality and hydrogen bonds in their backbone, peptoid-based systems are extremely designable, with less side chain chemistry-dominated characteristics and structural complexity compared to proteins and peptides. Yan *et al.*[184] created sequence-defined peptoids to regulate the synthesis of highly branched plasmonic Au-NPs. By engineering peptoid sequences, they hoped to develop a rule of thumb for designing peptoids that can potentially guide the morphological development from spherical to coral-shaped NPs. Cai *et al.*[185] employed amphiphilic peptoid sequences to control the nucleation, growth, and assembly of inorganic nanocrystals into superstructures. The functionalization of AgNPs by *N*-substituted glycine oligomer peptoids was reported by Maayan *et al.*[186] The capacity of functional biomimetic oligomers to facilitate the assembly of metal NPs, as well as peptoid/AgNP hybrids, has implications in sensing materials, biology, and catalysis. Tigger-Zaborov[187] used the characteristics of the peptoid mediators to design specific guidelines for the controlled aggregation of AgNPs in water at room temperature; they investigated the catalytic activity of Ag NPs in the reduction of 4-nitrophenol to 4-aminophenol. Peptoid-based approaches have demonstrated potential to synthesize predictive bioinspired NPs, NCs, and even highly complex designable hybrid materials; however, significant targeted research is critical to making such materials with programmable functions.

11.6.6 Molecular Simulation and Structural Insights at Atomic-scale Resolution

The understanding of the interactions at the peptide–inorganic particle interface that control the precise manufacturing of m-NP/m-NCs is crucial but yet ambiguous. Nonetheless, progress in our knowledge of organic/inorganic interfacial interactions in colloidal systems is crucial in paving the way for rational and predictable nanostructure design and manipulation in a wide range of applications. The structural insights at the atomic-scale resolution of such systems are difficult to get by experimental methods; however, molecular modeling[188] can offer much insight. To explain the adsorption of amino acid residues on Pt-crystallographic facets, Ramakrishnan *et al.*[189] used density functional theory (DFT) simulations to investigate the genesis of peptide facet-selective binding. They revealed that the presence of polar amino acid residues and the molecular architecture of peptides near the crystal facets have a major role in peptide surface recognition. Ramakrishnan *et al.*[190] demonstrated that specific surface recognition of amino acid side chains is caused by a combination of processes: (i) electron exchange, (ii) partial charge transfer, and/or (iii) dispersive effects after conducting a thorough examination of amino acid and peptide facet specific adsorption on Pt. Hong *et al.*[191] used DFT simulations to investigate the adsorption of a few single amino acids on Au and Pd cluster models to better understand amino acid adsorption at metallic surfaces.[191] Heinz *et al.*[192] conducted a quantitative investigation of the changes in energy and conformation upon adsorption on {111} and {100} to learn more about the molecular interactions involved. They investigated the selective binding of numerous short peptides (8–12 amino acids, omitting *Cys*) produced from phage-display methods to the surfaces of Au, Pd, and Pd–Au bimetal. They used a molecular dynamics simulation technique and proposed a soft epitaxy mechanism of binding. The strength of adsorption was projected to be in the range of 0 to −100 kcal per (mol peptide) and scale with the metal's surface energy, individual residue affinity *vs.* water affinity, conformational features, and polarization and charge transfer at the metal interface. The adsorption and desorption of coarse-grained peptide chains on typical Au and Pd surfaces were also investigated using Monte Carlo simulations[193] and the adsorption on Au and Pd was predicted to be mediated by aromatic anchoring amino acid residues such as *Tyr*, *Phe*, and *His*. Due to their lower (more negative) surface-interaction energy and weaker mobility on Pd surfaces, these residues were predicted to be slightly more likely to be adsorbed on Pd than on the Au surfaces. Through both theoretical and experimental studies, *Phe* has been identified as the primary motif to differentiate {111} from other facets in Pt-NPs. This understanding helps to the rational design of small organic compounds that show preferential adsorption to the {111} facets of both Pt and rhodium-Rh nanocrystals. Lin *et al.*[194] used the stochastic tunneling-basin hopping (STUN-BH) approach to generate the most stable peptide S7 configuration (*Ac-Ser-Ser-Phe-Pro-Gln-Pro-Asn-CONH*$_2$)

adsorbed on Au{111} facet. The simulation findings indicate that the morphology of *Pro* has an effect on the adsorption stability of *Phe* and plays a critical role in the facet selection for Au-atom stacking.[195] Indeed, the structural connections between a given peptide sequence and its binding affinity are yet to be fully understood. This lack of knowledge significantly inhibits practical progress in the field, leaving plenty of room for more study to fully exploit the promise of protein/interface manipulation.

11.7 Engineering Ultrasmall m-NCs

m-NCs, which are made up of only a few to a few tens of metal atoms, bridge the gap between isolated atoms and NPs. Their chemical, optical, electronic, and physical properties are vastly different from those of bulk metal, m-NPs, or the atom itself.[196] Many m-NCs synthesized in the gas phase with precisely controlled size have played a critical role in the development of a fundamental understanding of cluster properties and their applications.[197,198] Because m-NCs in the gas phase are unstable, prone to aggregation, and difficult to functionalize, solution-phase techniques for their production have been advanced.[199] The synthesis of atomically precise m-NCs in the solution phase, on the other hand, presents several unique synthetic and characterization challenges, including time-consuming and complex synthesis processes, application of extreme reaction conditions, and the use of an organic solvent; all of which may limit their biomedical applications. To alleviate the problem, a wide range of methods,[200,201] including the utilization of plant extracts,[202] DNA-based conjugation,[203] solid support-guided synthesis,[204] and the use of amorphous polymers[205] and dendrimers as hosts[206–208] have been proposed. Many native proteins, including BSA,[209] insulin,[210–212] lysozyme,[213] transferrin,[214] and egg white protein,[215] have shown their ability to be an efficient templating agent for the formation of Au-NCs. The efficiency of individual amino acid residues to synthesize and stabilize m-NCs has also been examined.[216–219] Glutathione (GSH), a ubiquitous tripeptide (*N*-γ-glutamyl-cysteinyl-glycine, with a gamma peptide linkage between the carboxyl group of the glutamate side chain and cysteine) in animal cells, has been extensively investigated as an environmentally friendly reducing/protecting regent for preparing m-NCs at physiological temperature.[220–222] GSH has also been utilized as a nucleation template for the synthesis of Pt-based peroxidase.[223] GSH-protected bimetallic Au-Pt NCs with NIR emission for the ratiometric determination of silver ions were reported by Gao *et al.*[224] Recently, a well-defined biomineralization process has been developed, in which a variety of biomolecules in a confined space were used to influence the size, shape, and distribution of NPs in a temporally and spatially controlled manner. Tang *et al.*[225] demonstrate the ability of metal–organic frameworks (MOFs) loaded with specific proteins for the growth of ultrafine metal AgNC and AuNC.

Rica and Chow[226] proposed a designer peptide *C*(*GRP*)$_3$ (*Cys-Gly-Arg-Pro-Gly-Arg-Pro-Gly-Arg-Pro*) to mediate the formation of Au-NC with a precise number of constituent metal atoms. Simply activating the reduction of Au-ions with $NaBH_4$ in the presence of *C(GRP)*$_3$ led to the growth of Au-NC. Conceptually, the *Cys* amino acid residue was inserted in *C(GRP)*$_3$ to concentrate metal ion precursors (due to the presence of –SH) and nucleate nanocrystal formation. In *C(GRP)*$_3$, the Pro-rich sequence *Gly-Arg-Pro-Gly-Arg-Pro-Gly-Arg-Pro* endowed the preferred conformation as a helical structure with a subnanometric cage that defines and limits of growth of m-NCs (Figure 11.16). The presence of positively charged *Arg* amino acid residue potentially facilitates the cell-penetrating properties. The resultant water-soluble Au-NC emits in the NIR range and is not toxic to cells. This research demonstrates that by changing the template and the size of the hollow cage for metal deposition, it is possible to create m-NCs of various sizes.

Wang *et al.*[227] rationally designed a bifunctional peptide *H_2N–CCYRGRKKRRQRRR–COOH* (*CCYTAT*) for the synthesis of designer m-NCs. Due to the presence of *tyrosine* residue (Y) in the peptide domain one, *CCY* may reduce Au-ions into cluster Au-NC and successively trap Au-NC using the –SH group of cysteine (C). The nucleus-targeting capacity is endowed by the second functional domain, which is constituted of *RGRKKRRQRRR* (a typical nucleus localization signal sequence derived from the protein HIV-1 TAT). They demonstrated that the fluorescence spectrum of the CCYTAT–AuNC exhibited red emission and specifically targeted the nucleus of cell lines without further modification. A cyclic peptide *c(RGDYC)* (Figure 11.17) as the template for the synthesis of fluorescent AuNCs was reported by Liang *et al.*[228] The cyclic peptide *c(RGDYC)* is composed of two functional elements that work together (Figure 11.17A). The arginine-glycine-aspartate (RGD) sequence in the first component can particularly bind the integrin avb3 receptor, which is well recognized as a tumor and angiogenesis marker. The second section contains *Tyr* and *Cys* residues, which can convert Au-precursors into AuNCs and attach NCs through the Au-S link. As a result, although having just five amino acid residues, the *c(RGDyC)* peptide has the potential to perform a threefold function in the synthesis as a reducing, stabilizing, and targeting agent (Figure 11.17B). Yu *et al.*[229] synthesized BSA–Au-NC with five distinct sizes: Au4, Au8, Au10, Au13, and Au25 in a

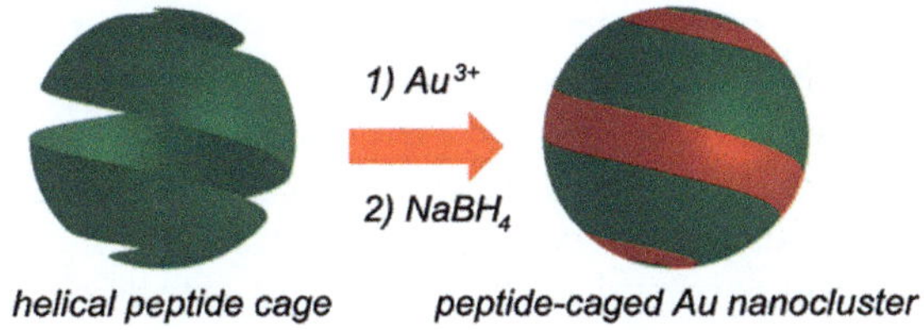

Figure 11.16 Schematic representation of the growth of Au nanoclusters with the peptide template C(GRP)3; the dimensional structure of the peptide can generate a cage that arrests the growth of Au nanoclusters. Reproduced from ref. 226 with permission from the Royal Society of Chemistry.

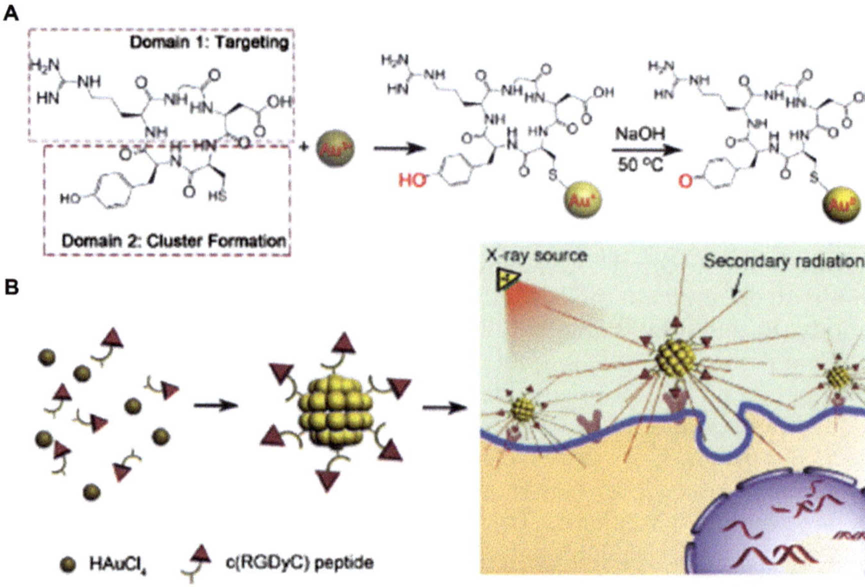

Figure 11.17 A schematic illustration of c(RGDyC)-AuNCs formation and application for enhanced radiotherapy. (A) The Cys and Tyr residues capture Au3þ ions and reduce it to Au0 under alkaline conditions, respectively. (B) c(RGDyC)-AuNCs accumulate in avb3 integrin-positive cancer cells and interact with incident radiation intensively, generating secondary radiation, and leading to radiation enhancement effect. Reproduced from ref. 228 with permission from Elsevier, Copyright 2017.

BSA–protein template with a predetermined shape using CO as the reducing agent. The conformation of BSA, which was influenced by environmental conditions, was the major determinant of the core size of as-synthesized Au-NCs. By adjusting the molar ratio of Au/BSA, the conformation of BSA in its natural state may be fine-tuned even further. This research is interesting not only because it demonstrates how to make protein-protected Au-NCs with distinct sizes but also because it demonstrates how to alter the size of Au-NCs by modifying the protein shape/confirmation.

Dutta and coworkers have recently used a multistimuli-responsive resilin-mimetic protein (RLP) with a consensus sequence *GGRPSDSYGAPGGGN* to demonstrate a facile aqueous-based method to synthesize Au-NC[15] and Pt-NC.[14] RLP concurrently plays a triple role as a directing, reducing, and stabilizing agent in the synthesis. Figure 11.18 shows the high-resolution TEM image of AuNCs, confirming the ultrasmall size of the AuNC formed. Figure 11.19 illustrates the fluoresce characteristics, atomic composition, and fluorescence lifetime of the Au-NC so formed. RLPs such as *rec1-resilin*,[230–233] AN-16,[234] and other designed RLPs[235] display multistimuli (pH, temperature, ions, mechanical stress, *etc.*) responsiveness, including dual-phase transition behavior (existence of both upper critical solution

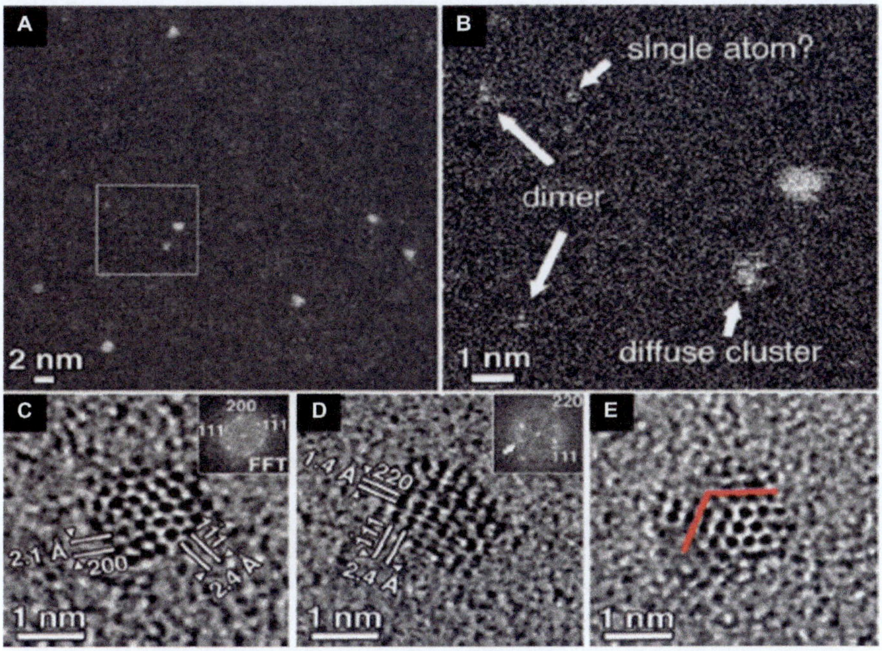

Figure 11.18 (A) HAADF-STEM image of AuNCs in green fluorescent Au–Rec1-resilin nanobioconjugates. (B) Enlarged view of area framed in (A), showing a variety of Au aggregates. (C) and (D) HRTEM images of AuNCs with FCC structure viewed along the[110] and[112] zone axes, respectively. (E) A AuNC with at least two[111] twins. Reproduced from ref. 15 with permission from the Royal Society of Chemistry.

temperature, UCST, and lower critical solution temperature, LCST). RLPs are also demonstrated to be intrinsically disordered protein[236,237] structures with rapidly interchangeable conformational ensembles in physiological conditions. The observation of unusual multistimuli responsiveness associated with the IDP nature of RLPs has garnered significant recent interest in RLPs and a large number of designer RLPs and modular RLPs have been advanced with a variety of potential applications (Figure 11.20). Maity *et al.*[238] propose an ingenious experiment that used a crystalline protein cage built by crosslinking a single crystal of a poferritin to structurally characterize the formation of NCs with reduction reactions and to clarify the detailed mechanisms of m-NC formation in protein environments. This study revealed that the Au-ions gradually moved toward the center of the three-fold symmetric channels of the protein cage to form NCs, with considerable conformational changes of the amino acid residues attached to Au ions. Indeed, the synthesis and use of fluorescent Au-NCs, as well as their fluorescence phenomena and reaction processes, have been of much interest.[239–245] Table 11.1 summarizes many protein/peptide compositions/consensuses that have been used to synthesis fluorescence m-NCs.

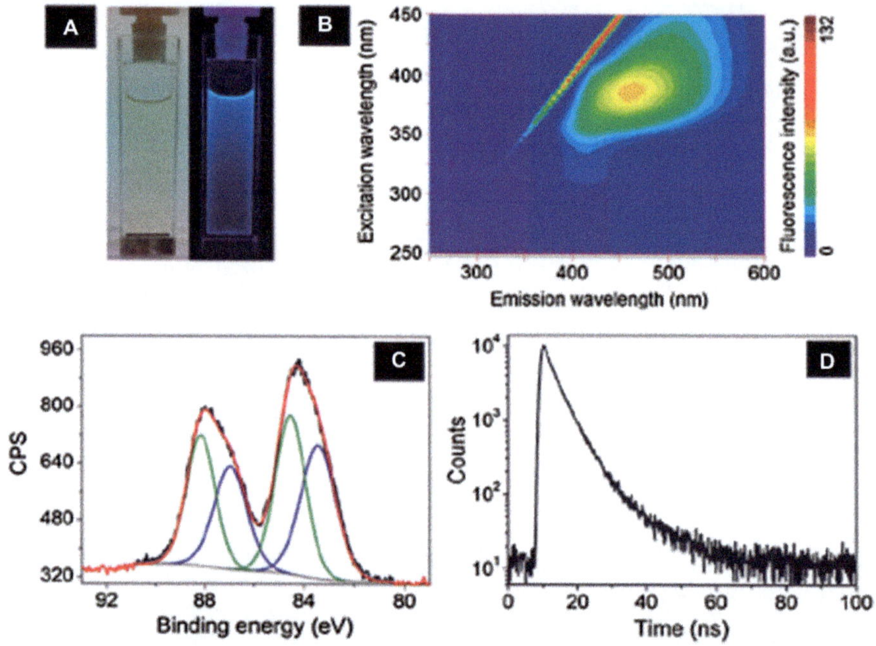

Figure 11.19 (A) Photographs under visible and 365 nm UV light, (B) 3D excitation/emission matrix contour plot, (C) Au 4f XPS spectra, and (D) fluorescence lifetime spectra of the blue fluorescent Au0 nuclei-Rec1-resilin nanobioconjugate. Reproduced from ref. 15 with permission from the Royal Society of Chemistry.

11.8 Applications of m-NPs and m-NCs

In the past several decades, researchers have been drawn to m-NPs and m-NCs because of their distinctive physical characteristics (such as particle aggregation and photoemission, fluorescence, and electrical and heat conductivities), chemical features (such as a catalytic activity), and potential applications. Their applications have been focused on wide-ranging areas, including advanced materials, biomedicine, immunoassay and cell bioimaging, biomolecule detection and sensing, pharmaceuticals, separation technology, purification and concentration, substrate coding, electronics, magnetics and optoelectronics, signal transduction and amplification, cosmetics, energy, catalysis, and environmental detection and monitoring.[267–279] They have the potential to improve the detection sensitivity of microbial monitoring, as well as the degradation and recovery efficiency of chemicals in environmental research.[280,281] Noble metal NPs and NCs have received the greatest attention among m-NPs and m-NCs. Au- and Pt-based NPs and NCs are the most celebrated, with significant inroads made in biomedical and catalytic applications. The size, shape, and surface functions of NPs/NCs substantially influence their characteristics and hence

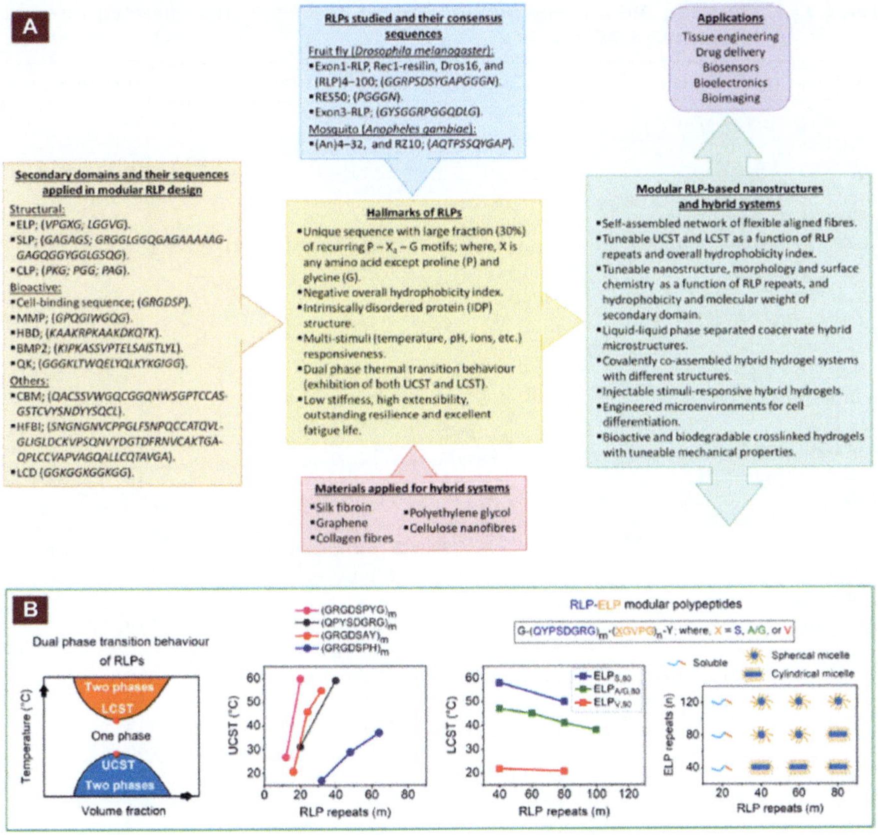

Figure 11.20 Hallmarks of RLPs, development of modular RLP-based materials and their structure–property relationship. (A) Summary of RLPs, secondary domains, and other materials applied for the development of modular RLP-based nanostructures and hybrids. The amino acid sequences of RLP repeats and secondary domains are presented as a single-letter code. UCST upper critical solution temperature, LCST lower critical solution temperature, ELP elastin-like polypeptide, SLP silk-like polypeptide, CLP collagen-like polypeptide, MMP matrix metalloproteinase cleavable peptide, HBD heparin-binding domain, BMP bone morphogenetic peptide, QK vascular endothelial growth factor-mimicking domain, CBM cellulose-binding module, HFBI hydrophobin protein domain, and LCD lysine crosslinking domain. (B) Schematics of dual-phase transition behavior and some of the structure–property relationship of RLP- and modular RLP-based systems. Reproduced from ref. 235 with permission from Springer Nature, Copyright 2021.

their application possibilities. Asymmetric m-NPs with varied ARs have been used in a variety of applications, including cancer cell imaging and hyperthermia phototreatment.[282,283] Similarly, the shape of Ag-NPs affects their catalytic capabilities significantly. In the oxidation of styrene, cubic Ag-NPs, for example, provide better yields.[284] The magnetic properties of cobalt oxide

Table 11.1 Protein, biomimetic peptide, and designer peptide-directed ultrasmall fluorescent MNCs.

Proteins and peptides	m-NCs	Applications	References
Thr-Leu-His-Val-Ser-Ser-Tyr (TLHVSSY)	Pt	Catalyst	154
YQPWKTQRELSV Pro-Trp-X-X-Gln-Arg-Glu-Leu-Ser-Val (PWxxQRELSV)	Pt	Catalyst	158
(i) Ac-Thr-Leu-His-Val-Ser-Ser-Tyr-CONH$_2$ (ii) Ac-Thr-Leu-Thr-Thr-Leu-Thr-Asn-CONH$_2$ (iii) Ac-Ser-Ser-Phe-Pro-Glu-Pro-Asp-CONH$_2$	Single-twinned seeds	Shape-controlled Pt	161
α-Chymotrypsin	Ag	Metal–protein interaction study	246
α-Lactalbumin	Au	—	247
β-Lactoglobulin	Au	—	247
BSA	Au, Ag, Cu, Au/Ag alloy	Hg(II), Cu^{2+}, Pb^{2+}, H_2O_2 sensing, bioimaging	248–255
Egg white	Au	Metal ion sensing	256
Fibrinogen	Au	—	247, 257
Histidine	Au	Biological labeling, biosensors	219
Horseradish peroxidize	Au	H_2O_2 sensing	258
Human serum albumin	Au	NO_x sensing	239
Insulin	Au	Bioimaging	259
Lactoferrin	Au	Cupric ion sensing	260
Lysozyme	Au, Pt	Hg^{2+} sensing,	213, 256, 261–263
Myoglobin	Au	—	247
Pepsin	Au	Hg^{2+} sensing, bioimaging	252, 264
Serum transferrin	Au	Bioimaging	260
Trypsin	Au	Hg^{2+} sensing	252, 265
Peptide – *CCYRGRKKRRQRRR–*	Ag	Cell imaging	266
Cys-Gly-Arg-Pro-Gly-Arg-Pro-Gly-Arg-Pro	Au,	Shape-controlled NPs	226
H$_2$N-Cys-Cys-Tyr-Arg-Gly-Arg-Lys-Lys-Arg-Arg-Gln-Arg-Arg-Arg-COOH	Au	Theranostics, stain cell nuclei	227
c(Arg-Gly-Asp-Tyr-Cys)	Au	Theranostics	228
Gly-Gly-Arg-Pro-Ser-Asp-Ser-Tyr-Gly-Ala-Pro-Gly-Gly-Gly-Asn	Au, Pt	Biosensors, theranostics	14–16
Glutathione (γ-Glu-Cys-Gly)	Au, Pt	Sensors, theranostics	220–223

NPs are also influenced by their shapes and sizes.[285] Due to their shape-dependent optical and physical characteristics, anisotropic metallic NPs, such as Au- and Ag-nanoprisms (NPSMs), have received a significant attention for their use in catalysis, molecular sensing, signal amplification, bioimaging, and therapeutic applications.[286]

In the vast majority of the applications of the core m-NPs and m-NCs, the particle is protected by multiple monolayers of inert material such as silica. The organic molecules that are adsorbed or chemisorbed on the surface of the particle might serve as biocompatible substances or linker molecules for further functionalization. A protective and biocompatible shell encasing the AuNP, as shown in Figure 11.21[271] with a typical Au-NP for biological applications, allows the attachment of various targeting vectors, imaging agents, therapeutic medicines, and biocompatibilizers such as polyethylene glycol (PEG) molecules. In biomedical applications, to shield NPs/NCs from the immune system, prevent reticuloendothelial system (RES) uptake, and decrease nonspecific binding, biocompatibilizer attachment is frequently a crucial parameter.[271] NPs/NCs are commonly derivatized with diverse functional groups, such as nucleic acid-targeted oligonucleotide probes, antibodies, and protein, to make nanoprobes. The m-NPs can also be in the shape of a nanovesicle with a membrane or a layer around it. Magnetic NPs are primarily used to help separate, purify, and concentrate various biomolecules. To accomplish this, various capturing molecules, including antibodies and oligonucleotide probes, can be fixed on the surface of magnetic m-NPs/NCs. The concentration of antigens from bulk solution,[287] the use of enzyme-entrapped NPs as nanocatalysts,[288] the separation/concentration of microbial cells from water samples,[289] and the purification of ultrapure supercoiled plasmid DNA[290] and *in vivo* tracking and recovery of progenitor cells[291] are all examples of applications for this type of magnetic particles or microspheres. The use of NPs for substrate coding, signal transduction, and amplification

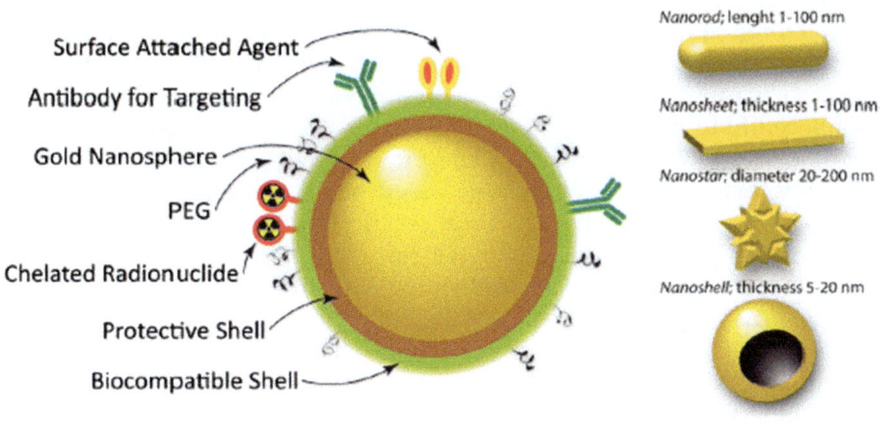

Figure 11.21 Image showing an example of a nanosphere of gold (Au). Gold nanoparticles (NPs) can also be based on, for example, nanorods, nanosheets, nanostars, or nanoshells, as indicated to the right with typical dimensions. The characteristics of AuNPs, and therefore their application possibilities, strongly depend on size, shape, and surface functionalities. Reproduced from ref. 271 with permission from MDPI, Copyright 2017.

has also been investigated.[292–294] It has been observed that having tight control over the average particle size of m-NCs, as well as a narrow distribution of sizes, enables the development of highly efficient fluorescent probes that can be customized to emit narrow light over a large wavelength range. These m-NCs help in the development of biomarkers in a range of colors.

The m-NC core itself might have multiple layers and be multifunctional; for example, combining magnetic and luminescent layers allows m-NCs to detect as well as manipulate particles. AuNCs, the most studied fluorescent m-NC, have a wide range of properties, including ultrasmall size, high water solubility, biocompatibility, photostability, and significant stokes shifts, to name a few. These characteristics make them promising fluorescent probes for biological labeling and imaging. AuNCs' produced light in the NIR range is capable of entering deeper tissue and detecting higher intensities due to poor biological autofluorescence, making them suitable fluorescent bioprobes for *in vivo* biological imaging. Because of their great adaptability, ease of manipulation, and minimal cytotoxicity, therapeutic nanocomplexes have been widely developed for the successful treatment of aggressive malignancies;[295,296] photothermal therapy (PTT), photodynamic therapy (PDT), gene therapy, and immunotherapy are some of the emerging therapies using such nanocomplexes. Jia *et al.*[297] described a simple approach for making protein-directed transition metal sulfide-NPs, which has potential as a photothermal agent in cancer treatment. They have great cellular absorption and biocompatibility, as well as strong photothermal conversion efficiency. Yang *et al.*[298] highlighted the importance of cysteine-rich protein and cysteine residue-related reactions. The fabrication of a multifunctional Gd_2O_3/Au hybrid nanoprobe employing BSA as the template in the biomineralization synthesis has been reported using a simple and one-step process.[299] The generated multifunctional nanoprobe's multimodal imaging capability was proven by successful NIR fluorescence and magnetic resonance blood pool imaging. Such BSA-Gd_2O_3/Au nanoprobe was further modified with RGD peptide to allow for targeted tumor imaging *in vivo*. Internally responsive nanomaterials based on m-NPs and m-NCs for activatable multimodal imaging of cancer, such as an activatable fluorescence/MRI bimodal platform for tumor cell imaging, have attracted significant recent focus.[300–304] Albumin-based nanoplatforms (MAlb-NPs) that adopt multiple imaging and therapeutic techniques have been attempted for cancer diagnosis and treatment[305] and antiglioma therapy.[306] Pt-NPs with protein coronas are autologously produced *in vivo* in human blood when a patient is treated with cisplatin (biosynthesized in the blood) and accumulate in tumors, according to Zang *et al.*[307] By consuming intracellular glutathione and inducing apoptosis, His-protein-coated Pt-NPs can not only function as anticancer drugs but also suppress chemotherapy-resistant tumor development.[308] When Pt-NPs are loaded *in vitro* with a chemotherapeutic agent and the peptide corona is GSH sensitive, the tumor inhibitory action of Pt-NP is considerably enhanced. Rationally designed, thermally responsive elastin-like peptides (ELPs) have recently been conjugated to multiple copies of a hydrophobic cancer drug for thermally targeting a tumor and therapeutic delivery.[309,310] Pt-NPs coated with albumin realize the same tumor-targeting properties as the FDA-approved paclitaxel–albumin complexes.

For monitoring drug delivery and distribution and the response of the targeted processes to the therapy, accurate tumor visualization is crucial. AuNCs with atomic precision have been created as effective NIR-II fluorophores for high-resolution brain imaging, kidney imaging, and metastatic tumor diagnostics.[311] The key feature of this dual-functional theranostic nanomedicine is that it can illuminate Pt transportation and maximize therapeutic efficacy by carrying the advantages of performing NIR-II imaging and GSH-scavenging tasks simultaneously. The long emission wavelength of AuNCs–Pt allows for deeper penetration and enables high resolution, NIR-II tumor imaging, which improves the visualization. Furthermore, the AuNCs–Pt effectively boosted Pt-drug efficacy by reducing tumor GSH levels, which is a preferred therapeutic strategy for multidrug-resistant tumors. Pt–AuNS (platinum-decorated gold nanostars) prodrugs are responsive to NIR light as ferroptosis-inducing agents in chemoresistant cancer cells and demonstrate their potential for use in future clinical applications.[312] AuNCs are also cleared from the body without toxicological consequences at high doses due to their ultrasmall size. It has been proposed to use as the multifunctional theranostic nanomedicine (AuNCs–Pt) to administer Pt(IV) drugs (Figure 11.22). This smart nanomedicine holds great potential for clinical imaging-based visualized therapy and platinum drug sensitization.[313]

11.8.1 Antibacterial Activity against Multidrug-resistant Bacteria

A wide range of metal and metal oxide NPs such as AuNPs, Fe_3O_2NPs, ZnO-NP CuO-NP, and TiONP, have been demonstrated to possess bactericidal effects and even antibacterial activity against multidrug-resistant bacteria (MDR). However, AgNPs and AgNCs are considered to be the most effective nanomaterial against bacteria.[314,315] Cell wall breakage, oxidation of cellular components, inactivation of respiratory chain enzymes, generation of reactive oxygen species (ROS), and breakdown of cellular components have all been hypothesized as methods for NPs to induce cell death. The use of peptides to immobilize NPs improves their stability and antibacterial characteristics, and it has the potential to be employed to combat antibiotic-resistant bacteria. The antibacterial activity of BSA–AuNCs against methicillin-resistant *Staphylococcus aureus* (MRSA) and vancomycin-intermediate *S. aureus* (VISA) has been established.[316,317] Pt-NPs biomineralized in the scaffold of BSA were observed to have excellent peroxidase-like activity. BSA–PtNPs are exceedingly stable and maintain activity in the presence of protein and even in human plasma samples, owing to the protective action of the protein shell. The antifouling ability of BSA–PtNPs makes them appropriate for a wide range of applications

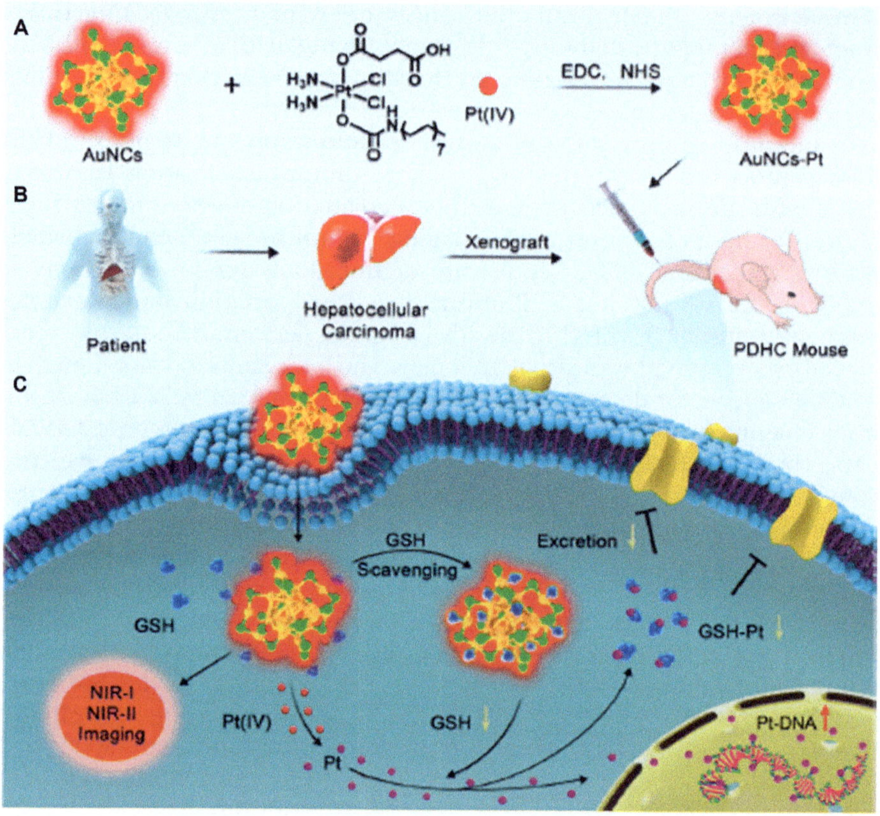

Figure 11.22 Schematic Illustration of a Theranostic Nanomedicine (AuNCs-Pt) with NIR-I and NIR-II imaging and GSH scavenging. Capacities for the eradication of a patient-derived tumor xenograft model of hepatocellular carcinoma (PDHC). Conditions: (A) Synthesis of AuNCs-Pt. (B) Illustration of the establishment of the PDHC model. (C) Dual-action mechanisms after endocytosis of AuNCs-Pt by cancer cells. The loaded Pt(IV) prodrugs can be reduced to toxic Pt, thereby inhibiting DNA replication and eventually apoptosis. In addition, AuNCs-Pt binds to GSH *via* Au – S bonds, resulting in the depletion of intracellular GSH, which reduces Pt detoxification and optimizes the therapeutic efficacy. Meanwhile, AuNCs-Pt with NIR-II tumor imaging capacity penetrates into deep tissues, allowing the visualization of the platinum transportation. Reproduced from ref. 313 with permission from American Chemical Society, Copyright 2020.

in biological sample sensors. For example, choline content in newborn formula and acetylcholine in plasma have been effectively identified using such sensors.[318] Vinayagam *et al.*[319] reported bioengineered Au-NPs using *Cynodon dactylon* extract and demonstrated its cytotoxicity and antibacterial activities. Kuo *et al.*[320] demonstrated that synthetic peptides containing arginine, tryptophan, and cysteine termini [such as (DVFLG)2REEW4C and

(DVFLG)2REEW2C] can be used to synthesize peptide–AuNPs, which are highly effective as antibacterial agents against staphylococci, enterococci, and antibiotic-resistant bacterial strains. It has been observed that superparamagnetic iron oxide NPs interact with microbial cells by penetrating the membrane and interfering with electron transport, as well as causing damage to macromolecules such as DNA and proteins by forming ROS. Peptide–AgNPs are emerging as one of the most promising inorganic NPs for treating bacterial infections, and they have emerged as a potent approach that has ushered in a new era.[321,322]

11.8.2 Application of NPs and NCs as Sensors

Song *et al.* showed that a label-free, sensitive, and facile method based on the selective aggregation of phosphorylated peptide–gold nanoclusters (peptide–Au-NCs) generated by Zr^{4+} ion coordination may be used to detect protein kinase.[323] The concentration of casein kinase II (CK2) in the range of 0.08–2.0 unit per mL with a detection limit of 0.027 unit per mL (3σ) was demonstrated. Park *et al.*[324] employed highly fluorescent Au-NCs made using a BSA-templated green synthesis method to create a new, label-free, fluorescent turn-on sensor for biological thiol detection. Xie *et al.*[325] developed a highly selective and ultrasensitive detection method for $Hg^{2+\bullet}$ based on fluorescence quenching. The specific and strong interaction between Au^+ and Hg^{2+} was used in this method, which effectively quenched the fluorescence of AuNCs. This technique can specifically detect and monitor Hg^{2+} ions at a minimal concentration of 0.5 nM. In a functionalized peptide-capped AuNC, the fluorescent AuNCs core may be used as a transducer, and the surface-functionalized molecules can be used as a recognition component.[326,327] Zang[328] proposed the use of dual-color peptide biomineralized m-NCs to determine protein kinase A and casein kinase II simultaneously. To make blue-emissive Cu-NCs and red-emissive Au-NCs, they specifically designed two artificial substrate peptides. Recently, for direct monitoring glucose in a whole blood sample that minimizes background absorption and fluorescence interference, the use of biomineralization to make a NIR fluorescent nanoprobe has been proposed.[329] In this study, NIR fluorescent protein (iRFP) and glucose oxidase (GOx) were used as templates for the biomineralization of Mn^{2+} to make an NIR fluorescent nanoprobe (iRFP-GO_x-MnO_2 NPs, iRGMs). The simplicity of manufacturing, observable color changes following exposure-related specific interaction, and quantitative read-out show that protein-facilitated biomineralization techniques have the potential to assist in the development of next-generation sensors for a variety of applications.

11.8.3 Application of NPs and NCs in Catalysis

Catalysts play a critical role in modern industry and >90% of all industrial chemicals produced in the world use catalysts within the manufacturing process. Due to their extremely large surface-to-volume ratio, the m-NPs and m-NCs have the potential to function as efficient catalysts in the production of chemicals. Therefore, extensive effort has been focused on the use of the biomineralization-inspired synthesis process for the colloidal synthesis of composition, size, and shape-controlled m-NPs and m-NCs. Due to their importance in the catalyst field, biomimetic synthesis of Pd,[330–343] Pt,[154,159] Au,[344–354] Ag,[355,356] and their alloys[357–359] have been of major focus. The development of electrocatalysts at solid–liquid interfaces has also been of significant research interest, because they are vital for driving the technological innovations that are essential to delivering reliable, affordable, and environmentally friendly energy.[360] By mimicking natural biomineralization, the production of shape-controlled metal nanostructures in aqueous media at room temperature under moderate conditions using carefully designed short peptides[154,159] of regulated sequence, including facet-selective binding segment and responsive peptides, offers not only an efficient and effective nanocatalyst but also potential flexibility in designing inorganic nanostructures with desired complex architecture and precisely oriented assembly. However, significant further research on the fundamental understanding of the interaction between protein/peptide and metal crystal faces in the amino acid sequence is critical to advance the frontier.

11.9 Conclusions and Perspectives

Biomineralization and biologically derived nanomaterials are exciting and emerging because they enable the creation of awe-inspiring functional nanomaterials, nanobioconjugates, NPs, and NCs from readily available resources under moderate, sustainable, and environmentally friendly circumstances. It has been recognized that proteins/peptides/peptoids have unique chemical compositions and sequence-level characteristics, responsiveness and environment-directed structural ensemble, specificity, and materials recognition properties that enable them to be used for rational design routes to synthesize programmed NPs and NCs with predictable properties and characteristics, which are difficult to achieve with other conventional NP/ligand combinations. Protein/peptide-directed synthesis, organization, and functionalization of NPs and NCs have advanced dramatically over the last two decades and are recommended for wide-ranging applications, including health, imaging, sensing, antibacterial, catalysis, and theranostics. Nonetheless, the greater part of the research has been directed toward the noble metal NPs and NCs; the prospect of other metals and metal oxides are less explored. While the design and synthesis of monometallic NPs and NCs utilizing the biomineralization technique has been the subject of major attention, relatively less focus has been on such nanostructures made of metal alloys, with only a few recent studies on bimetallic alloys[359] demonstrating activity enhancement. Even within the areas of alloy design, biomineralization can support the synthesis of 'high-entropy alloys' (multicomponent alloys, supported with more than five elements and the designs are based on near equiatomic molar ratio), which belong to a promising category with characteristics like cocktail effects with severe lattice distortion[361] but need further exploration. Indeed, the multidimensional compositional space that can be tackled with this approach is practically limitless.

Though the technology of biomimetic biomineralization is elegant and fascinating, and the possibilities are unlimited, the fundamental process involved is poorly understood and represents a grand challenge. The functional characteristics of NPs and NCs created using biomineralization are dependent not only on the characteristics of the NPs and the NCs but also on the structure, sequence, and ensemble of the individual protein/peptide involved and the collective biointerfaces at the NP/NC's material surface. The amino acid sequence, configurations, spacial organization, and/or conformational ensemble of the peptide/protein/peptoids are critical for directing the synthesis and achieving desired material functionality, and the binding motifs will vary with the peptide sequence and the environment. However, the detailed mechanism and kinetics of the peptide binding event and the impacts of the peptide sequence and binding motif on the surface characteristics and physicochemical properties of NPs/NCs remain poorly understood.[350] The nature of ligand–metal interactions on growing crystal faces has been shown to involve competitive ligand–solvent exchange equilibria and soft epitaxial matching of polarizable atoms (C, N, O) to hollow surface sites. Thus, the ability to expand the versatility of the peptide-mediated approach is dependent on our ability to elucidate and exploit structure–property relationships at the biointerfaces, and one key challenge to address this would be to advance both experimental and molecular simulation techniques to enable a clearer, more integrated characterization of the surface-adsorbed peptide structures.[362–365] Challenges include the inaccessibility of the peptide–particle interfaces to the available experimental detection techniques with high precision, as well as the complexity of polydisperse particle systems. This binding event of the peptide to the NP/NC surface provides key opportunities to control both the final overall morphology of the material and the resultant properties that arise from that specific structure. Such capabilities provide a powerful opportunity to tailor NPs and NCs through sequence engineering for targeted applications at levels that cannot be readily achieved with conventional nonbiological approaches.

Dedication

This chapter is dedicated to Mme Mireille Gouget and M. Gerard Derai, the 'Guardian Angels' in a very challenging but also exciting time (1991–1994).

Acknowledgements

The authors would like to acknowledge the financial support of ARC through ARC discovery projects (ARC-DPs) that helped develop the concepts of the chapter over a decade.

References

1. B. A. Krajina, A. C. Proctor, A. P. Schoen, A. J. Spakowitz and S. C. Heilshorn, *Prog. Mater. Sci.*, 2018, **91**, 1–23.
2. I. Chakraborty and W. J. Parak, *Adv. Mater. Interfaces*, 2019, **6**, 1801407.
3. A. M. Ealias and M. P. Saravanakumar, *Mater. Sci. Eng.*, 2017, **263**, 032019.
4. https://commons.wikimedia.org/wiki/File:Tutankhamun.jp.
5. E. C. Dreaden, A. M. Alkilany, X. Huang, C. J. Murphy and M. A. El-Sayed, *Chem. Soc. Rev.*, 2012, **41**, 2740–2779.
6. L. F. de Freitas, G. H. C. Varca, J. G. dos, S. Batista and A. B. volo Lugão, *Nanomaterials*, 2018, **8**, 939.
7. S. E. Skrabalak, L. Au, X. Li and Y. Xia, *Nat. Protoc.*, 2007, **2**, 2182–2190.
8. N. S. Abadeer, M. R. Brennan, W. L. Wilson and C. J. Murphy, *ACS Nano*, 2014, **8**, 8392–8406.
9. J. Zheng, C. Zhang and R. M. Dickson, *Phys. Rev. Lett.*, 2004, **93**, 077402.
10. E. Priyadarshini and N. Pradhana, *Sens. Actuators, B*, 2017, **238**, 888–902.
11. H.-H. Jeong, E. Choi, E. Ellisc and T.-C. Lee, *J. Mater. Chem. B*, 2019, **7**, 3480–3496.
12. S. Alex and A. Tiwari, *Nanosci. Nanotechnol.*, 2015, **15**, 1869–1894.
13. J. Li, J.-J. Zhu and K. Xu, *TrAC, Trends Anal. Chem.*, 2014, **58**, 90–98.
14. R. Balu, R. Knott, C. M. Elvin, A. J. Hill, N. Roy Choudhury and N. K. Dutta, *Biosensors*, 2019, **9**, 128.
15. R. Balu, L. Bourgeois, C. M. Elvin, A. J. Hill, N. Roy Choudhury and N. K. Dutta, *J. Mater. Chem. B*, 2015, **3**, 6580–6586.
16. S. Mayavan, N. K. Dutta, N. R. Choudhury, M. Kim, C. M. Elvin and A. J. Hill, *Biomaterials*, 2011, **32**, 2786–2796.
17. M. Faraday, *Philos. Trans. R. Soc. London*, 1857, **147**, 145–181.
18. G. Mie, *Ann. Phys.*, 1908, **25**, 377–445.
19. J. Turkevich, P. C. Stevenson and J. Hillier, *Discuss. Faraday Soc.*, 1951, **11**, 55–75.
20. G. Frens, Controlled Nucleation for the Regulation of the Particle Size in Monodisperse Gold Suspensions, *Nat. Phys. Sci.*, 1973, **241**, 20–22.
21. B.-K. Pong, H. I. Elim, J.-X. Chong, W. Ji, B. L. Trout and J.-Y. Lee, *J. Phys. Chem. C*, 2007, **111**, 6281–6287.
22. V. K. LaMer and R. H. Dinegar, *J. Am. Chem. Soc.*, 1950, **72**(11), 4847–4854.
23. M. M. Alvarez, J. T. Khoury, T. G. Schaaff, M. N. Shafigullin, I. Vezmar and R. L. Whetten, *J. Phy. Chem. B*, 1997, **101**, 3706–3712.
24. N. S. Abadeer and C. J. Murphy, *J. Phys. Chem. C*, 2016, **120**, 4691–4716.
25. K. L. Kelly and E. Coronado, *J. Phys. Chem. B*, 2003, **107**, 668–677.
26. S. Link and M. A. El-Sayed, *Annu. Rev. Phys. Chem.*, 2003, **54**, 331–366.
27. C. J. Murphy, T. K. Sau, A. M. Gole, C. J. Orendorff, J. Gao, L. Gou, S. E. Hunyadi and T. Li, *J. Phys. Chem. B*, 2005, **109**, 13857–13870.
28. K. Zheng and J. Xie, *Trends Chem.*, 2020, **2**, 665–679.

29. C. Han and W. Guo, *Small*, 2020, **16**, 2002621.
30. Y. Tao, M. Li, J. Ren and X. Qu, *Chem. Soc. Rev.*, 2015, **44**, 8636–8663.
31. D. Li, Z. Chen and X. Mei, *Adv. Colloid Interface Sci.*, 2017, **250**, 25–39.
32. X. Le Guével, C. Spies, N. Daum, G. Jung and M. Schneider, *Nano Res.*, 2012, **5**, 379–387.
33. X. Huang, Z. Li, Z. Yu, X. Deng and Y. Xin, *J. Nanomater.*, 2019, 6248725.
34. J. M. J. Santillán, D. M. Arboleda, D. Muraca, D. C. Schinca and L. B. Scaffardi, *Sci. Rep.*, 2020, **10**, 8217.
35. A. Mooradian, Photoluminescence of Metals, *Phys. Rev. Lett.*, 1969, **22**, 185–187.
36. J. P. Wilcoxon, J. E. Martin, F. Parsapour, B. Wiedenman and D. F. Kelley, *J. Chem. Phys.*, 1998, **108**, 9137–9143.
37. L.-Y. Chen, C.-W. Wang, Z. Yuan and H.-T. Chang, *Anal. Chem.*, 2015, **87**, 216–229.
38. T.-Q. Yang, B. Peng, B.-Q. Shan, Y.-X. Zong, J.-G. Jiang, P. Wu and K. Zhang, *Nanomaterials*, 2020, **10**, 261.
39. J. Zheng, P. R. Nicovich and R. M. Dickson, *Ann. Rev. Phys. Chem*, 2007, **58**, 409–431.
40. J. Zheng, C. Zhang and R. M. Dickson, *Phys. Rev. Lett.*, 2004, **93**, 077402.
41. R. Jin, *Nanoscale*, 2010, **2**, 343–362.
42. J. M. Forward, D. Bohmann, P. John, J. Fackler and R. J. Staples, *Inorg. Chem.*, 1995, **34**, 6330–6336.
43. Z. Lei, Z. Zhou, M. Liang, Y. Xiao and Z. Liu, *Front. Bioeng. Biotechnol.*, 2021, **8**, 628181.
44. J. Zheng, P. R. Nicovich and R. M. Dickson, *Annu. Rev. Phys. Chem.*, 2007, **58**, 409–431.
45. I. Rabin, W. Schulze, G. Ertl, C. Felix and C. Sieber, *Chem. Phys. Lett.*, 2000, **320**, 59–64.
46. J. W. Lichtman and J.-A. Conchello, *Nat. Methods*, 2005, **2**, 910–919.
47. S. Link, A. Beeby, S. G. Fitz, M. A. El-Sayed, T. G. Schaa and R. L. Whetten, *J. Phys. Chem. B*, 2002, **106**, 3410–3415.
48. Z. Luo, X. Yuan, Y. Yu, Q. Zhang, D. T. Leong, J. Y. Lee and J. Xie, *J. Am. Chem. Soc.*, 2012, **134**, 16662–16670.
49. Z. Wu and R. Jin, *Nano Lett.*, 2010, **10**, 2568–2573.
50. J. Liu, P. N. Duchesne, M. Yu, X. Jiang, X. Ning, R. D. Vinluan, P. Zhang and J. Zheng, *Angew. Chem. Int. Ed.*, 2016, **55**, 8894–8898.
51. J. Xie, Y. Zheng and J. Y. Ying, *J. Am. Chem. Soc.*, 2009, **131**, 888–889.
52. B. Shan, F. Huang, S. Yang, B. Peng, E. Yuan, P. Wu and K. Zhang, *Commun. Chem.*, 2019, **2**, 132.
53. M. B. Dickerson, K. H. Sandhage and R. R. Naik, *Chem. Rev.*, 2008, **108**, 4935–4978.
54. J. S. Evans, *Materials*, 2019, **12**, 581.
55. J. S. Evans, *Proteomics*, 2019, **19**, 1900036.
56. F. Reith, B. Etschmann, C. Grosse, H. Moors, M. A. Benotmane, P. Monsieurs, G. Grass, C. Doonan, S. Vogt, B. Lai, G. Martinez-Criado,

G. N. George, D. H. Nies, M. Mergeay, A. Pring, G. Southam and J. Brugger, *Proc. Natl. Acad. Sci. U. S. A.*, 2001, **106**, 17757–17762.
57. M. Sumper, E. Brunner and G. Lehmann, *FEBS Lett.*, 2005, **579**, 3765–3769.
58. R. T. DeVol, C. Y. Sun, M. A. Marcus, S. N. Coppersmith, S. C. Myneni and P. U. Gilbert, *J. Am. Chem. Soc.*, 2015, **137**, 13325–13333.
59. O. Paris, I. Zizak, H. Lichtenegger, P. Roschger, K. Klaushofer and P. Fratzl, *Cell. Mol. Biol.*, 2000, **46**, 993–1004.
60. R. R. Naik, S. J. Stringer, G. Agarwal, S. E. Jones and M. O. Stone, *Nat. Mater.*, 2002, **1**, 169–172.
61. S. Brown, *Nat. Biotechnol.*, 1997, **15**, 269–272.
62. S. R. Whaley, D. S. English, E. L. Hu, P. F. Barbara and A. M. Belcher, *Nature*, 2000, **405**, 665–668.
63. M. Suzuki, *Biosci. Biotechnol Biochem*, 2020, **84**, 1529–1540.
64. A. Arakaki, K. Shimizu, M. Oda, T. Sakamoto, T. Nishimura and T. Kato, *Org. Biomol. Chem.*, 2015, **13**, 974.
65. A. L. Boskey and E. Villarreal-Ramirez, *Matrix Biol.*, 2016, **52–54**, 43.
66. T. V. Avramenko, Y. N. Shkryl, G. N. Veremeichik and V. P. Bulgakov, *J. Nanopart. Res.*, 2020, **22**, 110.
67. H. Bönnemann, W. Brijoux, R. Brinkmann, E. Dinjus, T. Joupen and B. Korall, *Angew. Chem. Eng. Ed.*, 1991, **30**, 1312.
68. H. Bönnemann and R. M. Richards, *Eur. J. Inorg. Chem.*, 2001, **10**, 2455–2480.
69. T.-H. Lee, J. I. Gonzalez, J. Zheng and R. M. Dickson, *Acc. Chem. Res.*, 2005, **38**, 534–541.
70. F. Lu, H. Yang, Z. Yuan, T. Nakanishi, C. Lu and Y. He, *Sens. Actuators, B*, 2019, **291**, 170–176.
71. N. Xiao, J. X. Dong and S. G. Liu, *et al.*, *Sens. Actuators, B*, 2018, **264**, 184–192.
72. S. Ghosh, J. R. Bhamore, N. I. Malek, Z. V. P. Murthy and S. K. Kailasa, *Spectrochim. Acta, Part A*, 2019, **215**, 209–217.
73. X. Y. Wang, G. B. Zhu and W. D. Cao, *et al.*, *Talanta*, 2019, **191**, 46–53.
74. J. D. Aiken III and R. G. Finke, *J. Mol. Catal. A: Chem.*, 1999, **145**, 1–44.
75. Y. Zhang, C. Zhang and C. Xu, *et al.*, *Talanta*, 2019, **200**, 432–442.
76. X. Huang, Z. Li, Z. Yu, X. Deng and Y. Xin, *J. Nanomater.*, 2019, 6248725.
77. T.-H. Yang, Y. Shi, A. Janssen and Y. Xia, *Angew. Chem. Int. Ed.*, 2020, **59**, 15378–15401.
78. J. Xiao and L. Qi, *Nanoscale*, 2011, **3**, 1383–1396.
79. T. S. Rodrigues, M. Zhao, T. H. Yang, K. D. Gilroy, A. G. M. da Silva, P. H. C. Camargo and Y. Xia, *Chemistry*, 2018, **24**, 16944–16963.
80. Y. Xia, Y. Xiong, B. Lim and S. E. Skrabalak, *Angew. Chem. Int. Ed.*, 2009, **48**, 60–103.
81. Y. Xia, X. Xia and H.-C. Peng, *J. Am. Chem. Soc.*, 2015, **137**, 7947–7966.
82. Y. Xia, K. D. Gilroy, H.-C. Peng and X. Xia, *Angew Chem Int Ed.*, 2017, **56**, 60–95.

83. A. Ruditskiy, H.-C. Peng and Y. Xia, *Annu. Rev. Chem. Biomol. Eng.*, 2016, **7**, 327–348.
84. X. Peng, *Acc. Chem. Res.*, 2010, **43**, 1387–1395.
85. L. Xu, D. Liu, D. Chen, H. Liu and J. Yang, *Heliyon*, 2019, **5**, e01165.
86. A. Abedini, A. A. Bakar, F. Larki, P. S. Menon, M. S. Islam and S. Shaari, *Nanoscale Res. Lett.*, 2016, **11**, 287.
87. J. Sui, J. Yan, D. Liu, K. Wang and G. Luo, *Small*, 2020, **16**, e1902828.
88. G. Niu, A. Ruditskiy, M. Vara and Y. Xia, *Chem. Soc. Rev.*, 2015, **44**, 5806–5820.
89. S. E. Skrabalak and Y. Xia, *ACS Nano*, 2009, **3**, 10–15.
90. L. Bai, Y. Ouyang, J. Song, Z. Xu, W. Liu, J. Hu, Y. Wang and F. Yuan, *Materials*, 2019, **12**, 1497.
91. B. Lim and Y. Xia, *Angew. Chem. Int. Ed.*, 2011, **50**, 76–85.
92. H. Zhang, M. Jin, Y. Xiong, B. Lim and Y. Xia, *Acc. Chem. Res.*, 2013, **46**, 1783–1794.
93. H. Zhang, M. Jin and Y. Xia, *Angew. Chem. Int. Ed.*, 2012, **51**, 7656–7673.
94. J. Zhang, F. Huang and Z. Lin, *Nanoscale*, 2010, **2**, 18–34.
95. D. V. Goia and E. Matijevic, *New J. Chem.*, 1998, **22**, 1203–1215.
96. X. Y. Liu, *J. Chem. Phys.*, 2000, **112**, 9949.
97. B. Mutaftschiev, in *Handbook on Crystal Growth*, ed. D. T. J. Hurle, North-Holland, Amsterdam, 1993, p. 18.
98. Z. Peng and H. Yang, *Nano Today*, 2009, **4**, 143–164.
99. N. T. K. Thanh, N. Maclean and S. Mahiddine, *Chem. Rev.*, 2014, **114**, 7610–7630.
100. P. Tartaj, M. Morales, S. Veintemillas-Verdaguer, T. González-Carreño and C. J. Serna, *J. Phys. D: Appl. Phys.*, 2003, **36**, R182.
101. V. K. LaMer and R. H. Dinegar, *J. Am. Chem. Soc.*, 1950, **72**, 4847.
102. M. Ocaña, R. Rodriguez-Clemente and C. J. Serna, *Adv. Mater.*, 1995, **7**, 212.
103. C. J. J. Den Ouden and R. W. Thompson, *J. Colloid Interface Sci.*, 1991, **143**, 77–84.
104. H. Yang, X. W. Teng and S. Maksimuk, Metal Nanoclusters in Catalysis and Materials Science: The Issue of Size Control, in *Part II: Methodologies*, ed. B. Corain, C. G. Schmid and N. Toshima, Elsevier, Amsterdam, 2008.
105. A. R. Tao, S. Habas and P. D. Yang, *Small*, 2008, **4**, 310–325.
106. Y. Yin and A. P. Alivisatos, *Nature*, 2005, **437**, 664–670.
107. M. Volmer and A. Weber, *Z. Phys. Chem.*, 1926, **119**, 227.
108. D. L. Feldheim and B. E. Eaton, *ACS Nano*, 2007, **1**, 154–159.
109. C. W. Johnston, M. A. Wyatt, X. Li, A. Ibrahim and J. Shuster, *et al.*, *Nat. Chem. Biol.*, 2013, **9**, 241–243.
110. A. Malhotra, K. Dolma, N. Kaur, Y. S. Rathore, A. S. Mayilraj and A. Roy Choudhury, *Bioresour. Technol.*, 2013, **142**, 727–731.
111. C. H. Moon and E. D. Haberer, *MRS Symposium Proceedings*, 2012, **1461**, 1–6.

112. K. W. Pulsipher, S. Honig, S. Deng and I. J. Dmochowski, *J. Inorg. Biochem.*, 2017, **174**, 169–176.
113. S. R. Jones, T. D. Wilson, M. E. Brown, L. Rahn-Lee, Y. Yu, L. L. Fredriksen, E. Ozyamak, A. Komeili and M. C. Y. Chang, *Proc. Natl. Acad. Sci. U. S. A.*, 2015, **112**, 3904–3909.
114. H. Zhang, X. Liu, S. Feng, W. Wang and K. Schmidt-Rohr, *et al.*, *Langmuir*, 2015, **31**, 2818–2825.
115. K. Zhou, J. Zhang and Q. Wang, *Small*, 2015, **11**, 2505–2509.
116. Y. Hosomomi, T. Niide, R. Wakabayashi, M. Goto and N. Kamiya, *Anal. Sci.*, 2016, **32**, 295–300.
117. A. A. Glasgow and D. Tullman-Ercek, *Methods Mol. Biol.*, 2018, **1798**, 155–171.
118. A. V. Singh, T. Jahnke, V. Kishore, B.-W. Park and M. Batuwangala, *et al.*, *Acta Biomater.*, 2018, **71**, 61–71.
119. A. S. Schwartz-Duval, C. J. Konopka, P. Moitra, L. W. Dobrucki and D. Pan, *Nat. Commun.*, 2020, **11**, 4530.
120. W. A. El-Said, H. Y. Cho, C. H. Yea and J. W. Choi, *Adv. Mater.*, 2014, **26**, 910–918.
121. A. Shamsaie, M. Jonczyk, J. D. Sturgis, J. P. Robinson and J. Irudayaraj, *J. Biomed. Opt.*, 2007, **12**, 020502.
122. D. Drescher, H. Traub, T. Büchner, N. Jakubowski and J. Kneipp, *Nanoscale*, 2017, **9**, 11647–11656.
123. M. Ozaki, S. Yoshida, T. Tsuruoka and K. Usui, *Chem. Commun.*, 2021, **57**, 725–728.
124. J. M. Palomo, *Chem. Commun.*, 2019, **55**, 9583–9589.
125. S. K. Das, J. Liang, M. Schmidt, F. Laffir and E. Marsili, *ACS Nano*, 2012, **6**, 6165–6173.
126. K. T. Nam, Y. J. Lee, E. M. Krauland, S. T. Kottmann and A. M. Belcher, *ACS Nano*, 2008, **2**, 1480–1486.
127. D. He and J. Marles-Wright, *New Biotechnol.*, 2015, **32**, 651–657.
128. C.-L. Chen and N. L. Rosi, *Angew. Chem. Int. Ed.*, 2010, **49**, 1924–1942.
129. R. J. Mart, R. D. Osborne, M. M. Stevens and R. V. Ulijn, *Soft Matter*, 2006, **2**, 822–835.
130. M. Amiram, F. G. Quiroz, D. J. Callahan and A. Chilkoti, *Nat. Mater.*, 2011, **10**, 141–148.
131. R. L. DiMarco and S. C. Heilshorn, *Adv. Mater.*, 2012, **24**, 3923–3940.
132. R. Y. Tsien, *Angew. Chem. Int. Ed.*, 2009, **48**, 5612–5626.
133. J. Scheller, K. H. Gührs, F. Grosse and U. Conrad, *Nat. Biotechnol.*, 2001, **19**, 573–577.
134. C. M. Elvin, A. M. Carr, M. G. Huson, J. M. Maxwell, R. D. Pearson and T. Vuocolo, *Nature*, 2005, **437**, 999–1002.
135. B. Miroux and J. E. Walker, *J. Mol. Biol.*, 1996, **260**, 289–298.
136. K. Bolaños, M. J. Kogan and E. Araya, *Int. J. Nanomed.*, 2019, **14**, 6387–6406.
137. K. Karami, N. Jamshidian, A. Hajiaghasia and Z. Amirghofran, *New J. Chem.*, 2020, **44**, 4394–4405.

138. L. Van de Sande, S. Cosyns, W. Willaert and W. Ceelen, Albumin-based cancer therapeutics for intraperitoneal drug delivery: a review, *Drug Delivery*, 2020, **27**, 40–53.
139. Harnessing albumin as a carrier for cancer therapies E. N. Hoogenboezem and C. L. Duvall, *Adv. Drug Delivery Rev.*, 2018, **130**, 73–89
140. A. V. Singh, B. M. Bandgar, M. Kasture, B. L. V. Prasad and M. Sastry, *J. Mater. Chem.*, 2005, **15**, 5115–5121.
141. J. Xie, J. Y. Lee and D. I. C. Wang, *J. Phys. Chem. C*, 2007, **111**, 10226.
142. Z. Yang, S. Luo, Y. Zeng, C. Shi and R. Li, *ACS Appl. Mater. Interfaces*, 2017, **9**, 6839–6848.
143. X. Zeng, J. Sun, S. Li, J. Shi and H. Gao, *et al.*, *Nat. Commun.*, 2020, **11**, 567.
144. M. R. Hartings, N. Benjamin, F. Briere, M. Briscione and O. Choudary, *et al.*, *Sci. Technol. Adv. Mater.*, 2013, **14**, 065004.
145. A. J. Kumar, M. Kailasnath, J. Simon and S. Mathew, *Mater. Today: Proc.*, 2019, **9**, 111–115.
146. L. Lu, K. Ai and Y. Ozaki, *Langmuir*, 2008, **24**, 1058.
147. I. Chakraborty, N. Feliu, S. Roy, K. Dawson and W. J. Parak, *Bioconjugate Chem.*, 2018, **29**, 1261–1265.
148. S. Abe, B. Maity and T. Ueno, *Chem. Commun.*, 2016, **52**, 6496–6512.
149. R. Xing, T. Jiao, L. Yan, G. Ma and L. Liu, *ACS Appl. Mater. Interfaces*, 2015, **7**, 24733–24740.
150. G. J. Bedwell, Z. Zhou, M. Uchida, T. Douglas and A. Gupta, *Biomacromolecules*, 2015, **16**, 214–218.
151. L. Zou, W. Qi, R. Huang, R. Sue, M. Wang and Z. He, *ACS Sustainable Chem. Eng.*, 2013, **1**, 1398–1404.
152. I. Chakraborty and W. Parak, *J Adv. Mater. Interfaces*, 2019, **6**, 1801407.
153. W. Yang, W. Guo, J. Chang and B. Zhang, *J. Mater. Chem. B*, 2017, **5**, 401.
154. Y. Li, G. P. Whyburn and Y. Huang, *J. Am. Chem. Soc.*, 2009, **131**, 15998–15999.
155. S. R. Qiu, A. Wierzbicki, C. A. Orme, A. M. Cody, J. R. Hoyer, G. H. Nancollas, S. Zepeda and J. De Yoreo, *Proc. Natl. Acad. Sci. U.S.A.*, 2004, **101**, 1811–1815.
156. L. C. Ciacchi, M. Mertig, R. Seidel, W. Pompe and V. De, *Nanotechnology*, 2003, **14**, 840–848.
157. Y. N. Tan, J. Y. Lee and D. I. C. Wang, *J. Am. Chem. Soc.*, 2010, **132**, 5677–5686.
158. L. M. Forbes, A. P. Goodwin and J. N. Cha, *Chem. Mater.*, 2010, **22**, 6524–6528.
159. C. Y. Chiu, Y. J. Li, L. Y. Ruan, X. C. Ye, C. B. Murray and Y. Huang, *Nat. Chem.*, 2011, **3**, 393–399.
160. Y. J. Li and Y. Huang, *Adv. Mater.*, 2010, **22**, 1921–1925.
161. L. Ruan, C.-Y. Chiu, Y. Li and Y. Hu, *Nano Lett.*, 2011, **11**, 3040–3046.

162. R. Coppage, J. M. Slocik, B. D. Briggs, A. I. Frenkel, H. Heinz, R. R. Naik and M. R. Knecht, *J. Am. Chem. Soc.*, 2011, **133**, 12346–12349.
163. R. Coppage, J. M. Slocik, B. D. Briggs, A. I. Frenkel, R. R. Naik and M. R. Knecht, *ACS Nano*, 2012, **6**, 1625–1636.
164. R. Bhandari and M. R. Knecht, *Langmuir*, 2012, **28**, 8110–8119.
165. D. B. Pacardo, M. Sethi, S. E. Jones, R. R. Naik and M. R. Knecht, *ACS Nano*, 2009, **3**, 1288–1296.
166. R. Naik, S. J. Stringer, G. Agarwal, S. E. Jones and M. O. Stone, *Nat. Mater.*, 2020, **1**, 169–172.
167. Z. Tang, J. P. Palafox-Hernandez, W.-C. Law, Z. E. Hughes and M. T. Swihart, *et al.*, *ACS Nano*, 2013, **7**, 9632–9646.
168. C. J. Munro, Z. E. Hughes, T. R. Walsh and M. R. Knecht, *J. Phys. Chem. C*, 2016, **120**, 18917–18924.
169. J. Zong, S. L. Cobb and N. R. Cameron, *J. Mater. Chem. B*, 2018, **6**, 6667–6674.
170. R. Alvarez-Rodríguez, F. J. Arias, M. Santos, A. M. Testera and J. C. Rodríguez-Cabello, *Macromol. Rapid Commun.*, 2010, **31**, 568–573.
171. Y. Lin, X. Xia, M. Wang, Q. Wang, B. An, H. Tao, Q. Xu, F. Omenetto and D. L. Kaplan, *Langmuir*, 2014, **30**, 4406–4414.
172. N. K. Dutta, *et al.*, 'Template directed formation of metal nanoparticles and uses thereof', PCT/AU2013/001300.
173. W. Wang, C. F. Anderson, Z. Wang, W. Wu, H. Cui and C. J. Liu, *Chem. Sci.*, 2017, **8**, 3310–3324.
174. W. Wang, C. F. Anderson, Z. Wang, W. Wu, H. Cui and C.-J. Liu, *Chem. Sci.*, 2017, **8**, 3310.
175. N. M. Bedford, A. R. Showalter, T. J. Woehl, Z. E. Hughes, S. Lee, B. Reinhart, S. P. Ertem, E. B. Coughlin, Y. Ren, T. R. Walsh and B. A. Bunker, *ACS Nano*, 2016, **10**, 8645–8659.
176. J. M. Slocik, A. O. Govorov and R. R. Naik, *Angew. Chem., Int. Ed.*, 2008, **47**, 5335–5339.
177. C. Song, Y. Wang and N. L. Rosi, *Angew. Chem., Int. Ed.*, 2013, **52**, 3993–3995.
178. N. M. Bedford, A. R. Showalter, T. J. Woehl, Z. E. Hughes and S. Lee, *et al.*, *ACS Nano*, 2016, **10**, 8645–8659.
179. C.-Y. Chiu, L. Ruan and Y. Huang, *Chem. Soc. Rev.*, 2013, **42**, 2512–2527.
180. B. D. Briggs and M. R. Knecht, *J. Phys. Chem. Lett.*, 2012, **3**, 405–418.
181. T. R. Walsh and M. R. Knecht, *Chem. Rev.*, 2017, **117**, 12641–12704.
182. I. A. Banerjee, G. Muniz, S.-Y. Lee and H. Matsui, *J. Nanosci. Nanotechnol.*, 2007, **7**, 2287–2292.
183. M. A. Kline, L. Guo and R. N. Zuckermann, *Sequence-Controlled Peptoid Polymers: Bridging the Gap between Biology and Synthetic Polymers in Sequence-Controlled Polymers*, ed. J.-F. Lutz, Wiley-VCH Verlag GmbH, 2018, pp. 183–227.
184. F. Yan, L. Liu, T. R. Walsh, Y. Gong and P. Z. El-Khoury, *et al.*, *Nat. Commun.*, 2018, **9**, 2327.
185. B. Cai, Z. Li and C.-L. Chen, *Acc. Chem. Res.*, 2021, **54**, 81–91.

186. G. Maayan and L.-K. Liu, *Biopolymer*, 2011, **96**, 679–687.
187. T.-Z. Hagar and G. Maayan, *Colloid Interface Sci.*, 2019, **533**, 598–603.
188. L. B. Wright, J. P. Palafox-Hernandez, P. M. Rodger, S. Corni and T. R. Walsh, *Chem. Sci.*, 2015, **6**, 5204–5214.
189. S. K. Ramakrishnan, M. Martin, T. Cloitre, L. Firlej and C. Gergely, *Phys. Chem. Chem. Phys.*, 2015, **17**, 4193–4198.
190. S. K. Ramakrishnan, M. Martin, T. Cloitre, L. Firlej, F. J. G. Cuisinier and C. Gergely, *J. Chem. Inf. Model.*, 2013, **53**, 3273–3279.
191. G. Hong, H. Heinz, R. R. Naik, B. L. Farmer and R. Pachter, *ACS Appl. Mater. Interfaces*, 2009, **1**, 388–392.
192. H. Heinz, B. L. Farmer, R. B. Pandey, J. M. Slocik, S. S. Patnaik, R. Pachter and R. R. Naik, *J. Am. Chem. Soc.*, 2009, **131**, 9704–9714.
193. R. B. Pandey, H. Heinz, J. Feng, B. L. Farmer, J. M. Slocik, L. F. Drummy and R. R. Naik, *Phys. Chem. Chem. Phys.*, 2009, **11**, 1989–2001.
194. L. Ruan, H. Ramezani-Dakhel, C.-Y. Chiu, E. Zhu, Y. Li, H. Heinz and Y. Huang, *Nano Lett.*, 2013, **13**, 840–846.
195. C. H. Lin, S. P. Ju, J. W. Su and D. E. Li, *Sci. Rep.*, 2020, **10**, 2090.
196. H. Haberland, *Clusters of Atoms and Molecules: Theory, Experiment, and Clusters of Atoms*, Springer, Berlin/Heidelberg, Germany, 2013.
197. A. Halder, L. A. Curtiss, A. Fortunelli and S. Vajda, *J. Chem. Phys.*, 2018, **148**, 110901.
198. S. Vajda and M. G. White, Catalysis applications of size-selected cluster deposition, *ACS Catal.*, 2015, **5**, 7152–7176.
199. T. Huang and R. W. Murray, *J. Phys. Chem. B*, 2001, **105**, 12498–12502.
200. N. K. Dutta, A. K. Dutta and N. Roy Choudhury, 'Optically Coupled Hybrid Architectures Based on Fluorescent Proteins and Gold Nanoparticles, in *Encyclopedia of Metalloproteins*, ed. R. H. Kretsinger, V. N. Uversky and E. A. Permyakov, Springer, 2013, pp. 895–908.
201. X. Qu, Y. Li, L. Li, Y. Wang, J. Liang and J. Liang, *J. Nanomater.*, 2015, **2015**, 784097.
202. H. Duan, D. Wang and Y. Li, *Chem. Soc. Rev.*, 2015, **44**, 5778–5792.
203. A. Pandya, A. N. Lad, S. P. Singh and R. Shanker, *RSC Adv.*, 2016, **6**, 113095–113114.
204. H. Hu, S. Lu, T. Li, Y. Zhang and C. Guo, *et al.*, *Nanoscale Adv.*, 2021, **3**, 1865–1886.
205. J. He, S. Razzaque, S. Jin, I. Hussain and B. Tan, *ACS Appl. Nano Mater.*, 2019, **2**(1), 546–553.
206. J. Zheng, J. T. Petty and R. M. Dickson, *J. Am. Chem. Soc.*, 2003, **125**, 7780–7781.
207. Z. Zamani Nori, A. Landarani-Isfahani, M. Bahadori, M. Moghadam and V. Mirkhani, *et al.*, *RSC Adv.*, 2020, **10**, 33137–33314.
208. M. Rao Kotte, A. T. Kuvarega, S. N. Talapaneni, M. Cho and A. Coskun, *et al.*, *ACS Appl. Mater. Interfaces*, 2018, **10**, 33238–33251.
209. J. Xie, Y. Zheng and J. Y. Ying, *J. Am. Chem. Soc.*, 2009, **131**, 888–889.
210. C.-L. Liu, H.-T. Wu and Y.-H. Hsiao, *et al.*, *Angew. Chem. Int. Ed.*, 2011, **50**, 7056–7060.

211. A. R. Garcia, I. Rahn and S. Johnson, *et al.*, *Colloids Surf., B*, 2013, **105**, 167–172.
212. C. Guoand and J. Irudayaraj, *Anal. Chem.*, 2011, **83**, 2883–2889.
213. H. Wei, Z. Wang, L. Yang, S. Tian, C. Hou and Y. Lu, *Analyst*, 2010, **135**, 1406–1410.
214. J. Qiao, X. Mu, L. Qi, J. Deng and L. Mao, *Chem. Commun.*, 2013, **49**, 8030–8032.
215. D. Joseph and K. E. Geckeler, *Colloids Surf., B*, 2014, **115**, 46–50.
216. H. Chen, B. Li and C. Wang, *et al.*, *Nanotechnology*, 2013, **24**, 055704.
217. Y. Zhang, Q. Hu and M. C. Paau, *et al.*, *J. Phys. Chem. C*, 2013, **117**, 18697–18708.
218. X. Mu, L. Qi and P. Dong, *et al.*, *Biosens. Bioelectron.*, 2013, **49**, 249–255.
219. X. Yang, M. Shi, R. Zhou, X. Chen and H. Chen, *Nanoscale*, 2011, **3**, 2596–2601.
220. Y. Negishi, Y. Takasugi, S. Sato, H. Yao, K. Kimura and T. Tsukuda, *J. Am. Chem. Soc.*, 2004, **126**, 6518–6519.
221. Y. Negishi, K. Nobusada and T. Tsukuda, *J. Am. Chem. Soc.*, 2005, **127**, 5261–5270.
222. H. Zhang, Q. Liu and T. Wang, *et al.*, *Anal. Chim. Acta*, 2013, **770**, 140–146.
223. W. Li, H. Zhang, J. Zhanga and Y. Fu, *Anal. Methods*, 2015, **7**, 4464–4471.
224. Y.-C. Gao, C. Wang, C.-X. Zhang, H.-W. Li and Y. Wu, *Microchim. Acta*, 2021, **188**, 50.
225. L. Tang, J. Shi, H. Wu, S. Zhang and H. Liu, *Nanotechnology*, 2017, **28**, 365604.
226. R. de la Rica, L. W. Chow, C.-M. Horejs, M. Mazo and C. Chiappini, *et al.*, *Chem. Commun.*, 2014, **50**, 10648–10650.
227. Y. Wang, Y. Cui and Y. Zhao, *et al.*, *Chem. Commun.*, 2012, **48**, 871–873.
228. G. Liang, X. Jin, S. Zhang and D. Xing, *Biomaterials*, 2017, **144**, 95–104.
229. Y. Yu, Z. Luo, C. S. Teo, Y. N. Tan and J. Xie, *Chem. Commun.*, 2013, **49**, 9740–9742.
230. N. K. Dutta, M. Y. Truong, N. Roy Choudhury, M. Kim, C. M. Elvin and A. J. Hill, *Angew. Chem. Int. Ed.*, 2011, **50**, 4428–4431.
231. M. Y. Truong, N. K. Dutta, N. Roy Choudhury, M. Kim, C. M. Elvin, A. J. Hill, B. Thierry and K. Vasilev, *Biomaterials*, 2010, **31**, 4434–4446.
232. M. Y. Truong, N. K. Dutta, N. Roy Choudhury, M. Kim, C. M. Elvin, K. M. Nairn and A. J. Hill, *Biomaterials*, 2011, **32**, 8462–8473.
233. N. K. Dutta, M. Y. Truong, N. Roy Choudhury, C. M. Elvin, M. Kim and A. J. Hill, *Biomaterials*, 2009, **30**, 4868–4876.
234. R. Balu, N. K. Dutta, N. Roy Choudhury, C. M. Elvin, R. E. Lyons, R. Knott and A. J. Hill, *Acta Biomater.*, 2014, **10**, 4768–4777.
235. R. Balu, N. K. Dutta, A. Dutta and N. Roy Choudhur, Resilin-mimetics as a smart biomaterial platform for biomedical applications, *Nat. Commun.*, 2021, **12**, 149.

236. R. Balu, J. Mata, R. Knott, C. Elvin, A. Hill, N. Roy Choudhury and N. K. Dutta, *J. Phys. Chem. C*, 2016, **120**, 6490–6503.
237. R. Balu, R. Knott, C. M. Elvin, A. Hill, N. Roy Choudhury and N. K. Dutta, *Sci. Rep.*, 2015, **5**, 10896.
238. B. Maity, S. Abe and T. Ueno, *Nat. Commun.*, 2017, **8**, 14820.
239. L. Yan, Y. Cai, B. Zheng, H. Yuan and Y. Guo, *J. Mater. Chem.*, 2012, **22**, 1000–1005.
240. L. Shang, S. Brandholt, F. Stockmar, V. Trouillet, M. Bruns and G. U. Nienhaus, *Small*, 2012, **8**, 661–665.
241. L. Shang, R. M. Dörlich, V. Trouillet, M. Bruns and G. Ulrich Nienhaus, *Nano Res.*, 2012, **5**, 531–542.
242. T. S. Sych, Z. V. Reveguk, V. A. Pomogaev, A. A. Buglak and A. A. Reveguk, *et al.*, *J. Phys. Chem. C*, 2018, **122**, 29549–29558.
243. T. S. Sych, A. A. Buglak, Z. V. Reveguk, V. A. Pomogaev, R. R. Ramazanov and A. I. Kononov, *J. Phys. Chem. C*, 2018, **122**, 26275–26280.
244. Y. Li, Z. Tang, P. N. Prasad, M. R. Knecht and M. T. Swihart, *Nanoscale*, 2014, **6**, 3165–3172.
245. N. A. Merrill, E. M. McKee, K. C. Merino, L. F. Drummy and S. Lee, *et al.*, *ACS Nano*, 2015, **9**(12), 11968–11979.
246. S. S. Narayanan and S. K. Pal, *J. Phys. Chem. C*, 2008, **112**, 4874–4879.
247. S. Volden, S. M. Lystvet, O. Halskau and W. R. Glomm, *RSC Adv.*, 2012, **2**, 11704–11711.
248. J. Xie, Y. Zheng and J. Y. Ying, *J. Am. Chem. Soc.*, 2009, **131**, 888–889.
249. N. Goswami, A. Giri, M. S. Bootharaju, P. L. Xavier and T. Pradeep, *et al.*, *Anal. Chem.*, 2011, **83**, 9676–9680.
250. C. Guo and J. Irudayaraj, *Anal. Chem.*, 2011, **83**, 2883–2889.
251. X. Le Guével, B. Hötzer, G. Jung, K. Hollemeyer and V. Trouillet, *et al.*, *J. Phys. Chem. C*, 2011, **115**, 10955–10963.
252. J. S. Mohanty, P. L. Xavier, K. Chaudhari, M. S. Bootharaju and N. Goswami, *et al.*, *Nanoscale*, 2012, **4**, 4255–4262.
253. T. Das, P. Ghosh, M. S. Shanavas, A. Maity and S. Mondal, *et al.*, *Nanoscale*, 2012, **4**, 6018–6024.
254. Y. Xu, J. Sherwood, Y. Qin, D. Crowley, M. Bonizzoni and Y. Bao, *Nanoscale*, 2014, **6**, 1515–1524.
255. A. Mathew, P. R. Sajanlal and T. Pradeep, *J. Mater. Chem.*, 2011, **21**, 11205–11212.
256. P. L. Xavier, K. Chaudhari, A. Baksi and T. Pradeep, *Nano Rev.*, 2012, **3**, 14767.
257. Z. Suo, X. Hou, Z. Hu, Y. Liu, F. Xing and L. Feng, *Mikrochim. Acta*, 2019, **186**, 799.
258. F. Wen, Y. Dong, L. Feng, S. Wang, S. Zhang and X. Zhang, *Anal. Chem.*, 2011, **83**, 1193–1196.
259. K. Chaudhari, P. L. Xavier and T. Pradeep, *ACS Nano*, 2011, **5**, 8816–8827.
260. P. L. Xavier, K. Chaudhari, P. K. Verma, S. K. Pal and T. Pradeep, *Nanoscale*, 2010, **2**, 2769–2776.

261. A. Baksi, P. L. Xavier, K. Chaudhari, N. Goswami and S. K. Pal, *et al.*, *Nanoscale*, 2013, **5**, 2009–2016.
262. Y.-H. Lin and W.-L. Tseng, *Anal. Chem.*, 2010, **82**, 9194–9200.
263. C.-J. Yu, T.-H. Chen, J.-Y. Jiang and W.-L. Tseng, *Nanoscale*, 2014, **6**, 9618–9624.
264. H. Kawasaki, K. Hamaguchi, I. Osaka and R. Arakawa, *Adv. Funct. Mater.*, 2011, **21**, 3508–3515.
265. H. Kawasaki, K. Yoshimura, K. Hamaguchi and R. Arakawa, *Anal. Sci.*, 2011, **27**, 591.
266. Y. Cui, Y. Wang, R. Liu, Z. Sun, Y. Wei, Y. Zhao and X. Gao, *ACS Nano*, 2011, **5**, 8684–8689.
267. W.-T. Liu, *J. Biosci. Bioeng.*, 2006, **102**, 1–7.
268. M. De, P. S. Ghosh and V. M. Rotello, *Adv. Mater.*, 2008, **20**, 4225–4241.
269. B. R. Cuenya, *Thin Solid Films*, 2010, **518**, 3127–3150.
270. S. Guo and E. Wang, *Nano Today*, 2011, **6**, 240–264.
271. J. Elgqvist, *Int. J. Mol. Sci.*, 2017, **18**, 1102.
272. X. Zhu, R. Tang, S. Wang, X. Chen and J. Hu, *ACS Nano*, 2020, **14**, 2172–2182.
273. H. Deng, H. Liu, W. Kang, W. Kang and C. Lei, *et al.*, *Nanoscale*, 2020, **12**, 864–870.
274. Y. Fan, S. Liu, Y. Yi, H. Rong and J. Zhang, *ACS Nano*, 2021, **15**, 2005–2037.
275. K. Bian, X. Zhang, K. Liu, T. Yin and H. Liu, *et al.*, *ACS Sustainable Chem. Eng.*, 2018, **6**, 7574–7588.
276. J. Zhai, Y. Jia, L. Zhao, Q. Yuan, F. Gao and X. Zhang, *et al.*, *ACS Nano*, 2018, **12**, 4378–4386.
277. K. Bansal, M. Aqdas, M. Kumar, R. Bala and S. Singh, *et al.*, *Bioconjugate Chem.*, 2018, **29**, 1102–1110.
278. Y. Cui, J. Yang, Q. Zhou, P. Liang and Y. Wang, *et al.*, *ACS Appl. Mater. Interfaces*, 2017, **9**, 5900–5906.
279. B. Zhang, J. Wang, J. Yu, X. Fang and X. Wang, *et al.*, *Bioconjugate Chem.*, 2017, **28**, 330–335.
280. P. Gao, S. Wu, X. Chang, F. Liu and T. Zhang, *et al.*, *Bioconjugate Chem.*, 2018, **29**, 4140–4148.
281. C. Dai, C.-X. Yang and X.-P. Yan, *Anal. Chem.*, 2015, **87**, 11455–11459.
282. X. Huang, H. El-Sayed, W. Qian and M. A. El-Sayed, *J. Am. Chem. Soc.*, 2006, **128**, 2115–2120.
283. R. Weissleder, A clearer vision for in vivo imaging, *Nat. Biotechnol.*, 2001, **19**, 316–317.
284. R. Xu, D. Wang, J. Zhang and Y. Li, *Chem. – Asian J.*, 2006, **1**, 888–893.
285. Q. Song and Z. J. Zhang, *J. Am. Chem. Soc.*, 2004, **126**, 6164–6168.
286. X. Geng and T. Z. Grove, *Methods Mol. Biol.*, 2018, **1798**, 141–153.
287. A. Perrin, A. Theretz, V. Lanet, S. Vialle and B. Mandr, *J. Immunol. Methods*, 1999, **224**, 77–87.
288. H. H. Yang, S. Q. Zhang, X. L. Chen, Z. X. Zhuang and J. G. Xu, *et al.*, *Anal. Chem.*, 2004, **76**, 1316–1321.

289. K. C. Ho, P. J. Tsai, Y. S. Lin and Y. C. Chen, *Anal. Chem.*, 2004, **76**, 7162–7168.
290. C. L. Chiang, C. S. Sung, T. F. Wu, C. Y. Chen and C. Y. Hsu, *J. Chromatogr. B*, 2005, **822**, 54–60.
291. M. Lewin, N. Carlesso, C. H. Tung, X. W. Tang and D. Cory, *et al.*, *Nat. Biotechnol.*, 2000, **18**, 410–414.
292. J.-M. Nam, C. S. Thaxton and C. A. Mirkin, *Science*, 2003, **301**, 1884–1886.
293. J.-M. Nam, S. I. Stoeva and C. A. Mirkin, *J. Am. Chem. Soc.*, 2004, **126**, 5932–5933.
294. M. Li, Y. C. Lin, C. C. Wu and H. S. Liu, *Nucleic Acids Res.*, 2005, **33**, e184.
295. L. Zhang, X. Cai, L. Xu and K. Li, *J. Biomed. Pharmacother.*, 2020, **130**, 110585.
296. Z. Wang, Y. Wang, Y. Wang, C. Wei and Y. Deng, *J. Mater. Chem. B*, 2021, (9), 1781–1786.
297. Q. Jia, D. Li, Q. Zhang, S. Ye and Z. Xi, *J. Mater. Sci.*, 2019, **54**(20), 13255–13264.
298. G. Yang, Y. Lu, H. N. Bomba and Z. Gu, *Curr. Med. Chem.*, 2019, **26**, 1377–1388.
299. S.-K. Sun, L.-X. Dong, Y. Cao, H.-R. Sun and X.-P. Yan, *Anal. Chem.*, 2013, **85**, 8436–8441.
300. Z. T. Rosenkrans, C. A. Ferreira, D. Ni and W. Cai, *Adv. Health Care Mater*, 2021, **10**, 2000690.
301. Z. Zhao, H. Fan, G. Zhou, H. Bai and H. Liang, *et al.*, *J. Am. Chem. Soc.*, 2014, **136**, 11220–11223.
302. R. Yang, M. Hou, Y. Gao, S. Lu and L. Zhang, *et al.*, *Theranostics*, 2019, **9**, 6314–6333.
303. X. Cai, Q. Zhu, Y. Zeng, Q. Zeng and X. Chen, *et al.*, *Int. J. Nanomed.*, 2019, **14**, 8321–8344.
304. X. Zhu, R. Tang, S. Wang, X. Chen and J. Hu, *ACS Nano*, 2020, **14**, 2172–2182.
305. L. Chang, W. Xin, S. Hang, D. Shuaia and L. Wei, *Asian J. Pharm. Sci.*, 2020, **15**, 1–12.
306. T. Lin, P. Zhao, Y. Jiang, Y. Tang and H. Jin, *et al.*, *ACS Nano*, 2016, **10**, 9999–10012.
307. X. Zeng, J. Sun, S. Li, J. Shi and H. Gao, *Nat. Commun.*, 2020, **11**, 567.
308. E. N. Hoogenboezem and C. L. Duvall, *Adv. Drug. Delivery Rev.*, 2018, **130**, 73–89.
309. J. R. McDaniel, S. R. MacEwan, X. Li, D. C. Radford and C. D. Landon, *et al.*, *Nano Lett.*, 2014, **14**, 2890–2895.
310. J.-H. Choi, H.-J. Hwang, S. W. Shin, J.-W. Choi and S. H. Um, *et al.*, *Nanoscale*, 2015, **7**, 9229–9237.
311. H. Liu, G. Hong, Z. Luo, J. Chen and J. Chang, *et al.*, *Adv. Mater.*, 2019, **31**, 1901015–1901023.

312. A. C. del Valle, C.-K. Yeh and Y.-F. Huang, *Adv. Healthcare Mater.*, 2020, **9**, 2000864.
313. Y. Yang, Y. Yu, H. Chen, X. Meng and W. Ma, *et al.*, *ACS Nano*, 2020, **14**(10), 13536–13547.
314. Y. Chen, L. Ren, L. Sun, X. Bai and G. Zhuang, *et al.*, *NPG Asia Mater.*, 2020, **12**, 56.
315. T. C. Dakal, A. Kumar, R. S. Majumdar and V. Yadav, *Front. Microbiol.*, 2016, **7**, 1831.
316. A. Rai, S. Pinto, T. R. Velho, A. F. Ferreira and C. Moita, *et al.*, *Biomaterials*, 2016, **85**, 99–110.
317. S.-H. Kuo, C.-S. Chien, C.-C. Wang and C.-J. Shih, *J. Nanomater.*, 2019, **2019**, 4101293.
318. S.-B. He, G.-W. Wu, H.-H. Deng, A.-L. Liua and X.-H. Lin, *Biosens. Bioelectron.*, 2014, **62**, 331–336.
319. R. Vinayagam, M. Santhoshkumar, K. E. Lee, E. David and S. G. Kang, *Bioprocess Biosyst. Eng.*, 2021, **44**, 1253–1262.
320. Y.-L. Kuo, S.-G. Wang, C.-Y. Wu, K.-C. Lee and C.-J. Jao, *et al.*, *Nanomedicine*, 2016, **11**, 2497–2510.
321. P. V. Baptista, M. P. McCusker, A. Carvalho, D. A. Ferreira and N. M. Mohan, *et al.*, *Front. Microbiol.*, 2018, **9**, 1441.
322. S. Ramesh, M. Grijalva, A. Debut, B. G. de la Torre, F. Albericio and L. H. Cumbal, *Biomater. Sci.*, 2016, **4**, 1713–1725.
323. W. Song, Y. Wang, R.-P. Liang, L. Zhang and J.-D. Qiu, *Biosens. Bioelectron.*, 2015, **64**, 234–240.
324. K. S. Park, M. I. Kim, M.-A. Woo and H. G. Park, *Biosens. Bioelectron.*, 2013, **45**, 65–69.
325. J. Xie, Y. Zheng and J. Y. Ying, *Chem. Commun.*, 2010, **46**, 961–963.
326. S.-K. Sun, L.-X. Dong, Y. Cao, H.-R. Sun and X.-P. Yan, *Anal. Chem.*, 2013, **85**, 8436–8441.
327. J. M. Abad, S. F. L. Mertens, M. Pita, V. M. Fernández and D. J. Schiffrin, *J. Am. Chem. Soc.*, 2005, **127**, 5689–5694.
328. L. Zhang, W. Song, R.-P. Liang and J.-D. Qiu, *Anal. Chem.*, 2016, **88**, 11460–11467.
329. H. Deng, H. Liu, W. Kang, C. Lei, Z. Nie, Y. Huang and S. Yaoa, *Nanoscale*, 2020, **12**, 864–870.
330. R. Bhandari, R. Coppagez and M. R. Knecht, *Catal. Sci. Technol.*, 2012, **2**, 256–266.
331. A. Jakhmola, R. Bhandari, D. B. Pacardo and M. R. Knecht, *J. Mater. Chem.*, 2010, **20**, 1522–1531.
332. D. B. Pacardo, J. M. Slocik, K. C. Kirk, R. R. Naik and M. R. Knecht, *Nanoscale*, 2011, **3**, 2194–2201.
333. R. Coppage, J. M. Slocik, B. D. Briggs, A. I. Frenkel and H. Heinz, *J. Am. Chem. Soc.*, 2011, **133**, 12346–12349.
334. R. Coppage, J. M. Slocik, B. D. Briggs, A. I. Frenkel, R. R. Naik and M. R. Knecht, *ACS Nano*, 2012, **6**, 1625–1636.
335. R. Bhandari and M. R. Knecht, *Langmuir*, 2012, **28**, 8110–8119.

336. D. B. Pacardo and M. R. Knecht, *Catal. Sci. Technol.*, 2013, **3**, 745–753.
337. H. Ramezani-Dakhel, P. A. Mirau, R. R. Naik, M. R. Knechtc and H. Heinz, *Phys. Chem. Chem. Phys.*, 2013, **15**, 5488–5492.
338. R. Coppage, J. M. Slocik, H. Ramezani-Dakhel, N. M. Bedford and H. Heinz, *et al.*, *J. Am. Chem. Soc.*, 2013, **135**, 11048–11054.
339. R. Bhandari, D. B. Pacardo, N. M. Bedford, R. R. Naik and M. R. Knecht, *J. Phys. Chem. C*, 2013, **117**, 18053–18062.
340. D. B. Pacardo, E. Ardman and M. R. Knecht, *J. Phys. Chem. C*, 2014, **118**, 2518–2527.
341. N. M. Bedford, R. Bhandari, J. M. Slocik, S. Seifert and R. R. Naik, *et al.*, *Chem. Mater.*, 2014, **26**, 4082–4091.
342. B. D. Briggs, R. T. Pekarek and M. R. Knecht, *J. Phys. Chem. C*, 2014, **118**, 18543–18553.
343. H. Ramezani-Dakhel, N. M. Bedford, T. J. Woehl, M. R. Knecht, R. R. Naik and H. Heinz, *Nanoscale*, 2017, **9**, 8401–8409.
344. M. Sethi, W.-C. Law, W. A. Fennell, P. N. Prasad and M. R. Knecht, *Soft Matter*, 2011, **7**, 6532–6541.
345. R. Bhandari and M. R. Knecht, *Catal. Sci. Technol.*, 2012, **2**, 1360–1366.
346. Z. Tang, J. P. Palafox-Hernandez, W.-C. Law, Z. E. Hughes and M. T. Swihart, *et al.*, *ACS Nano*, 2013, **7**, 9632–9646.
347. Y. Li, Z. Tang, P. N. Prasad, M. R. Knecht and M. T. Swihart, *Nanoscale*, 2014, **6**, 3165.
348. L. B. Wright, N. A. Merrill, M. R. Knecht and T. R. Walsh, *ACS Appl. Mater. Interfaces*, 2014, **6**, 10524–10533.
349. B. D. Briggs, Y. Li, M. T. Swihart and M. R. Knecht, *ACS Appl. Mater. Interfaces*, 2015, **7**, 8843–8851.
350. N. M. Bedford, Z. E. Hughes, Z. Tang, Y. Li and B. D. Briggs, *J. Am. Chem. Soc.*, 2016, **138**, 540–548.
351. C. J. Munro, Z. E. Hughes, T. R. Walsh and M. R. Knecht, *J. Phys. Chem. C*, 2016, **120**, 18917–18924.
352. Z. E. Hughes, M. A. Nguyen, Y. Li, M. T. Swihart, T. R. Walsh and M. R. Knecht, *Nanoscale*, 2017, **9**, 421–432.
353. C. J. Munro and M. R. Knecht Solution, *Langmuir*, 2017, **33**, 13757–13765.
354. M. A. Nguyen, Z. E. Hughes, Y. Liu, Y. Li and M. T. Swihart, *et al.*, *J. Phys. Chem. C*, 2018, **122**, 11532–11542.
355. J. P. Palafox-Hernandez, Z. Tang, Z. E. Hughes, Y. Li and M. T. Swihart, *et al.*, *Chem. Mater.*, 2014, **26**, 4960–4969.
356. J. P. Palafox-Hernandez, C.-K. Lim, Z. Tang, K. L. M. Drew and Z. E. Hughes, *ACS Appl. Mater. Interfaces*, 2016, **8**, 1050–1060.
357. E. M. Zahran, N. M. Bedford, M. A. Nguyen, Y.-J. Chang and B. S. Guiton, *et al.*, *J. Am. Chem. Soc.*, 2014, **136**, 32–35.
358. N. A. Merrill, E. M. McKee, K. C. Merino, L. F. Drummy and S. Lee, *et al.*, *ACS Nano*, 2015, **9**, 11968–11979.
359. B. D. Briggs, J. P. Palafox-Hernandez, Y. Li, C.-K. Lim and T. J. Woehl, *et al.*, *Phys. Chem. Chem. Phys.*, 2016, **18**, 30845.

360. V. R. Stamenkovic, D. Strmcnik, P. P. Lopes and N. M. Markovic, *Nat. Mater.*, 2017, **16**, 57–69.
361. E. P. George, D. Raabe and R. O. Ritchie, *Nat. Rev. Mater.*, 2019, **4**, 515–534.
362. M. A. Nguyen, Z. E. Hughes, Y. Liu, Y. Li, M. T. Swihart, M. R. Knecht and T. R. Walsh, *J. Phys. Chem. C*, 2018, **122**, 11532–11542.
363. T. R. Walsh and M. R. Knecht, *Chem. Rev.*, 2017, **117**, 12641–12704.
364. R. Balu, J. Whittaker, N. K. Dutta, C. M. Elvin and N. Roy Choudhury, *J. Mater. Chem. B*, 2014, **2**(36), 5936–5947.
365. X. Lu, P. Xu, H.-M. Ding, Y.-S. Yu, D. Huo and Y.-Q. Ma, *Nat. Commun.*, 2019, **10**, 4520.

Subject Index